The Evolution of Avian Breeding Systems

Oxford Ornithology Series

Edited by C.M. Perrins

The Evolution of Avian Breeding Systems

J. DAVID LIGON
Department of Biology
The University of New Mexico

Original illustrations by
MIKE RAMOS

OXFORD
UNIVERSITY PRESS

OXFORD
UNIVERSITY PRESS

Great Clarendon Street, Oxford OX2 6DP

Oxford University Press is a department of the University of Oxford
and furthers the Univesity's aim of excellence in research, scholarship.
and education by publishing worldwide in

Oxford New York
Athens Auckland Bangkok Bogota Buenos Aires Calcutta
Cape Town Chennai Dar es Salaam Delhi Florence Hong Kong Istanbul
Karachi Kuala Lumpur Madrid Melbourne Mexico City Mumbai
Nairobi Paris São Paolo Singapore Taipei Tokyo Toronto Warsaw
and associated companies in Berlin Ibadan

Oxford is a trade mark of Oxford University Press

Published in the United States
by Oxford University Press, Inc., New York

A catalogue record for this book is available from the British Library

Library of Congress Cataloging in Publication Data
Ligon, J. David.
The evolution of avian breeding systems/J. David Ligon: original
illustrations by Mike Ramos.
(Oxford ornithology series; 10)
Includes bibliographical references (p.   ) and index.
1. Birds–Behavior.   2. Sexual selection in animals.   3. Courtship
in animals.   I. Title.   II. Series.
QL698.3.L494   1999   598.156′2–dc21   98–41860

ISBN 0 19 854913 X

Typeset by EXPO Holdings, Malaysia

Printed in Great Britain by Bookcraft (Bath) Ltd
Midsomer Norton, Avon

Preface

This book deals with two of the most conspicuous, and, to many students, two of the most interesting aspects of the reproductive biology of birds, namely the phenomena of sexual selection and mating systems. These two subjects are intimately related, in part because mating systems (together with parental care systems) seem to be influenced by, and to influence, the strength of sexual selection. For example, morphological traits (such as sexual dimorphism in body size) and extent of sexual dichromatism (such as plumage colour brightness) appear to be sexually-selected traits that, within a particular group, often correlate with mating system. Although sexual selection exerts strong effects on mating systems, and vice versa, they are also distinct and separate phenomena. Sexual selection is ubiquitous, occurring in plants as well as in animals; in contrast, the diversity of mating systems shown by birds is unique among vertebrates.

Because what birds do is relatively easy to see, and because they exhibit all major kinds of social and mating systems known to occur in vertebrate animals, they have proved to be an especially attractive group for studies dealing with mating and other social systems. Within the Class Aves there are monogamous (probably most of the approximately 9700 living species), polygynous (many), and polyandrous (relatively few) species, with many variations on these main themes. Birds may breed as isolated pairs, colonially (many pairs in close proximity), communally (several breeders, all using one nest), cooperatively (one female breeder per nest, plus one or more mates and non-breeding 'helpers'), or exhibit no parental care at all beyond laying eggs in the nests of host species (brood parasites). Overall, the Class Aves probably exhibits more diversity in both breeding behaviour and sexually-selected ornamentation than any other vertebrate group, even though it is the only class of vertebrates in which every species is oviparous.

My aims for this book are twofold. First, I hope to provide both an introduction to, and a review of, sexual selection and mating systems, as they pertain to birds. Ever since Darwin, birds have provided much of the stimulus for study of these subjects. No other single taxonomic group has been more studied or is better suited, all in all, for illustrating patterns and processes of these two interrelated phenomena. Until recently, except for comparative studies of closely related species (which usually presume evolutionary divergence from a common ancestor), the historical perspective was largely absent from considerations of both sexual selection and mating systems. Thus, a second goal for this book is to promote the idea that evolutionary or phylogenetic history is essential to an understanding of these essentially behavioural phenomena.

Recently, five excellent books have appeared in which sexual selection is a predominant theme. Malte Andersson's (1994) *Sexual selection* is very broad, as well as thorough. The others are more specialized, dealing either with a single species or a single theme. *Sexual selection and the barn swallow* by Anders Møller (1994) and *Polygyny and sexual selection in red-winged blackbirds* by William Searcy and Ken Yasukawa (1995) provide clear and thorough discussions of the theories associated with this subject, especially as they pertain to study species of the authors. The last two of the five books focus on a particular type of mating system that has figured prominently in the development of current thinking in the area of sexual selection: Paul Johnsgard's (1994) *Arena birds: sexual selection and behavior* and Jacob Höglund and Rauno Alatolo's (1995) *Leks.*

I have made no attempt to consider comprehensively the many theoretical papers on female mate choice, and I refer readers interested in a detailed treatment of sexual-selection theory to Harvey and Bradbury (1991), Andersson (1994), and Møller (1994), in particular. Johnsgard (1994) brings together detailed descriptions and numerous informative illustrations of courtship behaviours and male morphology associated with courtship. The illustrations in Johnsgard's book are especially useful as aids to visualizing the display patterns of many of the most strongly sexually-selected and spectacular birds in existence. Similarly, readers wishing to learn more about the theory behind the other main subject of this book, mating systems, and the related subject of parental care should see Davies (1991), Clutton–Brock (1991) and Clutton–Brock and Godfray (1991) respectively.

In addition to presenting ideas and discussion about sexual selection and mating systems in birds, I have another, more subtle, goal for this book. I have attempted to include studies of species and groups of birds that occur in regions where avian biologists are relatively scarce. Unfortunately, most of the world's birds, including a majority of its most interesting families (many of which are little-studied), occur in tropical regions where the environment is currently undergoing rapid and major alteration at the hands of humans. In view of the fact that the majority of avian field biologists live and conduct their research in the North Temperate regions, it is clear that most species of birds will never be the subject of even one detailed field study, at least in an environment that is more or less 'natural'. Thus, I hope that this book may help to encourage some ambitious, adventurous students to investigate some of the lesser known kinds of birds, especially those in tropical regions. (A study that my spouse, Sandy, and I conducted some years ago on a little-known African bird, the green woodhoopoe, was the most professionally satisfying project of my career, as well as the most fun.)

It is claimed that more than a thousand papers have been published on red-winged blackbirds. Although a thousand papers is not too many, I cannot help but think that some of the immense research effort those papers represent might have been more profitably used to investigate in depth some of the Neotropical relatives of redwings. In any case, it is clear that in the next hundred years, tropical regions, especially forested ones, of the planet Earth will be drastically changed, and many of the habitats of fascinating birds that may have existed there for millions of years will be gone.

One of the great benefits of an academic career is the opportunity to share one's enthusiasm for a particular subject with like-minded individuals. Several of the people whose help is gratefully acknowledged here are, or have been, graduate students who worked with me at The University of New Mexico. I want first to thank Geoffrey Hill, Sandra Ligon, and Richard Wagner, all of whom read every chapter, and suggested many improvements. Sandy also carried out the onerous task of checking text references against the literature citations. Rebecca Kimball critically read most chapters, helped greatly with organizing the literature citations, and made the initial compilation of the Appendix. Kristine Johnson and David Gray provided highly useful critiques of Chapters 2–7, and Brent Burt offered excellent advice on Chapters 10–12. Alan Feduccia's comments helped to improve Chapter 10. Joe Bennedict provided editorial suggestions for more than half of the chapters. Several of the participants in my graduate seminar 'Avian Social Systems' also provided useful input on specific chapters; these include Dan Albrecht, Julie Hagelin, Michele Merola–Zwartjes, Gary Miller, and Patrick Zwartjes. Over the past decade and a half I have profited greatly from discussions with Steve Zack, Peter Stacey, Jim Bednarz and Greg Farley about various aspects of avian biology, and from Randy Thornhill and Astrid Kodric-Brown about evolutionary biology in general, and about sexual selection in particular. I want also to acknowledge, with thanks, the mentors who patiently helped me to become an ornithologist: George M. Sutton, Pierce Brodkorb, William R. Dawson, Robert W. Storer, and Harrison B. Tordoff. Finally, I thank Dan and Rusty Ligon, and their mother, Sandy Ligon, for tolerating my disappearances, too many times, to 'work on the book'.

This project got underway in 1990, when I was hosted by Hugh Ford as a Visiting Research Fellow in the Department of Zoology at The University of New England in Armidale, New South Wales, Australia. Sandra Higgins and Viola Watt of UNE kindly typed the first drafts of several chapters, some of which Hugh critiqued. Beginning in fall 1991, four years as Chair of the Department of Biology at The University of New Mexico served to slow, but not completely extinguish, progress on the manuscript. Eventually, Anne Rice put the finishing touches on the semi-final and final drafts. I also extend my thanks to all of these individuals.

I acknowledge with thanks the University of Chicago Press for permission to reprint Figs 5.3 and 10.1 and the hierarchical classification in Chapter 13; Academic Press for permission to reprint Figs 5.2 and 8.2; the American Association for the Advancement of Science for permission to reprint Fig. 1.2;

the American Ornithologist' Union (AOU) and AOU Monographs for permission to reprint Figs 15.2 and 15.3; the National Academy of Science for permission to reprint Fig. 7.1; Ethology, Ecology and Evolution for permission to reprint Fig. 6.5; the Western Foundation of Vertebrate Zoology for permission to reprint Fig. 14.4; Oxford University Press for permission to reprint Figs 12.1 and 12.3; and *Oikos* for permission to reprint Fig. 1.3. I also thank the authors of these figures, whose names are given in the captions. Dover Publications, Inc. allowed me to use a number of illustrations by nineteenth-century artists, most of whom are unknown, that greatly enhanced the book. The figures from Dover Publications are also indicated in the captions.

University of New Mexico   J. D. L.
July 1998

Contents

Illustrations at chapter openings

These were prepared as wood engravings in the nineteenth century. Selected for publication by J. Harter and published by Dover Publications, Inc. (1979). Reprinted with permission.

1. Two pairs of kiwis engaged in their nocturnal activities. Paleognathous birds, which include the kiwi, other ratites, and their relatives, the tinamous, are thought to be the most primitive living members of the Class Aves.

2. Male and female booted racquet-tailed hummingbirds. In most hummingbird species, males are highly ornate, presumably as a result of strong sexual selection by female choice.

3. Male satin bowerbird with female in bower. The bowers of this and other bower-building species serve to attract females for mating.

4. King vulture. Natural selection favouring an unfeathered head apparently predisposed this and some other vulture species to sexual selection via development of fleshy ornaments located on the head and face.

5. The little-studied tree swifts, Family Hemiprocnidae, exhibit strongly forked tails, the length and symmetry of which may prove to reflect the male's condition or quality relative to other males.

6. The train of the Indian peacock is the most famous example of a sexually-selected trait.

7. The striking sexual dichromatism seen in Baltimore orioles and other icterids may have evolved from ancestors in which the males and females were both bright and similar in appearance.

8. Males of three species of birds of paradise. Members of this group of birds may represent the most extreme example of ornament differentiation due to sexual selection of any avian group. Although many of the bird-of-paradise species apparently are extremely closely related, male ornaments and displays are strikingly diverse.

9. In many kinds of birds, including the pink cockatoos shown here and other parrots, most coraciiforms (kingfishers, bee-eaters, etc.), and many alcids (the auks, puffins, auklets), females are as brightly coloured as males. Although mate choice by males has received relatively little study,

it is likely that in certain groups of birds, the ornaments of females will prove to be important in mate selection.

10. Male greater rhea with chicks. Like most other ratites and the tinamous, in rheas, parental care is conducted exclusively by the male. Sole paternal care in these birds may represent the earliest form of parenting in lineages leading to the living species of birds.

11. Although male resplendent quetzals are among the most spectacularly ornate of birds, they, like all other trogons and their nearest relatives, the coraciiforms, are monogamous, with extensive parental care by males.

12. Although most birds are socially monogamous, the frequency of extra-pair copulations that lead to fertilization varies greatly from species to species. The still-emerging picture suggests that in those species in which females require extensive parental care by the male in order to rear offspring, such as the golden eagle, extra-pair fertilizations are rare.

13. In the great argus pheasant, the spectacularly enlarged secondary feathers of the wing, with rows of colourful ocelli, provide a striking example of the relationship between the strength of sexual selection and polygynous mating systems.

14. The adaptive strategies of helpers of cooperatively breeding species vary, even within a single species. In the pied kingfisher, unrelated male helpers may eventually form a pair bond with the female whose chicks the helper provisions.

15. Several lekking species have received detailed study. One of these is the black grouse in Sweden.

16. Among neognathous birds, classical polyandry is found almost exclusively within the shorebird Order Charadriiformes. A few shorebird families are characterized by this mating–parenting system, the best known of which are the jacanas.

17. The striking tail-streamers of the African paradise flycatcher is presumably the result of strong sexual selection. However, this species is not only monogamous, but males participate fully in incubation.

*This book is affectionately dedicated to Sandy and my three sons,
Dave, Dan, and Rusty.*

1 Introduction

1.1 Sexual selection, mating systems, and parental care

This book is about two of the three most conspicuous aspects of the reproductive biology of birds, sexual selection and mating systems (see Sections 1.3 and 1.4 below); the third is parental care. Ever since Darwin, birds have provided much of the stimulus for study of these topics, and no other taxonomic group is better suited for illustrating patterns and processes of these inter-related phenomena. Although the Class Aves is unusually conservative in production of young, with every species being oviparous, it probably exhibits more diversity in breeding behaviour and sexually-selected ornamentation than any other vertebrate group. This diversity of reproductive strategies has even been compared to that of insects; for example, Wilson (1975) points out that non-breeding helpers of cooperatively breeding bird species are similar to the insect systems classified by entomologists as 'advanced subsocial,' and notes that birds also parallel insects in the repeated evolution of social or brood parasitism.

Sexual selection and mating systems, along with parental care, are intimately related, and evolutionary change leading to the reproductive patterns of contemporary species probably occurred simultaneously in each of these three areas. Sexual selection surely was present and important in the reptile precursors of birds as well as in the earliest birds, long before either the mating systems (beyond promiscuity) or parental care systems that characterize modern birds had appeared (Fig. 1.1). Similarly, the earliest form of parental care may have been restricted to egg-guarding, probably by males (see Chapter 10). Eventually, more complex parental care, and chicks more dependent on parental care, evolved, and biparental care evidently was strongly favoured in

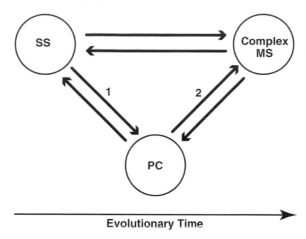

Fig. 1.1   An hypothetical evolutionary relationship between sexual selection (SS), parental care (PC), and complex mating systems (Complex MS). Sexual selection existed prior to, and affected, the evolution of parental care (step 1). The evolution of parental care promoted the evolution of the complex mating systems seen today (step 2). With the establishment of obligatory parental care and complex mating systems came the interplay between each of these factors; e.g. if parental care by males is essential, this will limit the strength of sexual selection. See text for elaboration.

the lineage producing neognathous birds (all living species except ratites and tinamous; see Fig. 1.2). Obligate parental care, whether, uniparental or biparental, would have strong effects on the evolution of more complex mating systems. Once a true mating system evolved (e.g. monogamy), this would have effects on the subsequent evolution of parental care and on the strength and form of sexual selection. Likewise, the evolution of obligate biparental care would affect the strength of sexual selection. Finally, the strength of sexual selection would also have effects on, as well as be affected by, the future evolution of both parental care and mating systems.

A key factor in the initiation of this scenario is the ability of a single parent to incubate the eggs and rear the young. Other things being equal, precocial hatchlings make the latter more likely to occur. This is the case in almost all paleognathous birds (ratites and tinamous), and uniparental care has also evolved independently several times in neognathous birds, most commonly in species with precocial young. Of the species in which uniparental care of altricial nestlings is the rule, almost all are either hummingbirds or members of the Order Passeriformes.

Once mating systems had evolved they could affect, as well as be affected by, sexual selection. For example, because parental care has unavoidable costs, it affects both mating systems and the strength of sexual selection. On the other side of the coin, sexual selection can also affect parental care. As an example, consider the pied flycatcher of Europe. Most male pied flycatchers

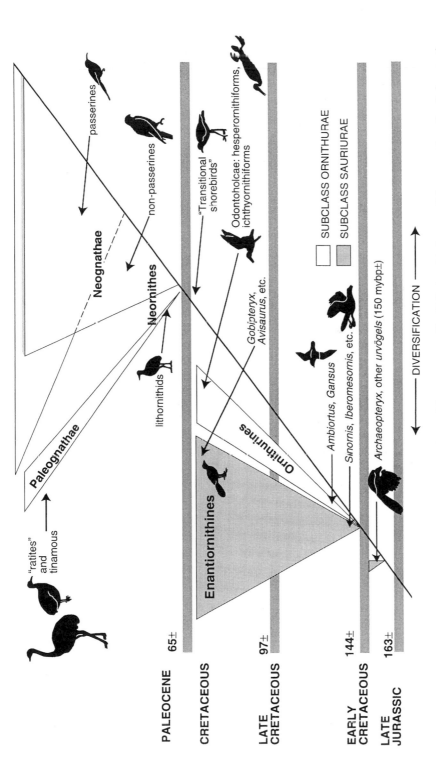

Fig. 1.2 A new view of avian evolution. The main points of interest are: (1) all living birds, possibly excepting the Paleognathae, are descended from 'transitional shorebirds'; (2) the neognathous non-passerines diversified rapidly during the Paleocene and Eocene; and (3) the passerine radiation occurred long after the non-passerine birds had reached an essentially modern degree of diversification. Reprinted with permission from Feduccia, A. (1995). Explosive evolution in tertiary birds and mammals. *Science*, **267**, 637–8. Copyright 1995 American Association for the Advancement of Science.

are monogamous, but some acquire a second mate. This shift in the mating system, which increases the strength of sexual selection in pied flycatchers, is possible because the second female can rear some young with little or no aid from the male. In most orders and families of birds, however, the male's contribution to the nesting effort is essential during one or more stages of the breeding cycle (see Section 10.3). In these species, parental care exerts a negative effect on sexual selection.

Although sexual selection has strong effects on mating systems, and vice versa, these are distinct and separate phenomena. Sexual selection is ubiquitous, occurring in plants as well as in animals (e.g. see M. Andersson 1994). In contrast, the diversity of mating–parenting systems shown by birds, combined with their universal oviparity, is unique among all major animal groups. Once monogamous mating systems had evolved, strategies involving matings outside the pair bond (extra-pair copulations) must also have appeared in both males and females. Extra-pair copulations increase the importance of sexual selection in monogamous species, and they might also be expected to have had an influence on parental care, especially by males (see Chapter 12).

Until recently, the historical perspective was largely absent from most considerations of both sexual selection and mating systems. Thus, another goal for this book is to encourage students of sexual selection and mating systems to consider explicitly the role of evolutionary history in the development of the reproductive patterns of the species or groups that they study.

This first chapter begins by briefly outlining the early evolutionary history of birds, as it is currently understood. It then introduces the subjects of sexual selection and mating systems, which are the two main themes of this book. The second part of the chapter addresses the complementary issues of adaptation and constraints. Until rather recently, the evolutionary history of many traits associated with the phenomena of sexual selection and mating systems has been largely ignored. In contrast, the subject of adaptation has received a great deal of attention. More recently, a related concept, often labelled phylogenetic constraint, has also been the subject of considerable discussion and debate. Much of evolutionary history of all organisms has been shaped by the complementary phenomena of adaptation and constraint.

1.2   Evolutionary history

Most considerations of sexual selection and parenting–mating systems have ignored, or paid only lip service to, phylogenetic influences, or the effects of evolutionary history on contemporary animals. However, over the past few years an appreciation of the role of history has developed, and many papers treating this issue have appeared (see Chapter 7).

1.2.1 Early birds

Alan Feduccia's (1996) *The origin and evolution of birds* provides a clear and thorough exposition of the evolutionary history of birds, which extends from the late Jurassic Period of the Mesozoic Era, some 150 million years ago, to the present. *Archaeopteryx*, the earliest known bird (defined first and foremost by the presence of feathers), is one of the most famous fossil vertebrates ever discovered. Subsequent to the discovery of *Archaeopteryx*, many additional Mesozoic birds have come to light. During the Cretaceous Period of the Mesozoic, there were two or more richly diverse radiations of birds.

In addition to the ancestors of modern birds there was a major radiation of a group known as the 'enantiornithines,' or 'opposite birds,' which are characterized by a unique form of fusion of the metatarsal bones. The enantiornithines and another avian lineage, the ornithurines, disappeared, along with the dinosaurs, at the close of the Cretaceous, about 65 million years ago, leaving behind the Neornithes, which gave rise to the surviving avian lineages—the Superorders Paleognathae (Infraclass Eoaves of Sibley and Ahlquist 1990) and Neognathae (Infraclass Neoaves, Sibley and Ahlquist 1990) (see Fig. 1.2). According to Feduccia (1995, 1996), the neognathous birds that survived the close of the Cretaceous were shorebird-like creatures that he refers to as 'transitional shorebirds.'. Their subsequent evolution progressed rapidly, so that by the early Eocene Epoch, some 5 to 10 million years later (about 50–55 million years ago), all modern orders of birds had appeared. Modern families had appeared by the late Eocene and early Oligocene (about 35 million years ago), and modern genera began to be clearly discernible by the Miocene about 14 million years ago (Feduccia 1996, p. 167; see Chiappe 1995) for some alternative views concerning the early evolution of birds. Some other recent estimates suggest that modern orders of birds arose considerably earlier than proposed by Feduccia, during the Cretaceous; e.g. Hedges *et al.* 1996; Cooper and Penny 1997).

These brief comments about the early history of birds are intended to make an obvious, but important, point: the Class Aves has had a very long and complex evolutionary history. Families and even genera of birds that we study today have been in existence for millions of years. Complex parental care behaviour and the altricial mode of chick development probably had evolved prior to the Eocene. If so, this suggests that parenting and mating systems much like those we see today were also present by no later than the Eocene, and, as mentioned earlier, sexual selection surely must have been important throughout the entire evolutionary history of birds.

During the almost unimaginably long period from the Eocene to the present, some 55 million years (a span of time that encompasses only about one-third of avian history), many evolutionary transitions occurred. Some of these, which may be counter-intuitive, are important to understanding sexual selection in contemporary birds. Consider, for example, the issue of sexual dichromatism in plumage brightness. It is usually assumed that species characterized by bright males and dull females evolved via sexual selection from an ancestor in which plumage of both sexes was dull. However, a case can be made that the ancestral condition in some living lineages characterized by plumage dichromatism probably was bright, colourful plumage in both sexes (see Section 7.2.5). In today's avifauna, some of the world's most brilliantly coloured species and groups of birds are sexually monochromatic (e.g. flamingos, scarlet ibis, parrots, coraciiforms), as also are most other, less spectacularly feathered birds. Bright, possibly ornate, plumage may have occurred in both sexes of groups currently characterized by striking sexual dichromatism (e.g. many ducks, galliforms, hummingbirds, and a number of passerine families). That is, sexual dichromatism may be derived from ancestors in which both males and females were brightly coloured, rather than having been derived from monochromatic, dull ancestors (see Section 4.6). In contrast to the neognathous birds, an opposite scenario may have occurred in the paleognathous ratites and tinamous. Nearly all living species in this group are monochromatic in feather colour, and plumage is bright in neither sex. The most striking exception is the ostrich, in which the plumage of the males is black and white and modified for display, while that of the female is a dull grey-brown.

A second example has to do with the relationship between parental care and mating systems (see Section 1.4 below). The complex parental—usually biparental—care that characterizes modern birds ultimately is derived from a condition of no parental care. Although monogamy is frequently taken as the starting point in discussions of the evolution of avian mating systems, this does not help us to understand how monogamy evolved from a state of no parental care in the first place. Field studies of mating systems can elucidate the costs and benefits of monogamy as compared to polygyny, for example, and how deviations from monogamy can be favoured under certain circumstances, but they do not really address the issue of *origin* (see Chapter 10).

In short, the species that we see and study today were constructed over many millions of years. Each adaptive change that appeared in a given lineage had subsequent effects on the evolutionary trajectory taken. Specific adaptations often affect the pathway of future evolution; such factors have been labelled 'phylogenetic constraints' (see below). Thus, it is important to remember that any adaptive modification usually will have a ripple effect on other aspects of a species' biology. The evolution of parental care, for example, must have had major effects on the development of offspring, both before and after the eggs were hatched. This, in turn, would have effects on the evolution of mating systems, which, in turn, would affect the intensity of sexual selection.

1.3   Sexual selection

In this section I provide a brief review of sexual selection, first identified as a specific phenomenon by Darwin (1859). Darwin recognized a fundamental difference between natural selection and sexual selection. Natural selection is based on differential survival and fecundity among individuals. In contrast, sexual selection is not directly based on a struggle for existence and reproduction; instead, sexual selection involves competition to attract and defend mates. As envisioned by Darwin, sexual selection includes two processes. First, there is competition between individuals of one sex, usually males, for possession of (access to) the other sex, usually females. For losers in this sort of competition, the result typically is few or no offspring. Darwin also assumed that certain characters typical of males of many species (e.g. large ornamental structures, often associated either with fighting or intimidating other males, or with attracting females) almost certainly had large survival costs associated with them. In short, such traits seemed to be related to increased mating success rather than increased prospects of future survival. Thus the functions of sexual and natural selection did not appear to be the same. Second, Darwin recognized sexual choice (often referred to as intersexual selection), usually of males by females. Ornamental traits of males that are thought to be used by females making a choice of mates are often referred to as epigamic characters.

Recently, several authors have provided modifications or extensions of Darwin's definition. Møller (1994, p. 3) recognizes the two components identified by Darwin:

> Sexual selection occurs as a result of a non-random association between a (secondary sexual) character and a component of mating success. Mating success includes the direct acquisition of mates, but also other components of success such as sperm competition, differential abortion and infanticide, and differential parental effort.

Andersson (1994, p. 7) writes that 'Sexual selection of a trait can therefore be viewed as a shorthand phrase for *differences in reproductive success, caused by competition over mates, and related to the expression of the trait.*' (Emphasis is Andersson's.) These definitions clearly are close to Darwin's view in that they explicitly include both intrasexual competition for mates and traits of one sex (typically males) that are used by the other (typically females) in making mate choice decisions.

On the other hand, Arnold (1994, S9) proposes that sexual selection be defined as '... ... selection that arises from differences in mating success (number of mates that bear or sire progeny over some standardized time interval).' Arnold points out that (1) this definition does not attempt to define selection itself, and (2) because this definition is cast in terms of fitness currency, it is clearly connected to formal evolutionary theory. Competition and choice, factors essential to Darwin's view of sexual selection, are not included in Arnold's definition, which by design is less precise than either Darwin's perspective, or those of Møller and Andersson. This is because, at least in part, Arnold intended that his definition be broad enough to include sexual selection in plants as well as in animals.

Darwin's view of a trade-off between naturally and sexually selected traits can be illustrated by viewing the ultimate 'goal' of natural selection simply as maximization of survival and fecundity probabilities. For example, Selander (1965) found that the presumably sexually-selected traits of large body size and out-sized tails of male great-tailed grackles (see Section 7.2.2) carried large costs, as reflected by decreased survival, as compared to the smaller-bodied and smaller-tailed females. This remains, more than 30 years later, one of relatively few well-documented cases among birds of a survival cost associated with sexually-selected traits. Another is cost of tail length in the barn swallow (Møller 1994). Exaggerated tails in the grackle and barn swallow thus are both costly (in terms of survival) and necessary (to obtain reproductive success). Thus, these two species support Darwin's contrast between sexual and natural selection.

Despite examples such as the great-tailed grackle and barn swallow, where ornamental traits of males carry a cost in terms of future survival, it is now recognized that, in fact, the ultimate effects of sexual and natural selection do converge (e.g. Searcy and Yasukawa 1995). In both, maximization of individual lifetime reproductive success is the ultimate or evolutionary 'goal,' although it may be reached via one or more of several possible pathways. Sometimes this convergence is easy to see. For example, for a variety of bird species the only significant correlate of lifetime reproductive success that has been identified is longevity (Clutton-Brock 1988; Newton 1989). In such species, behavioural or morphological traits that promote a long life-span also indirectly promote maximization of lifetime production of offspring (natural selection), which is the same ultimate goal of males as they compete for mates or matings (sexual selection).

Although the ultimate effects of natural and sexual selection may converge, in the eye of many biologists they remain separate phenomena. For example, behaviour designed solely to increase the prospects for continued survival or to increase fecundity (e.g. gathering food for nestlings with dangerous predators lurking in the area) is not generally considered to represent a component of sexual selection. Instead, following Darwin's conceptualization, sexual selection is that aspect of the biology of a given species by which males, typically, attempt to maximize matings and/or mates, and the factors used by females to choose mates. For males, sexual selection focuses on those aspects of reproduction involved with outcompeting other males, in one way or another, for paternity of young. For female birds, too, the competitive aspect of sexual selection may be conspicuous, particularly in species that are territorial and in which paternal care is essential for successful reproduction. In sex-role-reversed species, females compete for mates as vigorously as males of more typical species (e.g. Emlen *et al.* 1989). More commonly, competition by females is to obtain critically important partners in the nesting effort. Unlike the common situation for males, this competition is not primarily related to copulation *per se*. Although it is likely that female birds usually will have no trouble in locating a willing (and possibly high-quality) male for copulation, obtaining a territory-owning, potentially parenting male may be far more difficult, and can lead to intense female–female conflict, as described in Section 9.4.

Thus, to the extent that longevity is related to lifetime reproductive success, both traits that promote survival and traits that promote successful competition for matings can be measured by the same fitness currency—number of offspring that survive to breed. However, these two kinds of traits represent different phenomena within the overall life history strategy of an individual of a given species.

Finally, although a distinction usually is made between intrasexual competition and intersexual choice, it is important to note that the same trait or traits—morphological, physiological, or behavioural—may play an important role in both functions (Halliday 1983; Kodric-Brown and Brown 1984). In red junglefowl, for example, one morphological trait, the size of the male's comb, correlates positively with success of males in both intrasexual competition and in female choice (Ligon *et al.* 1990; Ligon and Zwartjes 1995*a*).

1.3.1   Sexual selection: early rejection and resurrection

In recent years the subject of sexual selection has assumed a leading position in the area of evolutionary biology; however, its development has had a very uneven history. Cronin (1991) provides an informative and interesting review of the origins and early history of the theory of sexual selection, and discusses in some detail the factors leading to its eclipse for many years. Briefly, following its recognition, naming, and elucidation by Darwin (1859, 1871), sexual selection was extensively discussed by Wallace (1889), who came to vigorously oppose it. In fact, it was opposed by most of the leading naturalists of the day. As Huxley (1938) remarked, 'None of Darwin's theories has been so heavily attacked as that of sexual selection.' Despite this opposition, Darwin maintained his belief in the significance of sexual selection. Only a few hours before his death Darwin read, at a meeting of the Zoological Society:

> ... I may perhaps be here permitted to say that, after having carefully weighed, to the best of my ability, the various arguments which have been advanced against the principle of sexual selection, I remain firmly convinced of its truth.' (see Cronin 1991, p. 249 for references).

With the demise of Darwin, the theory of sexual selection lost both its founder and its principal champion, and to the majority of evolutionists of the day, it was buried along with Darwin.

In 1915, however, 33 years after Darwin's death, Fisher (1915) provided the basis for the scientific resurrection of Darwinian sexual selection by developing a theory by which aesthetic mate choice by females could lead to the incorporation and elaboration of ornamental traits and displays of males. Fisher's work did not attract much interest at the time, or for many years thereafter, as the attention of the leading evolutionary biologists of the day had become focused instead on the issue of speciation. For a long period thereafter sexual selection was largely ignored by leading evolutionary biologists (e.g. Mayr 1963). Following this period, which Mary Jane West-Eberhard (1983) labelled the 'forgotten era of sexual selection,' sexual selection theory re-emerged in the 1970s (e.g. Campbell 1972; O'Donald 1973; Blum and Blum 1979), and, over the past 25 years, it has become generally recognized as fundamentally important to understanding many major issues in evolutionary biology, ranging from sperm competition and embryo sex ratio on the one hand, to the process of speciation on the other.

1.3.2   Future trends in the study of sexual selection

In view of the prominent position it now holds in evolutionary biology, the lack of interest in the issue of sexual selection for so many years is almost amazing. It is also a bit troubling. How could what seems so obvious to us today concerning the evolution of elaborate ornaments of male birds, and the importance to females of discriminating mate choice, have escaped the attention of the leading thinkers in evolutionary biology, many of whom surely were brilliant? Although, as this and the following chapters indicate, I believe that our current emphasis on sexual selection is on the right track, I recall a seminar that I participated in as a graduate student more than 30 years ago. The seminar was devoted to discussion of Mayr's (1963) then new book, *Animal species and evolution*. At that time none of us thought it peculiar that a 797-page book on evolution contained only a single passing remark concerning the subject of sexual selection (Mayr 1963, p. 201). The male ornaments that currently are thought to be so important in sexual selection were then considered by most evolutionary biologists to have evolved to prevent interspecific matings (see Chapter 8).

How will future evolutionary biologists view the perspective we employ in the late 1990s? Will the next generation discard the conceptual framework currently in vogue, as already recommended by Amotz Zahavi, one of the most insightful thinkers in behavioural evolutionary biology? Zahavi (1991, p. 502) writes:

> Sexual selection as defined by Darwin does not encompass a set of characters that pose a common problem to modern evolutionary biology. The only reason to continue using this term seems to be respect for Darwin's historical definition. But this respect blurs the main interesting problem within sexual selection. The central problem that sexual selection presents for evolutionary biology today is the selection for extravagance and waste; however, as already stated, these are not unique to

sexual selection. The common denominator of all those characters that display extravagance and waste is that they are all signals, whether they are included within the definition of sexual selection or not.

My general point is that ways of viewing phenomena and concepts change with time, and there is no assurance that this will not happen to the currently fashionable view of sexual selection, as unlikely as that may seem at present.

1.4 Mating systems

One of the important keys to understanding mating systems is recognition of the importance of parental care and type of young produced (e.g. precocial—semi-precocial—altricial). The benefits and costs of rearing precocial versus altricial young have critically influenced the evolution and maintenance of mating systems. For example, the large benefit of biparental care for the majority of neognathous birds is the basic factor that first favoured the evolution of monogamy in birds, and that has promoted its maintenance over most of their evolutionary history.

Another important approach to study of mating systems is appreciation of the influence of phylogenetic history. First, egg-laying, together with some other avian traits, forms the foundation for the evolution of complex paternal care in birds. Second, over evolutionary time it has been the opportunity provided by oviparity for extensive biparental care that has led to the different kinds of mating systems and mating strategies (including extra-pair copulations) observed and studied today. Presumably, prior to the evolution of parental care, mating in birds was affected solely by sexual selection. That is, in a promiscuous system, females were attracted to males solely for copulation. Once parental care behaviour had evolved in both sexes, however, selective benefits arose for the evolution of monogamy and, subsequently, other mating systems derived from monogamy (see Chapter 10, Fig. 10.3).

After distinctive mating systems had evolved, additional or alternative mating strategies, such as extra-pair copulations, also could develop. Extra-pair copulations can provide members of both sexes with important benefits. For attractive males, it is the opportunity to sire offspring in addition to those produced with the social mate. For females, it appears to be the opportunity to obtain the genes of desirable males for their offspring, while also having the parenting contributions of the social mate. The complex behavioural strategies associated with extra-pair activities of each sex could appear only subsequent to the development of stable pairs or polygamous units; i.e. for obvious reasons, the concept of extra-pair copulations is not applicable to promiscuous species.

Consideration of the maintenance and adaptive significance of the major categories of mating systems in birds requires that we address two even more basic questions about their reproductive biology, namely (1) Why are all birds oviparous, and (2) What is the evolutionary history of parental care? A short answer to the first question is that oviparity is an effective means of reproducing, with selection over the long term having favoured and refined this approach to reproduction (see Section 10.3). With regard to the second question, it appears that two traits, presumably characteristic of very early birds, or possibly even of their reptilian antecedents, favoured development of parental care; these are homeothermy or endothermy and the laying of eggs (Oring 1982). The combination of oviparity and endothermy set the stage for the evolution of parental care and monogamy in birds because these two factors make possible extraordinary benefits of intensive parental care. Oviparity and endothermy in conjunction meant that (1) as a result of an acceleration of metabolic processes only a short period of time elapsed between copulation and the appearance of the egg or eggs, and that (2) the rate of development of the zygote could be greatly accelerated by the application of body heat—incubation. These two points have important ramifications for the evolution of parental care by males, in particular, in that they provide a male bird both with an increased probability of paternity and the opportunity to contribute almost immediately to the welfare of its offspring-to-be, by guarding and/or incubating the eggs (see Sections 10.2 and 10.3).

1.4.1   Ecological classifications of mating systems

The subject of avian mating systems has been a popular one, and in the 30 years since the appearance of Lack's (1968) book, *Ecological adaptations for breeding in birds*, many reviews, focusing on different aspects of avian breeding systems, have appeared (e.g. Verner and Willson 1966, 1969; Orians 1969; Selander 1972; Wittenberger 1976, 1979, 1981; Emlen and Oring 1977; Wittenberger and Tilson 1980; Faaborg and Patterson 1981; Oring 1982, 1986; Erckmann 1983; Rowley 1983; Ford 1983; Rowley 1983; Murray 1984; Payne 1984; Gowaty and Mock 1985; Silver *et al.* 1985; Jehl and Murray 1986; Brown 1987; Stacey and Ligon 1987, 1991; Clutton-Brock 1988; Breitwisch 1989; Newton 1989; Stacey and Koenig 1990; Davies 1991; Emlen 1991; Wiley 1991; Koenig *et al.* 1992; Birkhead and Møller 1992; Koenig *et al.* 1992*a*; Johnson and Burley 1997).

Prior to the realization that avian mating systems might be influenced by ecological factors, their classification appeared to be straightforward. Species were classified either as monogamous (one male–one female), polygynous (one male, two or more females), polyandrous (one female, two or more males), or promiscuous (either or both sexes with two or more mates, and with no stable sexual–social bond). A change of mate over time, typically within a single breeding season, was indicated by terms such as serial monogamy, or serial polyandry. Mating systems of species and even sub-families and entire families of birds were characterized by one of these labels (e.g. Lack 1968). In the 1960s, however, it became apparent to some insightful avian behavioural ecologists that breeding systems of some kinds of birds were not fixed characters, but instead were dynamic and apparently correlated with certain kinds of environmental features (e.g. Orians 1961, 1969; Crook 1962, 1965; Verner 1964; Selander 1965; Verner and Willson 1966; Zimmerman 1966).

Subsequent to these ground-breaking papers, probably the most influential papers relating ecology and mating systems have been those of Stephen Emlen and Lewis Oring (1977) and Oring (1982). Previous traditional classifications were based on the number of mates obtained by males and females either simultaneously or serially. This approach lumps mating systems which function in very different ways, and it results in placing the outcomes of different selective factors into similar categories. Emlen and Oring (1977) and Oring (1982) present an alternative classification scheme based on the behavioural and/or ecological potential of individuals to monopolize mates and the means by which monopolization takes place. The concept of monopolization, '... the ability of a portion of the population to control the access of others to potential mates,' (Emlen and Oring 1977, p. 215), is a key aspect of their ecological classification of mating systems.

Other things being equal, polygyny should be a preferred strategy for males, as evidenced by the occurrence of facultative, occasional polygyny in a wide variety of usually monogamous species. Such observations suggest that while polygyny may benefit males, its manifestation usually is constrained; i.e., in most species there may be little, if any, 'polygamy potential' (Emlen and Oring 1977). The position taken here is slightly different: most birds are socially (and to varying degrees genetically) monogamous, because extensive biparental investment has been favoured throughout most of their evolutionary history, and, to this day, for most species, is usually the best available strategy for males as well as females, for reasons discussed in Chapter 11.

In some groups, however, mating systems are highly variable. Not only do mating systems of closely related species often differ, they also may differ within a species, both geographically and by sex. For example, male rheas mate with, and receive the eggs of, several females and thus are 'simultaneously polygynous' while female rheas mating first with one male, and a short time later with another, may be considered as 'sequentially polyandrous'. Similarly, male red-winged blackbirds with several females in their territories are considered to be polygynous, while females are socially monogamous, even though some females mate with more than one male.

Although the treatments of mating systems by Emlen and Oring (1977) and Oring (1982) are more precise than some earlier ones (e.g. Selander 1972), and have stood the test of time extraordinarily well, they have inevitably become somewhat dated. For example, Oring's (1982) comprehensive review of avian mating systems predated the evidence for extra-pair copulation as a widespread mating tactic. As originally applied, the term monopolization appeared to imply that members of one sex, usually female, copulate exclusively with their mate (i.e. that males monopolize fertilizations of their mates). In some populations or species, however, copulations frequently take place outside the social mateship bond, and they often lead to fertilizations (see Chapter 12). Therefore, for a number of avian species, mating systems based on ecological classifications frequently do not accurately or completely reflect the genetic situation. Patricia Gowaty (1985) may have been the first writer to specifically distinguish between social and genetic mating patterns. She suggested use of terms such as 'apparent monogamy,' or 'genetic monogamy' to emphasize the differences between social or apparent and genetic mating systems, respectively.

In addition to the issue of extra-pair copulations, simple labels like 'monogamy' encompass a variety of additional selective pressures on each sex, and thus they include a diverse array of reproductive strategies (Oring 1982; Mock 1985). In this book, I employ the terms monogamy, polygyny, polyandry, and lek-promiscuity to indicate the social number of mates that an individual has at one time. The unqualified term monogamy, for example, should not be assumed to imply anything about either the genetic parentage of a brood of nestlings or the duration of the pair bond. Rather, it is a label employed to indicate the overall social relationship of the male and female participants. At the other extreme, the term promiscuity implies an absence of discrimination by males. It is used here to refer to lekking species, in some of which males are demonstrably non-discriminating in their copulatory activities (e.g. Höglund *et al.* 1995); females of such species, on the other hand, are notably discriminating.

The ecological classifications of mating systems of Emlen and Oring (1977), Wittenberger (1979, 1981), and Oring (1982), among others, made important advances in the understanding of avian mating systems. By showing how the acquisition of mates was related to the temporal and spatial distributions of resources or members of the opposite sex, they demonstrated that mating systems could be flexible and adaptively modifiable, and that individuals often used resources to gain access to mates, or to monopolize mates, and thus to increase their own fitness. These approaches have been most useful in illustrating the kinds of ecological factors that can promote the development of 'facultative' polygyny from monogamy. Such purely ecological explanations of mating systems have been less effective in accounting for the original evolution of monogamy, and they are almost surely insufficient to explain the evolution of classical polyandry. These points are considered in later chapters.

1.5   Adaptations and constraints

In this first chapter I want also to discuss two fundamentally important concepts that permeate this book. The major topics of the book, sexual selection and mating systems, provide endless examples of both adaptations and constraints, which, in a sense, summarize the 'mechanisms of evolution' (Leroi *et al.* 1994). The concepts labelled adaptation and constraint have generated a great deal of discussion, in large part because the ideas these terms represent have been viewed in very different ways. Frederick Sheldon and Linda Whittingham (1997) provide an excellent review of these evolutionary phenomena from an ornithological perspective.

All readers of this book will be familiar with the word adaptation, and most probably think that they know what the term means in the context of evolutionary biology. However, 'adaptive' and 'adaptation' have proved not to be easy to pin down; e.g. 'Adaptation is a complex and poorly defined concept' (Futuyma 1986, p. 550). In large part, this is because these terms have meant different things to different evolutionary biologists. Reeve and Sherman (1993, p. 1) provide two quotes to illustrate this point: (1) 'The difficulty of the concept adaptation is best documented by the incessant efforts of authors to analysze it, describe it, and define it' (Mayr 1983, p. 324), and (2) 'Adaptation is considered a central, yet obscure, elusive, and controversial concept in evolutionary theory' (Krimbas 1984). Clearly, the issue of adaptation alone could merit book-length treatment (e.g. Williams 1966). Thus, the comments presented here should be viewed as no more than an introduction to the subjects of adaptation and constraint.

Before going into the differing viewpoints, I want to consider why the issue of adaptation is crucial to any comprehensive consideration of sexual selection. A recent definition of adaptation states that '... for a character to be regarded an adaptation, it must be a derived character that evolved in response to a specific selective agent' (Harvey and Pagel 1991, p. 13). This view of adaptation has special implications for students of sexual selection in birds, for at least two reasons. (1) It appears that many ornamental morphological traits and weapons (e.g. spurs) of male birds do fit the definition of adaptation offered by Harvey and Pagel (1991), in that they are indeed derived characters that evolved specifically in the context of mating success. (2) While many sexually-selected traits of male birds are adaptive for sexual selection (e.g. mate attraction), under natural selection they may be maladaptive in terms of the individual's survival. This trade-off provides a basis for assuming that costly ornaments did evolve in response to a single, identifiable selective agent, namely female preference.

A well-known example is the barn swallow's tail, which exhibits greatly elongated outer tail feathers (rectrices). Møller (1994) has shown a positive relationship between the length of the male swallow's outer rectrices and attractiveness to females, and he also has also shown that flight manoeuverability and general flight efficiency are negatively correlated with the length of the outer rectrices (see Section 5.6.1). In brief, males of this extremely aerial species could fly and survive better without the tail ornamentation that serves to attract females. Møller (1994) summarizes the trade-off between natural and sexual selection by pointing out that an individual male that grows too long a tail suffers in terms of natural selection, while an individual that grows a short tail is severely penalized in terms of sexual selection. Thus, the long outer rectrices are maladaptive in terms of their effects on the day-to-day existence of males, but this cost is more than counteracted by the benefits obtained in the sphere of sexual selection, and thus reproductive success. It is this trade-off that makes the evolution and maintenance of sexually-selected ornaments so especially interesting.

Although adaptation via natural selection is accepted by evolutionary biologists as the primary basis for the great diversity of functional traits exhibited by organisms, its potency does have limitations. Mayr (1983, p. 331), an adaptationist, states this point as follows: '... there are numerous factors in the genetics, developmental physiology, demography, and ecology of an organism that makes the achievement of a more perfect adaptation simply impossible.' Wilson (1975, p. 32) refers to the limits to evolutionary change or the relative difficulty of altering rates of evolutionary change as phylogenetic inertia, and points out that this inertia imposes constraints on the evolution of new adaptations.

Thus, for these reasons a consideration of both adaptation and phylogenetic constraints is important as we attempt to understand either the origins or the current adaptive significance of courtship ornaments or mating systems. Phylogeny has a major effect on many of the traits that are thought to be important in both phenomena. For example, the bright plumage of males of many North American passerine birds is strongly influenced by their phylogenetic history (Johnson 1991). Among other things, this means that if two or more species under consideration are closely related, each does not represent an independent data point. This is because a significant relationship among traits sampled across related species may be due to common inheritance rather than to convergent evolutionary changes based on similar selection pressures.

1.5.1   Adaptation: just what does this term mean and how do we recognize it?

As touched on above, a recurring and ever-current issue in evolutionary biology relates to the question of 'adaptation.' That this concept continues to be controversial may seem surprising 30 years after publication of Williams' (1966) classic book *Adaptation and natural selection*. However, the meaning of 'adaptation' continues to be re-analysed, defined (e.g. Gould and Vrba 1982; Thornhill 1990; Williams 1992; Reeve and Sherman 1993; Leroi *et al.* 1994; Sheldon and Whittingham 1997) and debated (e.g. Sherman 1988, and 1989 vs. Jamieson 1989*b*). The question of adaptation arises in almost every area of avian breeding biology, thus it is worthwhile to review what the word means to different evolutionary biologists, and some of the reasons that it continues to be a contentious issue.

There are at least two fundamentally different views concerning the biological meaning of the term adaptation. One of these employs a narrow use of the concept, the other perspective utilizes a broader definition, where only the effect of the trait is considered. Here I refer to these as the 'restrictive' and 'flexible' approaches. Because the meaning of adaptation has been viewed and interpreted differently by evolutionary biologists, several direct quotes are presented in the following discussion to provide the reader with a feel for this diversity of perspectives.

The restrictive approach

Williams (1966, p. 10) recognizes adaptation by attempting to determine '… whether a presumed function is served with sufficient precision, economy, efficiency, etc. to rule out pure chance as an adequate explanation.' He also suggests that we should speak of adaptation only when we can '… attribute the origin and perfection of this design to a long period of selection for effectiveness in this particular role.' Williams associates the terms goal, function, or purpose with adaptation to imply that the machinery involved was fashioned by a long history of selection for the goal attributed to it. Gould and Vrba (1982, p. 6) state that 'we may designate as an *adaptation* any feature that promotes fitness and was built by selection for its current role.' Thornhill's (1990, p. 32) view of adaptation is very similar to that of Williams: 'Recognition of an adaptation involves identification of a feature of an organism that is too complexly organized to be due to chance.' Finally, Williams (1992, p. 40) states that 'Adaptation is demonstrated by observed conformity to a priori design specifications.'

Most or all of these views of adaptation require that the term be restricted to traits where natural selection has shaped the character in question for its current function. Thus, in this usage, adaptation is not a label to be applied to each and every trait currently being maintained by natural selection.

The flexible approach

In the eyes of others, adaptation is viewed differently. Here I provide three quotes by authors who advocate a looser, more inclusive approach to recognizing adaptation. (1) 'Adaptation should be defined by its effects rather than by its causes as any difference between two phenotypic traits (or trait complexes) which increases the inclusive fitness of its carrier' (Clutton-Brock and Harvey 1979, p. 547). (2) 'An adaptation is, thus, a feature of the organism, which interacts operationally with some factor of its environment so that the individual survives and reproduces' (Bock 1979, p. 39). (3) 'An adaptation is a phenotypic variant that results in the highest fitness among a specified set of variants in a given environment' (Reeve and Sherman 1993, p. 9). A key to understanding the perspective of Reeve and Sherman (1993, p. 7) is their preference to '… not distinguish between adaptive traits and adaptation.' This, in effect, leads to labelling nearly all traits manifested by a species as adaptations. It also leads to the question: How could one convincingly test or falsify the prediction implied by their definition?

In short, according to Clutton-Brock and Harvey (1979), Bock (1979), and Reeve and Sherman (1993), effects define adaptation, whereas according to Williams (1966, 1992), Gould and Vrba (1982), Thornhill (1990), and Harvey and Pagel (1991), this is not correct usage of the term.

The publication dates of the quotes presented above indicate that the restrictive and flexible interpretations of adaptation have remained constant for quite some time. For example, there is virtually no difference in the definition of adaptation provided by Williams (1966) and that of Thornhill (1990), some twenty-four years later. Similarly, the views of Clutton-Brock and Harvey (1979), Bock (1979), and Reeve and Sherman (1993) are very similar. In short, over the past decade or so, the two major views of adaptation reviewed here have changed little or not at all; nor have they converged. The persistence of the two viewpoints appears to be based on individual preference for either one perspective or the other.

Clearly, these two major perceptions of adaptation can lead to different kinds of interpretations. As one example, consider the feeding of nestlings by non-breeding helpers in cooperatively breeding birds (see Chapter 14). Because this behaviour almost surely reflects a trait originally evolved in the context of parental care, rather than having evolved specifically in the context of helping, Williams (1966, p. 208), Jamieson and Craig (1987), and Jamieson (1989*a*) do not regard helping behaviour as an adaptation *per se*. Nor would Harvey and Pagel's (1991) framework classify helping behaviour as an adaptation. Instead, following the interpretations of these authors, helping is an effect of the existence of a behavioural response that evolved in another context (parental care). If helping has a beneficial effect, in the terminology of Gould and Vrba (1982), it is an 'exaptation'. Alternatively, following the definitions of Clutton-Brock and Harvey (1979), Bock (1979), and Reeve and Sherman (1993), if helpers increase either the direct or indirect component of their inclusive fitness, as a result of their helping activities (i.e. if it is adaptive), that behaviour represents an adaptation. In this book, I attempt to distinguish between 'adaptation,' as used by Williams and Thornhill, and 'adaptive.'

1.5.2   What is a phylogenetic constraint?

As with adaptation, consideration of 'phylogenetic constraint' requires a close look at definitions. Here I provide three recent definitions of phylogenetic constraint, all of which are similar (see also Sheldon and Whittingham 1997). Ligon (1993, p. 3) suggests that the term '... indicates that certain evolutionary pathways are not likely to be followed by a species or group of related species as a result of prior evolutionary history. In short, yesterday's adaptation may be today's constraint.' Figure 1.3 illustrates this perspective. McKitrick (1993, p. 309) defines phylogenetic constraint '... as any result or component of the phylogenetic history of a lineage that prevents an anticipated course of evolu-

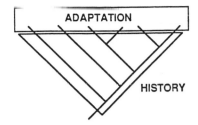

Fig. 1.3 Relative positions of adaptation and history (constraints). A key point is that all traits viewed as constraints are, or have been, adaptations. Reprinted with permission from Wanntorp *et al*. (1990). Phylogenetic approaches in ecology. *Oikos*, **57**, 119–32.

tion in that lineage.' Finally, Miles and Dunham (1993, p. 596) state that: 'Constraint (or inertia) implies the restriction of the range of variation associated with a trait in response to fluctuations in the environment or changes in the selective milieu.'

Animals clearly exhibit many traits that are less perfectly suited to the individual's requirements than we might imagine they could be, thus it should be kept in mind that natural selection does not work like an engineer; on the contrary, it works like a tinkerer (Jacob 1977). The result of the process of natural selection is that organisms exhibit traits reflecting their evolutionary history, and, in some cases, such traits prohibit or constrain the evolution of new or alternative responses to a particular environmental situation. Every student of birds is aware of the most conspicuous morphological and physiological adaptations of birds, and which simultaneously present constraints. The existence of wings, a wondrous adaptation, constrains the evolution of certain other kinds of adaptive uses of the forelimbs, and the cleidoic egg may constrain the evolution of internal development of embryos (Clutton-Brock 1991, p. 80). Gould and Lewontin (1979) argue that not only are constraints to adaptive change present, they should also be interesting in themselves to students of adaptation, in that they delimit or guide the kinds of adaptations that can occur.

To re-emphasize, the words adaptation and constraint should not be viewed as antonyms. Invoking the concept of constraints to help explain a trait or suite of traits should not be taken to imply that the species in question is in some sense handicapped by possession of inferior or imperfectly adapted traits. The ancient traits may well represent adaptations in the strict sense that they have been maintained over evolutionary time. However, those ancient traits, whatever their proximate bases, may also constrain some aspects of the future evolution of the species or groups of lineages possessing those traits.

1.5.3   Criticism and defence of the 'adaptationist programme'

Gould and Lewontin (1979) faulted adaptationists for their attempts to provide plausible but untested explanations for the phenomena they observe, and for ignoring the evolutionary history of the organisms they study (see also Prum 1994). These authors assert that the adaptationist programme breaks an organism into unitary traits and proposes a separate adaptive story for each. Gould and Lewontin suggest that constraints, which they believe are rarely recognized by adaptationists, can be more interesting and more important in delimiting pathways of change than the selective forces that may mediate change when it occurs. Finally, these authors feel that the adaptationist programme fails to distinguish current utility from reasons for origin. This point, earlier recognized by a Nobel Prize-winning adaptationist, Niko Tinbergen (1963), has recently been treated by Paul Sherman (1988), who refers to it as a 'levels of analysis' problem, and by Ian Jamieson (1989*b*), who criticizes Sherman's approach.

In his paper, *How to carry out the adaptationist program?*, Mayr (1983) provides a detailed response to the criticisms of Gould and Lewontin (1979). Mayr points out that post hoc explanations of a phenomenon should not be viewed as objectionable, because they provide hypotheses that can be tested and rejected. Mayr argues that so long as we recognize that constraints imposed by evolutionary history and developmental pathways prevent the achievement of perfect adaptedness and that, therefore, there is no perfect adaptation, the approach of asking hypothetical 'why' questions is legitimate, appropriate, and profitable.

The writing of Gould and Lewontin, on the one hand, and Mayr, on the other, indicate that whether one chooses to emphasize the useful or adaptive nature of traits found in living species, or to emphasize the limits to change built into organisms by their evolutionary history, is largely a matter of predilection. Attention to both, in so far as possible, is essential to obtaining a complete picture of the phenomenon of interest. Paleontologists like Gould are especially sensitive to what has occurred in the past, and are not intrigued by what in the long view of evolutionary time can be considered as minutiae. In contrast, field biologists try to understand the effects of traits they can see and measure on the survival and reproductive success of the individual animals they study, e.g. whether in a given year monogamy or polygyny is slightly more productive for male birds of a given species. Do such small variations tell us little about the evolution and maintenance of mating systems in contemporary species of birds, or is this the very stuff of evolution? (See also Tordoff 1967.)

1.5.4   Adaptation, constraint, and the comparative method

The comparative method is based on the premises that similar ecological–environmental pressures often will lead unrelated species to similar or convergent adaptive responses, and, conversely, that differing selective pressures will cause related species to diverge either behaviourally or morphologically. The comparative method is a powerful tool both for identifying adaptation, and, conversely, for recognizing constraints to adaptation. When, within a group of related species, such as a genus, one of them possesses a trait or character of interest that clearly differs from its relatives, we can be fairly confident that we are on the trail of an adaptive departure from the adaptive traits shared by the entire genus. On the other hand, when all members of a large group of related species share the same rare or unexpected trait, we should suspect that evolutionary history, and possibly constraints of some sort, are involved in the explanation of that pattern.

Many aspects of the breeding biology of birds can be viewed either as evolved adaptations, or, alternatively, as constraints on other, potentially superior ways to reach an adaptive solution to a particular aspect of a species' biology. This may seem surprising since the two concepts have often been viewed, erroneously, as if they were independent, opposing phenomena. In this section I provide examples to show that adaptation and constraint are complementary rather than conflicting concepts, and that both are involved in nearly all evolved responses to environmental challenges. To restate the view espoused here, adaptive changes via the evolutionary process often lead to constraints on future evolutionary pathways.

Fixed clutch size in gulls

Whether a trait is viewed as an adaptation or a constraint often depends on one's perspective. A particularly good ornithological example of this is provided by the response of McLennan *et al.* (1988) to the paper by Graves *et al.* (1984), 'Why does the herring gull lay three eggs?' Graves *et al.* (1984) demonstrated that the hatching of three eggs led to a great increase in the parents' reproductive effort, that the second-hatched chick suffered reduced weight gain, and that only 8 per cent of 74 pairs hatching all three eggs fledged three chicks. These data led Graves *et al.* to conclude that the third egg serves as insurance against loss of one of the first two eggs. However, because Graves *et al.* (1984) ignore clutch size in other *Larus* gulls, as well as in other members of the order Charadriiformes, McLennan *et al.* (1988) contend that the response of Graves *et al.* (1984) to their own question is incomplete. McLennan *et al.* (1988) suggest that since all members of the genus typically lay three eggs, the three-egg clutch is a phylogenetically-based trait that apparently is not readily susceptible to adaptive modification. These authors also argue that since the ancestral clutch size in the order Charadriiformes is apparently four, the appropriate question about clutch size in gulls is not 'Why three eggs?', but rather, 'Why not four eggs?' Their answer to the former question is because *Larus* species '... are descended from an ancestor that produced three-egg clutches' (McLennan *et al.* 1988, p. 2187).

Graves *et al.* (1984) answered the question they posed in terms of *current utility* in one species. In contrast, the response provided by McLennan *et al.* (1988) refers to the *origin* and maintenance of three-egg clutches in all *Larus* gulls. The differing interpretations of Graves *et al.* (1984) and McLennan *et al.* (1988) illustrate that it can also be difficult to determine whether a specific character, such as, in this case, clutch size, is an adaptation or constraint—if, indeed, it is not both (see Ligon 1993 for a more detailed discussion of this example). These differences also illustrate the alternative definitions of adaptation discussed above.

Incubation mounds in megapodes

The megapodes illustrate ways by which phylogenetic constraints can influence the evolutionary course of new adaptations. Among birds this monophyletic group is famous for the unique reproductive strategy of burying eggs underground, where embryonic development occurs. (See Jones *et al.* 1995 for a review of these fascinating birds.) Because incubation occurs underground in all living species of megapodes, most of which occur on islands, this trait can be viewed as an old adaptation that probably was present in the common ancestor of today's species. It is clear that phylogenetic history has played a major role in the reproductive biology of megapodes.

The apparently ancient and 'fixed' behaviour of burying eggs, along with other factors such as length of incubation and precocity of hatchlings, apparently has precluded the evolution of other, more typical forms of egg care and development. (It should also be noted that in most parts of the world, this mode of reproduction probably could not have evolved, or at least could not have persisted, due to predation by placental mammalian predators.; see Jones *et al.* 1995). Among the megapodes, the interspecific variation seen in kinds of incubation sites utilized, mating systems, spatial distribution patterns, etc. (Jones *et al.* 1995), can be viewed as more recent adaptations superimposed on the ancient trait (also adaptive) of burying eggs for their incubation. Thus, although the egg-burying reproductive pattern that characterizes all megapodes can be labelled as an adaptation, this one fundamental characteristic subsequently has set boundaries on the course of future adaptive modifications of species-specific reproductive strategies.

Nocturnal incubation in cuckoos and woodpeckers

Patterns of parental care can constrain the development of certain mating systems. Monogamy and biparental care, including shared incubation, are common mating and parenting patterns within the Class Aves. Often, however, the temporal distribution of incubation and brooding is unbalanced in that one sex incubates more than the other. For example, in most species females remain on the nest overnight. However, in two widespread groups, the non-parasitic cuckoos and woodpeckers, it is the male parent that does so (Skutch

1976, p. 157). Because nocturnal incubation by males is rare among birds overall, and because all non-parasitic cuckoos and all woodpeckers, so far as known, exhibit this trait, whatever their body size, geographic distribution, or ecology, it seems reasonable to ask whether this unusual and seemingly inflexible division of parental care may have imposed any sort of evolutionary constraint on members of these two groups.

Cuckoos. In cuckoos (Fig. 1.4), obligatory nocturnal incubation by males should preclude the evolution of multi-nest polygyny, since a male could perform night-time incubation at only a single nest. Within this primarily tropical and southern hemisphere group of about 143 species, all of the basic kinds of breeding patterns known for birds appear—except multi-nest polygyny. The absence of polygyny, so far as known, which is the second most common avian mating system overall, from this otherwise unusually diverse group (in terms of mating systems), supports the suggestion that the apparently fixed pattern of night-time incubation by the male is a constraint that prevents its occurrence.

The curious reproductive system of another cuckoo, the groove-billed ani, illustrates how this phylogenetically fixed trait of male night-time incubation can influence social–mating systems. In this species from one to four mated pairs jointly occupy a territory, and all females lay eggs in a common nest. The

Fig. 1.4 In the greater roadrunner, as in all non-parasitic cuckoos, the male incubates the eggs at night (e.g. Vehrencamp 1982). Reprinted with permission from Dover Publications, Inc.

dominant female begins to lay last and this bird tosses some of the eggs of the other females from the nest. Of special interest is the behaviour of her mate, the dominant or alpha male. This bird incubates all the eggs at night—his own future offspring, plus some offspring of other, subordinate pairs. This behaviour carries a great cost; breeding season mortality of dominant, incubating males is high, about 31 per cent, as compared to 15 per cent in other males (Vehrencamp 1978). In effect, the male that attains the alpha position has, at the same time, drastically jeopardized his future life expectancy as a result of nocturnal incubation.

Fig. 1.5   In woodpeckers, including the strikingly sexually dichromatic Williamson's sapsucker, the male makes major essential contributions to all aspects of the breeding effort: nest cavity excavation, incubation, brooding, and feeding nestlings. As in cuckoos, male woodpeckers also incubate the eggs and later brood the young at night.

Woodpeckers. The large (more than 200 species) and geographically wide-spread woodpecker group is highly conservative in terms of diversity of mating systems, with nearly all species breeding as simple, monogamous pairs. Why is there not more diversity in the reproductive biology of this major group of birds? As in the cuckoos, nocturnal incubation exclusively by males may prohibit the development of certain mating systems, such as multi-nest polygyny, since the male's incubation contributions are critical and cannot be divided or shared by females at two nests. In addition to the male's role as night-time incubator, male and female woodpeckers probably share the overall costs of reproduction about as equally as is possible. Thus nocturnal incubation exclusively by the male is just one critical factor that usually prevents the development of non-monogamous mating systems.

Male woodpeckers (Fig. 1.5) typically conduct most or all of the nest cavity excavation, and, following egg-laying, they also carry out about half of the daytime incubation. These male duties are perhaps as critical to nest success as night-time incubation, since in woodpeckers the female is not fed by her mate and must forage for her own nourishment. This point may be especially important in this group since many woodpeckers utilize hard-to-obtain food items which probably require a higher than usual amount of time to locate and procure.

In short, in woodpeckers the essential contribution of each sex to incubation, and later to the feeding of nestlings, may mean that neither the male nor the female can normally rear a brood of young without the contribution of the other. Thus, in woodpeckers there is a constraint against the evolution of multi-nest polygyny and multi-nest polyandry. However, breeding systems more complex than simple, single pair monogamy characterize two North American species, the red-cockaded and acorn woodpeckers, both of which are classified as cooperative breeders. In each of these species, as in other, more typical woodpeckers, night-time incubation is conducted by the dominant male parent and both sexes exhibit extensive care of eggs and nestlings. Thus, the social–mating systems of these two species are extensions of, rather than departures from, the basic woodpecker pattern of parental care.

1.6   Prospectus

This book consists essentially of two parts. The first considers sexual selection and related issues (Chapters 2–9) and the latter treats mating systems (Chapters 10–16). In Chapters 2–9, I provide a brief review of the general hypotheses that bear on the main theories of sexual selection, along with a review of empirical data based on studies of birds. These chapters may serve to bridge the theoretical work reviewed and summarized by Andersson (1994) and Møller (1994), on the one hand, and the detailed descriptive information of male displays presented by Johnsgard (1994), on the other. A few additional comments on this portion of the book may be useful.

Chapter 2 describes the main theories proposed to explain the basis of mate choice by females, which is the most controversial, and, to most students, the most interesting, aspect of sexual selection. This review provides the foundation for succeeding chapters. Next, a major aspect of sexual selection theory, namely factors promoting variation in reproductive success, is considered. Chapter 3 argues that a consideration of the principles of classical ethology can help us to better understand the proximate components of female mate choice, and provides a brief review of some still-relevant ethological phenomena and concepts. In Chapter 4 the roles of different types of morphological ornaments and vocalizations of males in mate choice are surveyed. Chapter 5 focuses on four critical aspects of the indicator or good genes perspective of sexual selection, namely (1) the role of testosterone in sexual selection, (2) carotenoid pigments as quality indicators, (3) the possible role of parasites in male ornamentation and female mate choice, and (4) the possible relationship between fluctuating asymmetry of bilateral ornamental traits, male quality, and female mate choice. Chapter 6 reviews a number of empirical studies of female mate choice in birds, along with an assessment and overview of the evidence to date that bears on ultimate explanations of female mate choice. Chapter 7 reviews some phylogenetically-oriented studies of traits associated with sexual selection. A necessarily somewhat speculative Chapter 8 considers the relationship between female mate choice and the process of speciation. Finally, Chapter 9 closes out the subject of sexual selection by briefly considering mate choice by males and intra-sexual competition.

The second half of the book, Chapters 10–16, considers the evolutionary origins and maintenance of avian mating systems, and the factors currently maintaining them. This requires consideration of three related aspects of avian reproduction: (1) oviparity and the evolution of parental care (Chapter 10), (2) social mating systems (Chapters 11, 13–16), and (3) extra-pair copulations, which often lead to extra-pair fertilizations (Chapter 12). The general theme of Chapter 10, in brief, is that egg-laying, together with some other critical traits characteristic of the Class Aves, forms the basic foundation for the evolution of male parental care in birds. Over evolutionary time it has been the opportunity provided by oviparity for extensive paternal care, more than any other single factor, that has led to the different kinds of mating systems and mating strategies (including extra-pair copulations, Chapter 12) that we study today. For example, once parental care behaviour had evolved in both sexes, and once young birds came to be utterly dependent on parental care, the opportunity arose for the evolution of monogamy (Chapter 11).

Subsequently, other mating systems were derived from monogamy (e.g. polygyny, cooperative polyandry, Chapter 13). After, or coincident with, the evolution of distinctive mating systems, additional or alternative mating strategies, such as extra-pair copulations, allowed members of both sexes to mate with individuals in addition to, or other than, their social mates; i.e. extra-pair activities, so fascinating to many students of reproductive strategies, could appear only subsequent to the development of pairs. Chapter 14 considers the phenomenon of cooperative breeding, which is characterized by the presence of non-breeding helpers. Strictly speaking, cooperative breeding is not a mating system at all; monogamous, polygynous, and cooperatively poly-androus mating systems occur among cooperative breeders. Lek mating systems, in which males are promiscuous and females often are genetically monogamous, are reviewed in Chapter 15. Chapter 16 attempts to provide insights into the most puzzling mating system of all, classical polyandry. Chapter 17, the final chapter, attempts to summarize, very briefly, the current status of our understanding of the subjects of sexual selection and mating systems as these subjects pertain to birds.

1.7 Conclusions and summary

Sexual selection and mating systems, the major topics of this book, have, along with parental care, evolved in parallel over the many millions of years since the close of the Cretaceous Era, and the subsequent evolution of the modern avifauna. Sexual selection, which is an inevitable consequence of sexual reproduction, probably predated the evolution of both the complex parental care and mating systems seen in contemporary birds; i.e. sexual selection is important in animals that exhibit neither parental care by either sex, nor any sort of social–mating relationship beyond copulation. For example, the evolution of young that absolutely required parental care led to biparental care and the development of complex mating systems.

Once all three of these factors became important, they came to exert reciprocal effects. Among today's birds, relationships can be seen between mating systems and the strength of sexual selection, and between the demands of parental care and mating systems. Studies of sexual selection, or of mating systems, should include consideration of the form of parental care required. For example, most types of birds exhibit monogamous mating systems and in these care by both parents is critical (see Section 11.3). Sexual selection in such species is thought to be less strong than in species in which a single parent, typically the female, can rear a brood of young; such species often are polygynous.

Many adaptive features, both morphological and behavioural, are involved in the phenomena of sexual selection, mating systems, and parental care. Although the great majority of traits that we see in birds serve adaptive func-tions, for many years evolutionary biologists have disagreed on the meaning of 'adaptation,' and the last words on the subject remain to be written. Similarly, the concept labelled 'phylogenetic constraint' has sometimes generated debate. Based on a definition of adaptation which emphasizes that the trait in question evolved specifically for its current function, the view taken here is that an adaptation at one point in the history of a lineage may be the source of a constraint at a later time. At any point in the evolutionary history of a lineage of birds, for example, each individual exhibits many adaptations *per se*, many adaptive traits that are not adaptations in the strict sense, and many constraints. For the Class Aves, one of the most distinctive, spectacular, and widespread morphological adaptations (apart from feathers, which define birds) is modification of the forelimbs and associated structures for flight. At the same time, wings apparently also represent major constraints on the evolution of the forelimbs for other functions.

Although most other traits of birds are less conspicuous than the wings, with regard to the interplay of adaptations and constraints, the same relationships hold. For example, production of crop milk and the associated behaviours for the feeding of young are clearly very special adaptations of pigeons and doves. The evolution of this unusual form of parental care also produces constraints, in that biparental care is critical (i.e. other mating systems, such as polygyny, apparently have not evolved, and possibly cannot evolve, in this group), and clutch size is limited to one or two eggs (i.e. there is likewise little scope for adaptive modification of clutch size).

2 Sexual selection: theories of female mate choice

2.1 Introduction

Sexual selection is based on variation in mating success. This variation has two components: intrasexual competition, which often is openly agonistic in nature, and competition to attract members of the opposite sex. In this chapter, the most controversial and thus most interesting aspect of sexual selection— mate choice by females—is considered. The key question here is the one that attracted Darwin's interest, namely, how and why are the amazingly extravagant ornaments exhibited by males of many avian species attractive to females? The major theories addressing this question are presented below. This is followed by a discussion of an issue critical to sexual selection theory, namely, variance in the reproductive success of males. This chapter is meant to provide the theoretical background for the issues addressed in Chapters 3–6, in particular.

Mate choice by females is the phenomenon that has led to so much interest in sexual selection among theorists and empiricists alike. For students of sexual selection, both the proximate and the ultimate factors influencing the choice by a female of one male over another have been perplexing, controversial, and sometimes contentious issues. As a result, mate choice by females has been, and continues to be, the most studied and argued aspect of the phenomenon of sexual selection. In monogamous species in which male parental care is the rule, one might expect females to choose a mate, at least in part, on the basis of traits related to indicators of parental quality. Provisioning of the female by the male during 'courtship feeding' (e.g. common tern, Nisbet 1973) is thought to serve this function. In other cases, it appears that the female may be interested in a particular territory, rather than in the male owner *per se*, especially when the territory holds resources of importance to her.

These are not the kinds of situations, however, that have generated the disagreements. Instead, it is those species in which males typically are polygynous or promiscuous, and in which males provide neither parental care nor territorial space for females to utilize, that have led to controversy (e.g. Bradbury and Andersson 1987). In many polygynous and lek-promiscuous species—e.g. most grouse, the ruff, most birds of paradise, bowerbirds, manakins, hummingbirds, and others—males contribute only gametes to the females' reproductive effort. Males of these species tend to be more highly ornamented and to display more striking courtship behaviour than do their closest relatives in which males are territorial and provide parental care. The extreme male ornamentations and displays seen in many polygynous and promiscuous species (e.g. Fig. 2.1) generally are assumed to function in the choice of mates by female birds. Since males provide neither resources nor parental care to chicks, the key question is, what is the significance of mate choice by females of such species?

In the following sections I provide a brief overview of each of the current major hypotheses or ideas proposed to account for the relationship between

Fig. 2.1 Perhaps excepting some of the birds of paradise, one of the most bizarre examples among birds of a sexually-selected ornament is exhibited by the male Bulwer's or wattled pheasant. The upper and lower lobes of the bright, sky-blue wattles, which contrast strongly with the red ring rimming the red eye, are extended at the onset of the courtship display by the injection of blood.

male ornaments and displays and mate choice by females. This review will allow the reader to evaluate the theories while considering the evidence discussed in subsequent chapters that are devoted to empirical studies of female mate choice.

2.2 Direct selection

The basic idea behind the direct selection hypothesis is that females are attracted to males for reasons having to do either with their own welfare during mating or nesting or for reasons reflecting aspects of their evolutionary history that are not directly related to mate choice (Kirkpatrick and Ryan 1991). In both cases, mate 'choice' may actually be based on natural selection rather than on sexual selection. Table 2.1 lists several factors potentially important to females that fall under the heading of direct selection.

For birds, male provision of resources, including the territory, is probably the most common (or at least best documented) type of direct selection for mating preferences. For some species, the evidence that females choose to mate with males controlling important resources, such as a territory, is convincing. Although direct selection clearly is an important factor in mate choice in many kinds of birds, particularly territorial, socially monogamous species, it is less obviously so in those species in which the male provides nothing but sperm to the female's reproductive effort, as in lekking species. In such species, neither parental care nor defended resources are provided by males, thus direct selection clearly does not provide an adequate, satisfactory explanation for such systems. To re-emphasize, it is those mating systems in which males usually exhibit highly ornate plumage, often in concert with bizarre behavioural displays, that provide the major evolutionary puzzles and controversies concerning the ultimate significance of mate choice by females.

Table 2.1 Direct fitness benefits associated with mate choice by female birds

Males provide territorial space and the resources therein (e.g. nesting and foraging sites)[1]

Males provide care and protection to females (e.g. 'courtship feeding')[1]

Males provide care and protection to the female's offspring[1]

Costs of search for mates

Avoidance of hybridization

Differences among males in sperm viability or quantity

Avoidance of disease or parasite transmission

[1]For the 90 per cent of avian species that are monogamous, one or more of these factors appear to be the primary direct benefits that females can obtain via selective mate choice.

2.3   Good genes

Over the past several years, two major theoretical categories of explanations for female choice have predominated. One of these, the 'good genes' school of thought, holds that females use male ornaments or displays to gauge the genetic quality of males that might sire their offspring. For example, the genes of chosen males may provide their offspring with inherited resistance to disease or parasites. In this case, the good genes hypothesis postulates that a female can enhance the well-being of her young (e.g. increase resistance to disease) by choosing to mate with a demonstrably healthy male, simply because the genes of the female are joined with the genes of a healthy male to produce her offspring.

The basic logic of the good genes hypothesis is simple, and to many students of sexual selection it is also compelling. Because birds and other animals exhibit so many kinds of precise and important adaptive responses to the environment (including the social environment), should not females attempt to maximize their own reproductive success and thus the health and vigour, and ultimately the reproductive success, of their offspring by choice of a demonstrably high-quality, healthy (i.e. parasite- or disease-resistant) mate?

By this view, if females are interested in judging or ascertaining the genetic quality of males, they should focus their attention on male features that honestly indicate physical condition. This would promote the evolution of those traits of males that signal their condition to prospective female mates. In brief, under the good genes scenario, ornaments or displays are viewed as visible indicators or markers of the underlying genetic quality of males. This is the general premise underpinning the handicap principle (Zahavi 1975, 1977), the truth-in-advertising hypothesis (Kodric–Brown and Brown 1984), the parasite hypothesis (Hamilton and Zuk 1982), the fluctuating asymmetry hypothesis (e.g. Møller 1993*a*), and other variations on the 'good genes' concept (e.g. Andersson 1982*a*, 1994; Borgia *et al.* 1985; Borgia 1987; Bradbury and Andersson 1987).

A general criticism raised by some population geneticists about the good genes hypothesis is that 'fitness' is not a heritable character. For example, Charlesworth (1987, p. 22) writes:

> … it is a well-known result of population genetics theory that natural selection tends to exhaust the additive genetic variance in fitness (Fisher 1930, 1958). The intuitive reason for this is that a positive correlation in fitness between parent and offspring will lead to an increase in the mean fitness of the population from one generation to the next, so that a population in genetic equilibrium under the action of selection alone cannot exhibit such a correlation. This creates a serious difficulty for the good genes hypothesis.

In other words, if fitness *per se* is genetically controlled, genetic variation, over time, will decrease to the point that all individuals within a population will have the same genetic make-up, and selection thus can no longer discriminate among individuals. In contrast to this expectation, Pomiankowski and Møller (1995) argue that sexually-selected traits can have high genetic variability as a result of strong directional selection in the past.

While it is almost surely true that overall fitness is not heritable (simply because it is made up of so many complex, subtle, and sometimes antagonistic components), it is also true that there is considerable evidence for the heritability of many important components of fitness. If a correlation exists between traits of males used by females in mate choice decisions and probability of inheritance from the male of traits beneficial to the females' offspring, such as resistance to the effects of locally and currently important parasites or diseases (Hamilton and Zuk 1982), or resistance to environmental perturbations during development, as indicated by the degree of fluctuating asymmetry (e.g. Møller 1994), then the good genes perspective may be correct. A discussion of the role of male condition in sexual selection is provided in Chapter 5.

2.4   Arbitrary mate choice

Arbitrary mate choice can be divided into two categories, Fisher's (1930, 1958) 'runaway hypothesis' and aesthetic mate choice (Burley 1986*a*).

2.4.1   Fisherian runaway

Over the past decade, the most commonly invoked alternative to the good genes hypothesis has been Fisher's (1930, 1958) 'runaway hypothesis'. This model is based on the idea that female choice of males can be based on male ornaments, with no underlying message about male quality being conveyed. An ornamental trait may be preferred by females simply because it is for some reason attractive, and the greater the development of the ornament, the more attractive it is.

Females that prefer an exaggerated male trait will, as a result of having mated with such males, produce sons also bearing the exaggerated trait, as well as daughters that have a preference for males bearing the exaggerated trait. As a result of the linkage between the gene(s) producing the favoured phenotype in males and the gene or genes leading to the female preference, referred to as linkage disequilibrium, this interaction will increase in intensity, leading to greater and greater elaboration of the male trait and stronger female preference for it, until further development is counteracted finally by natural selection on the trait (Lande 1981; Arnold 1983). Like the good genes hypothesis, Fisher runaway is an ultimate explanation, in that it accounts for the evolutionary effects of female mate choice.

Fisher (1930, 1958) suggested that the male trait initially must have some survival value in order to set the runaway process in motion. More recent models, however, indicate that, in theory, even the initiation of runaway selection can be based solely on arbitrary female preferences (Lande 1981; Kirkpatrick 1982, 1987). Under this latter view, females can be attracted to a particular male trait simply because it 'catches their fancy,' and this alone can lead to increasingly strong selection favouring that trait. (How it is that certain characters, such as a particular colour or movement, initially catch the female's eye is another question; a possible answer to this is briefly described below under sensory bias.)

To illustrate the hypothetical runaway process, let us consider a real species in which it certainly appears that this form of selection could have operated. Among the ornate pheasants, tail length is most extremely developed in the male Reeve's pheasants (Fig. 2.2), with lengths of up to 1.6 metres (Delacour 1977). Application of a Fisher runaway interpretation to this case would be more-or-less as follows. In the distant past, in the shorter-tailed ancestor of contemporary Reeve's pheasants, those females attracted to longer-tailed males produced sons with long tails. Over time, the gene(s) controlling male tail length and the gene(s) for the preference for long tails became linked, and females preferring long-tailed males produced daughters with a preference for the long tails, as well as long-tailed sons. The longest-tailed sons had the same mating advantage over rivals that their fathers enjoyed, or an even greater advantage, due to the increasing frequency of the gene(s) for preference for long tails in females. Across generations the daughters became increasingly more likely to prefer the longest-tailed males. As a result of the genetic linkage between tail length in males and preference for long tails in females, this pattern accelerated over time, or 'ran away', leading to the extreme elongation of tails seen today in Reeve's pheasants (see Section 6.4.1 for an assessment of the runaway hypothesis).

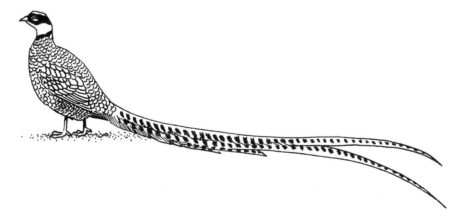

Fig. 2.2   Male Reeve's pheasants possess extraordinarily long tails (up to 1.6 m), even for a pheasant.

2.4.2   Aesthetics and sensory bias

Nancy Burley (1985) suggested that preferences for certain traits (e.g. colours or colour patterns) may exist within the brains of female birds prior to the evolutionary development by males of those favoured traits. This idea is similar to the sensory bias hypothesis, which was developed later and is described below. Such preferences, it was suggested, were 'aesthetic'. Aesthetic mate choice '... is a general term describing preferences for traits that are not indicative of male genetic or economic attributes, but are favoured simply because females find them attractive' (Johnson *et al.* 1993*a*, p. 138). Kristine Johnson *et al.* (1993*a*) consider the Fisher runaway and sensory bias hypotheses to fall under the general heading of aesthetic mate choice, and this is the arrangement followed here. The data used by Burley and Johnson and their colleagues to support the idea of mate choice based on aesthetic traits come from studies involving coloured leg bands. Some of these studies are reviewed in Chapter 6.

Other evidence that some birds may possess an aesthetic sense are the individually unique bower decoration characteristic of some species of bowerbirds. Based on the reasonable assumption that males decorate their bowers in a manner attractive to females, it appears that both sexes behave in a way consistent with the definition of 'aesthetic sense' presented by Johnson *et al.* (1993a). The extent to which the female bowerbirds prefer novel bower decor is not yet known (see Section 6.4.3.2).

The sensory bias hypothesis suggests that female animals may have a predisposition towards certain stimuli, and that when a male presents such a stimulus, the female responds in a positive manner. That is, the female's 'preference' predates appearance of the stimulating trait in the male. The best-known putative example of sensory bias (or exploitation) is the response of females of the tungara frog (*Physalaemus pustulosus*) to the 'chuck' call of males (Ryan *et al.* 1990; Ryan and Rand 1993, 1995; but see Pomiankowski 1994). Originally it was thought that female preference for the chuck had evolved before the call appeared. In other words, the sensory receptors of females were biased (for unknown reasons), so that they responded positively when the chuck call of male frogs appeared. A similar picture has been drawn by Basolo (1990, 1995*a,b*) for preferences by females of swords in swordtails (*Xiphophorus* sp., a fish) and some of their non-sworded relatives. The basic idea behind the sensory bias view of female mate choice is that a bias may evolve because it is adaptive in another context, because it was adaptive in the past, or it may not have an adaptive background at all (Basolo 1995*b*). Improved phylogenies suggest that the presumptive cases of sensory bias may be incorrect; females may retain a preference for certain male ornamental t raits even after the trait has been lost or modified (e.g. Hill 1994*a*). Thus, sensory bias explanations require accurate phylogenies, and this may be one of the greatest challenges to them (e.g. Meyer *et al.* 1994; Pomiankowski 1994).

With regard to sexual selection in birds, females may be attracted to males, and subsequently may mate with them, as a result of some sort of stimulus provided by males to which females are initially disposed to respond for some reason other than mate choice. If, for example, the attention of females of a given species is attracted to some environmental stimulus for reasons related to food (e.g. an attraction to red because red berries are an important dietary item), males may evolve a red morphological characteristic (along with a behavioural display that exhibits the red) to takes advantage of the tendency of females to approach the colour red (Fleishman 1992). In effect, some aspect of the male's ornamentation or display 'mimics' and exploits a pre-existing stimulus or attractant to the female.

It should be noted that, in contrast to the good genes and Fisher runaway hypotheses, which provide ultimate explanations for female mate choice, the sensory bias hypothesis is a proximate explanation of mate choice that, on its own, does not consider its evolutionary function (Hill 1995a). By this view, when elaboration of male ornamentation or behaviour occurs (in birds these usually are either visual or acoustical signals), it does so directly as a result of an already-present predisposition of females to respond to that type of stimulus. There need be no underlying message about male quality, although the sensory bias hypothesis does not preclude the possibility that either selection for good genes or a runaway process could promote further elaboration of the trait in question to convey such information (Ryan and Keddy–Hector 1992).

Probably the single most important general aspect of the sensory bias hypothesis is that it provides a plausible explanation for why most or all of the females in a population might be expected to respond in the same way to the appearance via mutation of a certain new male trait (i.e. why directional selection for a new mutant male ornamental trait might be likely to occur). That is, if some females were simply arbitrarily interested in any novel trait, while other females responded to other novel stimuli, there would be no mechanism for either rapid or consistent directional selection of some new mutant characteristics and not others throughout a species.

2.5   *Variance in reproductive success*

In addition to reviewing the main theories of male traits and female mate choice, it may be useful to review another important general issue, namely inter-individual variation in lifetime reproductive success. In this section, the relationship between such variation among males and the strength of sexual selection is emphasized.

The degree of male ornamentation often is related to mating system. Within a taxonomic unit, such as a family, a correlation between extent of polygyny and extent of sexual dimorphism in size, shape, or colour has been described for birds by Robert Payne (1984; see Chapter 7). The prevalent explanation for this correlation is that traits involved in mate choice often are especially exaggerated in polygynous or promiscuous species because of the great variation among males in mating success (Arnold 1983; Payne 1984), with the majority of females choosing to mate with the small minority of males that bear the most extremely developed ornaments. In such systems, most males may obtain no matings. For example, several studies of lekking species indicate that a small fraction of the males (10 per cent or even less) obtain 80 per cent or more of all matings (e.g. Vehrencamp *et al.* 1989). Under these circumstances, it is obvious that heritable traits of the highly successful males, including those associated with mating success, will spread in the population at a much faster rate than would the traits of any individual male in a genetically monogamous system.

When considering the relationship between the evolution of extreme male ornamentation and female mate choice, possibly the most important point to understand is the role of this variation among individual males in reproductive success. This is because:

> The intensity of sexual selection can be estimated from the variation in the breeding success among individual males in a population. ... The degree to which individuals vary in success sets limits on how rapidly selection can occur, assuming that genetic differences are associated with differences in success. (Payne 1984, p. 3.)

Great variation among males in reproductive success is thought to be most responsible for the extreme development of sexual dimorphism in body size, ornamental traits and/or displays in polygynous and promiscuous-lekking species. This is because those traits possessed by the few successful males are much more strongly favoured (are incorporated into the local breeding population much faster) than those of the reproductively less successful or unsuccessful individuals, simply because relatively few males in a population sire most of the next generation. Strong selection such as this presumably can rapidly promote the evolutionary development of extreme sexually-selected morphological characters—if there is additive genetic variance for the trait. In short, in polygynous species, morphological or behavioural display traits can become extremely exaggerated in an evolutionarily short time, in large part as a result of female preferences.

In addition, there is another factor that should further increase variance in male reproductive success in many species with strong sexual selection, e.g. lyrebirds, manakins, bowerbirds, and birds of paradise. Female lyrebirds and most birds of paradise lay a single egg, female manakins lay two, and female bowerbirds produce a clutch of either one or two eggs. For single-egg clutches, paternity obviously cannot be shared; it is all or nothing. That is, for males of species in which females lay a one-egg clutch, fertilizing part of the clutch is not a viable option; to a lesser degree this is probably also true for two-egg clutches. Thus, a minimal clutch size of one alone will increase variance in male reproductive success, relative to species that produce larger clutches.

Although the details differ, this general point is also true for some or all the lekking Nearctic grouse. In these species, females copulate only once prior to laying a clutch of eggs, with that single copulation sufficing to fertilize the entire clutch. Thus, as in the species that lay a single-egg clutch, paternity of a 'brood' is all or none. In both the lyrebird and the grouse, the effects of the extreme variance in male copulation frequency are further enhanced by the fact that one successful copulation completely removes a female from the pool of potentially available mates for all other males for the duration of the nesting cycle.

With regard to variation among individuals in mating success, Darwin (1859, 1871) first recognized the importance of inherent differences between the sexes in their reproductive potential: due to male–male competition and female mate preferences, and the ability of a single male to inseminate many females, many males will fail to sire any offspring, while a few others will be extremely successful. On the other hand, other factors being equal (which, in the real world they usually are not), female birds of a given species have, in general, the opportunity to produce about the same number of progeny as a result of the comparatively few eggs they lay and are able to incubate, the lesser intensity of female–female competition, and the greater likelihood that all females will have the opportunity to breed. These differences in the reproductive potential of the two sexes are thought to be reflected by sex-specific differences in reproductive strategies and in morphological traits.

By use of fruit flies, Bateman (1948) experimentally demonstrated and quantified Darwin's insight. He showed that a few males obtain many matings and many obtain none. In contrast, variation in reproductive success among females was comparatively slight. Bateman also found that a male's reproductive success was limited by its ability to coax females to mate, and that the number of offspring sired by a male was positively related to the number of his mates. Not surprisingly, the situation was different for female fruit flies. Females had numerous potential mates and the number of offspring they produced was not related to number of mates.

For birds, Payne (1984) analysed the relationship between mating system and variation in reproductive success by sex. Confirming Darwin's interpretation of the relationship between mating system and degree of sexual dimorphism, Payne presented detailed comparative data and analyses demonstrating that the variation in breeding success among individual males in a population (intensity of sexual selection) is highest in lekking species and lowest in monogamous species with shared parental care. Payne (1979, 1984) has also shown that for several species variance in breeding success among females is significantly less than for their male counterparts.

Using the relationship between mating system and his measure of intensity of sexual selection, Payne (1984) also examined the relationship between mating system and sexual size dimorphism to test the hypothesis that the extent of sexual size dimorphism is an evolutionary response to sexual selection. (Several other studies of the relationship between body size and ornaments and mating system are discussed in Chapter 7.) Payne compared lekking and non-lekking species with regard to degree of size dimorphism between the sexes, and in many cases found greater differences in those species with more intense sexual selection, although other factors also affected this relationship. For example: (1) within a taxonomic group, body size is positively related to both lekking behaviour and sexual size dimorphism; and (2) in some polygynous groups, such as manakins and hummingbirds, males are smaller than females. These birds display aerially, thus the small size of males may be associated with superior manoeuvrability during displays or male–male aerial confrontations, or both.

Although this discussion has emphasized the relationship between the intensity of sexual selection and mating system as the most critical factor promoting variation in male reproductive success, two additional points should also be briefly mentioned. First, in classically polyandrous species (see Chapter 16), greater variance in reproductive success may be shown by females than by males. In general, in such systems females should be larger than males (e.g. Oring *et al.* 1994). Second, the extent of such reversed size dimorphism might be expected to be related to the mating system, with those species in which variation among females in number of mates is greatest showing the most pronounced sexual size dimorphism. Even so, however, the highest intensities of sexual selection among females in classically polyandrous systems is less than for males in lek-promiscuous systems, simply because females are always more limited by the number of eggs they can lay than males are limited by sperm production.

2.5.1 Sexual selection and artificial selection

Employment of strong sexual selection based on variance in mating success to explain the development of extreme and sometimes truly bizarre sexual traits in wild birds, especially polygynous or lek-promiscuous species such as some cotingas, manakins, birds of paradise, bowerbirds, and pheasants, is analogous to what has brought about the outlandish development of morphological traits by artificial selection on domestic species (Fig. 2.3). Consider the variation among breeds of domestic chickens, pigeons, and ducks, for example, and the rapidity with which morphological characters can be selected and 'fixed' by artificial selection. It is not difficult to see that, in nature, circumstances that also promote intense selection, such as extreme skew among males in mating success, can produce spectacularly improbable birds. The basic process that produces a bizarre silkie chicken or a bizarre Bulwer's pheasant (Fig. 2.1) is the same: extreme variation in mating success based on strong selection for particular morphological characters. In the first case, this is provided by the human poultry breeder, and in the second, by the natural process of strong directional sexual selection.

Fig. 2.3   Male 'silkie' and male red junglefowl. All domestic breeds of chickens, including the silkie, are derived from the red junglefowl. The silkie illustrates well the powers of artificial selection with fine, hair-like feathers and other artificially-selected anatomical peculiarities, such as two halluces on each foot and a greatly reduced comb.

Although the processes of artificial and natural selection are fundamentally similar, modifications of domesticated species via artificial selection by humans is probably always more rapid than sexual selection in nature, for at least two reasons. First, under artificial selection, both sexes often are subjected to intense selection (i.e., there is great human-dictated variance in reproductive success of females as well as males). Second, most of the forces of natural selection that may retard the rate of directional evolution or limit the degree of change of sexually-selected morphological characters in wild animals (e.g. visually-oriented predators) are absent.

The parallels between the effects of artificial selection and directional natural/sexual selection can also be seen in comparisons between males of closely related species which differ strongly in ornamental plumage. Although males of such species may be extremely divergent in appearance, otherwise they appear to be very similar genetically, as are the breeds of domestic fowl. For example, males of the Lady Amherst and golden pheasants differ strikingly in appearance, yet all combinations of hybrids of these two pheasants are completely interfertile. Except for the genes (probably few in number) that control male colours and patterns, golden and Lady Amherst pheasants probably are extremely similar genetically (see Section 8.3).

2.6   Conclusions and summary

The significance of female mate choice is the most studied and most controversial aspect of sexual selection. This is particularly true for species in which
females obtain neither resources nor parental care from males. In such species,
sperm is the only male commodity of importance to females. Three broad categories of explanations have been proposed to account for female mate choice.

1. As its label indicates, *direct selection* is based on direct benefits to
the female. Examples include choosing a mate that occupies a territory of
high quality or reducing risk of predation associated with mating simply by
choosing the first male located.

2. The *good genes* perspective is based on the idea that females stand to gain
indirect benefits by choosing a demonstrably healthy male and thus obtaining
his genes for their offspring. Females might recognize healthy males by
assessing condition-dependent traits that reliably signal male vigour and health
(e.g. resistance to parasites and other pathogens).

3. To date, empirical evidence supporting *arbitrary mate choice* is limited.
Arbitrary or aesthetic mate choice includes three hypotheses. (a) The best
known of these is Fisherian runaway selection, which is based on the genetic
tie between a male ornamental trait and a female preference for the trait. The
critical factor driving such an evolutionary process may be the ethological phenomenon known as the supernormal stimulus (see Section 3.2.3), which provides a proximate mechanism for the attraction to females of ever larger or
more elaborate traits of males. (b) Evidence for aesthetic preferences is of two
types. First, some putative cases of aesthetic choice are based on the responses
of females to males with particular traits that do not appear on normal males
(e.g. coloured leg bands). Second, in at least one bowerbird, the Vogelkop
bowerbird, males exhibit novel colour preferences as they decorate their
bowers, and they arrange decorative items in novel patterns, for which no
functional explanation has been provided. (c) To date, the sensory bias hypothesis is not well documented among birds. This hypothesis posits that a
female is attracted to a particular male trait as a result of some aspect of the
species' evolutionary history apart from mate choice (e.g. attraction to a particular colour as a result of a prior association of that colour with food). Sensory
bias is a proximate explanation and thus does not address the question of
whether mate choice by females is functional.

An important aspect of sexual selection in birds is the potential for great
variation, usually among males, in mating success. Such variation is the
primary basis for the differences among species in the strength of sexual
selection. This factor is thought to account for the correlation between male
ornamentation, displays, etc., and mating system (see also Chapter 7).

3 Ethological concepts and sexual selection

3.1 Introduction

Sexual selection in birds is characterized most conspicuously by the extreme development of ornamental traits apparently designed to attract members of the opposite sex. In discussing the marvellous diversity of avian courtship displays and ornaments, Armstrong (1965, p. 66) writes:

> Nature uses the simple, primitive reactions, and not only formalizes them but gives them special, conventional and sometimes arbitrary meaning. She loves complexity, and experiments to discover into what strong designs she may weave her simplest patterns; she does not like to discard any organ or habit until she has played a prolonged game with it, trying to discover how it may be more amply employed—even to serve ends quite other than those for which it was designed.

Although Armstrong's phraseology is not considered scientifically acceptable today, the basic point he makes is valid. Specifically, existing characters or traits, whether anatomical or behavioural, are necessary precursors for the evolution of new traits. A complex anatomical ornamental structure or stereotyped display movement does not suddenly appear out of thin air. On the contrary, it is modified by selection from an already-existing trait, one which may originally have served an entirely different function. With regard to stereotyped behaviour patterns this important point has been especially well documented by the study of courtship displays of male birds, in which both anatomical structures and behavioural postures and movements often are derived and modified from pre-existing morphological and behavioural characteristics. Some examples are given in the following sections.

Development of ornamental traits sometimes seems to be in opposition to natural selection. Among birds, the trains of peacocks, the tails of widowbirds, and the bowers of bowerbirds provide well-known examples. Attempts to understand the evolution of these structures occur at two levels. We can ask how such an ornamental structure has the effect of attracting females (a proximate question) and we can ask why females choose from among several available males (an ultimate question). More specifically, just what might a particular male trait signify to females; i.e. what is the function of choice by females? It is important to emphasize at the outset of this chapter why this distinction is important in the area of sexual selection; it is because ultimate explanations (good genes, Fisherian runaway) have sometimes inappropriately been contrasted with proximate ones (sensory bias) (Hill 1995*a*).

More generally, why do we need to consider classical ethological concepts at the dawn of the twenty-first century? The short answer is that such concepts can contribute to a fuller understanding of the key issues of sexual selection, the proximate and ultimate bases of female mate choice. One of my colleagues has dismissed much of sexual selection as little more than a subset of classical ethology dressed up in newer terminology, while another has expressed the view that classical ethology is irrelevant to the modern study of sexual selection. As the existence of this chapter indicates, I do not agree with either of these extreme viewpoints.

Prior to and during much of the 1960s, the field of animal behaviour dealt primarily with attempts to explain phenomena by focusing on proximate issues. Two of the foremost pioneers of the discipline of ethology worked with birds. Nobel Prize winners Konrad Lorenz and Niko Tinbergen were among the first to propose that highly specific neural mechanisms may be responsible for the development of rules of behaviour commonly observed in animals. Examples include the specialized begging for food characteristic of nestling altricial birds, the following response of newly hatched precocial birds known as 'imprinting,' and the attraction to stimuli larger than life ('supernormal stimuli') (see below).

In birds, much behaviour appears to be largely 'hard-wired'. That is, the individual is programmed to respond to a particular stimulus in a rather stereotyped manner, although the manifestation of the behaviour may depend on species, sex, hormonal condition, etc. Specifically with regard to mating behaviour, the courtship behaviour of males and the responses of females to the males' displays have been explained proximately by use of these hypothetical neural concepts, some of which have recently been empirically confirmed (Balaban 1997). These displays and responses presumably have been designed by evolution to guide the animal to make appropriate responses to stereotyped, generally species-specific, stimuli. Such stimulus–response exchanges are often referred to as instinctive behaviour, and models of behaviour based on ethological concepts are presented in most textbooks on animal behaviour. At about two weeks of age, young male domestic turkeys will posture and strut in a manner like that of adult turkey gobblers (Fig. 3.1).

Fig. 3.1 Strutting turkey gobbler. Young male domestic turkeys commonly perform this display, sometimes while still in downy plumage, demonstrating the 'hard-wired' nature of certain types of behaviour.

Before the 1970s, ultimate explanations were used for some aspects of the reproductive activities of birds, especially clutch and brood size (e.g. Lack 1968). Little attention, however, was given to the ultimate factors underlying sexual selection, especially male courtship and female mate choice. The lack of interest in ultimate questions related to mate choice cannot be fully accounted for by the absence of theory to deal with them. Not only had Darwin first recognized the importance of sexual selection and considered its evolutionary implications in detail, but Fisher (1930, 1958) had provided a detailed theoretical explanation for the elaboration of male ornamental traits based on female choice. However, as discussed briefly in Chapters 2 and 8, for many years sexual selection was ignored by leading evolutionary biologists, and the significance of Fisher's work was not fully appreciated at the time. Also, as previously noted, the 1960s and 1970s marked the end of this 'forgotten era' of sexual selection, and to some extent social behaviour in general (West-Eberhard 1983). A number of important theoretical publications led the way (e.g. Hamilton 1963, 1964; Maynard Smith 1964; Campbell 1972; Selander 1972; Trivers 1972; Alexander 1974; Zahavi 1975, 1977). These papers, along with two extremely influential books (Williams 1966; Wilson 1975), opened the era of emphasis on ultimate factors in evolutionary biology. This shift in attention from proximate to ultimate explanations was also reflected by books on animal behaviour published at that time (e.g. Alcock 1975; Brown 1975). Thus, for about the last twenty years, most considerations of sexual selection have focused almost exclusively on ultimate explanations.

The premise upon which this chapter is based is that, with regard to sexual selection, particularly the issue of female mate choice, the pendulum perhaps has swung too far in the direction of ultimate explanations; that is, we need to pay more attention to proximate aspects of male choice. An explicit consideration of proximate ethological phenomena—together with evaluation of ultimate hypotheses—can contribute to a more comprehensive understanding of the workings of sexual selection in general, and female mate choice in particular, than can ultimate explanations alone. No one should be satisfied to ignore ultimate explanations, but simultaneous consideration of proximate factors will increase our appreciation and understanding of sexual selection. For example, the sensory bias hypothesis (Ryan *et al.* 1990), briefly described in Chapter 2, is a good example of how proximate explanations may enhance the understanding of the evolution of behaviour in general and sexual selection in particular.

Incorporation of ethological concepts into considerations of sexual selection may be particularly relevant in the case of birds, which exhibit many types of behaviours that are species-specific and highly stereotyped. Among vertebrates, this has been an especially useful group in the initial development of classical ethology. That is, birds have many behavioural (and presumably neurological) traits that may pre-adapt them to develop—probably relatively quickly—an orientation to a new stimulus, and a stereotyped response to it. Such a tendency could account for the commonly seen pattern of differences in colours or colour patterns of males of very closely related, but usually allopatric, congeners. Rapid, consistent response by females to a novel mutant colour or pattern of colour could explain the differences between males of two or more species or populations. In North America, a good prospective example is plumage differences of male rose-breasted and black-headed grosbeaks.

The contention here, then, is that although the subject of sexual selection has become a major subject in evolutionary biology, for the most part its students have not fully utilized the behavioural concepts that form much of the bases of classical ethology (see also West-Eberhard 1984). The stimulus-response component of courtship behaviour suggests that a consideration of 'old-fashioned' ethology is essential to development of a comprehensive understanding of male ornamentation and display and female response to these traits.

For example, one of the most prominent theories designed to explain female mate choice on an ultimate basis, the 'good genes' hypothesis, includes no explicit reliance on the proximate ethological mechanisms that are conspicuous components of most courtship behaviour: The good genes hypothesis as originally formulated (e.g. Zahavi 1975, 1977; Hamilton and Zuk 1982; Bradbury and Andersson 1987) does not explicitly address the question of what attracts a female bird to a particular colour pattern or display movement in the first place. This issue, however, can be addressed by a modification of classical ethology, such as the sensory bias hypothesis. As previously described, this hypothesis utilizes proximate phenomena to account for the responsiveness of females of a given species to certain kinds of stimuli provided by males. This is a significant contribution to sexual-selection thinking, in that it provides a plausible explanation for why many females of a population or species might respond in a similar manner to a novel male stimulus. The important point is that in order for a new male display trait to increase in a population, there must be some agreement among females in their preference for the trait, and sensory bias provides a mechanism for such agreement. In short, proximate and ultimate explanations are complementary and both are necessary to provide an adequate explanation of ornament and display evolution.

3.2   Ethological concepts and their relevance to mate choice by female birds

In this section I review some general ethological concepts and phenomena, originally developed in large part through research on birds, and suggest how they might be relevant to the process of female choice, especially the issue of attraction to new and novel stimuli. These generally have received little notice in most recent discussions of female mate choice (e.g. Bradbury and Andersson 1987; Andersson 1994, Møller 1994). From the ethological perspective, three questions immediately come to mind. First, what is the origin of courtship ornaments and displays? Second, why are they so diverse? Third, why are they usually unique to a single species, at least in fine detail? In this section, proximate elements of courtship, both of male display and of female mate choice, are emphasized by reviewing some major concepts of classical ethology. An attempt is then made to place them in the framework of modern sexual selection theory. Many 'rules' of behaviour, often first recognized and described many years ago by ethologists studying birds, can be employed in modern studies of both intrasexual selection and mate choice. For example, a key aspect of both the good genes and runaway views of female choice is the correlation over evolutionary time between increasing elaboration of certain male ornaments or displays and increasing responses among females to those characters. What are the proximate factors behind this presumed relationship between male ornaments and displays and their attractiveness to females?

Tinbergen (1952) pointed out that courtship fulfils a variety of functions: *orientation, persuasion, synchronization,* and *reproductive isolation.* The first two of these, orientation and persuasion, are probably most critical in species in which the male and female meet only briefly for mating, and in which males do not provide parental care. As previously noted, in such systems males often exhibit striking or bizarre ornamentation. (Synchronization is thought to be especially important in species with shared parental care and is discussed briefly in Section 11.5.1. Reproductive isolation is discussed in Chapter 8.) Orientation refers to drawing the attention of the female to the male, while persuasion involves increasing the sexual motivation of the female to the point that she solicits or at least accepts copulation with its attendant risks (e.g. sexually transmitted disease). In those species of birds where females and males may have no long-standing pair bond, or even any prior social interaction, the question arises: how (a proximate issue) does a trait or traits of males elicit or trigger copulatory behaviour in the female?

An older body of literature exists on the proximate origins of many discrete types of courtship behaviour, which are generally stereotyped and also often species-specific, at least in detail. Although we usually know little about the neurological bases either of species-specific stereotyped courtship behaviours or its precursors, the ethological terms and the behaviours they represent reflect widespread and repeated evolutionary phenomena. The behavioural categories briefly described here are mentioned in nearly all textbooks on animal behaviour, thus most readers will have some familiarity with them.

3.2.1 Displacement activities, intention movements, and related phenomena

Many courtship displays of birds apparently originated from either *displacement activities* or *intention movements*, which are types of behaviour originally evolved and exhibited in a context completely unrelated to courtship. Displacement activities in birds usually come from breathing movements, or preening, eating, or drinking, while intention movements usually develop from tendencies to attack or escape (e.g. Manning 1979). For example, postural adjustments made by birds of a particular species just prior to taking flight can come to be incorporated into a stereotyped courtship routine. Tinbergen (1952) referred to displays based on displacement activities or intention movements as *derived activities* to indicate their derivation from other types of behaviour. Once such behavioural movements have become completely separated functionally from their origins, and are modified to form social signals, they are said to have become *emancipated* from their original function, and the evolutionary process whereby they become modified to form social signals is called *ritualization* (e.g. Manning 1979, pp. 219–24).

Displacement activities may have originally appeared via the process of *disinhibition*. Hinde (1970, pp. 407–8) defines disinhibition as follows: 'When mutual incompatibility [of two behaviours] prevents the appearance of those types of behaviours which would otherwise have the highest priority, patterns which would otherwise have been suppressed are permitted to appear.' Coming close to a female to court her may produce conflict in the male between hor-monally-promoted sexual motivation and attack or escape tendencies, thus 'opening the door' to a seemingly irrelevant displacement activity, such as a particular type of preening movement (e.g. lovebirds, Dilger 1962). Eventually such signals come to be selected strictly in the context of male precopulatory behaviour; i.e., they have become emancipated from their original context(s), and ritualized to function in the new context.

Some of the most convincing evidence that the species-specific displays used by males in courtship originated from behaviour completely unrelated to male–female interactions comes from comparative study of closely related species. Lorenz's (1941) study of the courtship of ducks provides perhaps the most famous example. Courtship displays of male ducks of many species include patterns clearly derived from drinking and preening movements. This behaviour, not in itself related to sexual activity, may have originally appeared as a result of conflict between sexual and attack or escape tendencies, when males either approached, or were approached by, females. As described in the preceding paragraph, these internal conflicts may lead to displacement drinking or preening.

The assumption is that this behaviour, present in the common ancestor of contemporary ducks, arrested the female's attention and possibly stimulated her sexually (Manning 1979, p. 190). If males exhibiting such behaviour held even a slight mating advantage over males that did not respond to females in such a way, selection for (pseudo)drinking and preening movements by courting males would be strengthened. Often intimately associated with display movements are structures or colours apparently evolved in the context of emphasizing the effect of those movements (which may imply that the colours evolved subsequent to the evolutionary development of the movements; see Prum 1990). For example, in ducks of the genus *Anas*, males exhibit a courtship preening movement during which the drake's bill is drawn along the brightly coloured secondary flight feathers of the wing, the speculum, which usually is elevated to further increase its visibility (see Manning 1979, p. 190, Fig. 5.11). Bill and wing characters (i.e. colours) that further enhanced the signal effect of these movements would spread over evolutionary time through the population. In contemporary ducks, these displays presumably are no longer related to conflicting tendencies within the male, but instead are requisite components of courtship and usually are highly modified from their original form: They have become 'emancipated' from their original controlling mechanisms, and are now controlled by sexual mechanisms, e.g. hormones influenced by photoperiod and the presence of a female.

This explanation does not address the question of why males and females of the various species of *Anas* ducks possess specula that differ in colour. The issue of interspecific differences in colour patterns is considered in Chapter 8.

3.2.2 Sign stimuli or releasers, innate releasing mechanisms, and fixed action patterns

Stereotyped movements, such as courtship preening, and related physical structures, such as the speculum of male ducks, are referred to as *sign stimuli* or *releasers*. Releasers are sometimes likened to keys that 'unlock' the female's *innate releasing mechanism* (a hypothetical mechanism in the brain), which then permits the female to carry out a *fixed action pattern*, specifically, in this case, to solicit copulation. (See, for example, Tinbergen 1951, Hinde

1970, and Manning 1979 for discussion of these fundamentally important ethological concepts.) Releasers and fixed action patterns, which provide a predictable exchange of information and response between two individuals in many contexts apart from sexual selection, are probably also an important aspect of courtship behaviour and mate choice in all birds. I might mention parenthetically that this point may lend support to Zahavi's (1991; see Section 1.3.2) argument that sexual selection is but a subset of signal selection.

3.2.3   Supernormal stimulus

A *supernormal stimulus* is an exaggerated sign stimulus that elicits or 'releases' a specific response more readily or more strongly than the normal stimulus (Tinbergen 1948; Hinde 1970, p. 68). Prior to West-Eberhard's (1984) discussion of the relationship between ethological principles and sexual selection, the supernormal stimulus may have been the only classical ethological concept explicitly applied to modern sexual selection theory (Staddon 1975; O'Donald 1983). Staddon (1975) suggested that in the evolution of behaviour (in this case, response to a stimulus) favourable selection would be analogous to a reward and unfavourable selection to a punishment. With regard to female mate choice, the idea here is that those females that responded to a more extremely developed morphological trait would be 'rewarded' via enhanced fitness. For reasons discussed in Chapter 2, this would be true for both good genes and arbitrary selection scenarios. O'Donald (1983, p. 60) strongly endorsed the possible role of the supernormal stimulus phenomenon in sexual selection—the response of females to display traits of males—by stating: 'Response to a supernormal stimulus provides a mechanism for the relative expression of preference and thus completes the chain of inference that constitutes the theory of sexual selection by female preference.' Note that O'Donald appears to focus on a proximate, rather than ultimate, mechanism for female choice.

Recently, some classical ethological principles related to the supernormal stimulus have been used to help account for mate choice decisions by females. The sensory bias hypothesis (Ryan *et al.* 1990; see Chapter 2) is based on the premise that sensory systems of females, evolved in one context, can be exploited by males whose ornaments or displays provide similar stimuli to attract the females. Employing elements of classical ethology, based on basic rules of behaviour that pertain to response to stimuli, can help us to understand how a new, or apparently new, preference by females for a particular type of male ornament may originate.

3.2.4   Heterogeneous summation

In many bird species males exhibit an array of ornamental traits, both morphological and behavioural, that presumably evolved in the context of attracting females. This leads to the questions: why do males employ a number of different traits to advertise themselves, and are they all necessary or required in order to attract females? An ethological concept known as *heterogeneous summation* (Tinbergen 1951) may provide an answer. Heterogeneous summation refers to a stronger response to a number of simultaneously-presented stimuli than to a single stimulus (Hinde 1970, p. 69). In addition to a number of stimuli presented by a single male, such as the presence of multiple morphological ornaments, vocalizations, or other sounds, and behavioural displays, the enhanced response of females to multiple stimuli may account for group displays among males, such as leks. The fact that males of polygynous and lek-promiscuous species often display both many stimuli and extraordinarily striking or even bizarre stimuli lends some support to the suggestion that females prefer or require a number of stimuli. The additive stimulative effects of several males may also be related to lek mating systems in at least some species ('stimulus pooling'; see Oring 1982 and Chapter 15).

Ruffs provide an example of the possible relationship between leks and multiple stimuli. Many years ago, Stoner (1940) suggested that leks of male ruffs of diverse colours (see Fig. 15.1) may multiply the level of attractiveness and stimulation of females far beyond what any single male could achieve, regardless of its colour. Stoner suggested that a cluster of colourful males makes it easier for females flying in the vicinity to locate males and that the number and diversity of colours may be highly attractive to them. He also pointed out that when females are present at the lek, male–male conflict is rare and mild; e.g. males tend not to interfere with the mating opportunities of other members of the lek. That is, if males are dependent on the presence of other males to attract females, a high level of aggression among them would tend to discourage or drive away other males (especially the subordinate, pale satellite males which aid in attracting females initially) as well as females, with the result that a highly aggressive male would, as a result of its behaviour, actually decrease its mating opportunities. More recent studies of ruff leks (Hill 1991; van Rhijn 1991; Lank and Smith 1992; Höglund *et al.* 1993) have generally confirmed the benefits of lek size and diversity among males first suggested by Stoner. The relationship between groups of displaying male ruffs and their mutual stimulation of females is discussed in more detail in Chapter 15.

3.2.5   Habituation

Habituation is the waning of a response upon repeated exposure to a specific stimulus, and which is not followed by any kind of reinforcement (Hinde 1970, p. 577). In many species, males typically exhibit their displays many, many times in their efforts to attract females. Although repetitiveness of display, especially auditory displays, whatever their function, may be most apparent to the human investigator—e.g. song in the red-eyed vireo (more than 22 000 songs in a single day; de Kiriline 1954)—this is probably also true for

visual displays. One reason for the evolution of a diversity of display ornaments and movements by courting males may be to counteract the process of
habituation in female recipients of those displays. This suggestion also may
apply to diversity of songs rendered in species in which song is important to
attracting females (Searcy 1992).

3.2.6   Social facilitation

Social facilitation is the performance of a behaviour as a consequence of
performance of the same behaviour by other individuals (Hinde 1970, p. 582).
This term is also applied to situations where the behaviour, or even the mere
presence, of an individual increases the probability, rate, or frequency of a
specific behaviour pattern in another individual (Dewsbury 1978). With regard
to the issue of female mate choice (as opposed to mating behaviour), social
facilitation and copying behaviour may be most relevant in leks. For example,
if a female mates with a particular male as a result of having observed one or
more other females choose that male, this will have an effect on the success of
all of the males on the lek. Although social facilitation and copying do not
necessarily refer to the same phenomenon, their effects are clearly similar, in
that what one individual does may exert a pronounced effect on the behaviour
of another individual. Mate choice copying behaviour is discussed in Section
15.6.4.

3.3   Other ethological phenomena

In many species, courtship displays contain elements characteristic of the
parent–offspring relationship. Probably the most common such behaviour is
the feeding of the female by the male. As a part of their courtship, male galliforms of many species call females to be fed ('tidbitting' Stokes 1971; Stokes
and Williams 1971); in red junglefowl this male behaviour is very similar to
that of a female parent calling her chicks. Females of some species that
produce altricial young may assume the vocalizations and postures of begging
juveniles as they receive food from their mates (e.g. corvids such as the
pinyon jay, Marzluff and Balda 1992). This juvenile-type behaviour, first
adaptive in young animals of both sexes for obtaining parental care, is later
exhibited in sexual interactions, usually by females. However, during courtship
in pinyon jays, males are fed by females, as well as vice versa (Johnson
1988a).

Males of many species, especially polygynous ones, are typically larger than
females, and they tend to behave aggressively, sometimes directly to females
(Payne 1984, p. 13). Therefore, before presenting specific sign stimuli that
function to persuade females to copulate, male birds often must either overcome or utilize the fear they stimulate in females. This appears not to be a

trivial matter. In a variety of polygynous species in particular, the large size and aggressive behaviour of males appears to make females initially reluctant to approach them. Thus, a part of the courtship strategy of an individual male, perhaps especially on a lek where the male's movement is restricted by the presence of other males, involves (1) gaining a female's attention, (2) decreasing the female's fear response so that (3) the female is attracted to his proximity, and (4) inducing copulatory behaviour in the female.

Often there appears to be a lot of redundancy in courtship signals—plumage patterns, vocalizations, and displays—of birds (e.g. Smith 1977). This apparent duplication of the same message, usually via a variety of signals, may serve to ensure that the recipient of the signals clearly recognizes the message being conveyed. There are several additional plausible explanations as well. (1) An ethological explanation is that the large number of (presumably) sexually-selected traits displayed by male birds of many species may serve to provide diverse types of stimuli to females in order to elicit a stronger response by females than could be elicited by a single signal, as via the heterogeneous summation phenomenon. (2) Some of the ornaments and possibly even displays of males do not currently function to attract females (e.g. Ligon and Zwartjes 1995a). (3) Multiple signals may be required to provide a cumulative cost of courtship signals to females (Hill 1994b). (4) Different ornaments may signal different information. The significance of multiple ornaments (e.g. Møller and Pomiankowski 1993a) is discussed in Section 4.6.

In addition to sign stimuli or releasers, innate releasing mechanisms, fixed action patterns, supernormal stimuli, and heterogeneous summation, several additional ethological phenomena may be involved in female choice. One of these is the *attractiveness of novel stimuli*. A general feature of behaviour is that animals typically orient to a novel stimulus: responsiveness to such a stimulus is enhanced and readiness for action is increased (Hinde 1970, p. 145). Attraction to novel stimuli by female birds may be important in the initial development of preferences for male ornaments that appear by mutation. In the courtship repertoires of some bird species, males appear to take advantage of the attractive nature of a 'novel' stimulus.

Adult males of most species look much alike. How might novelty, or something similar to it, be produced in visual displays where all males possess the same types of ornaments (e.g. brightly coloured feathers arranged in distinctive patterns)? In a variety of species, male courtship includes an abrupt, possibly somewhat startling, presentation of some visual and/or vocal signal, and appears to be delivered in an unexpected or unanticipated temporal pattern, in conjunction with other features of the display. Suddenly opening the mouth to expose its brightly coloured lining is a feature of the courtship repertoire of several species of birds of paradise. A group of pheasants called tragopans quickly, and probably unexpectedly, jump out from behind a rock or log and expand and expose their unfeathered, brilliantly coloured, and intricately patterned throat sacs or lappets while simultaneously vibrating their

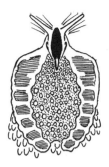

Fig. 3.2 Frontal view of displaying male Temminck's tragopan. Expansion and contraction of the huge lappet is under voluntary control.

'horns' (Fig. 3.2). Similarly, male argus pheasants present the brilliantly-coloured ocelli on the wings in a novel way (Fig. 3.3). In such species, concealment of elaborate colour or structure by the male until the overall courtship sequence has reached a certain stage suggests that a striking, but previously concealed, stimulus, presented abruptly, may lower the 'resistance' threshold of females. When the timing of the stimuli or releasers presented by the male or males is effective, they trigger or release the female's stereotyped precopulatory crouch, which appears to be a fixed action pattern.

Although morphological ornaments alone (e.g. feather colours and patterns) usually permit the exhibition of little individual novelty of visual stimuli presented to females, novel vocal displays occur conspicuously in many bird families. Due to the technology available for recording and precisely analysing vocalizations, these usually have been easier for the human observer to appreciate than subtle differences in plumage pattern or courtship display movements.

The mixing of visual and vocal stimuli and the timing of their presentation may substitute for novelty in the strict sense. The timing used by an individual courting male to reveal itself may comprise the unique or novel component of its visual display. Two very different passerine species, the tooth-billed bowerbird and Jackson's widowbird (Frith and Frith 1993; Andersson 1991), provide examples. In both species the male first attracts a female to its display site by use of vocalizations, which are emitted as the male remains hidden behind a small tree (bowerbird) or tuft of grass (widowbird). In each case, when the female approaches the hidden male, he suddenly pops into view and gives stereotyped vocalizations and postures that, if successful, promptly lead the female to solicit copulation. In both cases the male remains hidden for most of the duration of the display. For the bowerbird, the hidden vocal component took up 73 per cent of the total display time (averaging 75 seconds), while in the widowbird the vocalizations while hiding accounted for 85 per cent of the

Fig. 3.3   Male great argus pheasant displaying its specialized secondary feathers. Reprinted with permission from Dover Publications, Inc.

total display episode (averaging 70 seconds). Thus, the abrupt appearance of the male occurs rather late in the overall display.

The importance of novel stimuli *per se* may be most clearly illustrated by the individually unique decoration of their bowers by males of some bowerbird species. Use of physical objects to attract females may permit the greatest expression of individuality and novelty in courtship known in birds (see Section 6.4.3.2). In addition, once the ability has evolved to imitate—as well as to independently vary—the way the bower structure is decorated, or the way a set of decorative items is arranged (Diamond 1988), then each individual male in the neighbourhood can copy a successful, previously novel, pattern used by another male, as well as create new combinations of stimuli, which would also be novel.

3.4   Courtship in the red junglefowl from the ethological perspective

In this section, an ethological framework is used to describe male courtship and female mate choice in the red junglefowl, a species whose behaviour I have closely observed. Controlled experimental studies illustrate how some of the ethological concepts described above can help clarify proximate aspects of male courtship behaviour and female mate choice. Here I briefly describe the sequence of modes of behaviour that male junglefowl typically use, from their initial attempt to gain the female's attention until copulation is achieved. Wood-Gush (1954) provides a good description of several courtship displays of the fowl, along with their ethological interpretations (e.g. origins as displacement activity).

Mate choice trials were designed to enable a female to make a mating choice between two males without the complicating influence of male–male competition (e.g. Zuk *et al.* 1990*a*, 1992; Ligon and Zwartjes 1995*a*). Two test males were tethered in the centre of individual sections of a large enclosure separated by a wood partition. The males' tethers stopped them just short of the partition between the sections. A female was placed in a wire observation pen that ran the length of the enclosure, from which the hen observed the males before being released into the enclosure. Thus, both males were visible to the female, but not to each other. Because of the tether, it was necessary for a male to attract a female to him to obtain a copulation. Mature, free-ranging male junglefowl, especially high-ranking ones, typically attract females to them by use of the behaviour described below, rather than by approaching or pursuing the female; therefore, the tethers did not significantly affect normal male courtship and mating behaviour.

The courtship displays of male junglefowl appear to conform to Tinbergen's (1952) behavioural categories of *orientation* and *persuasion*, and to be designed to lead to the same ultimate goal—namely, triggering the release of the precopulatory crouch in the female. However, this occurs predictably only if the female's internal state is primed by hormones (i.e. if she is in egg-laying condition). Thus, a critical component of the male–female interactions described here is the physiological condition of the female. Given this pre-requisite for successful courtship, the specific male displays apparently serve somewhat different specific functions with regard to affecting the female's sexual behaviour.

Both before and after the female is released from the observation pen the cocks typically crow and engage in wing-flapping. When a female is out of sight, but known to be nearby, the frequency of crowing increases; when the female appears and approaches the male, crowing ceases. Thus, one function of the crow appears to be to attract the attention of females. When the female is within view of the tethered male, he typically attempts to draw her still closer by engaging in 'tidbitting' behaviour (Stokes 1971), which signals that the male has located a high-quality food item (whether or not he has actually done so) for the female to receive. That is, males use a stimulus associated with food with which the female is familiar: '... tidbitting and calling serve as powerful stimuli for a hen to approach the displaying male' (Stokes 1971, p. 27). The response of adult females to the tidbitting behaviour of males may be related to their early experience; newly hatched chicks soon begin following their mother and responding to her food-location calls. It appears that one aspect of parental behaviour of females has assumed a sexual context in males.

Once the female is close to the male, he may continue to tidbit or may shift to another type of behaviour, one in which special movements, together with tidbitting-like calls, are associated with showing potential nest sites to the female. In both red junglefowl and free-ranging domestic fowl, males often play a major role in nest site selection. (This behaviour was labelled 'cornering' by Wood-Gush (1954), in reference to the fact that the males he observed went to the corners of their enclosure to exhibit this behaviour. Within a pen, a corner is the site where a nest would almost surely be placed.) Neither the tidbitting nor cornering displays appear to serve primarily to induce the female to copulate; on the contrary, both attract the hen and encourage her to remain close to the male. Following Tinbergen's (1953) scheme, the crow, tidbitting, and cornering serve to orient the female to the male and to encourage her to remain in the male's close proximity (Stokes 1971; Stokes and Williams 1971). There is no clear relationship, however, between these orienting and attracting types of behaviour and which male is chosen. That is, both males use these modes of behaviour in their efforts to attract the female to them, but, at least under experimental conditions, they do not appear to play a significant role in the female's eventual mating decision.

When the initial aversion of the female to the male is overcome, Tinbergen's (1953) persuasion phase comes into play. After the female approaches and remains close to the male, he may then exhibit another display, the waltz. The waltzing male rapidly circles close to the female with a stilted gait and with one wing held extended downwards so that the primary feathers touch the ground. This appears designed to trigger, or release, the precopulatory crouch in the female. Evidence that the waltz serves as a threat comes from its function in male–male dominance interactions. As a dominant male approaches a subordinate, it often waltzes (Wood-Gush 1954). Typically, the recipient of this stereotyped signal assumes a submissive posture and steps away from the waltzer. Thus, in male–male interactions, the waltz appears to be a threat, and it probably serves a similar function in intersexual interactions in that it appears to trigger subordinate behaviour in the female at the same time she is attracted to the male.

Using the ethological concepts discussed above, the male and his behaviour create a conflict within the female, simultaneously eliciting attraction and fear. The female's hormonal condition, together with other male displays such as tid-bitting or cornering, attract the female. Once the female is close to the male, the waltz may threaten the female to some extent, creating a submissive response, and thus presumably fear. (Its effect on subordinate males is strong avoidance behaviour.) Apparently, this creates an internal conflict. If this is successful, the female crouches low to the ground with the wings partially extended and held very rigidly. This posture may have originally been a position taken just prior to flight. However, rather than actually indicating flight, in the context of mating it has been ritualized to produce the stereotyped precopulatory crouch—a specific fixed action pattern released by the male's display. In effect, the female's pre-copulatory crouch serves to counteract any fear and avoidance of the male by 'locking' her into the copulatory position prior to physical contact with the male. This allows the male to step onto the female's back, make cloacal contact, and transfer sperm, all within a few seconds. Thus, in this sequence of events, male behaviour and female responses produce an adaptive result for both parties. It may also be important to emphasize that females exercise choice in all of this. First, the female has the option either to approach a male or not, and second, once she has approached a male, she has the option either to assume the precopulatory crouch or not. When they are hormonally primed for reproduction, females approach a male and preferentially crouch for males bearing large combs (Ligon and Zwartjes 1995a).

The interplay of hormonally-induced sexual readiness, attraction to a particular male, and fear produced by the male's waltz combine to produce the crouch. This was made clear during studies in which several female junglefowl were repeatedly exposed to the same males for mate choice tests (Ligon and Zwartjes 1995b). Between tests, the females were maintained in small pens without males and were removed on alternate days for mate choice experiments. Most females were in the process of laying a clutch of eggs and thus were in a high state of sexual readiness, as indicated by the speed with which most of them approached a male and mated, when given the opportunity to do so. As the human investigators approached female junglefowl to place them in the mate choice pens, many hens responded to our approach by displaying the precopulatory crouch. The wings were maintained in the extended and locked position even as we picked them up!

In contrast, when females are not hormonally primed or when a male rapidly approaches or runs at a female for copulation without attempting to use the persuasion behaviour described above, female junglefowl typically will avoid and move away from the male, often running or even flying from a pur-suing male. This unambiguous avoidance of males that do not present courtship behaviour provides strong evidence for the importance in this species, and presumably many other species as well, of precopulatory display for overcoming the fear females have of males.

3.5   Bowerbirds

The courtship behaviour of bowerbirds provides some of the clearest examples of the importance of ethological principles in mate choice in that the males 'apply' these principles in their attempts to obtain matings. The bowerbird family, Ptilonorhynchidae, consists of 18 species that occur in New Guinea and Australia. Males of three members of the family do not construct court arenas or build bowers. The bower-building species are almost unique among birds in that the major physical traits used by males to court females are not parts of the males themselves (i.e. specialized plumage or other morphological structures), but instead are external structures that the males construct and decorate: the bowers.

The diversity of bower structure varies greatly among species, but bowers can be grouped into three or four categories (Marshall 1954; Gilliard 1969; Borgia 1986*a*; Diamond 1986*a*) that show a phylogenetic pattern (Kusmierski *et al.* 1993; but see Kusmierski *et al.* 1997). These are: (1) simple clearings on the ground, with the addition of green leaves turned upside down (toothbilled bowerbird); (2) the mat bower, a mat of lichens on the forest floor, with snail shells added (Archbold's bowerbird); (3) maypole bowers, built of sticks woven around a central pole, and of varying complexity depending on the species (several species); and (4) avenue bowers with parallel walls built of sticks and grasses (several species). Nearly all species add various kinds of decoration to the bowers, and, in addition, several even produce a coloured liquid composed of fruit pulp or charcoal masticated and mixed with saliva (Pruett-Jones and Pruett-Jones 1983) that is painted over the walls or floor of the bower. Females visit several bowers (Borgia 1985*a*; Pruett-Jones and Pruett-Jones 1982, 1983) and, so far as is known, all mating takes place at the bower. Thus, at least in the well-studied species, attracting females to and into the bower is essential to the reproductive success of male bowerbirds.

3.5.1   Transferral of the focus of selection

Bowerbirds beautifully illustrate many of the most basic ethological principles in that they gather and display objects in ways that have evolved to attract females for mating. Gilliard (1963, 1969) provided a fundamentally important insight, directly related to the concept of sign stimuli or releasers and fixed action patterns, when he suggested that many aspects of male courtship in bowerbirds have been transferred from the birds to another object, the bower, and that the bower thus is analogous to plumage in males of more typical species. Gilliard (1963, p. 43) writes:

> I believe that in these birds the forces of sexual selection have been transferred from morphological characteristics—the male plumage—to external objects and that this 'transferral effect' may be the key factor in the evolution of the more complex bowerbirds. This would explain the extraordinary development and proliferation of

the bowers and their ornaments: these objects have, in effect, become externalized bundles of secondary sexual characteristics that are psychologically but not physically connected with the males. The transfer also has an important morphological effect: once colourful plumage is rendered unimportant, natural selection operates in the direction of protective colouration and the male tends more and more to resemble the female.

Subsequent students of bowerbirds have concurred with Gilliard's interpretations, made more than thirty years ago, and view bowers as a very special example of the usually presumed, more general relationship between male ornamental traits and female choice (e.g. Borgia *et al.* 1985; Borgia 1986*a*; Diamond 1986*b*, 1988).

Comparative study of closely related species of bowerbirds provides evidence for the transferral effect. Gilliard (1963, 1969) first recognized that, in certain genera, an inverse relationship exists between degree of ornamental plumage and complexity of bower construction. In the four species of the Genus *Chlamydera*, males of two species possess bright, violet-coloured stripes on the nape that are used in displays to females, whereas in the other two species males possess no ornamental plumage at all. In one of the latter, males display to females by twisting their necks like males possessing colourful napes. This behaviour apparently was designed to exhibit the now non-existent nuchal stripe. The non-ornamented *Chlamydera* species construct more complex and brightly decorated bowers than do the ornamented ones, with the apparent transferral of the female's attention from male feather patterns to the physical objects displayed by the male (Gilliard 1963, 1969; Borgia *et al.* 1985; Diamond 1986*b*; Kusmierski *et al.* 1997). The presence of nape-exhibiting behaviour in the unornamented species suggests that with the transferral of female attention from bright plumage traits to the bower and its ornaments, natural selection has favoured the loss of the brightly coloured nuchal feathering (i.e. the ornamental plumage is gone, but the display associated with it has not disappeared). There is also an alternate explanation; the violet nape may have evolved in two species to enhance displays using neck movements. For example, in New World manakins, colourful feathers apparently evolved subsequent to the display movements associated with them (Prum 1990).

3.5.2 Sexual selection in bowerbirds

Bowerbirds are of special importance to the evaluation of theories of sexual selection for several reasons. First, as discussed above, they provide unusually clear examples of the role of ethological phenomena in sexual selection: releasers, fixed action patterns, supernormal stimuli, etc.

Second, the bower structures and ornaments gathered by males from the environment demonstrate unequivocally that different components of display may serve different functions, as suggested by Tinbergen (1952). For example, the male spotted bowerbird primarily uses two objects in decorating its bower, small bones and glass (Borgia 1995*a*). The conspicuous white bones are spread out over a wide area around the bower and probably serve to attract the attention of females from some distance (orientation of Tinbergen 1952), while the glass, which is mostly placed at the bower entrance and inside the bower, appears to help increase the readiness of the female to copulate (persuasion of Tinbergen 1952). The female's attraction to glass is interesting because it cannot be a specifically evolved response, since fragments of glass have been in Australia for no more than about a hundred years. Either the females are attracted to its novelty *per se* or this attraction is an example of sensory bias (Basolo 1990; Ryan *et al.* 1990), based on the similarity of glass to an object or objects that exist in the natural environment and that are attractive to these bowerbirds.

Third, bowerbirds illustrate in a unique way just how important precision of display may need to be in order to attract females. For example, male satin bowerbirds apparently make use of light in the bower presentation as the bower is oriented across the path of the sun (Marshall 1954, p. 40). Marshall took bearings on 66 bowers and found that the deviation from 360° was never more than 30°. He shifted a bower from 350 to 50° and found that its owner promptly demolished it and then rebuilt it in its former position. Similarly, Pruett-Jones and Pruett-Jones (1983) found that when they changed the position of certain decorations in bowers of Macgregor's bowerbird, the male owner would promptly return them to their original place.

Finally, Gerald Borgia (1995*b,c*) has recently produced the 'threat reduction' hypothesis to account for the evolution of complex bower structures, such as the stick or grass walls of 'avenue' bowers. The bower structure, in effect, provides a means by which females can escape sexually motivated advances of males. According to Borgia, these evolved as a means to encourage females to visit the highly-sexed males, by making it difficult for a male either to obtain an unsolicited copulation or to attack the female. Females may require high-intensity displays to assess males, but such displays may also be threatening. The presence of the walls of the bower increases the chances that a female bowerbird will visit a male's display area, simply because she can be confident that an unwelcome male advance, including copulation, can be avoided—as a direct result of the structure of the bower itself.

3.6   Conclusions and summary

The general theme of this chapter is that the origins of courtship behaviour of male birds and the response of females to them cannot be fully understood without a consideration of the behavioural phenomena comprising classical ethology: conflict behaviour, displacement activities, sign stimuli or releasers, fixed action patterns, and so forth. At the proximate level these phenomena form the basis for the evolution of complex displays and morphological ornamentation of males, and for the response of females to such behaviour and structures.

Much of the pre-nesting reproductive behaviour of birds is composed of the kinds of stereotyped behaviour that make up the core concepts of classical ethology. As illustrated by red junglefowl and bowerbirds, the courtship of males, which is itself stereotyped and to varying degrees innate, utilizes these basic ethological phenomena to attract females and to persuade them to copulate. This may be seen most readily in species such as junglefowl and the typical bowerbirds, in which pair bonds are absent. The key underlying factor is the hormonal state of the female, which is controlled in part by environmental cues such as photoperiod. Females not physiologically in reproductive condition are not sexually responsive to male courtship displays; likewise, males not in reproductive condition do not exhibit courtship displays. Thus, the releasers are presented by males and are effective in attracting females only when the central nervous system is hormonally primed. These two factors—hormonal condition and the interactions of releasers and fixed action patterns—clearly are necessary to produce courtship patterns of birds, both the displays of males and the responses of females to them. Hormonal priming, however, is not sufficient to account for the ultimate aspects of female mate choice. Specifically, why do females not respond equally to any and all males that present them with the appropriate ornaments and displays that serve as releasers; i.e. why do females choose among males, all of which exhibit similar (but not identical) ornaments and displays? One possible answer is that females respond to the most developed or largest traits of importance to them. At the proximate level, this may be driven by the supernormal stimulus phenomenon. If a 'runaway phase' of sexual selection existed over the course of evolution of male display traits and female response to them, then the runaway selection may be proximately driven by the supernormal stimulus phenomenon (O'Donald 1983). It is well established that females are not equally responsive to each male whose courtship they observe. Selectivity in mate choice by females on the basis of relative development of ornamental traits provides evidence that they are capable of (1) overriding a releaser–fixed action pattern response and (2) discriminating among males on the basis of the small quantitative differences in the traits they present (e.g. Andersson 1992*b*).

The sensory bias hypothesis (see Section 2.4.2) utilizes proximate elements in its explanation in that it assumes that males compete to most effectively stimulate certain sensory receptors of females. This could account for the fact that not all males, even though they possess the same basic set of releasers, are equally likely to attract females for mating. Incorporation of the releaser–fixed action pattern and supernormal stimuli phenomena of classical ethology, along with Ryan *et al.*'s (1990) perspective of sensory bias by males to most effectively stimulate the sensory receptors of females, provides a proximate explanation for the variance among males in mating success.

4 Morphological ornaments and song

4.1 Introduction

Among vertebrates, birds are a comparatively homogeneous group in their basic morphology and reproductive biology. They exhibit, however, mind-boggling diversity in those traits associated with courtship and mate choice. Apart from flight, the traits of birds that most excite the human imagination are those resulting from sexual selection. Armstrong (1965, p. 305) provides a colourful commentary on the spectacular and sometimes bizarre morphological and behavioural traits characteristic of the courtship displays of many birds:

> The element which more than any other is characteristic of sexual display amongst birds is strangeness. Attitudes and movements tend to be odd, exaggerated, or unwonted. Peculiar adornments are thrust into prominence: crests, wattles, ruffs, collars, tippets, trains, spurs, excrescences on wings and bills, tinted mouths, tails of weird or exquisite form, bladders, highly coloured patches of bare skin, elongated plumes, brightly hued feet and legs. … The display is nearly always beautiful; it is always striking.

This quote emphasizes the special and diverse nature of the ornamental and display traits generally believed to have evolved specifically in the context of female mate choice. In addition to their diversity, ornamental traits usually are, to some degree, novel, even among closely related species. Until recently, differences between closely related species in plumage colour patterns or eye colour, for example, were thought to be a manifestation of natural selection for 'premating isolating mechanisms' (see Section 8.2). It has now become more widely appreciated that such species-specific differences are more likely to be due to the effects of sexual selection in isolated populations, rather than serving to prevent mating between individuals of closely related species. Male birds of many species exhibit brighter plumage than their female counterparts. This is often true even when the plumage colour patterns or other structures of the two sexes are similar. In addition, males often possess ornamental traits not expressed by females at all (e.g. the specialized feathers of the peacock's train). These kinds of differences between the sexes are particularly conspicuous in species or groups of species in which polygynous or promiscuous lekking systems are the rule. Darwin was the first to assume that exaggerated ornamental traits and behaviour that characterize males of such species serve to attract females for mating. Although some studies of individual species have demonstrated preferences by females for males with the most elaborately ornamented plumage, such as the brightest or longest-tailed individuals (e.g. Hill 1990; Møller 1994), others have not detected a correlation between degree of development of certain ornaments and female mate choice (e.g. McDonald 1989*a*; Ligon and Zwartjes 1995*a*; see also Møller 1993*a* for a survey of such cases).

Certain display ornaments (e.g. tail length) are easily measured, and vocalizations and other kinds of auditory displays (e.g. drumming) are easy to quantify. The same is true of many visual displays (e.g. the number of stones carried by male black wheatears (Soler *et al.* 1996) or number of nests built within a wren's territory (Evans 1997). Other aspects of display, however, such as movements, coloration (e.g. iridescence), or non-linear morphological traits like air sacs, are more difficult to measure and assess. This disparity in the ease with which various traits can be quantified has led researchers to focus on certain kinds of ornaments, such as feather length, while often ignoring others.

For example, the perplexing studies of male traits and female preferences in the ring-necked pheasant by von Schantz and his colleagues (e.g. von Schantz *et al.* 1989, 1994; Göransson *et al.* 1990; Grahn and von Schantz 1994) ignored the possible role of the brilliant red wattles, which rapidly inflate during courtship (but see Hillgarth 1990*a* and Mateos and Carranza 1995). Similarly, several excellent studies describing mating success in sage grouse have neglected the air sacs and wattles (e.g. Wiley 1973; Hartzler and Jenni 1988; numerous publications by R. Gibson, J. Bradbury, and S. Vehrencamp, and their co-workers; but see Spurrier *et al.* 1991). In short, although anatomical 'soft parts' may often be of major importance to females as they make mate choice decisions (e.g. Zuk *et al.* 1990*a*; Buchholz 1995; Ligon and Zwartjes 1995*a*), because of the difficulty in obtaining accurate measurements of them, they have often been ignored in studies of sexual selection.

Possibly the most difficult aspect of bird displays to address is the issue of colour (but see Hill 1990). Bennett *et al.* (1994) have critiqued studies of the role of colour in sexual selection in birds. In brief, their main point is that researchers have assumed that human and other animals perceive colour in the same way, and this may be incorrect. Subsequently, Bennett *et al.* (1996) showed that in zebra-finches the ultraviolet waveband is used in mate choice decisions.

Since Darwin (1871) it has been assumed that song is a sexually-selected trait and that its evolution is causally related to mating success, serving both in male–male competition and to attract and stimulate females. The available evidence suggests that this supposition is correct (Payne 1983; Loffredo and Borgia 1986*a*; Searcy and Andersson 1986; but see Kroodsma and Byers 1991). Because of the contexts in which it occurs, and because it is so conspicuous, song clearly serves as a means of advertisement, although just what is being advertised, and to whom, may be less clear. Functions of song have been experimentally investigated in a large number of species. Indeed, the role of song in sexual selection has received considerably more detailed experimental investigation, and its importance is thus more firmly established, than that of morphological characters that appear to serve in visual displays. This is primarily because of the availability of technology that makes possible high-quality field experiments using playback of tape recordings, including the dissection and reconstruction of song components.

4.2   Morphological ornaments

4.2.1   Plumage elaboration and mating system

In some groups of birds, there is a relationship between mating system and degree of plumage elaboration of males. As a rule, in socially monogamous species, males and females tend to be more similar in appearance than in polygynous ones (e.g. Burley 1981; Payne 1984). In many, perhaps most, monogamous species, either males and females are alike in appearance or plumage patterns are generally similar, but with more intense coloration in males. However, numerous exceptions to this generalization exist. A striking counter-example is the eclectus parrot of the New Guinea region, in which females are bright red and blue, while males are primarily green. Two less exotic examples from North America are northern cardinals and Williamson's sapsuckers, both of which exhibit extreme sexual plumage dimorphism. (The plumage of males and females of Williamson's sapsucker is so completely different that the two sexes originally were considered to be different species; see Fig. 1.5.) Yet both of these latter species, and probably the parrot as well, are socially monogamous (Crockett and Hansley 1977; Ritchison *et al.* 1994).

In contrast to most monogamous species, in males of many polygynous species, plumage and soft parts (wattles, mouth lining, eye, etc.) ornamentation is often extreme, and males and females typically differ conspicuously in appearance. Such striking sexual differences raise two general questions: (1) why are males often brightly and even bizarrely ornamented in many kinds

of birds, especially polygynous ones, and (2) why do closely related species
often differ strongly in the details of ornamentation? The first question is
addressed in part in the next section, while the second is considered in detail in
Section 8.4. Interestingly, with the accumulation of data on a variety of
species, it turns out that both questions are more difficult to answer than
almost anyone would have guessed just a very few years ago, before careful
studies were initiated (e.g. the existence of apparently non-functional ornaments, discussed below).

4.2.2   Proximate control of plumage ornamentation

It has long been known that in the domestic fowl the ornate plumage characteristic of males develops fully in castrates (see Ligon *et al.* 1990). However,
the number of species studied in this regard is limited (one or a few members
of only four different avian orders, Galliformes, Anseriformes, Charadriiformes,
and Passeriformes). Owens and Short (1995) classify the control of sexually
dimorphic plumage development into three categories: oestrogen-dependent
control, testosterone-dependent control, and genetic control.

4.2.2.1   Oestrogen-dependent control

In some sexually dimorphic species of galliforms, the presence of oestrogen controls the development of sexually dimorphic plumage. Several general reviews
have described the proximate control of sexually dimorphic plumage development (Domm 1939; Witschi 1961; Vevers 1962; Owens and Short 1995). In
domestic fowl, if oestrogen is present at the feather follicle during moult, the
resulting plumage is female-like, regardless of the genetic sex. If oestrogen is
absent, regardless of the presence or absence of testosterone, the plumage is
bright and male-like (Stevens 1991). Less detailed studies of other galliform
species are consistent with this pattern. These data indicate that in these species,
in the absence of estrogen, male plumage develops as the default condition.

Oestrogen may be involved in the development of sexually dimorphic
plumage in other avian taxa as well. In the mallard duck, males alternate
between a bright cock plumage and a dull henny plumage. In these cases, it
appears that oestrogen is important in preventing the development of the bright
breeding plumage in the female. If females are ovariectomized, they begin to
alternate between bright and dull plumage, just like males (Thapliyal and
Saxena 1961; Witschi 1961). Similarly, in a passerine species, the red bishop
of Africa, males injected with oestrogen moulted into a dull henny plumage in
the breeding season (Witschi 1935, 1961).

4.2.2.2   Testosterone-dependent sexually dimorphic plumage

Owens and Short (1995) state that although it has been assumed that
ornamental plumage traits of male birds of sexually dimorphic species are

dependent upon testosterone, there is little support for this assumption. According to these authors, the ruff is the only species in which the presence of testosterone is known to be necessary for the development of male-specific plumage. Without testosterone, males do not develop the characteristic ruff around the neck, and females injected with testosterone do develop a small ruff (van Oordt and Junge 1934). In addition to the ruff, development of bright plumage in response to experimental treatment of testosterone propionate has been reported for two other charadriiform birds, the sex-role reversed Wilson and northern phalaropes (Johns 1964). In these birds, testosterone treatment causes feathers of both sexes become brighter. Similarly, testosterone may affect plumage colour in male willow ptarmigan (Stokkan 1979*a*,*b*), and it apparently accelerates the appearance of adult plumage in the satin bowerbird (Collis and Borgia 1992).

In addition, testosterone may be important in the development of plumage in monomorphic species. In herring gulls, testosterone has been reported as necessary for the attainment of adult plumage in both males and females (Boss 1943), and in the black-headed gull, castrated males do not moult into the breeding plumage that is characteristic of both sexes (van Oordt and Junge 1933).

4.2.2.3   *Non-hormonal control of sexually dimorphic plumage*

In some sexually dimorphic species, plumage is not dependent upon the presence of steroid hormones. It is unknown how common this is. Studies on house sparrows, involving both castration and hormone implants, have demonstrated that normal plumage dimorphism develops in the absence of either oestrogen or testosterone (Keck 1934). What controls the development of the sexually dimorphic plumage in this species is not known; i.e. sexual differences are 'genetic'.

4.3   *The significance of plumage ornamentation*

4.3.1   Sexual dimorphism in plumage colours

Many kinds of birds exhibit conspicuous differences in plumage colour and patterns between the sexes. Often such differences are of degree rather than being totally different colours. Several studies have implicated a relationship between male plumage 'brightness' and critical aspects of reproductive success in socially monogamous passerines. Here I briefly mention five such species.

1. In the great tit, the black breast stripe varies in size among males, and males with large stripes usually paired with females that laid large clutches (i.e. they appeared to attract 'high quality' females). Because there is no significant relationship between a male's stripe size and the quality of its territory,

it appears that females choose mates on the basis of the black breast stripe and the quality of male parental care (Norris 1990*a,b*, 1993).

2. Conversely, in adult male American redstarts, the amount of black on the breast is inversely related to reproductive success. Males with smaller bibs (less black) mated with females that laid more eggs and produced more hatchlings and fledglings. In addition, the earliest-arriving females at the breeding site mated with males exhibiting smaller bibs (Lemon *et al.* 1992). Although neither of these studies provided any evidence that variation in territory quality contributed to the differences in reproductive success, neither did they demonstrate that females actually choose males on the basis of plumage brightness.

3. Manipulative experimental studies have documented this critical point. Hill (1991) found that female house finches preferred brighter (more red) males and that male plumage colour correlated with paternal care and with overwinter survival. (See Section 6.3 for a more detailed treatment of Hill's work with house finches.)

4. In the cavity-nesting pied flycatcher, a variety of factors have been shown to influence male mating success, including nest-site quality, size and quality of the territory, and song rates (Lundberg and Alatalo 1992). In addition, experiments conducted by Sætre *et al.* (1994) have shown that female pied flycatchers prefer brightly coloured males to dull males. Thus, for this species, and probably most other sexually dimorphic passerines, a number of factors, including male plumage characteristics, are involved in mate choice by females.

5. Similarly, female yellowhammers prefer brighter males, which also tend to be older (Sundberg and Larsson 1994). More young fledge from nests of bright males, despite the fact that those males do not feed young at a higher rate than do duller males. Sundberg and Larsson (1994) suggest that older, more colourful males may be more experienced and of higher quality, and so may be preferred by females.

6. Investigation of mate choice in a non-passerine species, the European kestrel, makes much the same point (Palokangas *et al.* 1994). In an aviary experiment, female kestrels preferred brighter males. Under natural conditions such males spent more time hunting than dull males, and females mated to bright males reared more young than those mated to dull males. Palokangas *et al.* (1994) suggest that direct selection, specifically the correlation between bright feathers and provisioning effort, can explain the preference of female kestrels for bright males.

With regard to the major theories of mate choice reviewed in Chapter 2, the evidence suggests that in some, and perhaps all, of these cases choice of bright males may be related to both direct and indirect (genetic benefits to offspring) selection.

4.3.2   Tail length

The tails of birds usually serve an important aerodynamic function. Additionally, tails of males are often modified as ornaments. How might the natural selection function of the tail be co-opted for a role in sexual selection? Thomas and Balmford (1995) suggest that it is the ability of tails to maintain aerodynamic function while deviating from optimum flight morphology (unlike wings) that has led to their frequent elaboration through sexual selection. This idea also suggests that because tails are important in flight, structural modifications can be viewed as exhibition of a handicap as well as an ornament.

A common type of male ornamentation, presumably evolved in the context of female mate choice, is tail length. Winquist and Lemon (1994) analysed morphological characters and mating–parenting behaviour of a large number of bird species. They found that tail length was inversely related to the level of parental care. When males invest less in parental care, selection for exaggerated male tail length is more likely to occur. As might be expected, this relationship is also reflected in mating systems; males of polygynous species tend to have longer tails than males of related monogamous ones.

In some cases, elongated tails are also structurally complex or are coloured in a way that contrasts with other components of the plumage. This suggests that tail shape and colour, as well as length, may be important to females of such species. However, I am aware of no study that focuses on tail shape (as opposed to length), other than those dealing with the bilateral symmetry of the paired ornamental feathers (e.g. Balmford *et al.* 1993; Evans 1993; Møller 1994; Thomas and Balmford 1995; see Section 5.5.4). Probably the best examples of radiation of specialized and diverse tail ornamentation are to be found in the birds of paradise.

Among the long-tailed species that have received study, the significance of tail length in mate choice by females appears to vary from species to species (Møller 1993*a*). For example, females of the long-tailed and Jackson's widowbirds preferentially mated with longer-tailed males (M. Andersson 1982*b*; S. Andersson 1989, 1992). This is also true of female barn swallows (Møller 1994) and shaft-tailed whydahs (Barnard 1990). In contrast, male tail length was not correlated with mating success in three galliform species, the sage grouse (Gibson and Bradbury 1985), ring-necked pheasant (von Schantz *et al.* 1989; see also Mateos and Carranza 1995), and red junglefowl (Zuk *et al.* 1990*a*, 1992), or in the long-tailed manakin (McDonald 1989*a*), all of which possess elongated tail feathers. The significance of this interspecific variation in use of male tail length by females, among species in which the tail is elongated and in which it appears to the human observer to serve as an ornament, is discussed below in the section on multiple ornaments.

In those species in which tail length does influence female mating success, what is its significance? In Andersson's (1982*b*) classic study of the long-tailed widowbird, no correlation was sought between parameters of male physical condition and tail length. In Jackson's widowbird, however, tail length was positively related to body condition, specifically body mass, suggesting that preferred males were in better physical condition than less preferred ones (S. Andersson 1989, 1992).

Møller's (1994) studies of mate choice in the socially monogamous barn swallow make the same point more comprehensively. Møller found that female swallows prefer to mate with longer-tailed males, that females mated to such males are more likely to produce two broods per season, and that long-tailed males obtain more extra-pair copulations than do shorter-tailed males. Longer-tailed males also exhibit less fluctuating asymmetry than shorter-tailed males, which suggests that both tail length and degree of symmetry of the bilaterally elongated feathers indicate male quality (see detailed discussion of fluctuating asymmetry in Section 5.5).

In some other species, the relationship between male tail length and female choice may be more complex. In the socially parasitic pin-tailed whydah, individual males vary in the timing and rate of growth and abrasion of the four slim ornamental feathers (Barnard 1991); thus, during the extended breeding season, males vary considerably in tail length at any point in time. Barnard (1991) found that males that grew their long tails early in the breeding season were the dominant individuals in the local population. They were also the first to obtain call sites, from which they drove away later-developing males. Although all of the males in the localized study population had access to unlimited food, their tails nevertheless grew at very different rates, which suggests that onset and rate of tail growth do not simply reflect the male's nutritional condition, and that other aspects of overall male condition may influence tail development. The longest-tailed male whydahs were not the biggest, nor were they the most successful at obtaining copulations (Barnard 1989). This and other species of whydah, all of which are nest parasites, are atypical in that the timing of egg-laying by females is dependent on the reproductive cycle of the host species. Therefore, males that grow their tail ornaments most rapidly and maintain them over the longest periods are apparently more successful in obtaining matings than those males that may grow even longer tails later in the season.

In an experimental study of female mate choice in the shaft-tailed whydah, Barnard (1990) found that when given a choice of two males, oestradiol-primed females always preferred the longer-tailed individual. A complicating factor in this study was the level of display. Among pairs of experimentally modified males, the individual with the longer tail invariably displayed and vocalized significantly more than the shorter-tailed bird. This was true even for the same male: when its tail was artificially lengthened, it displayed more than its counterpart with the shorter tail; likewise, when its tail was shortened, it displayed less than its longer-tailed counterpart. This relationship between tail length and behaviour could be due either to some sort of self-awareness or, perhaps more likely, to feedback by the females. It is clear that male display, in addition to tail length, is important to females. When given a choice between a long-tailed stuffed male and a short-tailed live male, the females preferentially associated with the live bird, although they did not solicit copulations either from it or from the stuffed decoy. What do these studies of whydahs tell us about the major theories concerning the significance of tail length as it pertains to female mate choice? (1) The fact that the longest-tailed males do not receive the most matings indicates the absence of any simple relationship between female preference and maximum tail length. This finding is similar to that of several other studies as well (references listed above). (2) The fact that females mate with the first males at call-sites, and thus with the first males to grow the elongated ornamental tail feathers, suggests that females are mating with the first males to come into full breeding condition, rather than selecting the longest-tailed males *per se*.

This latter point, in turn, suggests a possible relationship between female preference and male physical condition or quality. Those males most resistant to, or not burdened with, an important parasite or other pathogen may be the first to begin the moult into breeding plumage and the first to obtain call sites. (Based on other studies that consider the relationship between moult and physical condition, this is a reasonable speculation, e.g. Zuk *et al.* 1990*a*.) Thus, male–male competition for call sites may dictate, to a large extent, which males will obtain most matings. This, in turn, indicates that females mating with male call-site holders may be acquiring matings from among the healthiest males in the local population. In choosing among the call-site holders, females may rely on morphological cues such as tail length at the time they are interested in mating, as suggested by the experimental manipulations of tail length described above. Other things being equal, the earliest-breeding females (that mate with the first males to complete the tail moult) are also likely to produce more surviving offspring because they parasitize the earliest breeders of the host species, which are likely to be the more experienced or more vigorous individuals in their population.

4.3.2.1   *Geographic variation in tail length as a test of Fisher's runaway hypothesis*

One prediction of the Fisher runaway hypothesis (evaluated in Section 6.4.1) is that if tail length in no way reflects male condition or quality, but is chosen by females purely as an arbitrary preference, then there should be a high level of geographic variation in tail length among different populations. This is because the point at which male tail length stabilizes will vary from population to population (Lande 1981, 1982; also see Arnold 1983 for discussion and illustration of these ideas).

To test empirically for this theoretical prediction, Alatalo *et al.* (1988) investigated the between-population variation in tail length of six species exhibiting elongated tails—the Neotropical fork-tailed flycatcher, the Asian paradise flycatcher, and four species of whydahs, all of which occur in Africa. They found that the elongated tail ornaments did not show higher geographic (inter population) variation than did body size characters, which suggests that arbitrary Fisherian selection is not going on in these species; i.e. ornament size is not an arbitrary result of random inter-population variation in female preferences.

4.3.2.2   Intra-population variation in tail length

Alatalo *et al.* (1988) also investigated the degree of intra-population variation in tail lengths, as compared to non-ornamental traits. These authors found that, within populations, coefficients of variation of ornament lengths were significantly greater than for any of several body size characters (bill length, tarsus length, wing length). In six out of ten geographic samples, there was a significant positive correlation between ornament length and overall body size. What this observation means is that bigger males produce longer ornaments, both absolutely and relatively.

Alatalo *et al.* (1988, p. 371) suggest that in these species tail length may be an indicator of male quality:

> ... [this] may indicate that bigger males are better capable of expressing particularly long ornaments. ... Superficially it fits well with the idea that ornaments are conditional viability indicators that exaggerate observable viability differences between males, but this and any other possible explanations need closer examination in future.

If larger body size reflects male quality in some sense, then ornament length may be a readily assessed indicator of such quality.

To summarize, females of some long-tailed species appear to use tail length in making mate choice decisions. This is true in both a socially monogamous species, the barn swallow, and in some socially polygynous or promiscuous species, such as the long-tailed and Jackson's widowbirds (M. Andersson 1982*b*; S. Andersson 1992). For some species, the degree of variation in tail length among individuals within a population exceeds variation in other mensural characters, and tail length is also an indicator of overall body size. Bigger individuals have more fully developed (longer) ornaments, suggesting that factors positively affecting body size also positively affect tail length. Finally, variation in ornament length between different populations is slight, which does not provide support for the Fisher runaway model of ornament exaggeration.

4.4   *Other structurally specialized ornamental plumage*

Males of many kinds of birds possess feathers that appear to have become structurally modified in response to female mate choice. Here I focus on modifications of the plumage other than colour or colour patterns. If we consider typical contour feathers, remiges, and rectrices as the 'ancestral' condition, many feather modifications that presumably evolved in the context of sexual selection come to mind. Elaboration of feathers of the head to form either a single crest or paired 'ears' is a common form of plumage decoration

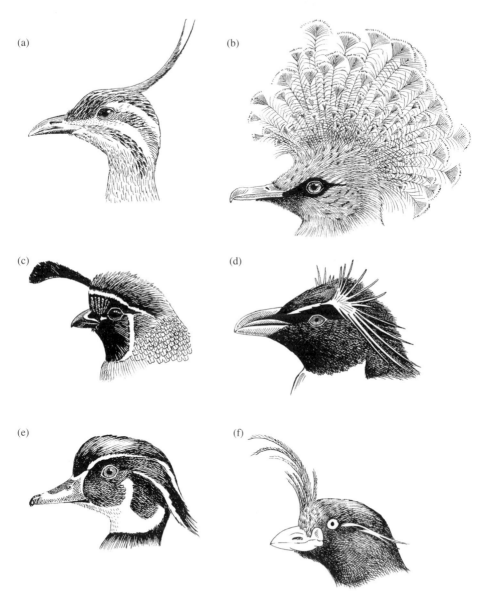

Fig. 4.1 Some examples of ornamental head plumage: (a) elegant tinamou; (b) Victoria crowned pigeon; (c) Gamble's quail; (d) rockhopper penguin; (e) wood duck; (f) crested auklet.

(e.g. Fig. 4.1). Another widespread form of feather ornamentation is iridescence. Although no studies to date have investigated the importance of patches of iridescent feathers in mate choice, it is likely that these structures function in social/sexual contexts.

Probably the best-known example of plumage ornamentation is the train of the peacock. The spectacular feathers forming the train are not tail feathers (rectrices) at all, but are highly modified contour feathers of the lower back. The ocelli, or eye-spots, located at the end of the elongated feathers of the train, are colourful and strikingly eye-like structures. The ocelli (up to 100–200 per male) are presented, 'shivered' in a frontal display, and the number of eye-spots exhibited by a male is correlated both with the symmetry of the train (Manning and Hartley 1991) and with its mating success (Petrie and Halliday 1994; see Chapter 6). Interestingly, eye-like ocelli also occur on the large secondary feathers of the great argus pheasant, as well as on most of the peacock pheasants (see Fig. 11.1 and the figure opening Chapter 13).

Within the Class Aves, it appears that, in one species or another, virtually every group of feathers has been used in the development of extreme ornamentation. Two members of the nightjar family Caprimulgidae provide a good example of this generalization. As a group, the plumage of the approximately 80 species of caprimulgids is noteworthy primarily for its extreme crypticity. (Anyone who has ever kept a poor-will in their living room, as I have, knows this.) With the exception of white patches in the tails or wings of several night-

Fig. 4.2 Male standard-winged nightjar in flight. The 'standards' are the amazingly modified second primaries.

jars that may serve in mate attraction, the plumage is notably cryptic. Yet two African species, the pennant-winged and standard-winged nightjars, exhibit certain flight feathers spectacularly specialized for visual sexual display (Fig. 4.2). In both, the second primary is tremendously enlarged and modified as an ornamental structure (Fry 1969; Fry and Harwin 1988). Not too surprisingly, the available evidence indicates that males of these two species, perhaps alone among the family Caprimulgidae, are polygynous or promiscuous.

In addition, males of many other polygynous species possess flight feathers (remiges) specialized to produce sounds used in courtship. Among North American birds, this is true of some hummingbirds, the common snipe, and the American woodcock.

4.5   Other kinds of morphological ornaments

In addition to feathers, a large array of other parts of the body of male birds have been specialized for a role in sexual selection (Fig. 4.3). Typically, many of these areas are referred to as soft parts (e.g. skin and structures of head and neck—including wattles, caruncles, snood, bill colour, eye colour, colour of mouth lining, knobs on head, cere, and tarsus and toes), and unusual colours or combinations of colours are usually their most conspicuous attributes.

Although feathers can indicate the bearer's physical condition at the time of moult, and the degree of wear or breakage may signal information about social status, because they are dead structures, feathers cannot reflect short-term change in the individual's physiological state. In contrast, soft parts often have the potential to change either colour or shape, or both, rapidly. This provides a means for the development of 'condition-dependent' traits that can signal up-to-date messages about a male's physical state (health).

The iris of the eye is probably an important signal in sexual selection in birds, but at this point it is extremely understudied. If one surveys eye colour of a wide variety of species, it turns out that in many the iris is brightly coloured, and it also often presents a strong contrast with the surrounding area of the head. According to Fox (1976), carotenoid pigments have been found in the eyes of 22 species of birds representing six orders. By displaying certain carotenoid pigments, the colour of the iris may provide a window into the individual's dietary background and health, if not into its soul.

4.5.1   Bare skin and sexual selection

Unfeathered areas of the body can also provide indication of condition in a fashion similar to that suggested for the eyes. In many cases display structures of this sort inflate and deflate rapidly, and they are, therefore, extremely difficult to measure in the usual ways. Some of the problems associated with

Fig. 4.3 Some examples of 'soft parts' ornamentation of the head and neck: (a) southern ground hornbill; (b) northern cassowary; (c) great frigatebird; (d) Andean condor.

studying these kinds of ornaments were discussed earlier in this chapter. Despite the challenges associated with measuring such traits, it should be recognized that for some, and perhaps many, kinds of birds, these sorts of traits may prove to be the ones most important in mate selection by females (Zuk 1991*a*).

Like ornamental plumage, brightly coloured and structurally specialized areas of bare skin have a history; i.e. at some point they evolved from a less specialized condition. With regard to origins of sexually-selected ornaments, these kinds of morphological structures are especially interesting to consider. In addition to a sexual-selection function, unfeathered heads and faces are sites of heat loss. For example, at high ambient temperatures, turkeys lose a considerable amount of body heat through the unfeathered head and upper neck (Fig. 4.4a; Buchholz 1996). Natural selection may first have favoured loss of feathering on and around the head in 'ancestral' turkeys for thermoregulatory reasons, with this area of bare skin subsequently becoming important in sexual selection. Supporting this interpretation of the turkey's unfeathered head is the probability that this lineage (two living species) may have originated in the tropics. (The range of the wild turkey extends from southern Mexico to the north-eastern United States, and the other living species, the ocellated turkey, is restricted to Central America.)

Expanding this line of thinking to the large Order Galliformes as a whole, it is worth noting that most families and subfamilies are largely tropical or subtropical in distribution (grouse being the most notable exception), and that in males of at least some species within every major taxonomic unit, bare skin exists on the face, head, and/or neck (Fig. 4.4). In a number of galliforms unfeathered areas of the head have been shown to function in sexual selection (e.g. Brodsky 1988; Boyce 1990; Spurrier *et al.* 1991; Buchholz 1995; Ligon and Zwartjes 1995*a*; Zuk *et al.* 1995).

Other groups provide parallel examples. The New World vultures are well known for their unfeathered heads and necks (Fig. 4.3d). The evolved function of bare heads in this group (as in their Old World counterparts) is usually

Fig. 4.4   Some examples of unfeathered areas of the head and neck that serve as ornaments, and that may be sites of considerable heat loss: (a) common turkey; (b) Australian brush turkey; (c) greater prairie chicken.

thought to be related to their mode of life—carrion eating. As suggested for the galliforms, once bare heads have evolved, they can be further modified to serve in communication of one sort or another (social and/or sexual selection). Among New World vultures, this may account for the red heads of the turkey vultures, the yellow heads of the lesser yellow-headed vultures, and the ornate heads of the Andean and California condors. The most spectacularly ornate member of this group is the king vulture with its bright orange-and-black face and orange wattles above the bill. In this regard, it is interesting, and perhaps significant, that the black vulture, one cathartid without colourful head ornamentation, is known to be genetically, as well as socially, monogamous (Decker *et al.* 1993; see Section 12.5).

4.5.1.1 Combs

Combs of the four species of junglefowl and their conspecific relatives, the domestic fowl, are brightly coloured, fleshy extensions of the integument of the head, which develop fully only in males. Vigorous, healthy roosters in breeding condition typically exhibit large, bright red combs and wattles. In the red junglefowl, the size of the male's comb (Fig. 5.1) is closely related to female mate choice (Ligon and Zwartjes 1995*a*; Zuk *et al.* 1995), and its structure and cellular composition have apparently evolved to signal its bearer's condition (Ligon *et al.* 1990). Specifically, the anatomy of the cock's comb suggests that one of its evolved functions is to convey information about testosterone level and thus the individual's physiological condition. (This is discussed in detail in Section 5.2.2.) Thus, whatever the initial function of the unfeathered head of junglefowl (e.g. heat dissipation), it appears that the comb has subsequently evolved to serve in sexual selection.

A similar relationship between testosterone and wattle- and comb-like adornments of the head probably occurs in other galliform birds. The combs of grouse and the wattles of pheasants come to mind (Fig. 6.2). Testosterone level has been shown experimentally to directly and strongly affect comb size in the willow ptarmigan (Stokkan 1979*a,b*), and a similar relationship may exist between testosterone level and relative development of the unfeathered structures of the head in other galliforms.

4.5.1.2 Wattles, lappets, and esophageal pouches

In some species (e.g. Bulwer's pheasant, Fig. 2.1), males can rapidly (within seconds) expand and contract their fleshy head/face ornamentation, apparently at will, while in others the size and shape of the ornamental structures are constant, at least during the breeding season. Among pheasants, rapid increase and decrease in the size of fleshy ornaments is a conspicuous characteristic of tragopans, Bulwer's pheasant, and the familiar ring-necked pheasant (Fig. 6.2), and it is also true of the turkey's snood (Fig. 4.4a). The ability to control the size of wattles, lappets, etc. should provide several benefits, among them an ability to increase and decrease the structure's conspicuousness in a situation-dependent manner (e.g. a 'coverable badge', Hansen and Rohwer 1986), and an ability to increase or decrease the rate of heat loss through the unfeathered skin.

The esophageal pouches of various North American grouse species—e.g. sage grouse and prairie chickens (Fig. 4.4c)—that are repeatedly inflated and deflated during the male's courtship displays provide well-known examples. While it is easy to appreciate the difficulty of measuring these structures—and especially the difficulty of precisely quantifying changes in their size and colour—to ignore inter-male variation in these traits is to risk overlooking a key (quite possibly *the* key) morphological character functioning in sexual selection in these species. The importance of the appearance of the pouches of sage grouse has been experimentally demonstrated by Spurrier *et al.* (1991), who modified their appearance in some males by adding artificial haematomas. Females avoided the manipulated males, thereby demonstrating that they do appraise the males' pouches. (See Section 5.4.2 for discussion of these studies.)

Less well known than the combs or pouches of grouse, but even more spectacular, are the wattles of male Bulwer's pheasants and the lappets and horns of male Temminck's tragopans (Figs 2.1 and 3.3; see also Johnsgard 1994, photographs 5 and 6). In these last two species, the bright colours of these structures are but one aspect of the males' display. In addition, there is a 'surprise' presentation of the wattles or lappets in specific settings, which is a key aspect of courtship display in both. The bright blue wattles of male Bulwer's pheasants apparently expand as a result of their filling with blood (Delacour 1977); the horns and lappets of male tragopans may undergo rapid enlargement in a similar manner.

In addition to galliforms, fleshy ornaments also are found in many other orders of birds (see Fig. 4.3), including cassowaries, gruiforms, charadriiforms, anseriforms, ciconiiforms, pelecaniforms, falconiforms, passerines, etc. The list of such species is a long one. Clearly, the evolution of structurally specialized areas of bare skin, often colourful, on the face has occurred repeatedly in unrelated groups, which in itself suggests two points: (1) for birds, the number of possible ways to exhibit bare skin is limited; and (2) the signal function of such structures is apparently important in species of many lineages.

4.5.1.3   Knobs and casques

A variety of birds, including curassows and guineafowl (Fig. 4.5), possess structures on the head that often appear to be ornamental (e.g. Buchholz 1991). Sometimes these are very hard, with a bony base. In the helmeted guineafowl, different structures on the head may serve different functions. This and all of the other guineafowl species possess bare, often colourful faces and heads; in addition, conspicuous paired wattles are present in some species (see

(a)   (b)

Fig. 4.5 Casques of the helmeted curassow (a) and the helmeted guineafowl (b).

Crowe *et al.* 1986, Plate 5). In these socially monogamous and plumage–monomorphic birds, the facial colours and wattles are likely to be employed by females in mate choice. In contrast, the style of fighting employed by male helmeted guineafowl suggests, as the bird's common name implies, that the casque on the head may serve as a defence against kicks from opposing males (personal observation). Although the significance of knobs and casques has not yet been demonstrated for any casqued or knobbed species, they probably are important in sexual selection (Buchholz 1991).

4.5.1.4 *Bill and tarsus colour*

Bills often are brightly coloured (Fig. 4.6), especially during the breeding season. The bright yellow bill of the European starling is a familiar example. The yellow colour is the result of carotenoid pigmentation, which becomes exposed in the spring, as black melanin pigments are withdrawn from the bill in response to androgens. (Bill colours of two other starlings, the red-billed and yellow-billed oxpeckers, also are likely to be based on carotenoid pigments.) Zebra finches provide another example of the use of bill colour in mate choice. In one of the few, if not the only, studies of the relationship between bill colour and mate choice, it was determined that female zebra finches prefer males with redder bills, while males prefer females with less red bills (Burley and Coopersmith 1987; Price and Burley 1994).

Burley *et al.* (1982) pointed out that the tarsi and toes of a variety of types of birds are often brightly coloured, and that colourful tarsi might well be important in the mate choice component of sexual selection. At least some species within many groups exhibit brightly coloured, often red or yellow, tarsi and toes. One conspicuous candidate, among many, for such selection is the blue-footed booby. Despite the common occurrence of brightly coloured tarsi, I am aware of no studies which demonstrate that tarsus colour is a sexually-

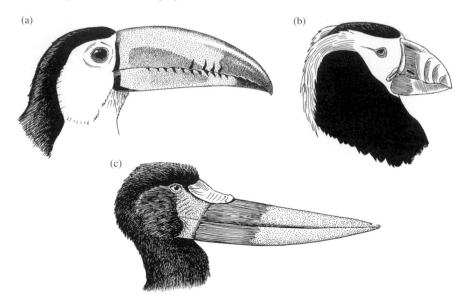

Fig. 4.6 Colourful, ornate bills of the (a) keel-billed toucan, (b) tufted puffin, and (c) saddle-billed stork.

selected trait, apart from those dealing with effects of colours and colour patterns of plastic leg bands.

4.6   *The significance of multiple ornaments*

Males of many kinds of birds, both socially monogamous and socially poly-gynous, exhibit from one to several sets of ornaments thought to be used in sexual selection. In fact, if we consider behavioural displays and vocalizations to be ornaments, all birds exhibit multiple ornaments. In addition to the issue of multiple ornaments, males of many species possess what appear to the human eye to be striking ornamental traits, some of which apparently are not currently used by females in mate choice decisions (Table 4.1).

Why many species bear more than a single ornament is an important issue with regard to theories about the evolution and maintenance of male orna-mentation and behavioural display, and it provokes several questions. (1) Why are some ornaments of males used by females in their mate choice decisions, while other, apparently equally complex ornaments currently appear not to be important to females? (2) Why do ornamental signals become increasingly elaborated? (3) Why might certain ornaments lose the ability over time to convey accurate information to females; i.e. why does a single, simple ornamental trait not suffice indefinitely?

Table 4.1 Some male plumage ornaments apparently *not* used by females in mate choice

Species	Ornament	Reference	Comment
red junglefowl	all plumage traits	Ligon & Zwartjes 1995*a* Ligon *et al.* 1998	Virtually the entire plumage of male red junglefowl is ornamental; females do not prefer normal ornate males over 'hen-feathered' ones
ring-necked pheasant	tail length brightness of body plumage	Wittzell 1991 Mateos & Carranza 1995	Recent manipulative studies by Mateo & Carranza 1995 suggest that tail length, but not body plumage, is important in female choice
turkey	all plumage traits, (beard, iridescent body plumage, large fan-like tail)	Buchholz 1995	As in red junglefowl, almost all of the male's plumage appears to be ornamental
sage grouse	tail length	Gibson & Bradbury 1985	
black grouse	tail length	Alatalo *et al.* 1991	
grey partridge	breast patch	Beani & Dessi-Fulgheri 1995	
long-tailed manakin	tail length	McDonald 1989	
Lawes' parotia	length of head wires	Pruett–Jones & Pruett–Jones 1990	

Møller and Pomiankowski (1993*a*) provided three hypotheses that could account for multiple ornaments. They restricted their analyses to plumage ornamentation, which, for many, and probably most, species encompasses only part of the overall ornamentation exhibited by a male bird.

1. The *multiple message hypothesis* ' ... proposes that each display reflects a single property of the overall quality of an animal. This is likely to be the case for ornaments that respond to condition on different time scales' (Møller and Pomiankowski 1993*a*, p. 167).

2. The *redundant signal hypothesis* ' ... suggests that each ornament gives a partial indication of condition. ... The redundant signal hypothesis predicts that (i) multiple ornaments should be particularly common among taxa with relatively uncostly and fine-tuned female choice, and (ii) females pay equal attention to the expression of all the secondary sex traits in order to obtain an estimate of overall male condition' (Møller and Pomiankowski 1993*a*, p. 167).

3. The *unreliable signal hypothesis* ' ... argues that some ornaments are unreliable indicators of overall condition and are only maintained because they

are relatively uncostly to produce and there is a weak female preference for them.' This predicts that (i) multiple sexual ornaments should be particularly common in taxa with the most intense sexual selection (i.e. lekking and other polygynous taxa), and (ii) there should be more evidence for condition dependence in ornaments of species with single as opposed to multiple ornaments (Møller and Pomiankowski 1993a, p. 167).

Møller and Pomiankowski (1993a) concluded that the evidence they considered supports the unreliable signal hypothesis, and that it does not support either the multiple message (which posits that each ornamental trait reflects a single aspect of the male's overall quality), or the redundant signal (each ornament gives a partial signal of condition) hypotheses. This led them to place male ornaments into two categories. If a particular ornament is unrelated to male condition, and is not currently used by females as they assess prospective mates, Møller and Pomiankowski (1993a) consider it to be a non-functional 'Fisher trait.' (It should be noted parenthetically that use of the term Fisher trait is misleading, in that it implies that the trait in question evolved via Fisherian runaway selection (Fisher 1958), which to date has not been documented (Kirkpatrick and Ryan 1991), and it may or may not even exist; see Section 6.4 for an assessment of the Fisher runaway hypothesis.) In contrast to 'Fisher traits', if the ornament is related to its bearer's condition, and is thus more costly for low-quality than for high-quality males to produce and maintain, the ornament will reliably reflect good genes. Such ornaments may become ever larger, more colourful, or more intricately designed, until further elaboration is stopped by the counteracting forces of natural selection (Møller and Pomiankowski 1993b).

Møller and Pomiankowski (1993a) argue that Fisher traits will turn over faster than traits that reliably reflect condition. These authors also suggest that females will tend to lose interest in Fisher traits, especially if female choice is costly; i.e. although females can be repaid for the costs of choosing if the male chosen provides good genes to their offspring, females are not compensated for costly choice if the male ornaments used provide no such genetic benefits. Following this line of reasoning, females should lose interest in Fisher traits and begin to choose mates by paying attention to other characters. Thus, over time the focus of choice of most females in a population or species will shift from one trait to another. For these reasons, Møller and Pomiankowski suggest that in highly polygynous taxa, especially lekking species, many secondary sex characters are likely to evolve, and be retained, and that many or most such ornaments in taxa with multiple sex ornaments will be Fisher traits.

Thus, following this scenario, over time female preference will switch from one trait or set of traits to another. Once a particular trait no longer provides accurate information about male quality (good genes), or no longer continues to become increasingly attractive and elaborated due to overriding costs (runaway halted by natural selection), then the attention of females will probably be directed towards either a completely new trait or a new kind of presentation of an old trait (which, in effect, makes it a new trait). Thus, among other things, investigation of the issues pertaining to multiple ornaments can provide insight into the process of shifts of preferences by females over time.

Møller and Pomiankowski (1993a) suggest that, in contrast to polygynous species, extravagant ornaments in monogamous species will usually demonstrate signs of condition dependence for two reasons: (1) males of many monogamous species provide extensive parental care and it will be particularly difficult to carry this out while bearing an exaggerated sex character; and (2) the aftermath of reproductive activity usually carries over to the non-reproductive season. Thus, although the Fisher process can occur in monogamous species, it will do so at a much weaker intensity than in polygynous species. For these reasons, Møller and Pomiankowski suggest that males of monogamous species are likely to evolve one or very few sex traits and that these are likely to be signals of 'good genes'.

To test their hypothesis, Møller and Pomiankowski (1993a) compared species with single and multiple sexual ornaments, with regard to ornament size, degree of variation, and degree of fluctuating asymmetry (see Section 5.6). In accordance with the unreliable signal hypothesis, species with single ornaments more often exhibit condition-dependent ornaments than do species with multiple ornaments. Basing their analyses and arguments primarily on patterns of fluctuating asymmetry, Møller and Pomiankowski (1993a, p. 175) conclude that '... multiple ornaments in birds do not appear to signal the quality of an individual; rather many of these ornaments are Fisher traits that do not currently signal male condition.'

Other evolutionary hypotheses also address the question of multiple ornaments. For example, females may shift their attention from one trait to another because of a fundamental conflict of interest between the sexes. Females may seek accurate indicators of male quality or condition, while the majority of males (the average and below-average individuals), on the other hand, may attempt to 'cheat', by attempting to produce signals which indicate (falsely) that they are of above-average quality (Hill 1994b). Hill's model provides an explicit mechanism for the continued elaboration of a male ornamental trait, and it also relates to the question of multiple ornaments by accounting for a shift by females from one ornamental trait to another (See Section 6.5).

Because males compete for females via ornaments, it pays a high-quality male to produce signals honestly indicating cost. This can be accomplished either by continued elaboration of a single trait or by producing a new ornamental trait attractive to females. Both evolutionary responses are driven by the importance to females of honest signals of male quality, together with the attempts of low-quality males to attempt to express those signals dishonestly. Hill (1994b) argued that extreme development of ornamental traits will occur only if average and below-average males can somehow match the ornaments of high-quality males. That is, if males of both low and high quality can produce traits of equal development, selection on females to be discriminating will strongly favour further elaboration of the same trait. Alternatively, females may change their focus to a new trait that more reliably signals quality. If average and low-quality males cannot match the signals produced by high-quality males, simple traits of high-quality males can function indefinitely as reliable indicators. Thus the females' demand for accurate indicators of male quality can either drive continued elaboration of traits, or lead to a shift by females from reliance on one male ornamental trait to another (Hill 1994*b*).

4.6.1   Multiple ornaments in red junglefowl

One of the species in which male plumage ornaments appear to be unimportant to females is the red junglefowl. In addition to ornaments on the head, the comb, paired wattles and ear lappets, male red junglefowl possess extremely ornate plumage, in terms of both structure and colour (see Figs 2.3 and 9.1). Although male jungle fowl possess a number of ornamental plumage traits— e.g. red or orange and structurally specialized hackle and saddle feathers, red shoulder feathers, blue or purplish greater wing coverts, iridescent tail coverts and rectrices—females appear to make little use either of individual plumage traits of males (Zuk *et al.* 1990*a*), or of all ornate plumage traits combined (Ligon and Zwartjes 1995*a*; Ligon *et al.* 1998) in their mate choice decisions. Experiments conducted by Ligon and Zwartjes (1995*a*) and Ligon *et al.* (1998) were designed to give females the opportunity to choose between two males, one with normal ornate plumage and one with hen-like plumage. Females mated with a hen-feathered male as frequently as with an ornate male. This suggests (1) that typical ornate plumage may not be used in mate choice decisions, and, therefore, (2) that relative development of male plumage traits is currently unimportant to females. In contrast, the same females showed significant discrimination between males when comb size was the manipulated variable.

Thus, although male junglefowl exhibit structurally specialized and highly colourful plumage that intuitively appears likely to be important in mate choice by females, this was not supported by studies of female mate choice that employed males with mutant female-type plumage. These results indicate that, although females do discriminate between males, the plumage is not the target of the female's attention. What then might be the current adaptive significance of the bright, showy plumage that characterizes present day male red junglefowl? One possibility is that, at least with regard to mate choice, the male's ornate plumage may have little or no current adaptive function, but that in the past it did function to attract females.

It is possible that in the ancestors of present day junglefowl, prior to the evolution of the specialized comb (which is unique to the Genus *Gallus*) the plumage of males may have played an important role in female choice. That is, male plumage traits once may have been important in female mate choice

decisions, but over evolutionary time this function of the plumage was superseded by the comb. This suggestion is strengthened by the fact that an evolved
function of the comb appears to be to convey accurate information about the
physical condition of its bearer, as discussed in Section 5.2.2. In contrast, the
plumage of the male junglefowl and its domestic relatives appears to provide
no information about the current health or condition of the male, and in fact, as
discussed earlier, the male-type plumage develops fully in castrates (Domm
1939; Ligon *et al.* 1990).

The absence of a relationship between ornamental plumage of male junglefowl and female mate choice, together with dependence by females on the
comb, provides circumstantial evidence that, over the course of evolutionary
time, females may shift their attention from one cue or set of cues to one or
more recently evolved signals, as hypothesized by Møller and Pomiankowski
(1993*a*) and Hill (1994*b*). This could be due either to loss of heritable variation in the former or to the development of new traits that provide more accurate indication of male condition. In short, when male traits do not provide
accurate information about their bearers, females may come to be attracted to
other, possibly newly evolved or conspicuous characters exhibited by males.

The fact that female red junglefowl do not prefer males exhibiting ornate
plumage over those bearing non-ornate plumage (Ligon and Zwartjes 1995*a*)
suggests that conspicuous, ornate morphological traits that appear to the
human observer to be likely to attract females of the species in question, may,
at present, serve no such function. The discovery that, in a number of species,
certain traits unarguably ornamental to the human eye are not used by females
in their assessment of mates has major general implications for the study of the
female choice component of sexual selection. For one thing, in the absence of
any supporting data, it is unwarranted to conclude that what appears to us to
be a conspicuous and attractive ornamental character is, in fact, currently
important to females of the species in question.

4.7   Sexually monochromatic bright or ornate plumage

The frequency and broad taxonomic distribution of monochromatic bright
birds suggest the possibility that monomorphic bright plumage was the ancestral condition in some avian lineages. Although species in which females are
as bright as males may seem unusual to those who are primarily familiar with
dichromatic species of the northern hemisphere (e.g. ducks, passerines), many
avian taxa, including entire families and even orders, are characterized by
having predominantly colourful (at least to the human eye), sexually monochromatic plumage. Among these are parrots (Order Psittaciformes),
kingfishers, bee-eaters, rollers, hoopoes, motmots and todies (O. Coraciiformes),
flamingos (O. Phoenicopteriformes), trogons (O. Trogoniformes), jacamars,
barbets and toucans (O. Piciformes), touracos (O. Musophagiformes), doves and

pigeons (O. Columbiformes), and a variety of passerine birds (O. Passeriformes), such as most New World jays, some cotingids (Trail 1990), and some starlings and icterines (Irwin 1994). Some of the best and most striking examples of brilliant plumage monochromatism occur in most of the families that make up the traditional order Coraciiformes. (Fry *et al.* 1992 provide a beautifully illustrated survey of all members of three coraciiform groups, the bee-eaters, kingfishers, and rollers.) Bright, monomorphic plumage also occurs in a minority of species in other avian taxa. The diversity of taxa in which this condition is found suggests that (1) selection for cryptic coloration may not be strong in many taxa or that (2) selection favouring bright plumage in both sexes may be common.

In most kinds of birds, males and females look much alike, whether or not their plumage is colourful. This, of course, correlates with the fact that the great majority of avian species are monogamous. Probably anyone reading this book can think of numerous species that are sexually monochromatic, monogamous, and cryptically or semi-cryptically coloured. It is intuitively easy to understand why both sexes of many species might possess cryptic plumage, regardless of mating system (e.g. species especially vulnerable to predators). Likewise, it is not difficult to provide *ad hoc* explanations for bright plumage in males and subdued plumage in females. In contrast, it is not so easy to explain why in certain groups both sexes are brightly, or even brilliantly, coloured. West-Eberhard (1983) put forth the idea that bright-plumage sexual monochromatism is related to the importance to both sexes of intra-specific interactions, and she uses the term *social selection* for selection that acts on traits that serve in behavioural interactions between individuals.

What might be the functional basis or current adaptive significance of brightly coloured males and equally brightly coloured females? Before addressing that question, it should be noted that there appears to be a phylogenetic component to bisexual brightness in the families mentioned above (e.g. sexual plumage monochromatism occurs in nearly all species of the Family Corvidae and the Order Coraciiformes). Interestingly, in these bisexually bright species—as in the great majority of birds—the only documented mating system is social monogamy with significant paternal care (cooperative breeding, which is relatively common in New World jays and Old World coraciiforms, is not, strictly speaking, a mating system at all; see Chapter 14). Thus, bright plumage in females is not obviously correlated with mating system, since more than 90 per cent of all birds are said to be monogamous (Lack 1968). However, the importance of the male's contribution to the female's breeding success may have led to increased female–female interactions compared to birds in which females do not compete for a critical resource. Here the suggested resource is a mate, especially one that holds a territory suitable for breeding and provides valuable parental care (also see Section 2.4).

One obvious possible functional explanation for female ornamentation is that males may choose mates on the basis of their appearance, as females have repeatedly been shown to do. This possibility has been experimentally tested in the crested auklet (Fig. 4.1f), a monogamous seabird in which both sexes are ornamented. Males prefer more decorated females, as well as vice versa (Jones and Hunter 1993). Somewhat similarly, Hill (1993*a*) found that in the house finch, a sexually dichromatic species, males preferred brighter females. However, in neither least auklets nor house finches is there any evidence that the ornaments of females serve to indicate individual quality (Jones and Montgomerie 1992; Hill 1993*a*).

4.7.1   Sexual disguise and intersexual aggression

Three studies suggest that in some cases plumage similarity between the sexes may serve to render sexual identification difficult, and thus serve in intersexual agonistic encounters. In most members of the Genus *Picoides*, to which the red-cockaded woodpecker belongs, males and females are alike in plumage except for a patch or stripe of red or yellow on the heads of males. In the red-cockaded woodpecker, however, the male's red is greatly reduced and is usually covered. (Like the epaulette of male red-winged blackbirds, the cockades of this woodpecker are 'coverable badges' Hansen and Rowher 1986.) Sexual dimorphism in body size is also slight in this species.

In the red-cockaded woodpecker, it appears that a function of the similarity between the sexes in size and appearance is related to production of a more unified and aggressive response by members of both sexes to intruders into the territory. Sexual similarity in appearance may also be related to living year-round in social groups. Reduced sex-identification characters may serve to reduce intra-flock aggression (Ligon 1970). Finally, it should be mentioned that this apparent convergence of the two sexes to a 'unisex' external morphology is probably not based on ecological factors, since male and female red-cockadeds, like other members of their genus, are ecologically segregated to some degree in that males and females forage at different sites (Ligon 1968*a*).

A somewhat similar situation and interpretation is presented by Whittingham *et al.* (1992) for red-winged blackbirds in Cuba. These authors compared song and sexual dimorphism of red-winged blackbirds from eastern North America and from Cuba. Male and female red-wings in North America are strongly dimorphic, with males being of larger size and possessing black plumage with red epaulettes, while females are streaky brown. Males also produce a song unlike that of females. In contrast, in Cuban red-wings, the sexes are more similar in size than in North America, and both males and females possess black plumage; however, female Cuban red-wings differ from males in that they do not possess a red epaulette. Finally, in the Cuban red-wings, the songs of males and females were very similar. Whittingham *et al.*

(1992) interpret the similarity of males and females of the Cuban red-winged blackbird in both appearance and vocalizations as associated with more equal roles between the sexes in territorial defence throughout the year. These authors suggest further that the similarities between the sexes may serve to allow each member of a territorial pair to defend its territory against an intruder of either sex, rather than only an intruder of its own sex.

Although their plumage is not bright, the sexual monochromatism and behaviour of Townsend's solitaires reflect a rather similar picture. Males and females are alike in appearance and size, and, during the autumn and winter, the two sexes behave similarly. Both female and male solitaires defend individual territories from individuals of both sexes by extensive singing and aggressive chases (George 1987). It appears that females are fully capable of defending a winter territory from male intruders.

To summarize, sexual monochromatism in these three species appears to be functionally related to agonistic behaviour between the two sexes. In each case, this appears to be directly related to defence of territorial space.

4.7.2   Intra-sexual interactions

In some other cases, intra-sexual interactions may be important for maintenance of ornate plumage in both sexes. Such social selection (West-Eberhard 1983) has been endorsed by Trail (1990), who compared two species of lekking cotingids and found that levels of social interactions among females were higher in the monomorphic bright species than in the sexually dimorphic one. Similarly, based on her survey of female brightness, Irwin (1994, pp. 901–2) concludes that, 'The association of bright female plumage and female aggression in diverse avian groups suggests that female aggression may be the general explanation for the evolution of bright female plumage.'

Although the suggestions of West-Eberhard, Trail, and Irwin may be correct in some cases, what about species in which female–female competition is intense and females are not bright? For example, fatal competition between cryptically-plumaged female black-headed grosbeaks is described in Chapter 9. Data relevant to this issue are scarce; however, I suspect that, in general, agonistic interactions among females are more intense among monogamous species with extensive paternal contributions (e.g. Arcese 1989) than among polygynous species with little or no paternal care, irrespective of female plumage colour.

Trail's (1990) observations on capuchinbirds of the Family Cotingidae provide an important exception to this generalization; this species is not socially monogamous, nor is there any evidence of paternal care. Thus, capuchinbirds appear to support the social selection interpretation of West-Eberhard. In short, while Irwin's (1994) suggestion, quoted above, may usually hold for species in which females are bright, it does not explain the flip side—those species either with strong female aggression, together with bright male and cryptic female plumage (e.g. black-headed grosbeak, Hill 1986), or those with strong female aggression and in which both sexes are cryptically coloured (e.g. song sparrow, Arcese 1989).

Finally, for the brightly-coloured and socially monogamous coraciiforms and parrots, all of which exhibit paternal care, an additional factor may be relevant: all species in these major tropical groups are cavity-nesters, which has at least two potentially important implications. First, because the nests are in cavities, incubating or brooding adults are not susceptible to visually-oriented predators. This might reduce selective pressures that would favour cryptic plumage. Second, because suitable cavities or sites for excavating cavities are frequently a critical limited resource, female–female aggression may occur for this reason, as well as for possession of a male mate.

4.8   Spurs: ornament or armament?

Among vertebrates, the best-known structures apparently evolved primarily in the context of intra-specific weaponry are the horns and antlers of ungulates. In birds, the tarsal spurs of many galliforms and the carpal spurs of several other groups (all three species of the Family Anhimidae, two species of Anatidae, two species of Jacanidae, 14 species of Vanellinae of the Charadriidae, and both species of Chionididae; see Fig. 4.7) represent partially analogous

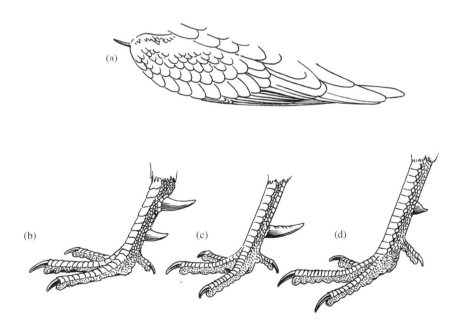

Fig. 4.7 Some examples of spurs: (a) a spur-winged plover; tarsal spurs of (b) Erckel's francolin, (c) red junglefowl, and (d) ring-necked pheasant.

structures in that spurs are thought to serve as weapons in male–male competition (see Davison 1985 for references). Until recently, students of sexual
selection would have been virtually unanimous in viewing spurs as armaments.
However, one set of studies, on the ring-necked pheasant, offers an alternative
explanation.

4.8.1   Spurs as ornament

Several recent studies by Swedish workers report that, in the ring-necked pheasant (Fig. 4.7(d), spur length is not important in male–male competition, but
instead is significantly correlated with female mate choice. von Schantz
et al. (1989) reported that, among free-living pheasants, male spur length was
strongly correlated with the success of males in attracting females to their
territories, where they nested and presumably also mated with the territory
owner. This was true both for unaltered males and for males whose spurs had
been artificially either lengthened or shortened. In addition, females that occupied territories of longer-spurred males produced more offspring than those
mating with shorter-spurred males. von Schantz *et al.* (1989) interpret these
results as an indication that spur length *per se*, rather than some correlate of
spur length, is the trait that females use in making mate choice decisions. These
results quickly generated controversy (e.g. Savalli 1989; Hillgarth 1990*b*).

Subsequent studies by Göransson *et al.* (1990) on the same population of
pheasants also reported that spur length was the most important predictor of
harem size, and that phenotypic condition and viability of males were
significantly related to spur length. Similarly, Wittzell (1991) concluded that
spur length was the most important factor for male reproductive success,
although he also acknowledged that spur length may have been correlated with
some other unmeasured character. Finally, von Schantz *et al.* (1994) showed,
via DNA fingerprinting, that long-spurred males sired more hatchlings and
surviving offspring, and that the females' production of surviving offspring
correlated with their mate's spur length.

In contrast to these results, using experimental approaches with captive
birds, Hillgarth (1990*a,b*) detected no relationship between male spur length
and female mate preference. Hillgarth provided captive female pheasants with
the opportunity to closely approach, and solicit copulations from, one-year-old
males. Many of the females did solicit copulation, while others simply
remained near certain males. (Proximity has often been used as an indicator of
female preference.) Although considerable variation in male spur length
existed, there was no correlation between this character and either indicator of
female mate choice (i.e. copulation solicitation or proximity).

The Swedish pheasant studies have been perplexing, in part because no
studies on ring-necked pheasants to date have described any behaviour that
suggests that males display their spurs to females, or that females attempt to
appraise the spurs as the males court. The courtship display of male ring-necks

is a lateral one, similar to that of many other phasianids and does not appear to involve exhibiting the tarsi, on which the short (*ca.* 15–20 mm) spurs are located (Taber 1949; see Fig. 4.7d). In fact, according to Davison (1985), no spur-revealing display has ever been described for any of the spurred galliform species. von Schantz *et al.* (1994, p. 523), however, continued to maintain that female pheasants do observe and assess the length of the male's spur:

> The length of the spurs is constantly expressed, even during locomotion and foraging, and cannot be enhanced or manipulated by behavioral mechanisms as can wattles and plumes. In addition, the spurs are exposed and clearly visible at territorial displays, during different stages of both intra- and inter-sexual displays, and during direct male–male combat.

My scepticism concerning the interpretations of von Schantz *et al.* (1989), Göransson *et al.* (1990), and von Schantz *et al.* (1994) is based largely on the fact that in their analyses of traits, these authors ignored the wattles, which develop under the influence of testosterone and which inflate during display. Nor do they consider other display characteristics of males. For example, Hillgarth (1990*a*) reports that female ring-necks preferred males that kept their wattles distended during the entire test session and which called repeatedly; spur length was not related to mate choice.

Recently, Mateos and Carranza (1996) experimentally studied the possible role of spurs in mate choice by female pheasants and, like Hillgarth, obtained negative results. Experimental manipulations of spur lengths of two matched stuffed ring-necked pheasants provided no evidence that females pay attention to this trait. However, their studies of live males showed that spur length did correlate significantly with wattle display and dominance in adult males. In male–male competition among ring-necks, both wattle size and length of ear tufts appeared to be important (Mateos and Carranza 1997).

Finally, to add to the perplexity, Grahn and von Schantz (1994) report that spur length is not as important when females choose among one-year-old males. Younger males have shorter spurs than do older birds, and these authors suggest that the short spurs of the young males are less reliable indicators of male phenotypic condition (see also Hillgarth 1990*b*). However, Mateos and Carranza (1996) report that in young male ring-necks spur length was positively correlated with physical condition and weight, which suggests that the spurs of one-year-olds do reflect phenotypic condition.

Apart from the controversial case of the ring-necked pheasant, the spurs of galliforms have unanimously been assumed to function exclusively in male–male competition (Davison 1985), and this assumption has been confirmed for the red junglefowl (Ligon *et al.* 1990; Zuk *et al.* 1990*a*).

4.8.2   Spurs as armament

In this section, using information based primarily on Davison (1985), I consider the tarsal spurs of galliforms. In red junglefowl and their domestic descendants, the game fowl, the spurs unquestionably are punishing weapons (Fig. 4.7c) and thus are critically important in male–male conflict, as evidenced by the ancient sport of cockfighting (Ligon *et al.* 1990). Moreover, unlike the situation reported by von Schantz *et al.* (1989, 1994) for ring-necked pheasants, experimental studies suggest that spur length is not associated with mate choice by female junglefowl. Females do not preferentially mate with mature, long-spurred males over short-spurred one-year-old cockerels; in fact, in 13 of 19 mate choice tests using a long-spurred older male and a short-spurred younger one, females mated with the short-spurred cockerels (Zuk *et al.* 1990a). Nor is spur length tightly correlated with dominance in adult male junglefowl. In this species, the spurs continue to grow throughout the male's life. As a result, the oldest males typically possess very long spurs; nevertheless, they often become subordinate to vigorous younger males that possess shorter spurs (personal observation); i.e. the vigour and persistence with which the weapons are applied are more important to the outcome of conflicts between males than the length of the weapons.

Davison's (1985) review of tarsal spurs in the Order Galliformes indicates that, depending on the species, spurs may be either single or multiple (Fig. 4.7(b–d)), and that they vary in shape from a short, blunt knob to a sharp-tipped scimitar. Within the galliforms, no member of the Megapodiidae (mound-builders), Cracidae (guans, currasows, chachalacas), Tetraoninae (grouse), or Odontophoridae (New World quail) bears spurs. In contrast, spurs are present in more than 113 species of the Phasianinae (pheasants, Old World quail, partridges, and francolins), the Meleagridinae (turkeys), and some Numidinae (guineafowl). In 16 of these species, females as well as males are spurred. Because of the intense male–male competition thought to be characteristic of polygyny, the presence of spurs might be expected to correlate with that kind of mating system. However, Davison's compilation and analysis does not clearly support this expectation. For example, sexual selection and male–male competition is very strong in many species of grouse, as well as in some megapodes and cracids (Buchholz 1991; Jones *et al.* 1995), yet no species in these groups possesses spurs.

On the other hand, of the 48 spurless species in the Phasianinae, Numidinae, and Meleagridinae, 46 are socially monogamous. Considering the spurred galliforms, 79 of 133 species are socially monogamous and 34 are 'polygamous' according to Davison (1985). Taken at face value, these numbers indicate that social monogamy is the more common mating system among both spurred and spurless species. A critical problem, however, with Davison's analysis of spurs and mating system is the fact that each species is treated as an independent data point. Many of the species he lists are closely related congeners, such as the 41 species in the Genus *Francolinus*, nearly all of which are spurred (and commonly labelled spurfowl) and thought to be socially monogamous (Crowe *et al.* 1986). Current phylogenetic analysis would treat these 41 species as a single data point rather than 41 independent cases of spurred and monogamous species, because they are probably descended ultimately from a single spurred common ancestor (Harvey and Pagel 1991). Even setting the francolins aside, Davison's (1985) data suggest a phylogenetic component to any relationship between spurs and mating system.

To summarize, excluding the francolins, among the Phasianinae most spur-less species are monogamous and most spurred species are polygynous. Within this group, the greatest stumbling block to a 'rule' concerning spurredness and mating system is the large genus *Francolinus*, most members of which are both spurred and monogamous. Moreover, both sexes of the ornate, sexually dimorphic and monogamous Congo peafowl possess spurs and, conversely, spurs are absent in the polygynous and spectacularly ornate great argus pheasant. Additional problems with attempting to make a causal connection between mating system and spurs in the galliforms as a whole include the presence of weakly developed spurs in two genera of guineafowl (all species thought to be monogamous) and, as mentioned previously, the absence of spurs in any species of grouse (many species promiscuous). Points such as these led Davison (1985) to conclude that mammalian weapons show a much closer association with polygamy than do avian spurs.

Sullivan and Hillgarth (1993) also tested for mating system correlates of tarsal spurs in the spurred galliforms. Their analyses, which attempted to control for lack of independence of data points, yielded ambiguous results. Thus, their results, like those of Davison (1985), do not provide clear evidence of a relationship between spur length or number of spurs and mating system.

4.9   Bird song and sexual selection

Birds, like humans, are acoustically as well as visually oriented organisms (Fig. 4.8). All birds, even those limited to hissing (e.g. New World vultures), use both sound and visual signals in communication. Loud vocalizations, perhaps especially song, are among the most conspicuous aspects of bird behaviour (Catchpole and Slater 1995). Other auditory signals, such as the drumming of woodpeckers, or specialized feathers that make distinctive noises in displays, appear to serve the same functions as song proper, namely advertisement of the presence and motivation of the signal emitter. For many birds, including both passerine 'songbirds' and a variety of non-songbirds, song may be a major part of the male's daily activities during much of the breeding season. As an extreme example, a red-eyed vireo in eastern North America sang 22 197 songs in a single day (de Kiriline 1954). (It may be worth noting that, unlike many other passerines, male red-eyed vireos forage while singing, thus perhaps making possible the incessant outpouring of song.)

Social signals manifested as song, because of their physical nature, have some important features not shared with fixed morphological traits. Individuals

Fig. 4.8   Bellbirds, including the bare-throated bellbird, are famous for the volume of their vocalizations. Reprinted with permission from Dover Publications, Inc.

of many species can produce a number of distinctly different songs. In some cases, variation in song also can provide graded signals or messages. For both of these reasons, song has the potential to produce a greater diversity of signals than can fixed plumage traits, or even behavioural displays, many of which are highly stereotyped. Within a species, individual differences among males in aspects of song can provide information about the singer's physical condition relative to other males (e.g. Cuthill and MacDonald 1990; Galeotti *et al.* 1997). In addition, because males of many species have song repertoires, a single male potentially can convey a greater diversity of stimuli to listeners of both sexes than could be emitted by display of a morphological character, such as a colourful patch of plumage. Finally, song can serve as a long-distance signal in habitats where visibility is obscured. Thus, for a variety of reasons song can be simultaneously important both in male–male competition and in female choice (Baker 1988; Kroodsma *et al.* 1989; Kroodsma and Byers 1991).

4.9.1 Energetic costs of song and crowing

During the breeding season, males of many species, both songbirds and others, spend a great deal of time vocalizing (e.g. tyrant-manakins, which spend about 90 per cent of the day calling; Snow 1961). It is intuitive that any activity involving so much time is likely to be of major significance to the actor. Moreover, if song serves to indicate male quality to females, we might expect its production to be costly (Zahavi 1975; Grafton 1990), and there is some indirect evidence that this is so. First, song output increases when the bird is provided with supplemental food (e.g. Strain and Mumme 1988). Second, cold ambient temperatures reduce singing rate (e.g. Santee and Bakken 1987). Eberhardt (1994) provided more direct evidence of a cost of song in Carolina wrens by measuring oxygen consumption during singing. During singing, energy consumption increased by from 2.7 to 8.7 times the basal metabolic rate, leading Eberhardt to conclude that song was more energetically costly than any other activity except flight.

Recently, Eberhardt's (1994) conclusions were challenged by Gaunt et al. (1996) on several counts. The main point of disagreement is not whether singing has any cost, but whether the level of physiological cost is as great as reported by Eberhardt, and, secondarily, the type(s) of costs incurred. With regard to the costs of singing, it does appear that at least some functionally song-like vocalizations, specifically the crow of the male red junglefowl and domestic fowl, are not energetically costly (Chappell et al. 1995; Horn et al. 1995). Whether or not there is a fundamental difference in the energy demanded by the cock's crow as opposed to the wren's song remains to be determined (Gaunt et al. 1996).

Eberhardt (1996) replied to the criticisms of Gaunt et al. (1996) by pointing out that even if singing is not costly in terms of elevated metabolic rates, it is often accompanied by active displays (e.g. northern mockingbird), it requires time that is unavailable for other activities, and there probably is some increased risk of predation. Thus, for most species the production of song has a cost, at least with regard to time budget considerations (see Eberhardt 1994 for references).

4.9.2 Functions of song in sexual selection

At any one time and place song may serve multiple functions and, conversely, the functions of song may also vary over time. For example, males of many migratory species of temperate zone passerines sing during migration, well before reaching their breeding grounds. Thus, at this time, song presumably does not function in territoriality. Quay (1989) showed that females of a variety of passerine species receive sperm prior to or during spring migration. One implication of Quay's discovery is that singing by males during migration may serve to stimulate females to copulate (but see Briskie 1996). Males often arrive at their breeding grounds well before any females appear and sing almost from the moment of their arrival, which suggests that at this time song serves exclusively in male–male interactions. A few days later, song draws the attention of newly-arrived females to the singers. These observations suggest that for migratory, north temperate passerines, the functions of song change over time and space.

If song influences mating success of individuals for any reason or in any manner, it may be considered to be important in sexual selection (Searcy and Andersson 1986). In some cases, song elaboration shows a correlation with mating system that parallels the relationship seen between plumage and size dimorphism and mating system. For example, among North American wrens the polygynous species have larger song repertoires than the monogamous ones (Kroodsma 1977), although this kind of interspecific relationship does not hold for all groups (Catchpole 1980).

Bird song and sexual selection are usually assumed to be causally related (e.g. Shutler and Weatherhead 1990); however, clear evidence that the conspicuous advertising song is important in female choice, beyond attracting the attention of females, is not common, given the large numbers of studies on song (Kroodsma and Byers 1991). If sexual selection has, in fact, shaped the evolution of bird songs, Payne (1983) lists six areas that should provide evidence concerning the relative importance of song in male–male competition and female choice. As Payne notes, most of these points or predictions will not clearly support one form of sexual selection while falsifying the other. These areas are:

1. *Context.* Under what circumstances is song used and is other behaviour associated with it?

2. *Development.* How does song develop? If male–male competition is most important, one might expect song to develop in a situation-dependent manner, such as, for example, mimicry to deceive other males. On the other hand, only if song structure is genetically determined can it provide information to females about the genetic make-up (quality) of the male in relation to other males.

3. *Repertoire size.* If repertoire size is important in sexual selection, song should correlate with some measure of male status, such as relative territory quality. If it is important in female choice, it should correlate with time of mating or, in polygynous species, with number of females attracted.

4. *Stereotypy and predictability.* This factor could be important in male–male competition if individual recognition is important to the singer. If females prefer certain males, these factors could make it easier for individual males to be located for mating by females.

5. *Conformity.* Males with unusual or atypical songs may be at a disadvantage both with regard to competition with other males and in attracting females.

6. *Dialects.* In species with song differences among local populations, males should have a better chance of establishing and defending a territory against other local males. Females should choose a mate that sings in a manner similar to their own song dialect.

Searcy and Andersson (1986) have also considered the means by which one could demonstrate the relative importance of song in influencing female choice versus its role in male–male competition. They suggest two criteria that together would make a strong case for female choice: (1) when songs are presented by playback in the absence of males, females respond preferentially to particular components of the song; and (2) those same song components or attributes are correlated with mating success under natural conditions when other likely influences on choice are controlled, either experimentally or statistically. Searcy and Andersson also propose that two sorts of evidence can demonstrate whether intra-specific variation in song affects male contests: (1) experimental documentation that song affects success in male contests; and (2) demonstration that male mating success in a natural situation is correlated either with success in male contests or the song attribute of interest. For example, it has been shown that when males were removed from their territories and replaced by speakers broadcasting song, other males were deterred from intrusion and occupation (see review by Searcy 1988), as compared to emptied territories. Male red-winged blackbirds trespassed less into territories from which song repertoires were broadcast rather than single song types (Yasukawa 1981a).

The following consideration of the role of song in sexual selection is modified from the criteria proposed by Payne (1983) and Searcy and Andersson (1986). Although it is probable that in many cases song serves simultaneously in male–male competition and female choice, these two categories are separated here in an effort to identify more specifically the way or ways that song functions in sexual selection.

4.9.3 Male–male competition

Song clearly is involved in male–male competition. Field ornithologists commonly use playback of song to attract and capture male territory-holding birds. With regard to intra-sexual selection, song probably is most important in conveying messages about ownership of space and willingness to defend it. However, because some aspects of song commonly correlate with other factors, such as age and dominance (Yasukawa *et al.* 1980; West *et al.* 1981a), it is often difficult to demonstrate that song characters alone are important in resolving male–male conflict. Here I consider three aspects of song that are relevant to male–male competition: context, development, and repertoire size.

Context. Many kinds of studies have shown that song is used in aggressive, competitive situations. Muting experiments demonstrate that males deprived of their ability to sing are less able to counter invasions into their territory, and usually lose part or all of the territory (Peek 1972; Smith 1979; Dufty 1986; M. V. McDonald 1989). Conversely, playback of song in a territory from which the male has been removed causes the territory to remain unoccupied longer than if song is absent (Searcy and Andersson 1986; Searcy 1988). Both kinds of manipulations show that song deters would-be invaders into the territory.

In addition to advertising, song is used by some species to indicate aggressive intent in an agonistic situation. Songs of male willow warblers contain a special syllable that is given more frequently as an attack nears (Järvi *et al.* 1980; see also Nelson 1984). In captive brown-headed cowbirds, dominant males sing songs not produced by subordinates; in this species, it appears that dominance is the basis for the song type, rather than vice versa. Rather than the more common pattern of using song to advertise a territory, male cowbirds advertise their within-group status by use of a specific song associated with dominance, and, at least in an aviary situation, subordinate males that attempt to produce dominant-type song are likely to be attacked by the dominant male (West *et al.* 1981*b*).

One of the best known vocalizations of male birds is the crow of the domestic cock. Leonard and Horn (1995) determined that the crows of dominant males differ from those of subordinate ones, and that males can distinguish between the crows of dominant and subordinate birds.

Development. Song in some passerines is acquired primarily by learning; young males learn to match the songs of older males or neighbours. Many studies have shown that young males learn the songs to which they are exposed (see Kroodsma 1979), and that they often change their songs to imitate neighbours. In fact, males may continue to alter their songs throughout their lives (McGregor and Krebs 1984, 1989). Payne (1982) suggests that young male indigo buntings match the songs of older, established males and that this involves competitive mimicry and deceit of other males through mistaken identity based on the older territorial male model's song. This is thought to benefit the mimic in establishing and maintaining a territory, and thus to be important in male–male competition.

In contrast to Payne's view of song as a strategy of deceptive mimicry, McGregor and Krebs (1982) found that although young male great tits also learn most songs from their immediate neighbours at 7–9 months of age, they do not share a greater number of songs with neighbours than expected by chance, nor do they selectively share songs with established neighbours. Moreover, McGregor and Krebs (1984) found no difference in reproductive success between males sharing song with established neighbours versus those sharing song with first-year neighbours.

These apparently conflicting results and conclusions may be explained by considering differences in the biology of male indigo buntings versus great tits. First, male buntings contribute little or no care to offspring, they are facultatively polygynous, and they may advertise to obtain additional mates over a period of many weeks. In contrast, male great tits provide critical care to a single large brood of young, thus the opportunity to obtain additional mates is limited. (Great tits do, however, practice extra-pair copulations; R. Wagner, personal communication.) Second, great tits require cavities for nesting, whereas buntings build cup nests in brushy areas. For male great tits in natural habitats (i.e. without nest boxes), the critical importance of nest sites and their distribution will affect the density and distribution of males, and thus the intensity of male–male competition. On the other hand, because female buntings may be attracted to an area on the basis of habitat quality, males compete to establish territories in the best available habitat. Third, great tits are permanent residents, retaining the same neighbours year-round, while indigo buntings are migratory and must re-establish territories and social relationships anew at the beginning of each breeding season.

Thus, for several reasons, the intensity of male–male competition probably is greater in indigo buntings than in great tits. If so, any behaviour of young male buntings that decreased aggression from at least some neighbours, such as song mimicry, might be strongly favoured; i.e. neighbouring males may be less inclined to attack a male that sounded like an older, well-established individual. In contrast, as a result of lesser male–male competition, such pressures should be weaker in great tits, and it may be of less advantage to a young male tit to copy the songs of older, established neighbours.

This speculative interpretation, that the tendency for males of a given species to learn a neighbour's song is related to male–male competition, is supported by observations of several types. (1) Captive, hand-reared indigo buntings imitate the songs of an adult male in a cage when they can interact aggressively with him (Payne 1981). (2) Individuals involved in severe interspecific territorial aggression may learn to sing the song of the opposing species (e.g. Bayliss 1982; Baptista and Morton 1988), including quite unrelated species.

Repertoire size. Within a species, song should be most diverse in individuals that are most successful in intra-sexual aggressive interactions. Howard (1974) found that male northern mockingbirds with large repertoires also possess territories of high quality, which reflects an important aspect of male–male competition. Similarly, in polygynous red-winged blackbirds, the size of the song repertoire is positively related to larger, better territories (Yasukawa 1981a). Artificial red-wing territories 'defended' by a tape-recorded repertoire of eight song types suffered fewer intrusions from non-neighbouring males than did territories defended by a single song type. The inverse relationship between repertoire size and frequencies of territorial intrusion also occurs in monogamous species (Krebs *et al.* 1978).

4.9.4   Female choice

The role of song in female choice appears to be more subtle and is perhaps, in general, less important than in male–male competition. Probably the clearest evidence that male song is important to females is found in the experiments showing physiological or behavioural responses (see below). Four lines of evidence that male song is important in stimulating female reproductive activities are presented here. All four suggest that differences in song among males could influence their reproductive success.

Context. 'Song often is used in sexual situations, with males singing to females and courting them with song, arousing the females' behaviour, endocrinology or ovarian development' (Payne 1983, p. 61). Several kinds of data suggest that male song is important in these ways. (1) In a strain of domesticated canary artificially selected by humans for its elaborate song, females exposed to tape recordings of large song repertoires built their nests more quickly and laid more eggs than did females exposed to smaller song repertoires (Kroodsma 1976). (2) Ovarian follicles of female indigobirds of two species became larger when they heard song mimicry by males of their own species as compared to song mimicry by males of a related species (Payne 1983). (3) Baker (1983) determined that female white-crowned sparrows responded with more copulation solicitation displays when hearing male song from their home dialect than when exposed to an alien dialect. The same was true of female yellowhammers (Baker *et al.* 1987*a*). In contrast, in female indigo and lazuli buntings song alone was inadequate to induce such sexual behaviour (Baker and Baker 1988). This difference between the buntings and the other two species provides an independent line of evidence supporting Payne's (1983) conclusion that in indigo buntings song is less important in female choice than in male–male competition.

There also is direct evidence that song can attract females. Stuffed male pied and collared flycatchers were placed next to nest boxes containing traps, along with loudspeakers broadcasting conspecific song. Many more females were trapped at boxes with broadcast song than at silent controls (Ericksson and Wallin 1986). These results indicate that song does attract females, but they can also be interpreted as a case of passive female choice (Parker 1983), since in this study the females were shown to discriminate only between conspicuous (by the presence of song) and inconspicuous (no song) nest boxes.

In captive brown-headed cowbirds, dominant males sing songs not produced by subordinates. Females in aviaries do not copulate unless the male first sings, and in playback experiments, female cowbirds court more vigorously for songs of dominants than for songs of subordinates (West *et al.* 1981*a,b*). Finally, female European starlings choose to mate with males that sing more complex songs (Mountjoy and Lemon 1996).

Development. Payne (1983) hypothesized that song development is directed by the genome in such a way that birds with different genomes develop different songs, even when developmental experiences are similar. This is a necessary assumption if females are to obtain information about the genetic quality of a male in relation to other males by hearing his song. As stated, this is a questionable assertion, and would be supported only if specific notes or sequences of notes of males stimulate the female, as described by Baker *et al.* (1987*b*).

By use of songs composed of elements from two dialect populations of white-crowned sparrows, Baker *et al.* (1987*b*) found (1) that captive females from one dialect responded to song that contained only half of the local song components, and (2) that some regions or elements of the song were more important than others in affecting female sexual response. These results suggest that different kinds of information are encoded in different parts of the song. In nature, female white-crowned sparrows do not sing; however, they can be experimentally induced to do so by treatment with testosterone. This has been a useful approach, since it permits the investigator to ascertain the song dialects to which the female was exposed in early life (to the extent that early experience is subsequently reflected by song type), and allows determination of, for example, the relationship between the song dialect that a female was born into and her later choice of a mate (see below).

A preference by females for the songs produced by their father might be expected under both the Fisher and good genes models; i.e. if preference is heritable, they should choose the song types preferred by their mothers. Several recent studies have investigated the possibility that females choose a mate by comparing their fathers' song to the songs of their potential mates, usually with negative results (e.g. Payne *et al.* 1987; Baptista and Morton 1988). For example, females of three species of Galápagos finches (large cactus finch, medium ground finch, and cactus finch) mate randomly with regard to their father's song type (Grant 1984; Millington and Price 1985).

Repertoire size. 'Repertoire size varies with the intensity of sexual selection, and is greater in species with non-monogamous mating systems than in species with monogamous mating, as a rule. Also, repertoire size within a species varies with the mating success of the male' (Payne 1983, p. 61). The idea that large song repertoires may have evolved through epigamic sexual selection was first suggested by Kroodsma (1977) and assumes that, other things being equal, females respond directly to the number of song types, and choose the male with the most song types. Baker *et al.* (1986) studied the sexual responses of female great tits to variation in size of males' song repertoires and found that the number of copulation–solicitation displays given by the females increased with repertoire size. Possible fitness benefits of such preferences to female great tits were reported by McGregor *et al.* (1981), who showed that in this socially monogamous species size of the song repertoire correlated with the most important components of lifetime reproductive success, survival, and production of breeding offspring. Heritability of repertoire size, however, was quite low.

Several other experimental studies have also reported that captive females court more for playback of multiple than single song types (e.g. song sparrow, Searcy and Marler 1981, Searcy 1984; swamp sparrow, Searcy *et al.* 1982; sedge warbler, Catchpole *et al.* 1984; red-winged blackbird, Searcy and Andersson 1986; yellowhammer, Baker *et al.* 1987*a*), which appears to support the hypothesis that sexual selection via female choice is involved in enhanced repertoire size. Using data from six species of North American passerines, Searcy (1992) showed a general relationship between median song repertoire size and strength of female preference.

Several field studies have indicated a relationship between song repertoire size and some measures of mating success. Howard (1974) reported that male northern mockingbirds with more diverse repertoires mated earlier in the season. However, the difference was probably caused by male–male interactions; males with larger repertoires apparently were more able to defend and hold the better territories against other males, and female mockingbirds apparently were attracted to these territories rather than to the song repertoires themselves. Catchpole (1980) found that male sedge warblers with large repertoires attract mates earlier in the breeding season than do males with smaller repertoires; other factors likely to affect mate choice were partly controlled. Female red-winged blackbirds also appear to prefer males with larger song repertoires. Again, however, males with larger repertoires tend to hold larger, better territories, which attract more females. Thus, song repertoires may in both of these cases be more directly related to male–male competition over territories rather than to attractiveness *per se* to females (Yasukawa *et al.* 1980).

In the polygynous great reed warbler, however, the evidence is convincing that song repertoire, male quality, and female choice are causally interrelated. Females base their mate choice on song repertoire size. Moreover, females obtain copulations from neighbouring males with larger song repertoires than their social mate, and post-fledgling survival of young was positively correlated with the size of their genetical fathers' song repertoire (Hasselquist *et al.* 1996).

Song rate has also been shown to correlate with male attractiveness in the pied flycatcher (Alatalo *et al.* 1990*a*). Alatalo *et al.* (1990*a*) suggest three possible reasons why females of this species might be attracted to males with high singing rates. High rates might signal (1) abundant food in the territory, since singing is limited by feeding demands, (2) that the male is unmated, since mated males tend to decrease their singing, or (3) male quality, in either a condition-dependent or Fisherian manner.

Dialects. 'In species with local song dialects, females respond differently to the different dialect songs and choose a mate with some constancy within their own song dialect' (Payne 1983, p. 62). As mentioned above, this prediction has received support from laboratory tests on female white-crowned sparrows (Baker *et al.* 1981; Baker 1983; Baker *et al.* 1987*b*) and yellowhammers

(Baker *et al.* 1981*a*). In these studies, the females responded with more copulation solicitation displays when hearing their home dialect than when exposed to an alien dialect. In addition, Tomback and Baker (1984) captured mated females from near dialect boundaries and induced them to sing by hormone treatment. They found that most females sang songs of the dialects from which they had come and which their mates sang.

In contrast, other studies of mated pairs have not found a relationship between male song and the hormonally-induced song of their mates (Baptista and Morton 1982; Petrinovich and Baptista 1984). Adult male white-crowned sparrows may (1) obtain mates even when singing a foreign song and (2) change their songs to match those of their neighbours (Baptista and Morton 1988) which argues against a tight link between song dialect and genetic differentiation of adjacent sub-populations of this species. However, in the brown-headed cowbird, females preferentially court for playback of their own subspecies' song (King *et al.* 1980), suggesting that song preferences of females could contribute to reproductive isolation in the manner suggested by Baker and co-workers (e.g. Baker *et al.* 1982).

4.9.5   The significance of song in some spectacularly ornate species

The relative importance of song versus other forms of male display varies from species to species and from group to group. Because song can be such an effective means of attracting the attention of females and conveying information about the singer, it is not too surprising that many species of spectacular songsters do not exhibit bright, showy visual characters. Although it is hardly a rule of thumb, the fact that among the passerines many groups containing spectacular singers are dull-plumaged and sexually monomorphic suggests that in various groups (e.g. New World mimids, wrens, and vireos, Old World sylviids) song is the primary means by which males engage in sexual selection. In other kinds of birds, such as many of the New World passerines (tanagers, wood-warblers, orioles, buntings, etc.), males exhibit both bright plumage and extensive song. A majority of the temperate zone species in these groups tend to be both socially monogamous and territorial, and the relative importance of song versus plumage ornamentation in female mate choice has not been determined.

In some passerine species of Australia and New Guinea, the lyrebirds and bowerbirds, courtship traits such as plumage ornamentation, bizarre behaviour, and vocalizations appear to have been pushed to extremes beyond what we would imagine, or be willing to believe, if these birds had become extinct prior to any description of their courtship. It appears that males of these species provide females with the greatest possible diversity of strong visual and vocal stimuli. In lyrebirds and bowerbirds, male song includes extensive mimicry, which, of course, provides the opportunity for vocal novelty. In these species, song may serve more importantly in mate attraction than in male–male competition.

In contrast, it appears that in another famous mimicking species, the northern mockingbird, complex song serves primarily in male–male competition (Howard 1974). Unlike the lyrebirds and bowerbirds, this mockingbird is socially monogamous, and males defend an all-purpose territory (Breitwisch and Whitesides 1987). Thus, it appears likely that in the former species mimicry does not serve the same functions as it does in mockingbirds.

4.10 Conclusions and summary

Morphologically, the Class Aves is an extremely homogeneous group. Within the restrictions of this basic uniformity, the structures used in, and modified by, sexual selection are amazingly diverse. In one taxonomic group or another nearly all feather groups have been modified as ornaments. One of the most improbable examples of this is the exaggeration of the second primary in the pennant and standard-winged nightjars. Although the modification of feathers of the lower back to become the train of the peacock also appears unlikely, contour feathers of that region have also become modified as ornaments in a variety of pheasants.

With regard to male ornamentation, a recent and surprising revelation is the apparent non-importance to females in a variety of species of certain ornate plumage characters. For several galliforms—turkey, sage grouse, ring-necked pheasant, red jungle fowl, as well as some other species—current evidence suggests that females do not rely on plumage traits when making mate choice decisions. Even within the galliforms, however, this is not a universal rule, with the peafowl providing a conspicuous exception to it (see Section 6.3).

Bare skin ('soft parts') also commonly serves a signal function, with some species in most orders and families of birds exhibiting this sort of ornamentation. Soft part ornaments include bills, combs, wattles, facial area in general, pouches, eye colour, colour of the mouth lining, etc. Traits such as these may provide a more precise or up-to-date means of providing information about the male's physical condition than could be conveyed by feathers.

Song is one of the most conspicuous attributes of birds. Considerable evidence indicates that songs, like morphological ornaments, are sexually selected traits. Song is important in the intra-sexual component of sexual selection, and is used by males of many species to deter intrusion into their territories. Males experimentally deprived of the ability to emit vocal displays are often unable to continue to defend their territories. In addition, the use of variations and gradations of vocalizations provide males with the means to convey precise information about their potential level of aggressiveness. With regard to the inter-sexual component of sexual selection, laboratory experiments in which song alone stimulates females to assume precopulatory postures or behaviour convincingly illustrate that song can be a powerful releaser of mating behaviour. Field experiments showing the importance of singing rate and possibly repertoire size on female mate choice provide additional evidence that characteristics of an individual's song can and do influence mate choice by female birds.

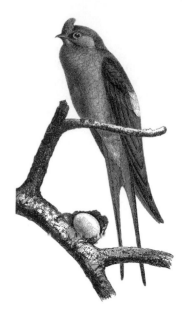

5 Male condition, parasites, and fluctuating asymmetry

5.1 Introduction

The physical condition or 'health' of male birds is important in all versions of the good genes view of female mate choice (e.g. handicap principle, Zahavi 1975, 1977; parasite resistance, Hamilton and Zuk 1982; truth in advertising, Kodric-Brown and Brown 1984). It is a given that healthy birds, by definition, will have more energy and vigour than unhealthy ones, other things being equal. Thus, healthy males will be better able to compete successfully with other males for females, whether by singing or fighting, for example, and females will be likely to be attracted to healthy males that have brighter ornaments or that display more vigorously than less healthy ones.

In the debate over the good genes view of sexual selection, the advantage to healthy males in displaying or fighting has not been an issue of contention. Disagreements have been based on three related questions: (1) does male health, or some of the numerous components of health, have a genetic basis; (2) what are the links between male physiological condition and the development or exhibition of male display traits, and between display traits and female mate choice; and (3) do females choose among males to obtain good genes for their offspring? With regard to question one, which is critical to the good genes perspective, few data are available for wild birds. For domestic fowl, however, the answer is that heritable resistance to a variety of diseases and parasites exists (Hutt 1949), and there is no reason to doubt that genetic resistance to pathogens also occurs in wild species. This chapter deals with the issues raised by questions two and three by focusing on factors affecting certain male ornamental traits.

Four major topics concerning the relationship between male condition and female mate choice are considered here: (1) the relationship between certain secondary sex characters and testosterone, and between testosterone and the immune system; (2) the relationship between carotenoid pigments, which must be ingested, and male condition; (3) the proposed relationship between parasite load and male ornaments, both within species and among different species; and (4) the relationship between physical condition of males and the degree of symmetry in paired bilateral traits.

5.2   *The role of testosterone in sexual selection*

It should be mentioned at the outset that this chapter does not deal with the subject of reproductive physiology. Specifically, the endocrinological and other physiological bases of courtship and mating, sperm production or ovulation, incubation behaviour, or parental care are not treated in this book. For general discussions of these topics, Marshall (1961), Witschi (1961), Lofts and Murton (1973), or Gill (1995) might be consulted. Although all of these subjects are certainly essential to the process of reproduction, which is the organism's ultimate *raison d'être*, most are not directly related to the most critical issues of sexual selection, as defined in Chapter 2, e.g. the basis for a female's choice of one male over another, or the ways by which a male bird may signal its phenotypic or genetic quality to members of its own and the opposite sex.

The testicular hormone testosterone, produced in males by the interstitial cells of the testes, influences female mate choice via its role in the development of certain male morphological ornaments and behavioural displays (e.g. see Ketterson and Nolan 1992 for references), as well as by its effects on the outcome of competition among males (e.g. Ligon *et al.* 1990). In contrast, this does not seem to be true for the ovarian hormone oestrogen. Although oestrogen is required as a proximate stimulus to evoke sexual interest in female birds (e.g. studies using injected estradiol to stimulate captive females to choose a male) and to produce female-typical plumage in many kinds of birds (Owens and Short 1995), among its many functions, differential production of oestrogen is not yet documented to influence female attractiveness, although this may eventually prove to be true for some species.

In addition to its inter- and intra-sexual roles in sexual selection *per se* (e.g. Wingfield *et al.* 1987; Ligon *et al.* 1990; Folstad and Karter 1992; Zuk *et al.* 1995), testosterone also indirectly influences mating systems via its influence on aggression, territoriality, etc. (Beletsky *et al.* 1995). For example, in the normally socially monogamous white-crowned sparrow, artificial elevation of testosterone in treated males resulted in polygyny (Wingfield 1984). However, similar treatment did not have this effect on dark-eyed juncos (Ketterson and Nolan 1992).

5.2.1   Effects of testosterone on secondary sex characters

Based on the few species of birds whose endocrine systems have been studied at all, it appears that the physiological bases for the development of male ornamental traits are variable. A common misconception, even among some highly distinguished students of sexual selection, is that bright, ornate plumage is produced by the effects of testosterone; e.g. 'In birds, plumage characters for sexual display are developed in the presence of the male sex hormone, testosterone ...' (O'Donald 1983, p. 54). However, contrary to this supposition, it appears that bright, decorative feathers are not usually under the influence of testosterone, and, in fact, they develop fully in its complete absence (see Section 4.2.2). Testosterone is important, however, for development of many other display traits. For example, song rate correlates with testosterone level in red-winged blackbirds (T. S. Johnsen, cited in Ketterson and Nolan 1992) and barn swallows (Saino and Møller 1995; Galeotti *et al.* 1997).

   In contrast, the typical dull or cryptic plumage of females of a variety of avian species develops in response to the presence of oestrogen. In the absence of oestrogen, females of a number of species moult into male-like plumage. In short, while testosterone is not required for development of the ornate plumage typical of males of many kinds of birds, oestrogen is required for normal plumage development in females. In the absence of any gonadal hormones (i.e. either testosterone or oestrogen), bright, male-like plumage is produced in both males and females.

   The report by Saino and Møller (1994) of a positive relationship between tail length and blood testosterone in barn swallows probably does not contradict the rule that testosterone does not control male plumage (see Section 4.2.2), because, in this case, the two are not directly causally related. Both tail length and testosterone level are related to male condition—the healthiest males grow longer tails and produce more testosterone than males in lesser condition.

   On the other hand, other types of ornaments, such as bill colour, wattles, combs, or other colourful areas of bare skin, are often directly controlled by testosterone. This leads to the prediction that male ornamental traits influenced by testosterone are likely to be important in female mate choice (see Beani and Dessi-Fulgheri 1995, Ligon and Zwartjes 1995*a*, and Zuk *et al.* 1995 for support for this suggestion). The converse, however, is not always true: ornamental traits thought not to be influenced by testosterone may also be important in female choice.

5.2.2   Testosterone and ornamentation in the red junglefowl

With regard to the role of testosterone on sexual selection, the red junglefowl and its conspecific domestic relatives provide relatively well-studied examples. In male fowl, testosterone has multiple effects on many aspects of morphology and behaviour that are directly related to sexual selection: the individual's social status, including its aggressiveness towards other males, its courtship behaviour, and certain secondary sex characters important to female mate choice, such as colour and size of the comb and wattles (Fig. 5.1). Vigorous, healthy male junglefowl typically exhibit large, bright red combs and wattles, and comb size of males correlates both with testosterone levels and with female preferences (Ligon *et al.* 1990; Zuk *et al.* 1995). In contrast, combs of sick or malnourished fowl become pale and rapidly decrease in size as a result of decreased testosterone production. Similarly, in mid-summer, following the breeding season, male red junglefowl may undergo a partial, 'eclipse' moult, and at this time their combs shrink significantly in size (Zuk *et al.* 1990*a*).

The anatomy of the cock's comb suggests that it has evolved, at least in large part, specifically to convey information about the testosterone level, and thus the internal state or health, of its bearer, as described above. The size and turgidity of the comb is due to the presence of large quantities of highly viscous intercellular mucoid, which is produced by fibroblasts of the connective tissue, only in the presence of testosterone (Hardesty 1931). When liberated from the fibroblast cells, the mucoid exerts considerable pressure that compresses the thin-walled veins of the intermediate dermal layer. This

Fig. 5.1   The comb of the male red junglefowl contains specialized cells that respond to testosterone. The comb thus has apparently evolved characteristics specifically to convey accurate information about the bearer's physical condition as reflected by testosterone level.

increases the pressure within the arterioles and capillaries and produces dilation in the latter. The enlarged capillaries in the peripheral layer of the dermis are pushed out to the epidermis, and it is the haemoglobin from the blood showing through the epidermis that gives the comb its red colour. Hardesty (1931) points out that among vertebrates the only other tissues that approach this condition are those adapted for a respiratory function. The unique histological nature of the cock's comb suggests that some of its features have evolved to indicate the bearer's physical condition, because comb size and colour are directly influenced by testosterone level, and because comb colour is due directly to red blood cells. Comb size and colour, which reflect the amount of androgen present, is an accurate predictor of the probable winner of fights among both hens and roosters (Allee *et al.* 1939; Collias 1943; Ligon *et al.* 1990). The same appears to be true for lekking black grouse, in which mating success was highly correlated both with winning fights and with blood testosterone levels (Alatalo *et al.* 1996).

In addition to its effects on specific anatomical traits, such as the cock's comb, testosterone has a direct effect on aggressive and courtship behaviour (Harding 1983), as well as on muscle and sexual development in general (e.g.

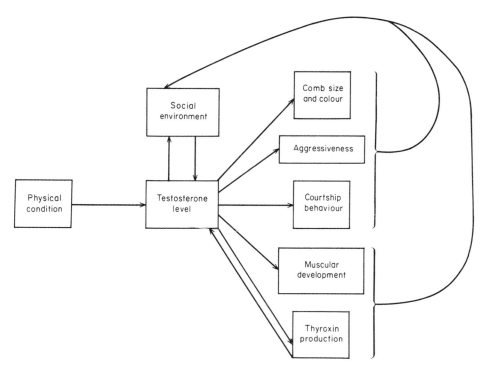

Fig. 5.2 Effects of testosterone in the red junglefowl. Reprinted with permission from Ligon *et al.* (1990). Male–male competition and the role of testosterone in sexual selection in red jungle fowl. *Animal Behaviour*, **40**, 367–73.

Turner and Bagnara 1971). These facts indicate that testosterone serves to promote both physical condition and behavioural traits critical to success in male–male competition, e.g. aggressiveness (Fig. 5.2).

5.2.3 Testosterone and the immune system

With regard to sexual selection theory, one of the most interesting aspects of testosterone is its relationship to the immune system (Fig. 5.3). It appears that testosterone depresses the immune response in at least some vertebrate groups (Grossman 1985; see Folstad and Karter 1992 for a review of this relationship specifically in the context of sexual selection). This suggests that male vertebrates exhibiting maximal development of those ornamental traits affected by testosterone are 'demonstrating' that they can survive and thrive despite suppression of the immune system. Thus, maximal elaboration of testosterone-driven ornamental traits, such as the comb of male red junglefowl (Ligon *et al.* 1990), simultaneously and inevitably increases the risk of infection by parasites and other pathogens (Zuk *et al.* 1995). This interpretation suggests that testosterone-mediated traits are honest signals of male condition (but see below).

By providing the cost required for reliability and evolutionary stability in signalling systems (Grafen 1990), the relationship between such traits and the immune system appears to provide a beautiful example of the handicap principle (Zahavi 1975, 1977). Although this relationship is not yet well studied, we can speculate on some possible mechanisms. First, it is known that in domestic fowl resistance to many pathogens has a heritable, genetic basis

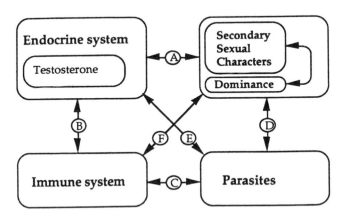

Fig. 5.3 A proposed relationship between testosterone and immunosuppression. Testosterone has a positive effect on the development of secondary sex characters and dominance, while hampering the immune response. Reprinted with permission from Folstad and Karter (1992). Parasites, bright males, and the immunocompetence handicap. *American Naturalist*, **139**, 603–33.

(Hutt 1949). A male with inherited resistance to a pathogen, such as coccidia (protozoan intestinal parasites), for example, may suffer less risk by elevating its testosterone level and thus maximizing the ornaments or displays, while at the same time suppressing its immune system, than a male without such inherited resistance. The issue of acquired immunity should also be acknowledged since it probably plays a role in combating many pathogens of wild birds. (Vaccines are available for many diseases of domestic fowl.) Genetic and acquired immunity provide two lines of defence against disease.

Why does testosterone have a negative effect on the immune system of male vertebrates (Folstad and Karter 1992)? At least two possible responses to this question exist. (1) This effect could be viewed as an interaction that has evolved specifically to demonstrate to females that males can contend with the handicap of a suppressed immune system: (that is, is testosterone-driven immunosuppression an adaptation *per se*?). This seems unlikely because females can only see the trait and not the testosterone level of the male and thus cannot enforce such a link. (2) The negative effect of testosterone on the immune system may be a non-adaptive cost of reproductive activity: that is, are males caught in the cruel bind of having to accept immunosuppression (for endocrine/physiological reasons that are not yet well understood) as a trade-off for production of the sexual traits mediated by testosterone, and that are required for the attraction of females? Although both interpretations view testosterone-mediated traits as true indicators of condition that also reflect a real cost to the individual, they are otherwise very different.

Testosterone, by enhancing ornamental signals while simultaneously reducing immunocompetence, is said to produce a 'double-edged sword' (Folstad and Karter 1992). This appears to be an example of the trade-off between natural selection (survival) and sexual selection. This interpretation may make it appear that the immunosuppression based on testosterone is maladaptive, or is vulnerable to cheating. Wedekind and Folstad (1994) respond to this concern by arguing that the relationship is adaptive, as follows. Animals may allocate limited internal resources to their most immediately demanding need. For example, substances required to produce testosterone go to that use in healthy male vertebrate animals during the breeding season, for obvious selective reasons. This may reduce the availability of those substances to other systems, including the immune system. By this view, the relationship between testosterone production and immunosuppression is one of adaptive allocation of limited resources rather than an unstable, maladaptive interaction; that is, under certain circumstances (e.g. illness during the breeding season) the individual has to make 'a hard choice': immune system or testosterone, but not both. The fact that testosterone production appears to decrease rapidly when an animal in breeding condition becomes ill (e.g. Zuk *et al.* 1990*a*), and thus, presumably, when its immune system is called into action, supports the idea of reallocation of resources.

A recent study provides some corroboration of the immunocompetence hypothesis. Zuk *et al.* (1995) determined that captive male red junglefowl with larger combs had higher levels of circulating testosterone, and they also exhibited fewer circulating lymphocytes, as expected if maintenance of traits such as the comb is costly to the immune system.

5.3   *Carotenoid pigments and sexual selection*

Many, many kinds of birds exhibit red or yellow colours in their plumage. Carotenoid pigments are often responsible for these colours. Alan Brush (1990, p. 2969) describes the significance of carotenoid pigments as follows:

> In animals, carotenoids are obtained ultimately from the diet; only plants can synthesize carotenoids. But once assimilated, carotenoids are modified, transported and deposited in surprisingly specific ways. This implies some level of genetic control, precise chemistry, and a support system that accommodates seasonal changes, sexual dimorphism, changes in plumage with age, specific differences important in social interactions and species identity. Each of these features is subject to natural selection and evolutionary change.

Via his studies of plumage colour of male house finches, Geoffrey Hill has been primarily responsible for bringing the condition-dependent nature of carotenoid plumage pigmentation of birds to the attention of sexual selectionists. In brief, Hill has demonstrated that the brightness of red pigmentation in the plumage of male house finches is used by females in mate choice, that redder males are better parents and survive better than paler males, that plumage colour of males is related to their dietary intake of carotenoid pigments, and that these carotenoids can be detected in the birds' plasma. In addition, there is geographical variation in the amount and distribution of carotenoid pigments utilized by males as plumage ornamentation, probably as a result of regional variation in the availability of carotenoid-containing foods (e.g. Hill 1990, 1991, 1992, 1993*b*, 1994*a*, and other references; also see below and discussion of house finches in Section 6.3).

The other major group of pigments, melanins, differ from carotenoids in several ways. Unlike carotenoids, melanin pigments are readily synthesized by animals. Another major difference between melanins and carotenoids is that melanins occur in feathers as small, rod-like bodies. In contrast, Brush (1990) states that carotenoids are deposited as amorphous substances with no clear structural features. In general, melanins produce the browns, black, and other less bright feather colours. The timing of changes in melanocyte activity during feather production is responsible for stripes, bars, and other patterns on feathers.

There is evidence that (1) melanin-based ornaments are often not important to females in their mate choice decisions, and, conversely, that

(2) melanin-based ornaments are used by females. Two examples supporting the former pattern are: (1) male grey partridges possess a brown, apparently ornamental, patch on the chest, which is probably coloured by melanin, and which is not important in female mate choice (Beani and Dessi-Fulgheri 1995); and (2) the red and orange colours of male red junglefowl are melanins (Witschi 1961), and male plumage is apparently not important to females (Ligon and Zwartjes 1995*a*). Two possible counter-examples include (1) the black bib of the male house sparrow, the size of which Møller (1988), but not Kimball (1996), found to be used by females in mate choice, and (2) great tits, where males with the largest black breast stripes attracted mates that laid large clutches (Norris 1990*a*).

In an analysis demonstrating a relationship between carotenoid-based plumage colours and sexual dichromatism in passerine birds, David Gray (1996) points out that, in contrast to internally synthesized or produced signals, carotenoids can, on their own, provide an honest indicator of quality. Melanins, on the other hand, provide an honest signal only if coupled with a second factor, such as a true handicap trait. Gray also suggests that the types of birds most likely to use carotenoids as sexually-selected signals of condition are those for which carotenoids are available in the diet, but are not abundant (e.g. insectivorous and granivorous species). In contrast, carnivores may not acquire sufficient carotenoids for a signalling function to evolve, while, at the other extreme, frugivorous species may ingest so many carotenoids that these pigments would not be an accurate indicator either of health, or of foraging, or of competitive abilities.

5.3.1 Plumage colour as a signal of nutritional condition

Hill and Montgomerie (1994) compared the brightness of carotenoid-based plumage coloration with feather growth rate and time of moult in male house finches from four different populations, and reported several significant findings. (1) Among the four populations, there was no significant difference in mean feather growth rate, thus interpopulation variation in plumage brightness did not reflect differences in access to food *per se*. Growth rates of tails of captive birds provided with *ad libitum* food was significantly greater than the growth rates of males in any of the wild populations, indicating that all of the house finch populations studied were food-limited to the same extent. (2) Within populations, there was a positive relation between growth rate of a male's tail feathers and brightness of his plumage, which suggests that brighter males are in better physical condition. (3) Within populations, a negative correlation existed between the onset of moult and plumage colour; i.e. brighter males started moulting earlier. These results support the proposition that carotenoid-based plumage coloration is an indicator of nutritional condition during moult. Individual differences in nutritional condition may reflect differences in either ability to locate foods high in carotenoids or in the presence of gut parasites that affect carotenoid uptake.

Hill *et al.* (1994*a*) also compared the colour of blood plasma and plumage colour in three populations of house finches and found a significant positive correlation between hues of plasma and plumage. Males, but not females, of the populations with brighter plumage also had plasma of brighter hue.

5.3.2   Carotenoids and the appearance of new ornaments

With regard to carotenoid signals, Brush (1990) describes a fascinating situation relevant to a critical question in the study of sexual selection, namely: what is the origin of morphological change in ornaments? Historically, all cedar waxwings of North America have had a yellow terminal band (a carotenoid pigment) on the tail (Fig. 5.4). Within the past thirty years, however, individuals with red terminal bands (red-tipped tails) began to be recorded in the north-eastern United States, and up to 20 per cent of the individual waxwings within a population may now be of the red-tipped morph. The carotenoid pigment responsible for the red tips, rhodozanthin, occurs in a species of honcysuckle introduced to the region (Witmer 1996).

This situation is extremely interesting for a variety of reasons. First, there is apparently no evidence of selection against red-tipped birds. The fact that they are relatively common only thirty years after having first been recorded suggests just the opposite. Second, cedar waxwings are attracted to small red fruits (McPherson 1988), and the waxy red ornaments on the birds' wings (Fig. 5.4) appear to be important in sexual selection (Mountjoy and Robertson 1988). Thus, at least some individual waxwings may be attracted to red tail tips. Third, the red pigment appears on an area that is genetically disposed to be coloured by carotenoid pigments; i.e. the new pigment occurs at a site that is already 'programmed' to exhibit carotenoids and that may have a signal function.

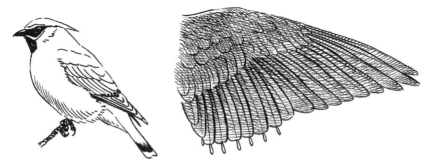

Fig. 5.4   Cedar waxwings often possess waxy red growths at the ends of the inner secondaries. Sometimes tiny waxy growths also appear at the tips of the rectrices.

These points suggest that the new red-tipped morph may come to be favoured by sexual selection. For the sake of discussion, let us assume that this is the case. Why might this be so? Cedar waxwings are apparently disposed to be attracted to small, red fruits (McPherson 1988) and to the red waxy wingtips, as a result of their prior evolutionary history, thus red-tipped individuals of both sexes probably are viewed as desirable mates (Mountjoy and Robertson 1988). At a proximate level, this scenario implicates the sensory bias phenomenon. This suggestion does not argue against a functional, ultimate basis for the attraction to red as an indicator of quality, just as waxy wingtips may do. In short, a shift in preference to a new colour pattern, the red-tipped tail, may currently be evolving in some populations of the cedar waxwing. If so, as more becomes known, this case should become an especially instructive one for students of sexual selection, particularly with regard to questions about the origins of new morphs and new preferences.

To summarize, because carotenoid pigments may provide a signal of an individual's prior foraging history, they, like structures influenced by testosterone, have the potential to be indicators of quality. This point is relevant to the issue of parasites and sexual selection (see next section), because carotenoids and carotenoid-pigmented structures can be strongly affected by parasites (Zuk 1992). For example, in red junglefowl and their domestic relatives, coccidia and a nematode gut parasite appear to decrease the colour intensity of carotenoid-dependent tissues thought to be important in mate choice. Thus, with regard to parasites, carotenoid-based ornamental plumage colours can signal the health of their bearers.

5.4   Parasites and sexual selection

Over the past decade and a half, since the landmark publication by William Hamilton and Marlene Zuk (1982), the possible relationship between the ornamental display traits of male birds (including song and other vocal displays) and their parasite loads has received a great deal of attention, including a recent book devoted exclusively to the subject of birds and their parasites (Loye and Zuk 1991). The Hamilton and Zuk hypothesis (hereafter referred to as H–ZH) is a particular version of two broader concepts, namely the good genes model of female mate choice (e.g. Bradbury and Andersson 1987) and the handicap principle (Zahavi 1975, 1977). Specifically, females should choose to mate with males that exhibit good health, or condition, based on a genetically-based resistance to parasites or other pathogens. The 'handicap' is represented by the bright plumage putatively evolved in response to high parasite loads.

The H–ZH makes two main categories of predictions—interspecific and intraspecific. The interspecific prediction, which has proved to be both controversial and difficult to assess (as discussed below), is that species faced with major problems in terms of pathogens (especially prevalence of parasite infection) should evolve more exaggerated ornamentation (e.g. brighter plumage, more complex song) than species less threatened with infection by parasites, as a result of strong selection on males to provide accurate cues to females about their health or condition. That is, among a suite of species, those most threatened by parasites should exhibit the greatest development of ornamental characters, such as bright plumage. The basis of this prediction is the assumption that interspecific differences in pressure from parasites lead to differing intensities of sexual selection among different species, which gives rise to variation in the degree of development of secondary sex characters.

Stated in another way, a greater probability of being infected with parasites has led to increased intensity of sexual selection, because of the increased significance of discriminating mate choice by females. This is because females attempt to obtain genes for their offspring from males whose appearance suggests a resistance to locally important parasites or other pathogens. This effect of mate choice on the fitness of females via the health of their offspring is what drives the development of male display characters that accurately reflect the health and vigour of their bearers. For this system to work, there must be a real cost associated with the display structures or behaviour of males (Zahavi's handicap principle). In short, the H–ZH suggests nothing less than a causal, evolved relationship between ornamentation of males and frequency of parasite infection: 'Across species, then, Hamilton and Zuk (1982) predicted that the degree of ornamentation should be positively correlated with the proportion of individuals infected with parasites' (Zuk 1991*b*, p. 317).

The more straightforward intraspecific prediction is that among males of a given species, the less parasitized, non-parasitized, or most fully recovered males should possess and display ornamental traits that are more developed than those of parasitized males. In addition, to obtain genes for parasite resistance for their offspring, females should prefer to mate with those males physiologically most capable of producing the brightest plumage or the most energetically-demanding displays, while simultaneously coping with and resisting the deleterious effects of parasites.

A hypothetical problem with the good genes interpretation of female mate choice is that when sexual selection is strong, as in lekking species, most females will mate with only a small proportion of the males in the population, thereby reducing the genetic variation available within the population and reducing the advantage of choosy mate choice. This situation has been referred to as the 'paradox of the lek' (e.g. Kirkpatrick and Ryan 1991). The H–ZH responds to this paradox by assuming that the specific kinds of parasites affecting the hosts change over time and that, as a result, genetic resistance to the effects of parasites will not come to be fixed in all individuals of the host species or population. This idea is based on the assumption that if females choose males based on evidence suggesting good health or vigour, a continuing source of heritable genetic variation in fitness must be present in the population, or else the genes responsible for the phenotypic expressions of vigour will become fixed; i.e. all individuals in the population will have the same alleles and there will be no variation for selection to act upon. To date, there is no empirical evidence on this critical point. However, there is incredible genetic diversity in the immune system alleles of mammals (Brown and Eklund 1994), and presumably this is true of birds as well.

In the following sections, I provide a brief review of some of the evidence pertaining to each of the major predictions of the H–ZH.

5.4.1   Interspecific tests of the Hamilton and Zuk hypothesis

5.4.1.1   Positive results

Several tests of the interspecific predictions of the H–ZH by use of birds have been attempted. First, using a method that scored plumage brightness and overall 'showiness' of display characters in North American passerine birds, Hamilton and Zuk (1982) found that a higher incidence of individuals afflicted with blood parasites (protozoans and nematodes) had been reported among species in which males were bright than among those species characterized by dull-coloured males. These results supported their hypothesis.

Subsequently, Read (1987) published a paper entitled 'Comparative evidence supports the Hamilton and Zuk hypothesis on parasites and sexual selection.' Read analysed data both from European passerines and from a larger data set on North American passerines. He considered several factors not treated by Hamilton and Zuk, such as behavioural, ecological, and taxonomic variables. With regard to taxonomic variables, he looked at intrageneric variation in the relationship between the species' colour and parasite prevalence. Read (1987, p. 69) reported that '… in all cases, the intrageneric correlations between colour and parasite prevalence in European passerines become significantly positive when other factors are held constant shows that these variables slightly obscure the parasite-colour relationship.'

Using data for 526 species of tropical birds, Zuk (1991b) conducted an analysis of the relationship between male brightness in a given species and the proportion of individuals carrying blood parasites. Female brightness was also scored and a 'sexual dichromatism' score calculated for each species. (Appropriate controls for phylogeny were conducted.) An unspecified prediction here is that the intensity of sexual selection might be expected to correlate positively with the degree of plumage dimorphism between the sexes, and that the most dimorphic species should thus show greater frequencies of parasite infection. Zuk also divided the birds into resident and migratory categories to test the prediction that resident species which live year-round with the same set of parasites should show a stronger relationship between brightness and parasite level than would migrant species. Both male and female brightness scores were significantly associated with parasite load; however, among the 75 migratory species alone, brightness was not significantly correlated with parasite level. In contrast, for the 451 residents brightness scores for both males and females remained significantly associated with parasite load. Sexual dichromatism scores were unrelated to parasite level in both migrants and residents.

Two other sets of brightness/showiness and parasite prevalence data provide additional, if equivocal, support for the H–ZH. Pruett-Jones *et al.* (1990) studied blood parasite infections in 10 species of birds of paradise, and in 64 other passerine species (plus 15 non-passerines) in New Guinea (Pruett-Jones *et al.* 1991). In both studies, a significant positive correlation emerged between relative parasite intensity and showiness in males. Moreover, parasite infections also correlated across species with degree of sexual dimorphism and varied with mating systems. Promiscuous species were showier and had significantly higher parasite prevalences than monogamous species. However, male showiness did not correlate with either the variance in parasite load or the opportunity for selection on parasite load, which argues against a causal relationship between parasites and sexual selection. Pruett-Jones *et al.* (1991, p. 237) conclude:

> We view the equivocal nature of interpreting our findings in terms of the Hamilton and Zuk hypothesis as less important than the fact that the correlation between showiness and parasite burden has now been demonstrated in another avifauna with an evolutionary history different from the other avifaunas previously studied. The association between parasites and showiness appears to be a general phenomenon, whether the mechanism underlying the association is as yet determined.

5.4.1.2   Negative results

Read's (1987) corroboration of the H–ZH was retracted two years later when Read and Harvey (1989) reassessed the comparative evidence purporting to support it. Using six different ornithologists to score for brightness and showiness in North American and European birds, Read and Harvey compared plumage colour scores and parasite prevalence. Although some subsets of the data showed a significant relationship, others did not. The correlation between prevalence and the averages of the six new brightness scores for each species was 0.18 ($P = 0.06$), a figure generally considered to be at the margin of statistical significance. Other tests appeared to more strongly support the H–ZH. For example, across European species, there was a significant positive correlation between parasite prevalence and male brightness, and a positive correlation was found within more taxa than expected by chance. An interesting finding is that the less commonly sampled European species (10 or fewer individuals) showed a highly significant relationship when tested alone, and that when these species are removed from the overall analysis the significant association between parasite prevalence and colour disappears. The upshot of this, as Hamilton and Zuk (1989) point out, is that the analyses presented by Read and Harvey (1989) do not convincingly discredit the H–ZH, and in fact, the former authors interpret the significant correlations as providing further support for it. This appears to be a case of viewing the proverbial cup either as half empty (Read and Harvey 1989) or half full (Hamilton and Zuk 1989); i.e. refutation or corroboration appears, to at least to some extent, to be in the eye of the beholder.

Two other studies similarly failed to detect the predicted relationship. Weatherhead *et al.* (1991) investigated the prevalence of blood parasites in 10 species of wood-warblers to test for a correlation between plumage brightness and parasitism and between the level of sexual dimorphism and parasite prevalence. These authors failed to find a correlation between any index of plumage and prevalence of parasites. Similarly, S. G. Johnson (1991) failed to detect any relationship between hematozoan parasites and male brightness in a large sample of North American passerines.

Finally, following the assumption that bird song probably evolved as a result of sexual selection through female choice, Read and Weary (1990) tested the interspecific component of the H–ZH by use of quantitative data on bird song, as it relates to blood parasites. Songs of 131 species of European and North American passerine birds were analysed for song duration, intersong interval, song continuity, song rate, song versatility, and song and syllable repertoire size. Prevalence of haematozoa were based on data used by Read (1987). Two song variables were significantly associated with blood parasites. There was a negative relationship between haematozoa prevalence and song continuity, contrary to the direction predicted by the H–ZH, and a positive correlation with song versatility. Both relationships came about through taxonomic associations. None of the other song variables correlated with haematozoa prevalence. Read and Weary (1990) conclude that there is no evidence of an association between song elaboration and parasites.

5.4.2   Intraspecific tests of the Hamilton and Zuk hypothesis

In general, the intraspecific component of the H–ZH has been less controversial, and therefore also possibly less interesting, than the interspecific aspect. After all, it is intuitively logical that females should choose to mate with healthier, more vigorous males, whatever the reason for their vigour. An extension of the intraspecific hypothesis adds the prediction that male morphological ornamental traits used by females in mate choice should be more affected by parasites than other, non-ornamental morphological characters. That is, sexually-selected characters, presumably evolved in the context of conveying information about male condition, should be proportionately more adversely affected by parasites than traits unrelated to sexual selection. In one manipulative study this appears to be the case (Zuk *et al.* 1990*b*, see below; see also Møller 1992*a*).

Intraspecific tests of the H–ZH using birds have investigated either external parasites, internal parasites, or both. Studies involving external parasites include the following species: sage grouse (Boyce 1990; Johnson and Boyce 1991; Spurrier *et al.* 1991); rock dove (Clayton 1990); barn swallow (Møller 1990*a*, 1992*a*); satin bowerbird (Borgia 1986*b*; Borgia and Collis 1989, 1990); and zebra finch (Burley *et al.* 1991).

With regard to internal parasites, studies to test at least some aspects of the intraspecific component of the H–ZH have been conducted on the ring-necked pheasant (Hillgarth 1990*a*), red junglefowl (Zuk *et al.* 1990*b*), sage grouse (Boyce 1990; Gibson 1990; Johnson and Boyce 1991; Spurrier *et al.* 1991), common grackle (Kirkpatrick *et al.* 1991), Lawes' parotia, a bird of paradise (Pruett-Jones *et al.* 1990), and red-winged blackbird (Weatherhead 1990).

5.4.2.1 External parasites

The available evidence suggests that in general females of the species listed above prefer to mate with males carrying few external parasites. Male sage grouse bearing lice were significantly less likely to obtain matings than were louse-free individuals. Of special interest with regard to female choice, lousy male grouse often had readily visible haematomas on their expandable esophageal pouches. These males were avoided by females (Spurrier *et al.* 1991).

Similarly, in the rock dove, females significantly preferred to mate with louse-free males. Clayton (1990) suggests that these results can be viewed as support for either the H–ZH or selection on females to avoid receiving parasites from their mates. Although feather lice do not directly affect the surface appearance of the hosts' plumage, and infected males do not groom more frequently than uninfected ones, parasitized males display significantly less to females. This could be due to the energetic drain associated with the decreased thermoregulatory capabilities of the feathers; thus by attending to the display rates and durations of displays by males, females could be avoiding lousy males. Such avoidance may be of major importance in that ectoparasites can be major vectors of disease. Male pigeons, like females, incubate and brood the nestlings, and Clayton (1990) suggests that females may choose unparasitized males, not to obtain genetic resistance to lice for their offspring, as postulated by the H–ZH, but to avoid the direct transmission of ectoparasites to their offspring as well as to themselves. Avoidance of parasite infection by being choosy is a direct selection explanation of mate choice, whereas choosing males for parasite-resistant genes is not.

Møller (1994) reports that in barn swallows tail length of males varies with ectoparasite burden and that females preferentially choose long-tailed males both as mates and as extra-pair copulation partners. Møller also found that the blood-sucking mite he studied affected nestling growth and that there appeared to be heritable resistance to the effects of the mites. Finally, Møller reported a negative relationship between mite load and length of the ornamental tail feathers of males in the following year, leading him to conclude that heavy mite loads apparently affect the ability of male barn swallows to grow large secondary sexual ornaments. All of these points provide support for the H–ZH.

The relationship between louse infection and male plumage and attractiveness in the satin bowerbird was examined by Borgia and Collis (1989, 1990), who evaluated three predictions: (1) the bright male hypothesis of H–Z, that females should prefer to mate with bright and parasite-free males; (2) the correlated infection model, which suggests that females avoid males with ectoparasitic infection because the presence of ectoparasites indicates a low overall male resistance to disease; and (3) the parasite avoidance model suggests that females avoid males with ectoparasites to reduce the likelihood of infection by such males. Of these three models, only the parasite avoidance model was consistently supported by the data. In short, although females do tend to mate preferentially with less parasitized or unparasitized males, the selective basis for this discrimination does not seem to be a manifestation of a good genes strategy (Borgia and Collis 1990).

Burley *et al.* (1991) investigated the relative frequency of ectoparasites on sex and age classes of the socially monogamous zebra finch. In adult finches ectoparasite counts tended to correlate positively and strongly with bill colour (a variable character used by both sexes in mate choice), contrary to the key intraspecific prediction of the H–ZH. Burley *et al.* (1991) recognize that many potentially important pieces of information, especially about the parasites, are not available, e.g. information on the pathogenic effects of the various parasites on their hosts.

5.4.2.2 Internal parasites

Hillgarth (1990*a*) experimentally investigated the relationship in ring-necked pheasants between infection with various endoparasites, especially several species of coccideans (some of which cause considerable mortality in chicks), and mate choice. Hillgarth demonstrated that pheasants possess a heritable resistance to parasites, and that females prefer to mate with males carrying low levels of coccidia. The latter result is explained proximately by the fact that males with low coccidia levels exhibit more display vigour (e.g. prolonged extension of wattles and more persistent calling) than do heavily parasitized males. These results are consistent with the H–ZH.

Zuk *et al.* (1990*b*) experimentally infected male red junglefowl with an intestinal nematode. The parasitized birds grew more slowly than controls and at maturity had smaller combs, but not shorter tarsi or bills. The comb, an ornamental character used by females in mate choice decisions, was more strongly affected by parasites than were non-ornamental traits, and females mated significantly more often with unparasitized males than with parasitized ones. All of these results are consistent with the H–ZH.

In Wyoming, Boyce (1990) and Johnson and Boyce (1991) determined that male sage grouse with malaria were significantly less likely to appear at the lek on a given morning, and that they obtained significantly fewer copulations than uninfected individuals. Moreover, when males with malaria did obtain copulations, they tended to do so later in the breeding season. This also contributes to their relatively poorer reproductive success, because older, experienced hens and those in good condition breed early. Yearling sage grouse hens breed later, lay smaller clutches, and are less successful at rearing young than older females. Thus, for several reasons male grouse infected with malaria are less reproductively successful than are uninfected males. In addition, Spurrier *et al.* (1991) provided an antibiotic to captive male grouse to counter low-grade bacterial infections. Treated birds strutted more frequently and were more often chosen by females than were controls.

In contrast, Gibson (1990) found that sage grouse in California appeared not to be affected by the single blood parasite identified. Courtship display was not affected by the presence of the parasite and, in fact, Gibson was unable to detect any relationship between presence or absence of the parasite and any measure of male courtship behaviour or reproductive success. In comparing his results with those of Boyce (1990) and Johnson and Boyce (1991), Gibson (1990, p. 277) concludes that '... blood parasite incidence can, but need not, be related to differential mating success at leks in this species.'

Kirkpatrick *et al.* (1991) investigated the presence and diversity of blood parasites in the common grackle in an effort to test one prediction from the H–ZH, that the parasite must alter the appearance of a sexually dimorphic trait of importance in female choice. Overall, Kirkpatrick *et al.* conclude that the large number of variables, including seasonal variation in parasite loads, effects of various species of parasites, plumage colour variation of hosts, and so forth, makes it unlikely that field studies will yield clear answers to many of the questions posed by the H–ZH.

Pruett-Jones *et al.* (1990) sampled blood parasites in Lawes' parotia. Males clear, maintain, and defend their display sites or courts (Pruett-Jones and Pruett-Jones 1990). Parasite intensity of males showed negative relationships with all phenotypic traits examined that were associated with mating effort. Based on a small sample, Pruett-Jones *et al.* (1990) determined that the variable most strongly affected by parasites was court attendance, but that this was not the factor females appeared to use in discriminating among males, which was display probability (Pruett–Jones and Pruett–Jones 1990). Pruett-Jones *et al.* (1990) point out that this result can be viewed as a contradiction of the intraspecific prediction of the H–ZH. They conclude that their data suggest that females may be avoiding highly parasitized males, perhaps through the proximate effect of parasites on court attendance by males, rather than actively choosing resistant males. This interpretation neither assumes nor predicts a relationship between preferences in females and traits in males that indicate resistance, which forms the basis of the H–ZH.

Finally, Weatherhead (1990) analysed the possible relationship between blood parasites and fitness and secondary sexual traits of male red-winged blackbirds. Parasitized and unparasitized males did not differ either in their ability to acquire a territory or to survive from one year to the next; nor was there a relationship between parasitism in males and the number of females that nested in their territories. Similarly, parasitized and unparasitized females did not differ with regard to how early they started nesting, how many eggs they laid, or their year-to-year survival. Thus, no fitness costs associated with being parasitized were detected. However, parasitized males differed morphologically from unparasitized ones, and they were significantly less aggressive. Overall, apparent mating patterns were unrelated to either the males' or the females' parasite status. Weatherhead (1990) concluded that only if unparasitized males realize significantly greater success than parasitized ones in obtaining extra-pair copulations would the H–ZH be supported in this species.

An alternative idea is that parasitized individuals of both sexes might be fitter than non-parasitized ones; i.e. selection might favour resistance at the secondary level of living with the parasites, rather than avoiding them. Davidar and Morton (1993) found that in purple martins adult females that were infected had higher reproductive success than uninfected ones. They suggested that the infected females were those that survived the effects of the parasites because they were robust individuals. In contrast, the uninfected females included some weak females that probably had never been exposed to the parasites.

5.4.3   Cause and effect: testosterone, mating system, plumage showiness, and parasites

Several authors, including Pruett-Jones *et al.* (1990), Read (1990), and Burley (1991), are not optimistic that the correlational approaches usually employed to study the relationship between parasite loads and degree of development of secondary sex characters will yield clear, unambiguous tests of the H–ZH. For example, Burley *et al.* (1991, p. 373) writes:

> ... the Hamilton–Zuk hypothesis is essentially non-falsifiable by the correlation procedures typically used to address it. A lack of (or positive) correlation may simply mean that the wrong parasite was selected for study, that the parasite's effects were felt at a different life stage, or some other variable intervened.

Pruett-Jones *et al.* (1990) and Burley *et al.* (1991) provide an intriguing alternative to the H–ZH, one that reverses the cause-and-effect postulated by Hamilton and Zuk (1982). These authors point out that the development of elaborate secondary sexual characters may cause individuals to be more susceptible to parasites; e.g. 'Thus bright coloration, previously evolved in response to social factors, may impose a physiological cost that ultimately increases susceptibility to parasite infection' (Burley *et al.* 1991, p. 360). This interpretation turns the scenario of Hamilton and Zuk (1982) completely around: instead of parasites increasing the intensity of sexual selection favouring ornamental plumage via the H–Z mechanism, a greater intensity of sexual selection is postulated to increase vulnerability to parasites!

If we consider the negative relationship between testosterone level and immunosuppression, information that was not available to Hamilton and Zuk (1982), this idea appears to have merit. Specifically, polygynous species appear to maintain testosterone levels at or close to their physiological maxima for the entire breeding season. Wingfield *et al.* (1990) suggest that in polygynous species selection has favoured males that can maintain maximal levels of testosterone over the longest periods of time. In contrast, males of monogamous species maintain high levels of testosterone for only relatively brief periods during the course of the breeding season, which probably allows the immune system to respond to most of the pathogens that the individual encounters.

Consider, for example, two related species, one highly polygynous and with colourful wattles that reflect testosterone level, and the other monogamous and less showy. The respective mating systems of these hypothetical species have developed in response to ecological factors (e.g. see Emlen and Oring 1977). If males of the colourful polygynous species produce testosterone at a maximal level over the course of several months in order to attract additional females, one might expect to find decreased resistance to parasites—and thus more parasites—compared to the less ornate monogamous one, as a result of the more complete, and more extended, inhibition of the immune system of the former.

This interspecific argument is similar to the intraspecific one discussed above. Recall that variance in the reproductive success of males of polygynous species is especially high (see Chapter 2). It is likely that the proximate physiological or endocrinological factor most important in ornaments or displays, which is usually testosterone, should be produced at the highest possible levels, and probably for longer periods of time, in the most successful individuals within the polygynous species. This may occur, however, at the cost of decreasing the individual's immune response. In short, we might expect both an intraspecific and an interspecific relationship between intensity and duration of mating effort and risks associated with immunosuppression, such as increased susceptibility to pathogens.

5.5   *Fluctuating asymmetry of ornamental characters*

An aspect of female choice that recently has received a lot of attention is based on the phenomenon of fluctuating asymmetry (Van Valen 1962; Palmer and Strobeck 1986; Parsons 1990; Watson and Thornhill 1994). Fluctuating asymmetry is a population phenomenon. It occurs when bilateral symmetry is the normal condition and when there is no consistent tendency for the trait on one side of the body to be more developed (e.g. larger or longer) than on the other side; i.e. within a population either the right or left side of individuals can be bigger. Fluctuating asymmetry occurs when both sides of any bilaterally symmetrical trait are under the control of the same gene or genes, and when an individual is unable to undergo identical development of the trait on both sides of its body. Considerable evidence exists to the effect that a large degree of fluctuating asymmetry in a morphological character indicates that an individual has been unable to cope with stress (environmental or genetic) during development of that trait (e.g. Palmer and Strobeck 1986; Parsons 1990). Many diverse factors can contribute to fluctuating asymmetry, such as nutrition (Nilsson 1994; Swaddle and Witter 1994) and social interactions (e.g. Witter and Swaddle 1994). Thus, the degree of fluctuating asymmetry may represent a measure of the sensitivity of a given trait to environmental stress during its development. However, because not all morphological traits develop at the same time, or are equally sensitive to a particular environmental stress, not all of the bilaterally symmetrical traits on an individual animal are expected to show the same degree of fluctuating asymmetry (Watson and Thornhill 1994; Hill 1995*b*).

5.5.1   Fluctuating asymmetry and sexual selection

Specifically with regard to sexually-selected ornaments in birds, Anders Møller and his colleagues have brought the phenomenon of fluctuating asymmetry to centre stage (e.g. Møller 1991, 1992*a,b,c*, 1993*a,b,c*; Møller and Höglund 1991; see Møller and Pomiankowski 1993*b* for references to many additional papers by Møller). The bases of many of their analyses and interpretations are founded on the relationship between size of the ornamental character of interest (in most of the studies to date this has been tail length; see Fig. 5.5) and degree of fluctuating asymmetry. In an ordinary (i.e. non-sexually selected) trait under strong stabilizing selection, the degree of fluctuating asymmetry exhibited by a population or species, and thus the extent of developmental disruption, is greatest at the upper and lower extremes of the trait's size distribution. In contrast, for exaggerated ornamental traits, which are thought to be costly to produce, there is a negative relationship between trait size and the degree of fluctuating asymmetry: the larger the trait (e.g. the longer the tail feathers), the more symmetrical they are expected to be.

In birds, most sexual ornaments, but not all, are paired and bilaterally symmetrical. (Conspicuous exceptions are the medially-located single comb of the four species of junglefowl, the single medial wattle of the green junglefowl, and the single snood of turkeys.) Often the paired ornaments are expressed or presented in a manner such that any deviation from a high degree of symmetry should be readily apparent (Møller and Höglund 1991). That is, the male's mode of presentation or display behaviour appears to provide a means by which choosing females could readily assess the degree of deviation from bilateral symmetry of an ornamental morphological trait. To date, although female response to male asymmetry has been documented in nature only for the barn swallow (Møller 1994), it has also been shown to occur in captive zebra finches (Swaddle and Cuthill 1994*a,b*). The displaying male Temminck's tragopan provides a striking possible example of this suggestion (Fig. 3.2). The male tragopan's lappet may provide information based on its colours or

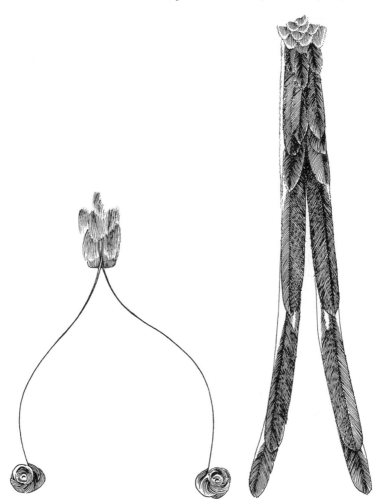

Fig. 5.5 Elaborated tails of the scissor-tailed flycatcher and the king bird of paradise illustrate well the potential of tails to indicate symmetrical as well as ornate development.

colour contrast, its size, and also on the symmetry of the colour patterns. Whether such potential sources of information about the displaying male are used by female tragopans as they choose a mate remains to be determined.

Because sexual ornaments often show greatly exaggerated size, shape, or colour, Møller and Höglund (1991) suggest they are likely to be very costly to produce in their full expression. Moreover, the great intricacy that many ornaments show in design of structure and coloration, such as the ocelli of the peacock's train, suggests that a variety of factors (e.g. mechanisms which causally mediate fitness, such as those underlying metabolic, immunological, and developmental competence; Watson and Thornhill 1994) can disrupt the developmental pathways that otherwise would lead to the production of perfectly symmetrical, exaggerated secondary sex characters.

Møller and his co-workers have considered the significance of fluctuating asymmetry in birds from a number of angles. Keeping in mind that the study of fluctuating asymmetry is in an early phase of its development, that its significance in sexual selection is currently controversial, and that many aspects have not yet been investigated (Watson and Thornhill 1994), a brief review of major findings and interpretations to date, based on studies of birds, is provided. First, the Møllerian view of the role of sexual selection is presented. Then, in Section 5.5.4, alternative interpretations are considered.

5.5.2   Sexual selection and fluctuating asymmetry in feather ornaments

Møller and Höglund (1991) predicted that sexual ornaments should show a larger degree of fluctuating asymmetry than either other morphological traits on the same individual or homologous traits in females of the same species or related non-ornamented species. Following the logic of these authors, if ornaments honestly indicate the quality of individuals, the bilaterally symmetrical ornamental characters of high quality individuals should be at the upper end of the size spectrum and exhibit little asymmetry. This leads to two predictions: (1) a negative relationship should exist between the degree of asymmetry and the size of the ornaments; and (2) this relationship should not hold for non-ornamental traits or for homologous traits in conspecific females, or in either sex of related non-ornamented species.

To test these ideas, Møller and Höglund (1991) used elaboration of feather ornamentation. In 15 of 16 species, the measured trait was tail length; the other species, the standard-winged nightjar, has a structurally specialized and elongate flight feather on each wing (see Fig. 4.3). Møller and Höglund compared fluctuating asymmetry in tail length and wing length (a non-ornamental trait) of males of related ornamented and non-ornamented species, and between males and females of ornamented species. They also compared ornamental and non-ornamental traits of the same individuals. Their predictions were confirmed in all comparisons. Møller and Höglund (1991) conclude that the large degree of fluctuating asymmetry in ornaments, and the negative relationship between ornament size and degree of asymmetry, do indeed suggest that production of ornaments is costly. That is, the extent of fluctuating asymmetry in ornaments reliably reflects the genetic quality of males, and thus is consistent with the good genes models of sexual selection.

In a follow-up paper, Møller (1993a) analysed fluctuating asymmetry in two sets of species for which some male traits associated with female preference had been identified. In one group of six species, empirical studies had concluded that females prefer males with the greatest degree of elaboration of the measured trait (species with a 'female preference'). In the other group, also of six species, the data indicated that females do not to rely on the measured ornament in making their mate choice decisions (species with 'no female preference'). For several comparisons of the two groups, no significant differences were found: relative size of the ornaments did not differ; degree of variability did not differ; and absolute degree of fluctuating asymmetry in ornaments did not differ.

In contrast, several other comparisons did reveal significant differences between the two groups. (1) The relationship between the degree of fluctuating asymmetry and size of the ornament differed between taxa with and without a female preference. (2) The negative relationship between absolute degree of fluctuating symmetry and size was statistically significant for three of the species with a female preference, but for none of the species without a female preference. (3) The negative relationship between relative degree of fluctuating asymmetry and size was significant for five of the species with a female preference, but for none of the species without a female preference. (4) The mean standardized regression coefficient for the relationship between asymmetry and trait size was negative for taxa with a preference, while this was not the case among species without a preference. (5) There was very little overlap in standardized regression coefficients between taxa with and without a female preference. These analyses imply that museum specimens can reveal which male ornamental traits females of a given species utilize as they make mate choice decisions.

5.5.3   Developmental stability and ornamentation

In another thought provoking paper relating fluctuating asymmetry to sexual selection, Møller (1993c) discusses, in general terms, the genetic bases for the finding that fluctuating asymmetry tends to be greater in ornamental than in non-ornamental traits. He suggests that a causal relationship exists between differences among traits in their functional importance and their degree of 'developmental canalization' (degree of resistance to modification based on their genetic underpinnings; see Møller 1993c for relevant references), and that this relationship may explain why some characters of a given individual or of a species show high degrees of fluctuating asymmetry, while others do not. It may also explain why certain characters often become exaggerated secondary sexual ornaments, while others never do so. Stated in another way, a prediction based on this perspective is that secondary sex traits are often derived from structures that do not appear to have great functional significance and that, as a result, are less strongly canalized than are functional traits. Non-functional traits, because of their relatively greater developmental flexibility, are more likely to become exaggerated secondary sexual ornaments than are functional traits.

As an example of this perspective, consider the wattles and tarsi of male red junglefowl. Symmetry of the legs is probably of extreme functional importance—the legs and feet are used to walk and run, to spring into the air while fighting with other males and prior to flight, to scratch for food, and to mate with females, among other essential functions—while the bilaterally symmetrical wattles, which appear to be ornamental, are not known to serve any functional role. Consistent with Møller's view, the wattles of male red junglefowl exhibit significantly greater fluctuating asymmetry than do the tarsi (Kimball *et al.* 1997).

Another question addressed by Møller (1993*c*) relates to how females come to focus on certain characters very early in their evolution as sexual ornaments, i.e. before they have evolved the degree of specialization—length, colour, shape, etc.—currently seen. In brief, Møller's scenario, which is a specific version of the good genes perspective, is as follows. (1) Females of the ancestral population should have been interested in males exhibiting a low degree of fluctuating asymmetry, for the reasons described above. (2) The females' attention should focus on one or several of the functionally less important morphological characters because fluctuating asymmetry will be greatest in those characters, as a result of their lesser developmental stability. (3) The size of an exaggerated morphological trait is an honest signal of male quality, simply because larger characters cost more to produce and maintain than smaller ones. Stated in another way, costly signals reliably reveal quality because high-quality individuals pay a lower cost for a given level of signalling than low-quality individuals. Thus, males may compete indirectly for females in an uncheatable manner by increasing (evolutionarily) character size over time. (4) The degree of fluctuating asymmetry differs relative to the size of the character; by producing a large character, fluctuating asymmetry is also increased. (5) Therefore, if females attend to both ornament size and fluctuating asymmetry, males will attempt to attract females by producing ornaments with two characteristics—large size and bilateral symmetry—that provide complementary signals of quality to females. (6) These hypothetical evolutionary steps can be reinforced by the sensory apparatus of females, for the simple reason that larger ornaments are more readily perceived than smaller ones; e.g. they can be seen from a greater distance. (7) Once females have developed an attraction to the ornament, assuming that they obtain fitness benefits as a result of choosing males with large, symmetrical ornaments, they may be attracted to ever larger ornaments. This would occur up to the point at which further increase in ornament size is counteracted by natural selection. At a proximate level, this process may be accelerated by the supernormal stimulus phenomenon.

To summarize briefly, the importance of male symmetry to females can direct the attention of females to certain types of morphological features and initiate its evolutionary development as a sexually selected ornament. The chosen feature is likely to be one of little functional significance (the tails of barn swallows appear not to support this point), because development of such a character is not rigidly canalized, and therefore it is vulnerable to perturbation during development and thus to a high degree of fluctuating asymmetry. The developmental flexibility of such a trait allows it to change (e.g. to become larger) readily, relative to more functional and more highly canalized traits. This can account for the findings that (1) ornamental traits have higher fluctuating asymmetry than non-ornamental ones and (2) females appear to prefer males with both the largest (and thus costliest) and most symmetrical ornaments. Although both size and degree of symmetry may be important to females, in Møller's scenario it is the instability of a weakly canalized trait or traits (as indicated by high fluctuating asymmetry) that initiates the process whereby females focus on a particular morphological character. This leads to both the elaboration of that trait and a reduction of fluctuating asymmetry in individual males of high quality.

Fluctuating asymmetry in spur length
Møller (1992a) also investigated patterns of fluctuating asymmetry in spurs of birds. Spurs appear to be sexually-selected traits, but, unlike the feather ornaments discussed above, they are usually thought or known to be important in male–male competition, rather than in female mate choice (see Section 4.7.1). Spurs, like the ornamental feathers discussed above, show a higher level of fluctuating asymmetry than do non-ornamental traits of the same individual. Paralleling the relationship described above between ornamental feather traits and fluctuating asymmetry, there is a negative relationship between spur length and degree of fluctuating asymmetry. These results indicate that any sexually-selected trait may show this relationship, whether the trait is used in female mate choice or in male–male competition.

Sexual selection and fluctuating asymmetry in the barn swallow
Møller (1994) presents comprehensively analysed and synthesized evidence on the importance of fluctuating asymmetry in the reproductive biology of barn swallows. In this species, the length of the outermost tail feathers is important in mate choice; the longest-tailed males mate earlier and receive more extra-pair copulations than short-tailed individuals. Tail length shows fluctuating asymmetry, and longer-tailed males show less asymmetry than shorter-tailed ones; i.e. short-tailed males fail to grow symmetrical tails more frequently. In contrast, for wing and tarsus length, the degree of asymmetry is significantly related to the absolute size of the traits, with high levels of asymmetry in extreme phenotypes and low levels in between. As discussed above, this is the common pattern for morphological traits (Palmer and Strobeck 1986).

Møller (1990b) suggests that tail length in male swallows serves as a reliable advertisement of male quality. This is because tails are more costly to produce for short-tailed than for long-tailed individuals per unit length, as demonstrated by the greater degree of fluctuating asymmetry in shorter-tailed males. This line of reasoning is based on the premise that fluctuating asymmetry in secondary sexual traits honestly signals the ability of individual males to cope with environmental conditions during growth and development of the ornament. High-quality males are capable of growing both long and symmetrical tails, whereas if low-quality males attempted to grow long tails, they would be conspicuously asymmetrical. Thus, the level of fluctuating asymmetry provides a means of condition-dependent quality control of the male traits used by female swallows as they make mate choice decisions.

One of the most inspired of Møller's many thought-provoking papers on fluctuating asymmetry is the study on traits of barn swallows before and after the Chernobyl, Ukraine, nuclear radiation disaster (Møller 1993*b*). After the accident, the degree of fluctuating asymmetry in male tail length was increased and deviant tail morphology appeared in males. This result suggests, but does not conclusively prove, that mutations caused by the radiation released in the Chernobyl area affected tail elongation and tail structure. If this is correct, however, it supports the idea discussed above, that ornamental traits are likely to controlled by relatively few and relatively unbuffered genes.

Fluctuating asymmetry and female mate choice in the red junglefowl
Male red junglefowl possess three types of specialized, fleshy ornaments on the head—the comb, wattles, and ear lappets. The comb is a single, medially-located structure that previous manipulative experimental studies have indicated is used by females for choosing between males (e.g. Ligon and Zwartjes 1995*a*). In contrast, the wattles and ear lappets are fleshy, bilaterally paired structures. Ligon *et al.* (1998) conducted mate choice tests in which the degree of symmetry of these paired structures was experimentally manipulated. These investigators obtained no evidence that female red junglefowl were sensitive to asymmetry of either of these paired ornaments.

5.5.4   Alternative explanations for the relationship between tail length and degree of fluctuating asymmetry

Several publications have appeared that challenge the Møllerian view of fluctuating asymmetry, two of which will be mentioned here. Balmford *et al.* (1993) analysed fluctuating asymmetry in 63 long-tailed bird species and found that fluctuating asymmetry in wing length was less in highly aerial species than in others. This suggests that symmetry of feathers used in flight may be related more to aerodynamic demands than to other factors. With regard to fork-tailed species, such as the barn swallow, Balmford *et al.* (1993) found no sexual differences in the degree of fluctuating asymmetry; i.e. there was no tendency for fluctuating asymmetry to be greater in the longer-tailed sex. Moreover, species with forked tails had unusually low fluctuating asymmetry in their outermost tail feathers. Balmford *et al.* suggest that this is due to strong natural selection to minimize asymmetry, where it impairs aerodynamic efficiency more than in other tail types. In short, these authors suggest that the general negative relationship between tail length and fluctuating asymmetry is based, not on sexual selection, but on natural selection.

Evans *et al.* (1994) also provide alternative explanations for the relationship between tail length and fluctuating asymmetry. Briefly, the longer the tail, the more deleterious fluctuating asymmetry is, with regard to flight efficiency. When tails of red-billed streamertails (a hummingbird) were experimentally shortened, asymmetry had no effect on the ability of the birds to negotiate a maze. Conversely, when tails were lengthened, asymmetry had a pronounced effect on flight agility. These authors suggest that individuals producing long tails must also produce highly symmetrical ones to minimize flight costs. This point can be viewed as complementing, rather than replacing, a signal function of quality based on tail length and symmetry (see also Evans 1993).

5.5.5   Other evidence concerning the possible role of fluctuating asymmetry in sexual selection

A study of the relationship between degree of elaboration and fluctuating asymmetry in the Indian peacock's train (Manning and Hartley 1991) provides independent support for Møller's suggestion that the degree of male ornamentation should be positively correlated with degree of symmetry. These authors found that the number of highly specialized and conspicuous ornaments of the train, the eye-like ocelli, were significantly and strongly associated with train symmetry. More recently, Petrie and Halliday (1994) experimentally removed eye-spots and determined that this led to decreased mating success. This study did not investigate the possible role of symmetrical arrangement of the ocelli, as opposed to their number.

Oakes and Barnard (1994) experimentally manipulated tail symmetry in the paradise whydah, a species in which the tail is both long and structurally ornate (see Fig. 6.4), and found that females did not prefer males with symmetrical tails. Somewhat surprisingly, females significantly chose to associate with males bearing asymmetrical tails. As a result of the way the tails were altered, the asymmetrical tails also appeared to be longer, and the female whydahs may have focused on length rather than symmetry. It should also be noted that in this study there was no negative relationship between fluctuating asymmetry and length of the tail. This may support Møller's (1993*a*) suggestion that traits important to females will show a relationship between length and fluctuating asymmetry, while traits not important in sexual selection will not do so.

Finally, one study supports the suggestion that a basic bias towards symmetry exists, independent of evolved ornamental traits. Using zebra finches and an experimental design similar to the one described by Burley *et al.* (1982), Swaddle and Cuthill (1994*a*) tested for the effects on female mate preferences of coloured bands placed either symmetrically or asymmetrically on the legs of males. All test males wore two green and two orange leg bands, thus all birds exhibited the same total amount of orange (O) or green (G) on the legs. (In other studies, green bands were clearly not attractive to female zebra finches; Burley *et al.* 1982.) The critical variable was the specific placement of the bands in relation to each other. Bands were arranged either asymmetrically or symmetrically. Female zebra finches spent significantly more time near symmetrically banded males, regardless of the arrangement of the colour bands. Several additional points should also be mentioned: (1) females displayed no colour band position preference within symmetry categories (e.g. O/G, left and right, did not differ from G/O, left and right); (2) it is bilateral symmetry (left and right legs the same) that is important, rather than within-leg vertical symmetry (upper ring same as lower); and (3) since cross-asymmetrical colour patterns (e.g. O/G left, G/O right) were not preferred, it is not just the average colour of the two legs that must be similar, but the specific colour pattern.

Moreover, these results indicate that any inherent preference for symmetry can be generalized, as these manipulations did not alter the symmetry of any natural secondary sexual trait. Although neither the proximate nor ultimate basis for the preference for symmetrically banded male zebra finches is yet known, these results may prove to have important general implications. For example, it has been argued that animals (including *Homo sapiens*) are attracted to symmetrical patterns, initially at least, for reasons having nothing to do with ascertaining the quality of a prospective mate. Kirkpatrick and Rosenthal (1994) discuss papers by Enquist and Arak (1994) and Johnstone (1994) that present more general explanations for the attractiveness of symmetry.

5.6 Conclusions and summary

This chapter considers various aspects of the good genes hypothesis of mate choice. Signals affected either by testosterone, carotenoid pigments, parasites, or symmetry of bilateral ornamental traits can provide information about male condition and thus potentially about the male's genetic quality.

Testosterone and sexual selection
The male gonadal hormone testosterone plays many roles in the process of sexual selection. Certain ornamental traits reflect testosterone level; these kinds of ornaments probably provide accurate information about the health or condition of males. In contrast, other ornaments are not indicative of testosterone and, in fact, they develop fully in its absence. In some species, ornate plumage develops in the absence of testosterone and feather traits are not currently important to females in mate choice. The negative relationship between testosterone level and the immune system suggests an important trade-off between sexual selection and natural selection. A male maximizes its efforts to

attract mates or matings by maximizing output of testosterone. On the other hand, testosterone also suppresses the immune system, which increases the risk of serious damage or even death due to pathogens. Although understanding of this complex interaction is far from complete, it is probably best to view the testosterone–immunosuppression relationship as one in which the male has only so much of certain critical physiological resources, and that when these resources are allocated to reproduction, there is a price to be paid elsewhere.

Carotenoid pigments and sexual selection
Carotenoid pigments provide red and yellow colours in the plumage of many kinds of birds. Unlike melanin pigments, carotenoids must be ingested, thus they may be condition-dependent indicators of some aspect of male quality. That is, differences among males in the extent of carotenoid pigmentation can reflect differences among individuals in their ability either to locate foods containing the required pigments or to assimilate them. With regard to the use of carotenoid as signals, the house finch is the best-studied case. Male house finches vary both within and between populations in the amount of pigmentation they exhibit, and females from all populations prefer the most colourful males. More colourful male house finches provide more parental care and survive better than do less colourful ones. These facts suggest that female house finches obtain direct benefits by choosing a colourful male and, since brighter males survive better, that they may also obtain good genes for their offspring.

Parasites and sexual selection
The Hamilton–Zuk hypothesis (H–ZH) consists of interspecific and intraspecific components. The interspecific hypothesis posits that avian species more challenged by parasite infections will possess brighter plumage than species with less risk of being parasitized. To date, the data gathered to test this suggestion provide a mixed picture, with some studies apparently supporting it, while others do not. (One can only wonder if the publishing bias towards positive results has affected this picture.) Despite the great interest that it has generated, evidence for the interspecific element of the Hamilton and Zuk hypothesis, which suggests that bright plumage and other ornamentation have evolved in response to parasite loads, remains ambiguous, 15 years after its publication. In large part, this is due to the scanty knowledge of the effects of most kinds of parasites on wild birds; for example, the parasites usually investigated may not be the relevant ones.

   In addition, there is the possibility that the cause-and-effect suggested by the H–ZH is backwards. That is, under strong sexual selection, as seen in polygynous systems, males may tend to be more ornate (brighter) than in monogamous species. Due to the stresses associated with polygyny, males may also be more susceptible to parasites and other pathogens than their less ornate, socially monogamous relatives. By this view, a positive relationship between male showiness and parasite load is an *effect* of male mating effort (as measured by testosterone-dependent morphological traits and energetically costly displays), rather than the *cause* of the mating effort often associated with males of ornate species. Both the H–ZH and this alternative interpretation predict a positive relationship between showiness and prevalence of parasites; thus, differentiating clearly between these two explanations will not be easy. However, this cause-or-effect issue must be assessed and evaluated before the Hamilton–Zuk interpretation of the relationship between degree of ornamentation and prevalence of parasites can be accepted.

The intraspecific component of the H–ZH predicts that ornaments of males will be affected by parasites, and that the least parasitized, or most parasite-resistant, males will have the greatest development of ornamentation. Overall, this prediction appears to have received stronger support than the interspecific prediction. Several studies have documented (1) an effect of parasites on one or more aspects of courtship display or ornamentation, and (2) a preference by females for the unparasitized males. Some experimental studies provide evidence that unparasitized (healthier) males obtain more matings than do parasitized males. This is also predicted, however, by the alternative hypothesis that females avoid matings with parasitized males for reasons other than genetic benefits.

A key additional point in support of the intraspecific component of the H–ZH is the finding that traits directly influenced by testosterone are more affected by parasite infection than are other, non-ornamental traits such as tarsus length and bill size. One interpretation is that certain specific traits may function as cues or indicators of male health or vigour, and that they have evolved to convey this information to conspecifics. Alternatively, the effects of parasites on certain ornamental traits may mean that, in the face of infection, the physiological response is reduced production of testosterone. In brief, the strongest evidence in support of the intraspecific component of the H–ZH are experiments in which (1) parasite loads of males were experimentally manipulated, (2) their effects on the phenotypes of males were measured, and (3) female preferences for unparasitized males were demonstrated via mate choice tests unconfounded by male–male interactions.

Sexual selection and fluctuating asymmetry

A number of studies on birds have made a case for the still-controversial proposition that the degree of fluctuating asymmetry of ornamental traits, especially tail length, is related to male quality, and that females use the degree of asymmetry in mate choice decisions. This is because the degree of fluctuating asymmetry reflects environmental or genetic stress during development of the trait. Several patterns have emerged from studies of ornamental traits and fluctuating asymmetry. The most important of these is a negative relationship between trait size and fluctuating asymmetry in ornamental traits used, or likely to be used, by females. To date this has been documented in nature in only one avian species, the barn swallow. It is probable that, as is so often the case, in some species fluctuating asymmetry will turn out to be a major factor in female choice, whereas in other species it will not be important. Finally, it has also been suggested that responses to symmetrical ornaments may occur for reasons having little or nothing to do with assessing differential quality among males. For example, in fork-tailed species, the negative relationship between tail length and fluctuating asymmetry may be causally related to aerodynamic factors. Because the study of the relationship between fluctuating asymmetry and sexual selection is still in its early stages, it is too soon to predict how generally important fluctuating asymmetry will ultimately prove to be in sexual selection.

6 Empirical studies of the major hypotheses

6.1 Introduction

In the latter half of this chapter each of the major hypotheses proposed in
Chapter 2 to account for mate choice by female birds is briefly reviewed and
evaluated. First, however, to give the reader a feel for the kinds of data that
have been obtained to assess these hypotheses, I summarize a number of
empirical studies of individual species. Most of these provide clear evidence
that females use certain male traits in making mate choice decisions; i.e. in a
diverse array of species, directional selection, based on certain characteristics
of the males (Ryan and Keddy-Hector 1992), occurs. Traits of males may be
either morphological or behavioural in nature; most commonly, both appear to
be important. This is not surprising, for two reasons. First, specific behaviour
is almost always used to display or exhibit the morphological ornaments of the
male, and second, males of most kinds of birds have several mechanisms for
self-promotion. In general, more attention has been paid to morphological than
to behavioural traits of males, largely because they are often easier to measure
accurately and to manipulate experimentally. In contrast, quantifying all of the

varied and often subtle behaviour associated with male courtship is far more challenging. Thus, the conclusions of most studies may be biased towards overemphasis of male morphological characters in mate choice decisions by females.

Several species that have been studied with regard to sexual selection are not included in this chapter, or are treated briefly, simply because they have been recently and extensively reviewed elsewhere. These include the black grouse (Höglund and Alatalo 1995), pied flycatcher (Lundberg and Alatalo 1992), barn swallow (Møller 1994), and red-winged blackbird (Searcy and Yasukawa 1995).

6.2   Direct selection

In their review of the major hypotheses proposed for female mate choice, Kirkpatrick and Ryan (1991) suggest that direct selection may be foremost (see Chapter 2, Section 2.2). For birds, this is probably true for the great majority of species, and may, in fact, be true for all species in which parental care by males is important. In addition to paternal care *per se*, the quality of the territory controlled by the male is often also of great importance to reproductive success. Apart from the parental care provided by males, the most common resource of value to female birds is territorial space and the resources that it holds (e.g. food required to support the female and her offspring, safe nest sites, and/or roost sites). For some species, it is the quality of the territory, rather than the male territory-holder, that attracts the female. This is indicated in those cases where reproductively experienced females appear to be more faithful to former breeding sites than to former mates. In such cases, a male may move from one territory to another one nearby, while his mate of a previous nesting cycle returns to the original territory rather than joining her previous mate on his new territory.

One line of evidence that females may focus on territory quality rather than on male quality (bearing in mind that the two are probably correlated) comes from an analysis of mating patterns of female red-winged blackbirds in Washington, where Searcy (1979, p. 96) concluded that, in general, territory quality is the most important factor influencing female choice in redwings. In other populations of redwings, where males provide more parental care, it has been more difficult to ascertain the relative importance of territory quality and male quality (e.g. Yasukawa 1981*b*; Searcy and Yasukawa 1983).

An apparently common strategy of female birds is to pair and occupy a territory with one male, the 'mate' while seeking copulations from another, possibly superior, individual that is unavailable as a social mate (e.g. S. M. Smith 1988; Kempenaers *et al.* 1992; see Chapter 12). This strategy, too, indicates that both the territory and the identity of the provider of sperm are important to females.

Many other species, both polygynous and socially monogamous ones, also clearly demonstrate resource-based mating systems based on male ownership and defence of some sort of critical resource. For cavity-nesting species, suitable nest sites are probably the most common critical limited resource. In other cases, the resource may be more specialized. The polygynous male orange-rumped honeyguide defends beeswax and mates with females that come to partake of the wax (Cronin and Sherman 1976). Similarly, the male purple-throated carib (a hummingbird) defends flowering plants and mates with females that come into his territory for nectar (Wolf 1975).

Among socially monogamous species, male coloration also may be important in direct selection, in that brightness often correlates with parental quality. A relationship between male ornamentation, female choice, and direct benefits to the female via male parental care has been shown for the house finch (Hill 1991), European kestrel (Palokangas *et al.* 1994), and yellowhammer (Sundberg and Larsson 1994), among others.

6.2.1   Empirical studies

Mate choice in several of the species discussed in this section probably involves both direct selection and male quality, 'good genes'. Thus, categorization of these studies under the heading of direct selection is somewhat arbitrary. However, in the cases treated here, it is clear that direct benefits are an important aspect of the benefits females may obtain by their mate choice decisions.

Mallard

Over the past several years, a number of studies have appeared that deal with sexual selection in mallards (e.g. Klint 1980; Bossema and Kruijt 1982; Williams 1983; Holmberg *et al.* 1989; Weidmann 1990; Omland 1996*a*,*b*; see also Johnsgard 1994, pp. 74–6). With regard to the issue of mate choice, mallards, like most other north temperate zone dabbling ducks, are especially interesting. This is because they exhibit the unusual combination of social monogamy, with pair bonds extending over many months, but no paternal care or exclusive territoriality. Mallards pair on the wintering ground, thus females must assess males on the basis of traits other than their parenting ability or resources they may hold. This could be due either to direct selection, in that females may choose males that provide the best security from predators and sexual harassment by other males, or they may choose on the basis of the males' genetic quality. The available data suggest that both direct selection and good genes selection are plausible, complementary explanations for mate choice by female mallards.

Holmberg *et al.* (1989) identified several variable traits of males that were significantly correlated with female choice: social display activity, plumage status, and body size. Specifically, females preferred males with (1) a high rate of display activity, (2) completely moulted, more perfectly developed plumage (green head, white neck ring, brown or chestnut chest, grey flanks, and tail curl), and (3) small body size. Weidmann (1990) corroborated some of these results. By experimentally manipulating male plumage, he showed that females preferred unblemished males. (See Alatolo *et al.* 1991 for a similar finding in black grouse.) Flawed plumage may signal that a male is of low social status, which, in turn, might indicate an inability to prevail in intrasexual conflicts and thus to effectively protect the female from sexual harassment by other males.

Although several studies have provided evidence that some aspect of male plumage is important to female mallards, others have suggested that, in this regard, its usefulness is limited. In a recent investigation of unmanipulated plumage characters and bill colour of males, Omland (1996*a*) found that only bill colour correlated significantly with mate preferences of females. Omland (1996*b*) also conducted studies in which plumage traits and bill colour were experimentally manipulated. Removal of certain plumage colours and patterns did lower mating success as compared to control birds. Taken together, these studies suggest that, of morphological traits, bill colour of males is the primary factor in female mate choice decisions, but that certain aspects of the plumage also have a weakly significant effect.

Grey partridge
Dahlgren (1990) investigated the basis of mate choice by females of the socially monogamous grey partridge. Males and females are similar in size and appearance, and the plumage of both sexes is primarily cryptic. A male contributes to parental effort by standing watch as its mate forages and incubates. After the chicks hatch, the male expends a great deal of time vigilantly scanning for predators. As discussed in Chapter 2, females of many socially monogamous species are likely to make mate choice decisions on the basis of either resources controlled by the male or contributions made by the male to the reproductive effort. Apparently consistent with this generalization, Dahlgren found that females chose between two males on the basis of vigilance behaviour exhibited prior to pair formation.

Beani and Dessi-Fulgheri (1995) conducted further tests of mate choice in grey partridges. By use of experimental manipulations, these authors determined that the main morphological plumage trait distinguishing males, the brown belly patch, was not important to choosing females. Injection of testosterone had no effect on the size of the breast patch. These experiments suggest that patch size, which is highly variable in size among males, and which also occurs uncommonly in females, probably does not signal condition. The red tissue behind the eye is brighter in males than in females, and is also brighter in testosterone-treated individuals than in controls. Thus, this morphological trait, unlike plumage, may signal condition. The male trait identified by Beani and Dessi-Fulgheri (1995) as most important to females is rate and acoustical structure of certain vocalizations, which is influenced by testosterone level as is vigilance behaviour (Fusani *et al.* 1997).

The results of Dahlgren (1990) and Beani and Dessi-Fulgheri (1995) should be viewed as complementary. Both rate of calling and vigilance behaviour may signal important information about a male's condition (i.e. good genes). For example, the ability to remain vigilant for prolonged periods throughout the day might reflect the energy balance of the male. Presumably, only a well-nourished male, possibly with fat stores, could afford to remain vigilant a majority of the time. The fact that females assess male vigilance *per se* suggests the possibility that they use this trait to obtain mates in top physical condition.

In addition, vigilance can send an additional message about the male's quality as a mate and parent. Prior to incubation, the presence of the male allows the female to spend more time foraging and less time scanning the environment for predators (Dahlgren 1990). The exhibition of sentinel or vigilance behaviour by males could be used by the female as an indicator of male parental quality, particularly since the male will later provide the same 'resource' for his foraging offspring. In short, the ultimate basis for female choice in the grey partridge could be both direct selection—procurement of a resource (a sentinel for the female as she forages, and later a caretaker for her offspring)—and good genes selection by obtaining a mate that has come through the winter in top physical condition.

Pinyon jay

Johnson (1988a) conducted experimental studies of female mate choice using captive pinyon jays. Pinyon jays are highly social, living year-round in integrated flocks (Marzluff and Balda 1992). These jays breed in colonies and are socially monogamous. Males provide an abundance of parental investment. Early in the breeding cycle, they extensively feed their mates as the females lay and then engage in almost continuous incubation, as a result of low environmental temperatures during the late winter–early spring breeding period. After hatching, the male feeds both the young jays and the female, which broods the chicks almost continuously for the first two weeks of their nestling lives. The male-to-female sex ratio of adult pinyon jays is about 1.5:1.0 (Ligon and White 1974; Marzluff and Balda 1992), which suggests that sexual selection may be strong in this monogamous species.

In choosing among males, females did not use traits identified as important in male–male encounters in captivity (see Section 9.3): i.e. neither body size nor bill length of males was important in female choice. Chosen males typically had larger testes and brighter feathers than non-preferred ones (Johnson 1988b). Although it is reasonable to speculate that these two traits are causally correlated—i.e. a relationship between testis size and testosterone production and between testosterone production and plumage—this is unlikely to be the case (see Section 5.2.1). Instead, both testosterone level and plumage brightness may be related to condition.

In the choice tests, both dominant and subordinate males were chosen by females. Johnson did not study courtship and pairing in nature, thus it is not known just how precisely her experimental findings relate to mate choice in free-living birds. However, they seem to be consistent with what occurs under natural conditions—high-speed, prolonged flights, with several males pursuing a female, apparently as part of the pairing process (Marzluff and Balda 1992). The fact that experimental females did not prefer large, dominant males is consistent with the possibility that the normal pairing behaviour of free-living females leads them to select superior fliers, which may not be the largest males.

Why might flight ability or efficiency of male pinyon jays be important to females? As mentioned above, female pinyon jays are completely dependent on their mates for food, both for themselves and for their small offspring, and males often forage at great distances from the nest colony. Therefore, morphological traits related to efficient, long-distance food delivery by males may be sought by females. Average-sized male jays may be superior to larger individuals in providing food, by virtue of (1) requiring less nutrition for themselves on an absolute basis and (2) possibly delivering more food per trip or making more trips to the nest per day. An increase in food delivery could be due to more efficient wing-loading. (Similar arguments have been made for those birds of prey in which males are smaller than females; in such species, as in pinyon jays, males deliver food to the female and nestlings.) Further support for these suggestions is provided by Johnson and Marzluff (1990), who report that pairs of pinyon jays composed of relatively small males and large females are more successful in reproduction than are other size combinations of mated pairs.

Pied flycatcher
Pied flycatchers are small, migratory, hole-nesting passerine birds that are widespread in Europe. Mate choice and sexual selection have been studied exhaustively in this species; e.g. '... the Pied Flycatcher has become somewhat of a European equivalent of the North American Red-winged Blackbird in attracting an extraordinary amount of research interest.' (Lundberg and Alatalo 1992, p. 2.) Male plumage is highly variable, ranging from distinctly black and white to the other extreme of female-like dull grey. Although males are usually monogamous, a minority obtain two mates by defending two separated territories that contain suitable nest cavities. The male provisions young of his first brood, but usually does not contribute to the rearing of the second female's brood. The interested reader is referred to the book by Lundberg and Alatalo (1992) for detailed summaries of most of the subjects considered here, plus many others.

With regard to female mate choice, it is clear that direct selection is of great importance, in that male mating success is affected by size and quality of the territory and quality of the nest cavity (e.g. Alatalo *et al.* 1986; Slagsvold 1986; Dale and Slagsvold 1990). In addition, two traits of the males themselves, song rate and brightness of plumage, have been shown to affect mating decisions by females (Alatalo *et al.* 1990*a*; Sætre *et al.* 1994).

House sparrow

Møller (1987*a*,*b*, 1988, 1989) described the relationship between certain traits (especially the size of the black bib, referred to as a 'badge') of male house sparrows in Denmark, and their effects on various aspects of both male–male competition and female mate choice. The male's badge, which is produced by melanin, serves to signal social status. The 'honesty' of the signal is apparently maintained through social interaction. Body weight and testis size also correlated with bib size in male sparrows. Møller concluded that a positive relationship exists between bib size and both dominance relationships among males and mate choice among females. Presumably related to this dominance, large-bibbed males were more likely to obtain territories with cavity nest sites (safer sites) and more nest sites than other males. With regard to mate choice, large-bibbed males attracted mates earlier in the season than did small-bibbed males, and females implanted with oestradiol (to stimulate sexual behaviour) chose large-bibbed males over small-bibbed ones.

Recently, Rebecca Kimball carried out somewhat similar studies on house sparrows in New Mexico, USA. In some cases, Kimball's methodologies differed from those of Møller, and her findings likewise did not always parallel the results he obtained. First, using a mate choice apparatus similar to those described by Burley *et al.* (1982) and Hill (1990), Kimball (1996) failed to detect any relationship between bib size of males and female preference for large-bibbed males (in these studies females were not artificially primed by oestradiol treatment). The only morphological trait that correlated with female preference was bill depth. Møller (1989) also found a significant relationship between male bill depth and female mate preference in yearling male house sparrows. In neither study was the biological significance of this relationship explained.

Second, Kimball (1995) discovered a significant negative relationship between the degree of fluctuating asymmetry of the bib and bib size, which in itself may indicate a relationship between bib size and male phenotypic quality (see Section 5.6). Additionally, males with low fluctuating asymmetry began nesting earlier and successfully fledged more young over a breeding season. Kimball also found that the reproductive effort expended by females was correlated with the bib symmetry of males: specifically, females mated to males with more symmetrical bibs laid more eggs over the course of the entire reproductive season than did those mated to less symmetrical males (more clutches per season rather than more eggs per clutch). Females mated to symmetrical males did not, however, invest more heavily in parental care, as measured by feeding rates.

Finally, Kimball (1995) compared directly her major findings pertaining to mate choice and reproductive success in free-living house sparrows with those of Møller. In both studies, large-bibbed males obtained territories and mates sooner, and had higher annual reproductive success, than small-bibbed males. In contrast to Møller, Kimball did not find that larger-bibbed males acquired territories of higher quality, as measured by number of protected nest sites. These similarities and differences may be explained as follows. In Møller's (1989) study, large-bibbed males held higher quality nest sites and possibly obtained more extra-pair copulations, whereas in Kimball's studies the greater reproductive success of large-bibbed males was due to the fact that their mates began nesting earlier in the spring, thereby producing more broods over the course of the breeding season than did females mated to small-bibbed males.

To summarize, direct selection clearly is a major aspect of mate choice for females of many species of birds. In fact, it may be the single most common factor for females making mate choice decisions. However, studies of such systems are not usually very informative concerning the most controversial issue of sexual selection, namely the basis of female mate choice in those species in which the only contribution by the male to the female's reproductive success is the sperm he provides. For such species, direct selection in any of its forms does not appear to be the answer.

6.3   Good genes

The basic premise of good genes or indicator hypotheses is that display traits of males are used by females to assess relative health and vigour of prospective mates, and that such traits are indicative of a genetically-based resistance to at least some of the environmental vicissitudes to which animals are exposed. Male traits of importance to females should be condition-dependent; that is, they should accurately reflect the physical condition of the male (e.g. flight displays of male bobolinks; Mather and Robertson 1992), and thus they should be honest indicators of male health or viability. The following studies support the premise that male traits used by females in their mate choice decisions have the potential to indicate male quality.

6.3.1   Empirical studies

Most of the studies considered in this section are of species in which males provide no parental care, and most do not provide territorial space. Thus, it is especially likely that females of such species assess males for their intrinsic quality. Several of the species discussed in this section exhibit either polygynous (e.g. red junglefowl, ring-necked pheasant, turkey) or promiscuous mating systems (e.g. the lekking Indian peafowl, sage grouse, great snipe, Jackson's widowbird, and the non-lekking satin bowerbird, Vogelcop bowerbird, long-tailed widowbird). Two of the species considered here to support the good genes hypothesis are socially monogamous (barn swallow, house finch), although in both direct selection probably also plays an important role in female mate choice.

Indian or common peafowl
The prime example of male ornamentation and behaviour presumably having evolved in the context of female mate choice is the Indian peafowl. A number of careful studies of male ornaments and female mate choice have been conducted on feral populations of this species in England. The peafowl is famous primarily for the large elaborate train, which consists of three types of feathers: (1) the longest feathers, which end in a fishtail-like structure, (2) the curved feathers that form the lower border of train when it is erect, and (3) the elongate feathers that end in an 'eye-spot' or ocellus structure. The number of ocelli is related to the age of the male (Manning 1989), but apparently not beyond four years (Petrie 1993). In addition, there is a significant positive relationship between the number of ocelli and the symmetry of their distribution (Manning and Hartley 1991; see Section 5.6). As might be expected, certain behavioural traits of males, especially the length of calls given during display, also correlate with mating success (Yasmin and Yahya 1996).

With this brief background in mind (and ignoring the role of male behaviour in female choice), several additional findings can be better appreciated. First, peahens preferred males with more elaborate trains: a significant correlation exists between the number of eye-spots a male exhibits and the number of females he mates with (Petrie *et al.* 1991). Subsequently, Petrie and Halliday (1994) experimentally altered the ocelli of peacocks tails and showed that males with a reduced number of eye-spots also had significantly less reproductive success than they had previously enjoyed. Second, males with low mating success suffer higher predation than do males that are more reproductively successful, which suggests that the males that obtain little mating success are also inferior in other ways (Petrie 1992). Third, females mated to males with the more elaborate trains lay more eggs (Petrie and Williams 1993). Finally, of great significance with regard to ultimate explanations of mate choice, Petrie (1994) has shown that offspring of the males most successful in attracting females grow and survive better under semi-natural conditions than do young birds sired by males with less well-developed trains. Specifically, the offspring of males with larger ocelli were significantly larger at 84 days of age and significantly more of these young birds survived for at least 24 months under semi-natural conditions. (There was also a significant positive correlation between weights of young birds at day 84 and their weights at the time of their release into the wild.) Thus, by virtue of this last point—the superiority of offspring of preferred males—a good genes explanation of mate choice has been convincingly made for this classic example of strong sexual selection.

Sage grouse
The question of female mate choice in sage grouse is especially interesting for several reasons, many of which are associated with their lek mating system (see Chapter 15). Different workers have come to different conclusions regarding the cues used by female sage grouse to assess prospective mates. Wiley (1973) proposed that proximity of male territories to central locations in the lek was used by females in mate choice. The idea that females prefer centrally located males, and that males therefore compete for central locations, has since become firmly established (e.g. Wittenberger 1981, p. 496; Halliday 1983; Partridge and Halliday 1984). In contrast, Hartzler (1972) identified only one factor, display rate, that was significantly related to mating success. Thus, for some time the strategy of female choice in sage grouse remained unclear.

Subsequently, Robert Gibson and Jack Bradbury (1985) analysed the relationship between male mating success and a variety of traits possibly used by females to assess males. In this study, 13 of 27 males obtained copulations, and three male characters were correlated with male mating success (whether or not a male obtained any matings). These were lek attendance, strut rate (numbers of strut displays given per minute), and one acoustic component of the display. Of the 13 males that mated, copulations were distributed as follows: three mated three times, four mated twice, and six mated once. Among these males, only one independent variable, another acoustical component, was related to number of matings. Gibson and Bradbury tentatively suggest that one set of factors may be necessary for mating status (i.e. potential acceptability as a mate), whereas another may be used to choose among the pool of acceptable males. Gibson *et al.* (1991) further analysed components of displays of male sage grouse and ascertained that the time between the two popping sounds, the 'interpop interval', was most consistently associated with male mating success. Their analyses also suggested that the relative importance of specific sound signals varied markedly from year to year and from lek to lek.

Equally interesting is the list of characters not found to be important in female choice. Mating success of males was unrelated either to size measures, including wing length, tail length, and body weight, or to age (Gibson *et al.* 1991). Various other morphological characters of possible importance to females, such as size and colour of the inflatable oesophageal sacs, were not measured (see Chapter 5 and Johnson and Boyce 1991).

What might be the significance of female choice based on display performance? Sandra Vehrencamp and her co-workers found that displays of male sage grouse are apparently costly, with the metabolic cost for vigorous displays being 14–17 times basal metabolic rate, an expense similar to that of flight (Vehrencamp *et al.* 1989). Vehrencamp *et al.* (1989) calculated that on a 24-hour basis, actively displaying male sage grouse expend energy at the typical maximum rate sustainable by homeotherms. Thus, females preferentially mate with males that are capable of expending a lot of energy over extended periods of time. Vehrencamp *et al.* (1989) also made the counter-intuitive discovery that males expending the most energy via display lost the least weight. Those males apparently ingested more and perhaps higher quality food by flying farther from the lek to forage than both the weak displayers and the non-displayers. Vehrencamp *et al.* (1989) obtained no evidence that either blood parasites or other diseases, as determined by haematocrit levels, were associated with display effort (cf. Section 5.4.2.2). Malaria, however, seems to be an important disease of sage grouse in Wyoming (Boyce 1990).

Turkey
Male turkeys are highly ornate, with iridescent body plumage, colourful tail feathers that are fanned during the male's courtship displays, and several additional specialized features that appear to serve as ornaments. These include the beard, specialized hair-like feathers projecting from the chest, tarsal spurs, and an unfeathered head and neck that is brightly coloured blue and red (Fig. 6.1). Red bumps on the side of the neck are called side caruncles and large growths labelled frontal caruncles occur at the base of the neck. The skullcap on the crown of the head varies in colour from red to white and light blue and appears to change with motivational state. Finally, there is a distensible, finger-like process at the base of the upper mandible called the snood. In addition to these morphological specializations, turkeys exhibit a well-known and distinctive display that involves erecting the body plumage, fanning the tail, and dragging the primary wing feathers along the ground, while emitting the famous 'gobbling' vocalization (Fig. 3.1). Thus, the turkey appears to be a strongly sexually-selected species.

Using captive-reared turkeys of wild stock, Richard Buchholz (1995) conducted mate choice tests and found that among those females that solicited copulation, two traits of males, and, surprisingly, only two, were significantly correlated with female preference. Females exposed to live males chose individuals with longer snoods and wider skullcaps; females also spent significantly more time with males that they subsequently solicited for copulation. Buchholz (1995) also treated females with oestradiol and conducted mate choice experiments using models of male turkeys that differed only in snood length and skullcap width, and again found that females significantly preferred the model with the longer snood and wider skullcap.

Thus, the conclusion, based on two different experimental approaches, is convincing. By choosing males with longer snoods and wider skullcaps females also mated with males that (1) strutted at a higher rate (see sage grouse, above), although under the experimental conditions used no direct

Fig. 6.1   Non-displaying turkey with snood at minimal size.

relationship between strut rate and female choice emerged, and (2) were in better body condition (as measured by weight/tarsus length). It is also worth noting that the bright plumage did not figure significantly in female preferences.

Red junglefowl
The red junglefowl, a member of the pheasant group, is the conspecific ancestor of the domestic fowl (Hutt 1949; Stevens 1991). Males are elaborately ornamented, with both a colourful orange, red, and black plumage and a fleshy red comb and wattles on the head, while the plumage of females is drab and cryptic. Red junglefowl are thought to be polygynous (Collias and Collias 1985; Thornhill 1988) and, like the peafowl, sage grouse, and turkey, males typically provide no parental care. Males routinely assist females, however, in choosing a nest site, and during the period of egg-laying, they frequently provide high-quality (high protein) food items to females, as well as protection by serving as sentinels while the gravid females forage (personal observation). Thus, in this galliform species, direct selection appears to play a role in female mate choice.

Marlene Zuk and co-workers (e.g. Zuk *et al.* 1990*a,b,c*, 1992, 1995) studied the role of male ornaments in mate choices of female red junglefowl. In choice tests using two one-year-old males, comb length (which was highly correlated with overall comb and wattle size) and colour, and eye colour were correlated with male mating success. In another set of choice tests, females chose between two males of different ages—either cocks two years old or older or one-year-old cockerels. In 13 of 19 trials, females chose the younger birds (not significant). This result was somewhat unexpected, because it has often been suggested that older birds should be preferred as mates, and because the older junglefowl were significantly more ornate in terms of sickle feather (longest ornamental feathers of the tail) and saddle feather lengths. They also possessed conspicuously longer spurs and were significantly larger. The two groups, however, did not differ significantly in either comb size or eye colour (Zuk *et al.* 1990*a*).

One of the most counter-intuitive outcomes of the studies of mate choice was the finding that the highly ornate plumage of male junglefowl appears not to be important, and perhaps is not used at all, in mate choice decisions by females. To test directly for a role of male plumage in mate choice, Ligon and Zwartjes (1995*a*) presented pairs of males to female red junglefowl. One male possessed normal, wild-type plumage, while the other had mutant cryptic, brown plumage identical to that of females. The females showed no preference for the normal-plumaged males. Other manipulative choice tests, however, showed that the same females strongly preferred males with large combs, regardless of plumage type. These results, taken together with the findings of Zuk *et al.* (1995) and Ligon *et al.* (1998), indicate that females discriminate between males on the basis of their combs, and that the plumage is not the target of the females' attention.

In the red junglefowl, comb size, a sexually-selected, condition-dependent male trait, correlates with rate of development of the males' chicks. Chicks of large-combed males grow faster than do those of small-combed roosters, for at least their first three months of life (K. Johnson *et al.* 1993*b*). Rapid growth in this species is an indicator of general vigour, and it probably reflects resistance to pathogens, such as coccidia and certain nematodes, that affect chick development (see Zuk *et al.* 1990*b*). In the conspecific domestic fowl, resistance to parasites such as these is known to be heritable (Hutt 1949).

Ring-necked pheasant
Male ring-necked pheasants attempt to attract a number of females (Fig. 6.2). An assessment of the claim that male spur length is the foremost factor in mate choice decisions of female pheasants was presented in Section 4.8.1. Experimental studies of male ornaments and female choice in ring-necks have been conducted by Hillgarth (1990*a,b*) and Mateos and Carranza (1995, 1996). Hillgarth (1990*a*) reported that females preferred males which displayed most consistently; specifically, those males that extended their wattles and elevated

Fig. 6.2 Male and females of the ring-necked pheasant. The male traits used by females in their selection of a mate have proved to be controversial. Reprinted with permission from Dover Publications, Inc.

their feathers for long periods received significantly more solicitations to mate. Mateos and Carranza (1995, 1996) conducted experimental manipulations of traits of both live and stuffed males. Their results, like those of Hillgarth (1990*a*), disagree with those of von Schantz *et al.* (1989, 1994), who concluded that the males' spurs were of most importance to females (see Section 4.8.1). Mateos and Carranza (1995) reported that tail length, length of the ear tufts, and presence of black points (tiny black feathers on the wattles) positively influenced the females' choices, but that neither wattle size nor colour nor brightness of the plumage did so.

The negative conclusions concerning the role of wattle size in mate choice by female ring-necks are perplexing. I suspect that the males' wattles are important to females. This suspicion is based on two points. (1) Both wattle

(a)   (b)

Fig. 6.3   Heads of male ring-necked pheasant with (a) wattles contracted and (b) wattles inflated during display.

size and length of ear tufts, which are correlated, are important in male–male interactions, and both traits appear to signal male condition (Mateos and Carranza 1997; in the related red junglefowl the comb is important both in female choice and male–male competition (Ligon *et al.* 1990; Ligon and Zwartjes 1995*a*), and the same is true for the snood of turkeys (Buchholz 1995, 1997). (2) Hillgarth's (1990*a*) finding that females preferred males that maintained fully extended wattles for long periods suggests that females are sensitive to wattle size (Fig. 6.3), plus the length of time that they are extended.

Three additional points might also be mentioned. First, the spacing of the black points, which is affected by wattle expansion, may reflect the turgidity of the wattles; if so, the points may be used to assess wattles; or, the black points make simply enhance the conspicuousness of the wattles. Second, the experimental protocol used by Mateos and Carranza (1995) may have led them to an erroneous conclusion (that wattle size is not important to females). Mateos and Carranza (1995) pasted artificial wattles on the faces of live and dead male pheasants. Zuk *et al.* (1992) also placed artificial combs on live male red junglefowl and found that females were not attracted to males bearing them, even though a number of studies, both experimental and correlational, have indicated that the comb is the primary cue used by females of this species (e.g. Ligon and Zwartjes 1995*a*; Zuk *et al.* 1995; Ligon *et al.* 1998); for unknown reasons, female junglefowl did not respond to the artificial combs, and the same may be true for the female pheasants. Third, the data presented for this manipulation (Mateos and Carranza 1995, p. 740 and Table II), apparently based on only two pairs of live males, indicate that one or two of these males was chosen twice as often as the other, whether large or small artificial wattles were attached to it. With manipulation of so few males, the possibility cannot be dismissed that one of them was more attractive for unrecognized or unstudied reasons.

Great snipe
The great snipe is a lek-breeding species in which males display at night by emitting a burst of song while the wings are lifted and the tail is spread. Mating success of individual males is highly variable. Höglund and Lundberg (1987) took morphological measurements of individual males and recorded their songs and displays in an effort to discover which male traits were correlated with reproductive success. Three factors proved to be associated with male success—display location, number of displays per unit time, and amount of white on the male's tail. (1) The males most successful in attracting and mating with females were the ones located nearest the centre of the lek. (Subsequent work suggests that this position effect is the result rather than the cause of the attractiveness of central males; Höglund and Robertson 1990.) (2) Those males that gave the greatest number of displays per unit time, irrespective of their lek position, were preferred. (3) More recently, Höglund *et al.* (1990) concluded that females use the amount of white, which is correlated with male age (Höglund and Lundberg 1987; Höglund *et al.* 1990), as a cue in mate choice. Central males were older than peripheral ones and were present more often on the lek. These points indicate that female great snipe tend to choose older males, both because of their location and because they have a larger number of white tail feathers.

Höglund and Lundberg (1987) suggest that the white tail feathers may be of special significance, in part because of their greater visibility at night; i.e. the older males may be more attractive simply because they are more conspicuous as they spread their tails in display. In addition to the mating benefits provided by whiter tails, there also may be greater costs in that whiter-tailed birds are probably more conspicuous to predators, such as owls.

Satin bowerbird
In a series of papers on the satin bowerbird, Gerald Borgia and his co-workers (e.g. Borgia 1985*a*,*b*, 1986*a*,*b*; Borgia *et al.* 1985; Borgia and Gore 1986; Loffredo and Borgia 1986*b*) presented several lines of evidence indicating that female satin bowerbirds prefer to mate with males owning high-quality bowers, and that males both compete for resources that attract females to bowers and attempt to interfere with each other's mating success by destroying bowers. Well-built bowers with certain kinds of relatively rare decorations (e.g. blue feathers) attract more females, and individuals possessing and defending such bowers tend to be the older and more dominant males. In addition, male bowerbirds use vocal mimicry—a second form of creativity in addition to bower decoration—in their songs, and older males are also superior mimics; in one of two years, song features were correlated with male mating success (Loffredo and Borgia 1986*b*).

Males are about six years old before attaining adult plumage (Vellenga 1980), which suggests that certain demographic features, such as extremely low mortality of adult males, are important, or that males require a very long time to perfect their bower-building and song-mimicry behaviour, or that both factors have favoured a retardation of maturation in males. Young male satin bowerbirds apparently must learn both to construct high-quality bowers and to sing complex songs; thus these traits can be viewed as cultural ones that require time to perfect. This demographic fact of prolonged male immaturity, which is associated both with subordinate behaviour and poor bower-constructing abilities, dictates that older males will (1) build better bowers, (2) defend them more successfully from theft and destruction, and (3) sing more complex songs than younger males. Since bowers and song do attract females, it is the dominant, older males that obtain most matings.

Lawes' parotia

Female mate choice has been studied in one bird-of-paradise species, Lawes' parotia by Stephen and Melinda Pruett-Jones 1990 (see also Johnsgard 1994, pp. 243–9), a typical representative of this group. Males clear terrestrial courts as display sites, where all intersexual and intrasexual interactions take place. Some male courts are clumped in their distribution, while others are well isolated. Males are highly ornate, with elongate, specialized head 'wires', an iridescent breast plate, and bright blue irises. In addition to the specialized plumage and a variety of displays (Pruett-Jones and Pruett-Jones 1990, Table 3; see Figs. 61 and 62 in Johnsgard 1994), males collect a variety of objects for their courts (Pruett-Jones and Pruett-Jones 1988), a behaviour reminiscent of most bowerbirds.

Females actively choose among males for mating. In this species the mate choice decision by females appears to be a lengthy process. Of several male characters identified and quantified, the one most clearly associated with male mating success was probability of display, which is associated with a male's presence at his court during the time of female visits. Individually marked females visited the same males repeatedly at their display courts, thus providing the females with a means of measuring the amount of time the male is available to present displays.

If a female loses her one-egg clutch, she returns to mate again with the same male. This fact, together with repeated visits to the same set of males, suggests that the female has made a true choice, and one that is important to her. Interestingly, however, morphological traits (e.g. length of head wires), the characters most conspicuous in the polygynous bird-of-paradise species, were not correlated with mate choice by female parotias. Thus, mate choice in an ornate bird of paradise is not unlike mate choice in some other strongly sexually dimorphic and polygynous/promiscuous species, such as the long-tailed manakin (McDonald 1989*a*) and the red junglefowl (Ligon and Zwartjes 1995*a*), in which the most extreme development of feather ornaments also fails to correlate with male mating success.

Long-tailed widowbird

In the first experimental field study of the role of male morphological traits in female choice, Andersson (1982*b*) manipulated tail lengths of territorial male long-tailed widowbirds, which occupy open grassland in equatorial Kenya. Breeding males are black with red epaulettes on the wings, and they possess extremely long (*c*. 0.5 m) tails. When the male is engaged in flight display over his territory, the tail is extended and expanded vertically below the slowly flapping bird.

Andersson elongated tail lengths of one group of males, shortened tails of another, and used two other groups as controls. (Tails were snipped and glued back together without changing their length in one set of controls, and in the other, the birds were simply handled with no tail manipulation.) To control for variation in territory quality, the number of nests initiated in the territory of each male was determined before and after the tail-length manipulations were carried out. Similarly, because the same male was compared in the same territory before and after the tail-length manipulations, each male served as its own control. The experimental design was such that neither re-nesting of already present females nor the possibility that some females nested in territories of males with whom they did not mate should bias the outcome in the direction of the hypothesis. Possible changes in male behaviour as a result of the manipulation were randomized by the experimental design. Changes in display rate did occur—males with artificially shortened tails displayed more frequently, but not significantly so.

After treatment, mating success of males changed as predicted. Males with elongated tails had highest success. Conversely, although males with shortened tails increased their rates of display, they had least success in attracting females. Hence, male tail length apparently did influence female mate choice decisions as Andersson predicted. These results provide strong evidence that female choice has driven the elaboration of a highly developed ornamental plumage character.

Might the long tail indirectly convey information about its bearer's energetic capabilities? To the casual human observer, it looks as though displaying male long-tailed widowbirds expend considerable effort to remain airborne. The fact that males with most of their tails removed displayed more frequently than longer-tailed birds supports the idea that long tails do indeed increase the metabolic costs of display. Supporting evidence comes from an interspecific analysis of wing length in widowbirds and their congeners. Male long-tailed widowbirds have unusually long wings relative to females, and this appears to be a response to flight costs associated with the long tail (Andersson and Andersson 1994).

This interpretation implies that the tails represent a handicap. This may be true, in terms both of energy expended to grow and carry the tail, and with regard to the males' safety, since the huge tail probably makes them more vulnerable to both ground and aerial predators. Alternatively, it is also possible that males with shortened tails increased their display efforts in response to a perceived lack of interest by females.

Jackson's widowbird
S. Andersson (1989, 1991, 1992) studied male display traits and female choice in Jackson's widowbird, a congeneric relative of the long-tailed widowbird

discussed above. Like that species, male Jackson's widowbirds possess long, structurally specialized, and conspicuous tails. However, unlike the long-tailed widowbird, males of this species display on leks and their displays are not aerial in nature (see Johnsgard 1994, pp. 273–6). Instead, males perform stereotyped jumps from small display courts or 'dance rings' constructed by the males in open grasslands (see also Section 3.3). Also unlike long-tailed widowbirds, female Jackson's widowbirds do not construct their nests near the males' display sites; on the contrary, they nest away from the lek area, with no male participation. Thus, females of the latter species visit the lek solely for mating.

Andersson's (1989) first analysis indicated that two components of the males' displays were significantly correlated with female preference—the rate of the jump display and tail length. He also found a positive relationship between condition (body mass) and tail length, with longer-tailed males being heavier. The success of males in obtaining matings was apparently based on three factors. Both display rate and lek attendance affected the number of visits by females, and number of visits was related to male mating success. Second, tail length seemed to be important in inducing a female to copulate once she had come to the male's display site. Subsequent experimental manipulations of tail length demonstrated that, within the natural range of variation, those males with longer tails were significantly more likely to be chosen by females (Andersson 1992).

Barn swallow
Møller's investigations of sexual selection in a European population of the migratory barn swallow may be the most comprehensive series of studies of sexual selection in a single species ever produced. His book, *Sexual selection and the barn swallow* (Møller 1994), consolidates the results of this amazing effort. According to Hill (1995c), between 1987 and 1994, Møller published 76 papers on sexual selection in the barn swallow!

Briefly, with regard to female mate choice, male barn swallows attract females to their territories by displaying their tails in a fan-shaped fashion and by singing vigorously. The tail is made conspicuous as females examine the male. (1) Tail length is related to male mating success. Long-tailed males arrive early in spring, and tail length is related positively to mate acquisition and early onset of breeding (see also Smith and Montgomery 1991). This leads to higher reproductive success than that of shorter-tailed individuals, due largely to the greater frequency of second broods by pairs with a long-tailed male. Møller also experimentally manipulated the lengths of the elongated tail streamers and experimentally confirmed both that males with long tails obtained mates more quickly than males with shorter tails and that they had greater reproductive output over the breeding season. (2) The longest-tailed males were also preferred by females seeking extra-pair copulations, particularly those females mated to males with experimentally shortened tails (cf. Smith *et al.* 1991). (3) Infestation of feather mites has a large effect on tail length; by choosing a male with parasite resistance, females gain an indirect benefit for their offspring and they have a mate with few mites to transmit to nestlings. (4) Females were sensitive to the fluctuating asymmetry of the males' tails, as well as to tail length, and there is a negative relationship between tail length and degree of asymmetry of the tail streamers. (5) Long-tailed males survived better than short-tailed individuals, demonstrating that the former are in better condition. (6) Males given artificial, long tails return the next year with significantly shorter tails, demonstrating a foraging cost associated with tail length; thus tail length is a handicap trait. One important general implication of this classic study is that sexual selection can be a significant selective agent on male morphology even in socially monogamous species (see also house finch, below).

Zebra finch
Male zebra finches have red or red–orange bills and legs. In a well-known and novel study, Burley *et al.* (1982) determined that female zebra finches preferred to associate with males that had red bands on their legs. Not only do female zebra finches prefer males wearing red leg bands, but under conditions of captivity, males wearing such bands (and their mates) enjoy enhanced reproductive success (Burley 1986*a*). Males randomly assigned red bands had about twice the reproductive success of males assigned unattractive band colours. Burley attributes this to increased reproductive effort by the females mated to the red-banded males; i.e. females mated to males perceived to be of high quality are willing to invest extraordinary amounts of parental care. Finally, band colour also affects mortality patterns, in addition to mate-acquisition abilities and reproductive success. Given that fitness (number of descendants) is the most critical measure of quality, the behaviour of females that are mated to males exhibiting the greatest amount of red virtually ensures that such males will have greater than average fitness. Burley's remarkable findings suggest that the perceived quality of a mate can have several ramifications.

Recently, Collins and ten Cate (1996) reviewed a number of studies on male choice in zebra finches, calling into question the conclusions of Burley and her co-workers concerning the importance of the male's red bill in female mate choice. These authors pointed out that some recent studies have failed to find any effect of bill colour in female preference, and that some of them suggest instead that females use male display rate to choose a mate (see Collins and ten Cate 1996 for references). Their review led Collins and ten Cate (1996) to provide some alternative suggestions: (1) there may be a hierarchy of choice criteria, with display rate being more important than beak colour; (2) female preference for beak colour may be influenced by the experience of females during their maturation; and (3) it is possible that males rather than females respond strongly to beak colour differences between the sexes.

An important point in this debate is the fact that males with the reddest bills also have the highest display rates (Houtman 1990, reference in Collins and ten Cate 1996). Thus, both traits—one morphological, one behavioural—may signal the same message about male condition or quality. Although the questions and issues raised by Collins and ten Cate (1996) merit careful consideration by future researchers, the basic point that bill colour of male zebra finches is important to females appears to remain on solid ground.

House finch

Sexual selection and male traits have been intensively studied in the North American house finch (e.g. Hill 1990, 1991, 1992, 1993*b*, 1994*a*), a small, sexually dichromatic, and socially monogamous passerine. Typically, the feathers covering the face and chest of males are brightly coloured. Within a population, the colour of males varies, ranging from pale yellow to bright red. This variation reflects the carotenoid pigments in a male's plumage (see discussion of carotenoid pigments and house finches, Section 5.3). A male's colour is unrelated to intrasexual dominance (Belthoff *et al.* 1994), nor does it reflect age. Unlike many other passerine species, male house finches do not defend territories during the breeding season, which eliminates the issue of territory quality as a confounding variable in field studies of mate selection by females. Therefore, if females actively choose among males in nature, they probably do so on the basis of male traits or characteristics *per se*.

An analysis of the relationship between plumage pigmentation and pairing success of male house finches provides circumstantial evidence for a tie between female mate preference, male plumage, and male quality. Hill (1991) compared the mean coloration of pigmentation of a number of males with that of males known to have paired successfully, and found that the paired males were significantly more colourful than the male population as a whole. This suggests a causal relationship between the degree of plumage pigmentation and ability to attract a mate. Males with greater amounts of carotenoid pigments also showed greater nest attentiveness and higher overwinter survival than did those males with less pigmentation.

To test for the importance of male coloration to female house finches, Hill (1990) used an experimental apparatus similar to the one described by Burley *et al.* (1982). Male plumage colour was altered by dietary manipulations at the time of moult, and, in one experiment, by dyeing birds to create variation among four males that were naturally similar in colour. Hill conducted four sets of mate choice experiments, all of which led to the same conclusion: females invariably chose the reddest males available. He also conducted a test in which females rather than males were used, to test for any sort of non-sexual association preference, with negative results, indicating that the strong female preferences obtained did indeed reflect mate choice.

Extending Brush and Power's 1976 study on feather pigmentation in house finches, Hill (1992) also conducted feeding experiments, which revealed that (1) on a standardized diet, all birds converge on a similar appearance, and (2) in the wild, bright males are not better at utilizing carotenoids than their dull-plumaged neighbours. This means that variation in colour among males is a result of variation in ability to obtain carotenoids in the diet. Thus, plumage coloration is a condition-dependent trait, probably related to the males' foraging efficiency (Hill and Montgomerie 1994).

Hill (1994a) also determined that house finches from an introduced population in Hawaii, where males are on average duller in colour than those in either the eastern or western US, moulted into plumage as red as that of brightly coloured males from the eastern US when provided with a high-carotenoid diet. Thus, as indicated both by female choice tests and tests of the effects of diet on plumage colour, the Hawaiian birds appear to differ from their mainland ancestors only in the availability of the carotenoids involved in synthesis of plumage pigmentation.

6.3.2   Mechanisms by which male condition may be signalled to females

Male birds have evolved many different means of signalling their condition to females. In this section, several mechanisms for indicating condition are reviewed.

6.3.2.1   *Prolonged, energetically costly displays, including vocalizations*

Males of many diverse species provide vigorous and/or prolonged displays. Such displays can be either vocal or visual, or they may utilize both means of signalling. Displays often take the form of more or less stereotyped, species-specific movements (e.g. 'dances'), along with simultaneous production of sounds. To the extent that these displays are metabolically costly, they may serve to provide information concerning a male's physical condition relative to other males. For example, male sage grouse use both dances and auditory signals in their displays, which can be conducted repeatedly in a single morning and over an extended period of several weeks only by those males in the best physical condition (Vehrencamp *et al.* 1989).

6.3.2.2   *Structures or behaviours directly reflecting testosterone levels*

Males of many species are able to signal physical condition to females via either morphological structures or types of behaviour that are directly indicative of their testosterone levels relative to other males (see discussion in Section 5.2). Testosterone is thought to negatively affect the individual's immune system (Folstad and Karter 1992), thus traits directly influenced by that hormone appear to be honest signals about the condition of the male displayer. A high level of testosterone may also indicate that the male displayer has survived and maintained good health despite the handicap of reduced immunocompetence (see Section 5.2.3). The combs of male galliform birds of a variety of species reflect testosterone level (e.g. Stokkan 1979*a,b*; Brodsky 1988; Ligon *et al.* 1990) and appear to send this kind of signal. In some other species, both sexes possess conspicuous morphological traits affected by androgen level; e.g. the black colour of the mouth lining of rooks and pinyon jays during the breeding season (Marshall and Coombs 1957; Ligon 1978). In both of these corvids, open mouths are an important component of male–female interactions (e.g. reciprocal courtship feeding).

Other ornamental structures not directly affected by testosterone can also correlate with body condition. For example, tail length in Jackson's widowbird is not controlled by testosterone (Witschi 1961); however, S. Andersson (1989, 1992) found that females preferred long-tailed males and that such males were in the best condition (relative body mass).

6.3.2.3   Ornaments coloured by carotenoid pigments

Males of many species of passerine birds, in particular (e.g. house finch), may signal their quality via plumage colours based on carotenoid pigments, which form the basis of most of the yellow, orange, and red colours (see Section 5.3). In addition to plumage, certain other, apparently sexually-selected traits are under the influence of both testosterone and carotenoids. For example, the yellow bill colour of male and female European starlings depends on both androgens and carotenoid pigments for development of the bright yellow bills seen during the breeding season (Witschi 1961). Similarly, the yellow bills of male American goldfinches is based on carotenoids, as is the yellow portion of the breeding plumage. As in starlings, carotenoid pigments in the goldfinch's bill are exposed at the onset of the breeding season as a result of testosterone-mediated loss of melanin (Mundinger 1972; K. Johnson *et al.* 1993*b*). In these examples, testosterone and carotenoids affect the same trait—the bill—and it is likely that in many kinds of colourful birds, information about condition is provided via simultaneous exhibition both of carotenoid pigments and testosterone-mediated traits.

6.3.2.4   Bilateral symmetry of ornamental feathers

One of the most recently proposed and most controversial of the various good genes arguments relates to the phenomenon of fluctuating asymmetry. This is the relationship between male mating success and the degree of deviation of paired ornamental characters from perfect bilateral symmetry (see Section 5.6). To review, consider the relationship between fluctuating asymmetry and length of paired elongate, ornamental tail feathers. The degree of asymmetry of the paired elongate feathers reflects the degree of developmental homeostasis during feather growth. The greater the perturbation of those feathers during their development, the more asymmetrical they will be. In the socially monogamous barn swallow, males with both the longest and the most symmetrical tails are thought to exhibit good genes, and they are preferred by females, both as mates and for extra-pair copulations (Møller 1994).

The proposed relationship between fluctuating asymmetry and length of tail feathers as a signal of genetic quality has been challenged on the basis of aerodynamic and mechanical principles related to forked tails (e.g. Balmford *et al.* 1993; Evans and Hatchwell 1993; Evans *et al.* 1994; Thomas and Balmford 1995). Another possible challenge to the fluctuating asymmetry–good genes connection is based on the argument that sensory systems are built to be sensitive to symmetry for reasons having nothing to do with mate quality, and that this form of sensory bias could account for the reports that female animals seem to prefer males with more symmetrical ornaments (see Enquist and Arak 1994, Johnstone 1994, and Kirkpatrick and Rosenthal 1994).

The arguments invoking degree of fluctuating asymmetry as a signal of male quality are based on the good genes and handicap perspectives of female mate choice. Because of the relationship between degree of asymmetry and amount of environmental stress during development, demonstration that females of a number of species do attend to the degree of symmetry of ornamental traits of males will provide another line of evidence that females are indeed 'shopping' for good genes when making mate choice decisions.

6.3.2.5   Size of morphological ornaments

Perhaps surprisingly, in a number of studies, size of feather ornaments has not proved to be important in female mate choice (see Chapter 4.3.2). In some cases, however, feather size does relate to male mating success, e.g. the Indian peacock's train. In addition to the outer tail feathers of barn swallows (Møller 1994), tail length is possibly important in the ring-necked pheasant (Mateos and Carranza 1995) and two species of African widowbirds (M. Andersson 1982*b*; S. Andersson 1991). In the Jackson's widowbird, tail length correlates with male body condition, which suggests that it is a condition-dependent indicator. Size of certain other kinds of male ornaments (e.g. snood of turkeys, comb of red junglefowl) are apparently also important in female mate choice.

6.3.3   Convergence of the evidence for good genes

In all of the empirical studies of individual species reviewed above, plus other studies considered in Chapters 4 and 5, male traits identified as important to females in mate choice decisions have the potential to convey either direct or indirect information to females about the physical condition of prospective mates. The fact that such information can be signalled in so many different ways suggests that any purely arbitrary system of female choice is unlikely to be the common basis for the diverse types of male displays that exist. Although a good genes or indicator interpretation of male traits and female preference is consistent with the kinds of male traits used by females in mate choice, the evidence remains for the most part circumstantial.

Several studies indicate that more attractive males also produce more offspring (e.g. zebra finch, yellowhammer, house finch, European kestrel). While this result could be due to the males' good genes, alternative explanations are also viable. For example, it is possible that brighter birds are more experienced at parenting (e.g. Lozano and Lemon 1996), hold more productive territories, or that females expend more parental effort when mated to bright males. If so, then genetic quality of the male may or may not be the correct explanation for any correlation between male ornamentation and reproductive success.

A few of the studies reviewed in this chapter do, however, provide more direct evidence for the good genes hypothesis in that they make the critical connection not only between male traits and female preference, but also between female mate preference and number or quality of offspring.

For example, under controlled, standardized conditions, offspring of male Indian peafowl that are most successful at attracting females grow and survive better than offspring of less successful males (Petrie 1994). Because most complicating environmental factors were eliminated (e.g. parental quality of the mother), these results clearly suggest that females that mate with the most attractive males obtain good genes for their offspring. Similarly, in the red junglefowl, the size of the male's comb, a sexually-selected, condition-dependent trait, correlates with rate of development of the males' chicks. Chicks of large-combed males grow faster than do those of small-combed roosters, for at least their first three months of life (Johnson *et al.* 1993*b*).

6.4 Arbitrary female choice

Evidence supporting either of the two well-known hypotheses for arbitrary mate choice—Fisherian runaway and the aesthetic choice–sensory bias perspectives—is extremely limited, and, moreover, is susceptible to alternative interpretations.

6.4.1 Runaway selection

As discussed in Chapter 2, for a period of several years, Fisherian runaway selection was the favoured explanation for male ornamentation. A number of theoretically-oriented biologists strongly championed Fisher's model, or modifications of it, as the theory most likely to account correctly for the evolution of extreme ornamental traits of male animals (Lande 1981; Kirkpatrick 1982; West-Eberhard 1983). However, the popularity of the runaway hypothesis appears to be on the wane. In their review of the evolution of female mating preferences, Kirkpatrick and Ryan (1991, p. 34) concluded that the attention the runaway process now receives '… comes more from its historical importance as the first modern hypothesis for preference evolution than from any empirical support for it.'

The current lack of support for the runaway hypothesis, however, probably should not be taken as the final verdict on the subject. Even if the runaway phenomenon exists as postulated by Fisher, several factors work against its documentation. (1) If a runaway-like process has played a role in the elaboration of the extreme ornamental traits in some species of male birds, it probably did so within a small number of generations and over a brief period of evolutionary time. The ephemeral nature of the runaway process could explain, in part, why to date no good empirical evidence for it exists (Kirkpatrick and Ryan 1991). The runaway stage could occur so rapidly, in an evolutionary sense, that the likelihood biologists will encounter an ongoing case of it is small. (2) In view of the relatively small number of highly ornate bird species (most of which are both difficult to access and difficult to study) and the existence of careful studies of only a handful of these, together with the (presumably) brief period of evolutionary time over which the runaway process might operate, it would be far more surprising to discover a case of ongoing runaway selection than to fail to discover such a case. (3) The improbability of discovering a clear-cut case of runaway selection is also due to the fact that less than one per cent of the ornate bird species have been, or probably ever will be, studied in a way designed to document it, which, even under ideal conditions, will be anything but easy, and may not even be possible. Much of this difficulty is due to the fact that the key predictions of runaway are similar to those of the good genes hypothesis (e.g. longer-tailed males preferred under both hypotheses). Borgia (1987, pp. 62–4) discusses difficulties associated with testing the runaway models.

In short, the failure to date of locating and studying a good example of runaway selection could be due to one or more of the following factors. (1) Very few highly ornate birds (the most likely candidates) have been studied in an appropriate manner. (2) Runaway may occur only over a brief period of evolutionary time. (3) Most importantly, a means of distinguishing between the Fisher runaway and good genes hypotheses has not yet been devised.

I am aware of only one possible example among birds of the occurrence of rapid elaboration of an ornamental trait that could be a case of 'runaway'. Alatalo *et al.* (1988) reported a statistically significant increase ($r = 0.78$, $P < 0.001$) over time in tail length in specimens of the northern paradise whydah (Fig. 6.4) collected over a period of 50 to 60 years. Alatalo *et al.* (1988) point out that this finding could be coincidental, since of 11 statistical tests, only this one is significant. Their caution is well taken. However, since a runaway process would start and stop independently in each population or species, and at different points in time, there is no reason to assume that it would necessarily be occurring simultaneously in two or more related, but independently evolving, species. That is, an isolated case of tail elongation over time within a group of congeners, such as reported by Alatalo *et al.* (1988), might be an expected aspect of Fisher runaway.

This dubious, but possible, example of runaway illustrates how difficult it will be to identify confidently a case of Fisherian selection. Even when a significant change in size of an ornament is detected over a brief period of evolutionary time, it will be next to impossible to ascertain that the increase is not

Fig. 6.4 Paradise whydahs surrounded by other, short-tailed, species of African finches. Reprinted with permission from Dover Publications, Inc.

a statistical coincidence. Moreover, even if it is concluded that the increase is real, there is no way to eliminate the alternate possibility that the increase has been driven by environmental change, which might be related to some form of the good genes phenomenon. For example, Evans (1991) found that ornaments of male scarlet-tufted malachite sunbirds—the size of pectoral tufts and tail length—changed substantially between low- and high-rainfall years, which suggests that they are condition-dependent traits.

6.4.2 Sensory bias

A brief description of the sensory bias hypothesis (Ryan *et al.* 1990) was presented in Section 2.4.2. Pomiankowski (1994) points out that analyses of this hypothesis require an accurate phylogeny. For example, a new phylogeny opened the possibility that in *Physalaemus* frogs (discussed in Section 2.4.2), the male 'chuck' call, rather than a female preference for the call, may be ancestral (Ryan and Rand 1993, 1995). If so, it appears that a more complex 'ornament' (the chuck call in male frogs) was lost in the species that do not exhibit it, and that females of this lineage have retained a preference for the display. In short, the best putative evidence for the original sensory bias explanation of female choice has undergone fundamental revision (see Pomiankowski 1994).

This does not mean, however, that responses to a particular sort of stimulus could not have affected the evolution of male ornaments and female preferences. For example, Kirkpatrick (1987, p. 69) speculated that attraction to food of some particular colour could lead to preferences in females for males or male ornaments of a similar colour. I am aware of one possible case in support of this hypothetical scenario. McPherson (1988) studied food preferences of the highly frugivorous cedar waxwings. The waxwings prefer fruits that are small and red over those of other sizes and colours. McPherson (1988, p. 967) anticipates the point I am attempting to make here: '... red ... colour serves as a conspicuous cue or 'orienting stimulus'... to guide birds to a potentially good food source.' The most conspicuous morphological ornaments of cedar waxwings are the small, red waxy extensions of the secondaries (Fig. 5.4), and the waxy red tips are apparently used by both sexes in mate choice (Mountjoy and Robertson 1988). As discussed in Section 5.3.2, the waxwings' red tips are composed of carotenoids, thus they may serve as condition-dependent indicators of the quality of individual birds. The initial attraction of these birds to small, red food objects possibly led to the evolution of the waxy wing tips, and these structures, because of their carotenoid nature, may have become condition-dependent indicators of their bearers' quality.

In addition, if the supernormal stimulus phenomenon is viewed as a form of sensory bias, then there is considerable additional evidence for a role of sensory bias in the evolution of male ornamental traits in birds and female preferences for them. The sensory bias–supernormal stimulus phenomenon may prove to be a common proximate explanation for mate choice in female birds, perhaps especially for species in which males provide nothing in the way of resources or parental care. The response of females to male tail length in long-tailed widowbirds (M. Andersson 1982*b*), for example, may be based on the supernormal stimulus phenomenon.

One empirical study on an avian species invoked the sensory bias hypothesis to explain a puzzling female preference. Searcy's (1992) experimental study of song preferences in female common grackles suggested to him that the disparity between what males sing (only one of four song types) and what females prefer (all four song types) might mean that female preference for repertoires may pre-date their evolution. This idea, however, has subsequently been refuted (see Section 7.2.3).

There is one convincing case in birds for a pre-existing bias of the sort outlined above, namely loss in males of an ancestral ornamental trait, with retention of the preference for that trait in females. Experimental studies of mate choice in house finches revealed that, like their eastern counterparts in Michigan, females from Hawaii and some areas of California—areas where males tend to be unusually pale—prefer the reddest males available (Hill 1994*a*). Preferred males are considerably redder than any males those females normally encounter. Because birds from all of these areas belong to the same subspecies, and thus are very closely related, perhaps this is not too surprising. However, in males of a distinctive subspecies of house finch in south-central Mexico, *C. m. griscomi*, the colourful carotenoid pigments are restricted to a small patch of feathers under the chin. When given a choice, females of this population also prefer males with more extensive red than occurs in males of their own subspecies. A phylogenetic analysis indicates that *C. m. griscomi* is derived from ancestral house finches with more red, thus the preferences shown by females of *C. m. griscomi* may reflect a 'pre-existing bias' originally evolved in their ancestor.

Although it is usually assumed that the direction of evolutionary change, particularly of male ornaments, is from less complex to more complex, this may not always be the case (see Section 7.2). Hill's (1993*b*) studies of male colouration in different populations of the house finch suggest the reverse; i.e. in some contemporary species, male display characters may have evolved from a more ornate to a less ornate condition, even though females exhibit a preference for what is presumably a stronger stimulus than that currently exhibited by males of their species or population (Hill 1994*a*). Geographic variation in the colour pattern of male house finches supports the notion that male display traits should be costly (Hill 1994*b*); i.e. because females of *griscomi* prefer larger patches of colour, males presumably would produce them if they were able to do so. The fact that female preference is not necessarily altered in tandem with the change in male characters is also predicted by Hill's (1994*b*) view of sexual conflict as a driving force in the evolution of signals of male quality (see Section 6.5 below).

6.4.3   Aesthetic mate choice

Two lines of evidence may support the existence of truly aesthetic mate choice. One deals with the effects of coloured legs on mate choice by female zebra finches, while the other is based on the use of novelty in bower decoration by male bowerbirds, perhaps especially the Vogelkop bowerbird.

6.4.3.1   *Zebra finches and coloured leg bands*

Nancy Burley's work on the effects of leg-band colour in mate choice in zebra finches provided the first evidence for birds that coloured bands could increase the sexual attractiveness of individuals. One set of experiments (Burley 1985), in particular, appears to document the existence of a female preferences for novel male characters, apart from their preference for orange or red bands (Burley *et al.* 1982), which are similar to the males' bill and tarsus colour. Female zebra finches strongly preferred males wearing yellow bands over their orange legs to males wearing other colours found naturally on the body surface of males (black and white). Because the colour yellow is not found on zebra finches, this result cannot be accounted for via the supernormal stimulus phenomenon, at least not in its usual form. Here, then, appeared to be evidence that males exhibiting a novel colour attracted females more strongly than did males that differed little or not at all from the normal phenotype of the species.

Burley (1985) then conducted a follow-up experiment to determine whether it was the colour yellow *per se* that attracted females or whether the attraction was based on the contrast between the yellow bands and the orange legs. Two groups of males were normal leg (unbanded) and normal leg with a yellow band. The tarsi of a third group of males were painted yellow, and yellow bands were added ('super-yellow' males). Finally, Burley painted the toes and tarsi of a fourth group red to make the legs appear dark orange–red, and then added yellow bands to produce a 'super-red–yellow' phenotype. Females perched next to super-red–yellow males significantly more frequently than next to males of any other phenotype. 'Super-yellow' males were least preferred, demonstrating that the colour yellow was not attractive to females, contrary to the results of the first experiment.

Burley interprets these findings as follows: female zebra finches are attracted to the contrast provided by the orange–red and yellow colours, and this may be related to the fact that the body surface of male zebra finches is covered with contrasting colour patterns, e.g. the vertical black and white stripes under the eyes. Whatever the proximate basis for the preference for yellow and orange–red, these results suggest that novel colours (here yellow), or novel patterns of presentation, which red-and-yellow legs unarguably are, can produce a preference by females upon initial exposure. They also suggest that the course of future morphological change driven by sexual selection may be unpredictable.

6.4.3.2 Bowerbird bowers and the question of an aesthetic sense

Because objects brought by males to their bowers represent visual signals produced by particular and sometimes unique arrangements of colourful objects, it is not a great mental stretch for the human observer to see similarities between bower ornamentation and human art. Jared Diamond (1982, 1986*a*) raised this intriguing question in two articles entitled *Evolution of bowerbirds' bowers: animal origins of an aesthetic sense* and *Animal art: variation in bower decorating style among male bowerbirds, Amblyornis inornatus.* Diamond studied bower structure and design in three populations of the Vogelkop bowerbird. This species builds the 'most elaborately decorated structures erected by an animal other than humans ...' (Diamond 1986*a*, p. 3042).

Geographic and individual variation exists in both bower structure and preferred items of decoration. Specifically: (1) there are marked differences in decorations used by individuals whose bowers are located only a few kilometeres apart; (2) colour preferences for ornamental objects vary from population to population, as do the objects chosen as decorative items; (3) within a local population differences exist between immature and adult males in bower construction and design, and such differences also occur even between neighbouring adult males; (4) in two separate cases, two neighbouring bowers contained items not used by the remainder of the local population, suggesting that one male might be copying a unique preference from its neighbour; (5) one population paints its bowers, whereas others do not (Diamond 1987); and (6) by use of poker chips, Diamond (1988) also experimentally confirmed individual variation in preference of colours of decorative items used by males. Provided with poker chips of a variety of colours, different individuals within a population varied in their selection criteria and in the way they arranged the chips chosen to decorate their bowers. Decorating decisions involved trials and 'changes of mind'.

Diamond (1987, 1988) presents arguments, based both on observational data and on his poker chip experiments, that preferences for certain items and certain colours are learned, and that the similarities in decorative patterns among adult males within a local population reflect a cultural tradition, rather than instinctive behaviour. All of the patterns of variation listed above suggested to Diamond that many aspects of bower design by male Vogelkop bowerbirds may be arbitrary; i.e. the patterns of decoration may have no specific adaptive function. However, individually unique patterns of decoration by males imply that females of this species are attracted to novel visual stimuli, and possibly that bower decor reflects male quality (Diamond 1988).

In this species, it is likely that any genetic basis for preference for certain items, or for unique arrangements, is present in females as well as males; i.e. the preferences of each sex—items gathered and placed in bowers by males and reaction to those items by females—are probably influenced by the same genes. This point, together with a consideration of (1) the role of ethological factors in female choice, (2) the presumed evolutionary history and current functional significance of bowers, and (3) theories that predict episodic shifts in female preferences for male signals (e.g. Møller and Pomiankowski 1993*a*), lead to the conclusion that the putative aesthetic and artistic behaviour of bowerbirds can, like more mundane ornaments, be attributed to the process of functional female choice. In a more recent paper on bowerbirds, Diamond (1988) agrees with Borgia *et al.* (1985) that bowers probably serve as 'markers' of various aspects of male quality, and that female preference for well-constructed, decorative bowers (and thus for males producing such bowers) has driven the evolution of bowers towards increased complexity and ornateness.

The intraspecific, interindividual variation in bower structure and decor exhibited by the Vogelkop bowerbird appears to provide an example of what may prove to be a common, if generally difficult to recognize, aspect of courtship display among birds, namely the use of variation or novelty as an aspect of courtship display. Individual variation is perhaps most obvious in the songs of certain birds, most conspicuously those 'mimicking' species (including bowerbirds) that routinely incorporate sounds from the environment to produce individually unique signals.

6.4.3.3   *The current significance of bowers*

The 'marker hypothesis' (Borgia *et al.* 1985) is designed to account for the current functional significance of bowers. This hypothesis posits that males compete for females by the attractiveness of their bowers. Attractiveness of bowers is determined by quality of construction and the number and specific nature (e.g. colours) of decorative items associated with it. Much male–male competition in the satin bowerbird is effected by reciprocal bower destruction and theft of display items. Those males that receive the most matings are the ones most successful in building high-quality bowers, in decorating them with rare attractive objects, and in preventing the destruction of their own bowers, while attempting to destroy the bowers of neighbouring males. Also supporting the marker hypothesis is the fact that females visit several bowers, thus providing them with the opportunity to make comparisons of bower quality. Thus, males compete for females in large part via their bowers. This has probably provided the selective pressure for the evolution of the almost unbelievable development of bowers in some species, along with their decorative objects.

The specific role of bowers across species in mate attraction appears to be variable; i.e. the specific ways in which bowers are important differ from species to species (Diamond 1986*a*, 1988). In the satin bowerbird, males compete for females by stealing preferred decorative items and by destroying each other's bowers (Borgia 1985*b*), whereas in the spotted bowerbird, mate attraction appears to be largely via displays at the bower (Borgia 1995). In the

Vogelcop bowerbird, males compete for females in part by creating individually novel patterns of decorations (Diamond 1986*b*, 1987). The characteristics of decorations also vary. In the satin bowerbird, favoured display items are similar to parts of the males' morphology (e.g. blue objects), whereas in others they appear to provide a strong contrast with the male's plumage. In still others, the objects gathered may be both common and dull in appearance.

As in satin bowerbirds, male MacGregor's bowerbirds compete via bower destruction and theft of attractive objects, and by interfering in mating attempts as well (Pruett-Jones and Pruett-Jones 1983). Thus, among the bowerbirds that have been studied, there is strong evidence for female choice based in part on bower quality in satin and spotted bowerbirds, as well as for male–male competition via stealing and/or bower destruction in these two species, plus MacGregor's bowerbird. As previously mentioned, in the Vogelkop bowerbird, males vary in favoured decorative items; presumably this is correlated with variability in female preferences (Diamond 1987, 1988).

If females are attracted to novel visual stimuli—and, given the diversity of decorative objects brought to bowers, it is reasonable to assume that this might be the case—then the regional variation in bower decoration in Vogelkop bowerbirds (Diamond 1986*b*, 1987, 1988) may have an arbitrary origin; perhaps some females were initially attracted to a novel bower decor and the preferences of those females may subsequently have been copied by other females.

A role for novelty in female choice is suggested with regard to the origin of each of the major categories of bowers and the items used for decoration. For example, even though older and more dominant male satin bowerbirds gather and defend blue objects in their bowers, the preference by females for the colour blue may initially have been arbitrary (or it may reflect a sensory bias of the sort discussed by Hill 1994*a*; see above). Whatever criteria females use, males must compete on the basis of those criteria. It is this intermale competition, reflected in differential bower quality, which suggests that high-quality bowers are indicative of high quality males, and which thus provides support for the good genes interpretation of current function in this species.

On the other hand, in Vogelcop bowerbirds, an aesthetic function of bower decor is suggested by the great intraspecific variation in bower design, and especially in the diversity of objects used in decoration. Specifically, the variation both between populations and among individuals within a population in choice of decorative items suggests that selection for novelty *per se* is present. Female preferences in some populations of the Vogelkop bowerbird may currently be undergoing transition; that is, selection on male traits via female choice may currently be shifting from one set of stimuli to another (see Hill 1994*b* and Møller and Pomiankowski 1993*a*). Alternatively, novelty of decor may in itself reflect relative quality among males.

It is also possible that the novelty apparently preferred by females of some populations of the Vogelcop bowerbird (Diamond 1987, 1988) may be an example of the 'unending nature of change' (West-Eberhard 1983). In this species, bower ornamentation can continue to evolve for as long as some females continue to respond to novel stimuli, which appears in this case to have driven bower structure and ornaments to amazing extremes. (It is assumed here that selection on males for novel displays is the basis for the incorporation into the bowers of foreign novel objects, e.g. poker chips.) In practice, exaggeration of bower display by use of new decorative objects will at some point be limited simply by the availability of locally obtainable objects. In contrast, new patterns of presentation of available items could presumably continue 'unendingly' (see West-Eberhard 1983). For example, the facts that in several species (1) paint is manufactured and used and (2) decorative objects may be continuously rearranged mean that the potential for future change of an idiosyncratic nature is virtually unlimited. In addition, male bowerbirds (probably all species) are accomplished vocal mimics; thus, in addition to their amazing bowers, they also have the ability to provide novel vocal displays important in female choice (Loffredo and Borgia 1986b).

It should be clear that the two major views of the bases of female choice espoused for bowerbirds—good genes (e.g. Borgia 1985a,b, 1986a; Borgia et al. 1985; Borgia and Gore 1986) and arbitrary preference (e.g. Diamond 1982a, 1986a, 1987)—need not be mutually exclusive, if the distinction between origin of a preference and current function of that preference is kept in mind.

6.5   *A hypothetical scenario for the appearance and subsequent elaboration of male traits important in female choice*

One of the greatest challenges for proponents of the good genes viewpoint of female choice is how to account for the mind-boggling diversity of ornamentation and display patterns found in birds, not to mention other animals. If the only message conveyed by ornaments or display behaviour is male quality (e.g. resistance to pathogens, as indicated by health and vigour), why do we see so much diversity in display signals even within closely related groups? Among birds there are many cases where closely related species differ strikingly in feather colour. It strains belief to suggest that males of closely related (e.g. congeneric) forms must generally convey a message of health or vigour so distinctive and specific that it can only be transmitted with a specific colour or colour pattern (although this might be true in some cases).

The great diversity of colours and patterns of both feathers and 'soft parts', and behavioural displays, provides the best circumstantial evidence for the involvement of chance factors in the initial evolutionary appearance of male ornaments and displays. 'The diversity of sexually selected signals exhibited

by birds probably reflects first the random nature of mutations and the idiosyncratic nature of evolution.' (Payne 1984, p. 43.) This sentence is quoted to
emphasize both the source—mutation—and the unpredictability of new morphological and behavioural characters that can come to be important in mate
choice. Over time, new traits that have a heritable basis appear in individuals
of natural populations by chance mutation.

In addition to sexual and natural selection, mutations also form the raw
material for artificial selection (see Section 2.5.1). With regard to plumage
colours and colour patterns, consider, for example, the varieties of domesticated breeds of chickens, turkeys, guineafowl, ducks of two separate lineages
(mallard and muscovy), peafowl, golden pheasants, coternix quail, bobwhite
quail, budgerigars, and rock doves, among others. Even the ring-necked pheasant, which, unlike the chicken, is not a species with a long history of domesti-
cation and selection for phenotypic variation, exhibits a large number of
plumage colours chosen and 'fixed' by artificial selection. Gamebird farm
brochures offer a variety of mutant-plumaged, ring-necked pheasants: black-
neck, buff, buff Isabel, Alaskan snow, green, black, white, etc. Thus, over
even short periods of evolutionary time, the raw material for phenotypic
change—mutation—is not a rare event. All of the many phenotypic variations
found in domestic birds presumably originated as the result of random
mutations, with subsequent selection of those characters favoured by poultry
fanciers.

These and many other possible examples indicate that the source of new
variants in plumage or traits such as the cock's comb is pretty well understood.
The key questions to address are: (1) how do females of a population or
species initially come to prefer a newly-appeared trait, and (2) how does a new
trait and the preference for it spread and become fixed through an entire population or species? The following scenario, which, for simplicity, focuses on
species in which males provide no resources to females, considers the appearance of a new male ornamental trait via mutation, and transformation
of that trait into an accurate reflection or indicator of male health or
vigour.

Within the limits set by phylogenetic history and morphological possibilities, new traits, which appear as a result of random mutation, are almost
always derivatives or extensions of pre-existing traits (e.g. brighter colour of
the bill). Some females may be strongly attracted, while some others may be
neutral. Moreover, other females may copy the mate choice of those females
initially strongly attracted to the new trait, thus increasing the absolute and
relative mating success of the mutant males. The key point is that at the time of
its original appearance, a new male trait attractive to some females can initially
be unrelated to any form of functional female choice (i.e. unrelated to either
direct selection or good genes selection). A preference by some females for a
newly-appeared trait in a few males revises 'the rules of the game' for all
males in a population, because matings with those females that favour the new

trait, as well as a portion of the neutral females, will be lost to the males that do not possess it (O'Donald 1983).

Because male birds typically present a number of morphological ornaments, together with a number of display postures or movements, the appearance of a red-billed mutation in a black-billed population or species, for example, will not totally alter the overall appearance of the mutant males. Only one trait— red bill colour—is new. Most of the suite of characters (e.g. plumage colours and colour patterns, postures and movements, sounds) used in the male's display will be familiar to females. This, in itself, further increases the likelihood that a mutant male will attract some females for mating. That is, the appearance of a new phenotypic trait by mutation often will not alter the appearance of a 'mutant' male so drastically that it will be avoided by females. This suggestion is similar to what has been referred to as '... the "optimal discrepancy hypothesis"—the idea that the most attractive object is one that differs just a little from a familiar standard' (Cronin 1991, pp. 173–4). This kind of phenomenon has previously been documented by Bateson (1983) to be important in mate choice.

In this scenario, the new trait is not immediately tied to male quality in terms of either health or general physical condition; however, over time such a link may develop. One way for this relationship to be initiated is via male–male competition. Mutant males that most fully utilize (develop or exhibit) the new trait will attract a portion of the female population. This almost automatically means that the mutant males will encounter and be in conflict with other males; i.e. males that are attractive to females will also have other males in their vicinity. The more attractive the mutant male, the more intense male–male competition will be. Of course, male–male competition also continually tests the condition of non-mutant males as well.

The proximity of the mutant male to females that it attracts will in itself lead to increased challenges from other males. As a result, only those males with the attractive new trait *and* with the ability to compete successfully with other males will be able to take full advantage of the drawing power of the new trait. Thus, any of many different kinds of newly mutated and attractive traits will also come to be correlated with a male's physical condition relative to other males. That is, newly mutated ornaments or display behaviour can become 'honest' indicators of male physical condition via male–male competition. The suggestion here, then, is that the increasingly tight relationship between a newly-appeared and originally arbitrary trait and the evolutionary development of its function as an honest indicator of male condition or quality is the result of concurrent female choice and male–male competition. The essential point is that male–male competition, either directly or indirectly, may play an important role in creating the initial link between an attractive new ornamental trait and its eventual evolved function as a reliable signal of condition. This scenario stresses (1) an arbitrary, non-functional origin of a new trait via mutation, and (2) transformation of that trait into an accurate reflection or indicator of male health or vigour.

Possibly the best empirical evidence to date for this hypothetical sequence of events is the male zebra finch, for which the addition of red bands confers both sexual selection and natural selection benefits to its bearers (Burley 1986*b*, 1988). This result supports the idea that a male trait attractive to females may generate its own costs and benefits, independent of its physiological basis (i.e. although the red bands do not increase the intrinsic quality of the male, female zebra finches perceive red-banded males as being of high quality, and they respond to such mates with increased parental effort). Recall that this suggestion reverses the sequence of events postulated by Fisher (1958). Fisher suggested an initial phase of functional mate choice, followed by arbitrary female preferences for more and more elaborately decorated males—his 'runaway selection'. However, such an arbitrary phase of attraction to new or novel traits would probably not last long (Kirkpatrick and Ryan 1991), especially if the condition of males bearing the new trait are tested by male–male competition. Once a link develops between the new trait and male vigour, then female choice is based, more or less inevitably, in part on the male's success in male–male competition. When the tie is established between indicators of male quality and female choice, a good genes basis for female mate choice has evolved.

This situation can continue indefinitely, for as long as the trait accurately reflects heritable variation in male quality. If, at some later time, the trait in question no longer conveys such information about males, another trait, possibly initially attractive to females for purely aesthetic reasons, may come to provide functional information, as described above, and the scenario can be repeated. My intent is to suggest one possible mechanism—competition among males—by which an originally non-costly trait could come to be correlated with male quality. Once a new trait comes to reflect male quality, variation in development of the trait can reflect variation in quality (e.g. the turkey gobbler's snood; Buchholz 1995, 1997).

Hill (1994*b*) presents a model to account for continued elaboration of a functional ornamental trait via pressure by females on high-quality males to provide honest signals, and on low-quality males to express those signals dishonestly (Fig. 6.5). The issue, in brief, is this: how does evolution proceed to produce a trait or traits that are extremely elaborated? Hill argues that extreme development of ornamental traits will occur only if males (specifically males in something less than tip-top condition) can come up with a means of reducing the cost of character expression (i.e. if such males can 'cheat'); if this does not take place, simple traits can remain reliable indicators indefinitely. Once all males, both high-quality and low-quality individuals, are able to produce equivalent traits, selection on females to be discriminating will strongly favour further elaboration of either the same trait or a change of focus by females to a new trait that more reliably signals quality.

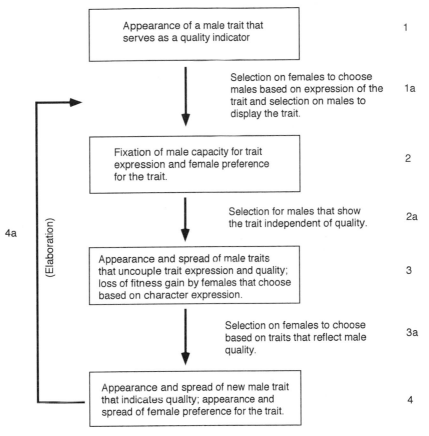

Fig. 6.5 A model of trait elaboration in males via sexual conflict. Evolutionary events are within the boxes and the forces of change are indicated by the arrows. Reprinted with permission from Hill, G. E. (1994*b*). Trait elaboration via adaptive mate choice: sexual conflict in the evolution of signals of male quality. *Ethology Ecology and Evolution*, **6**, 351–70.

High- and low-quality males compete for females, in part, via ornaments. It pays a high-quality male to produce signals honestly indicating cost, which forces low-quality males to attempt to match those signals. Thus, the solution for high-quality males is to produce signals that low-quality birds cannot copy. It is the females' demand for accurate indicators of male quality, plus the ability of lower quality males to find ways to produce signals indicating (falsely) their high quality, that drives either continued elaboration of traits or a shift from use of one trait to another.

According to Hill's model, if a trait is not costly, it should not be important in female choice (although it may have been in the past). This interpretation can help to account for that fact that females of species such as the red jungle-fowl apparently do not attend to ornate plumage of males when making mate choice decisions (Ligon and Zwartjes 1995a, Ligon *et al.* 1998), while females of other species, such as the house finch (Hill 1990, 1991), do so. This difference is based on the origins and costliness of male plumage colour. As described previously, the plumage of red junglefowl is coloured by melanins produced within the male's body and it develops fully in the complete absence of testosterone. Moreover, male junglefowl can produce the normal conspicuous ornate plumage even on a low-quality diet (Zuk *et al.* unpublished). In contrast, plumage colour of male house finches is highly dependent on environmental factors.

Hill's perspective also allows one to make predictions about the role of particular traits in sexual selection. As an example, consider the northern cardinal, a common and strikingly dichromatic species. Although the red plumage of male cardinals is based on a carotenoid pigment, plumage coloration may not be important in female choice if the particular carotenoid responsible for the red colour is very common. The fact that all young males (insofar as known) moult into a completely red plumage when only a few months of age suggests in itself that in this species the red plumage coloration may not be very costly. Alternatively, it is possible that subtle variation in the red colour of males may signal condition, as in the house finches. This remains to be determined.

6.6   Conclusions and summary: an attempt to integrate the hypotheses

As documented in this chapter, many studies have made it clear that female birds have the ability to discriminate between males on the basis of one or more ornamental and/or behavioural traits, even though all males of a given population or species possess the same basic set of releasers (e.g. colourful plumage or other ornaments, which are often associated with stereotyped displays). The critical ultimate issue, however, has been whether the discriminating females gain fitness advantages by obtaining good genes for their offspring. Although many biologists have found it logical to assume that females can gain such benefits, obtaining unequivocal evidence that this hypothesis is correct has proved to be difficult, in large part because good genes, runaway, and sensory bias scenarios make many of the same predictions; for example, all predict directional selection on male ornamental traits (Ryan and Keddy-Hector 1992).

Moreover, even if female preference for a trait is initiated in a purely Fisherian fashion, which can happen in theory (Lande 1981), it will almost inevitably come to be affected by the relative physical condition of males. To illustrate this point, consider two male birds that are genetically identical with regard to alleles for long tails. Feather development is known to be influenced by nutritional state or by general health; thus, unless the two males are also identical in physical condition, one will probably grow a somewhat longer tail than the other, due to the greater amount of energy it can afford to put into tail growth. Fisherian females choosing long-tailed males are thus also choosing healthier males, and their offspring receive the benefits of having a healthy sire. Once a connection is made between a trait such as tail length and the health of its bearer, in this or any other fashion, then good genes selection is off and running.

It is possible, and perhaps probable, that more than one mechanism (direct selection, good genes, Fisherian runaway, sensory bias) has played a role in the evolution and maintenance of male display characters of contemporary birds. If the relative importance of each of these processes varies from species to species at a given point in time, it is altogether unlikely that any one of the major theoretical hypotheses will ever provide a satisfactory or complete, across-the-board explanation for mate choice in female birds.

In the past, the question of female choice led to something of a polarization of opinions (Bradbury and Andersson 1987): *either* females make an adaptive, or functional, choice based on some aspect of the quality of males, i.e. good genes, *or* the basis for female choice is arbitrary (e.g. Fisherian runaway), with no relationship assumed between female choice and male quality apart from the relative development of the trait or traits of males to which females respond. It appears, however, that the debate over the ultimate significance of male ornamentation may be, in part, an example of a 'levels of analysis' problem (Tinbergen 1963; Sherman 1988) in that two separate issues may be involved. First, there is the question of the *origin* of ornamental characters, and second is the question of the *current* adaptive or functional significance of the male ornaments, which may or may not tell us anything about their origins.

The origins and genetic control of female preferences are among the most fundamental issues in sexual selection and the most difficult to address empirically (see discussion by Borgia 1987). This is largely because the species and traits that we see today originated in the distant past and subsequently have been modified many times. At best, we can make inferences based on comparative studies involving circumstantial evidence. The origins of certain ornamental traits could originally have been based on arbitrary choice, for example, while currently selection for 'good genes' may be operating—or vice versa, as originally suggested by Fisher (1958). Moreover, to add to the difficulty of resolving these issues, many ornamental traits that may have been important formerly in female mate choice decisions may be retained, but no longer used by females (e.g. Møller and Pomiankowski 1993*a*; Ligon and Zwartjes 1995*a*; Ligon *et al.* 1998).

Thus, there is no logical necessity to characterize arbitrary or aesthetic (runaway and sensory bias) female choice at some point in time prior to the present and good genes (current use of certain male traits by females as indicators of male physical condition or quality) as mutually exclusive alternatives. In fact, it appears that either could have been important in the initial development of new male traits and preferences of females for them, following the appearance of the traits by mutation. That is, the first or initial use by females of a newly evolved male trait could have been either functional (i.e. provided information about the quality of the male), as described by Fisherian and good genes models, or arbitrary (females were attracted for non-functional reasons, e.g. sensory bias).

This suggestion is not merely an attempt to end on a conciliatory note with regard to the merits of one theory versus another. Instead, each major hypothetical phenomenon—good genes, Fisher runaway (ultimate explanations), and sensory bias, including 'aesthetic' choice (proximate explanations)—may have played a role in the development of traits and preferences exhibited by contemporary birds. The evidence also indicates that for whatever reason a new male trait initially becomes attractive, choice by females eventually becomes functional. That is, over time females come to focus on male traits indicative of quality or condition. When that takes place, further trait elaboration can occur, as described, for example, by Hill (1994*b*).

7 Phylogenetic studies of reproductive patterns

7.1 Introduction

The classic studies by Lorenz (1941) on ducks first demonstrated that behavioural traits of birds may conform closely to morphological ones, and that behaviour, like morphology, can reflect phylogenetic history. A recent example is seen in the study by Kennedy *et al.* (1996) of the relationship between social displays and phylogeny in the Order Pelecaniformes. David Winkler and Frederick Sheldon (1993) make this point in a novel way by demonstrating that nests of swallows parallel their phylogeny (Fig. 7.1). As a group, swallows show a great diversity of nest types, from burrows to cavities to nests built of mud. Winkler and Sheldon superimposed nests on a phylogeny based on DNA-hybridization and found that nest type is tightly linked to the inferred evolutionary history of swallows. Nest sites and modes of construction are a physical manifestation of complex, largely stereotyped behaviour, and in this respect they are similar to the courtship displays of many birds.

Phylogenetic studies that focus on sexual selection and mating systems are still rare. Nevertheless, one should keep in mind that all traits have

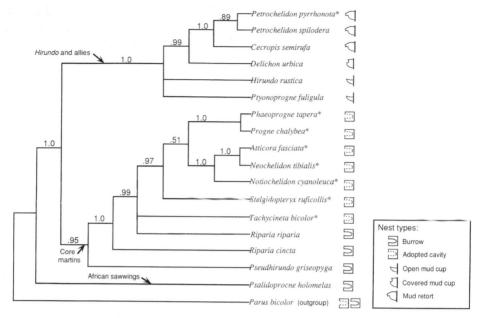

Fig. 7.1 Phylogenetic patterns of nest construction in swallows, Family Hirundinidae, based on DNA hybridization. Nest type and phylogeny closely correspond. Reprinted with permission from Winkler, D. W. and Sheldon, F. H. (1993). Evolution of nest construction in swallows (Hirundinidae): a molecular phylogenetic perspective. *Proceedings of the National Academy of Sciences, U.S.A.*, **90**, 5705–7. © 1993 by the National Academy of Sciences.

precursors, and that to more fully appreciate the reproductive patterns of contemporary species and groups of species, we should be aware of their evolutionary history. The ancestors of contemporary birds were just as complex as living species, and over much of the evolutionary history of this group, diversity was probably considerably greater than that exhibited by today's avifauna (Brodkorb 1960; Feduccia 1996). In brief, all aspects of the biology of birds, including the main topics of this book—sexual selection and mating systems—have a long and complex history. No study of such topics can be complete in the absence of any consideration of the history of the lineage or the trait in question. Of course, study of the behaviour of long-extinct birds is, in a strict sense, impossible. However, modern phylogenetic approaches provide important insights into some of the behaviour of the ancestors of today's species. In certain cases they can also show the evolutionary pathway and progression of behavioural traits. Two phylogenetic analyses discussed below that produced unexpected and fascinating results are the history of social parasitism in the cuckoos and sexual dichromatism in New World blackbirds.

Until very recently, most avian behavioural ecologists attempted to account for the patterns they studied by current behavioural, ecological, and demographic variables. As pointed out by Richard Prum (1994, p. 1657), 'Most models of the evolution of social behaviour, particularly in vertebrates, are essentially ahistorical.' However, as he and Steve Zack (1995, p. 37) have noted, 'History matters'. In a review of cooperative breeding in African shrikes of the Genus *Lanius*, Zack (1995) provides a valuable example of the perils of ignoring phylogenetic history (see Section 14.4.1.2). In this short chapter, I briefly review a number of phylogenetically-based studies that have explicitly explored the history of sexually-selected traits and other aspects of avian reproductive biology.

7.2 *Sexual selection and phylogenetic history*

The studies reviewed in this section deal with ornaments of male birds that are thought to be important in sexual selection, particularly female mate choice.

7.2.1 Courtship displays and social organization in manakins

The New World manakins exhibit some of the most spectacular and complex display behaviour known for birds. Prum (1990) investigated the stereotyped courtship behaviour of male manakins and confirmed a close relationship between behaviour and phylogeny. Prum's phylogenetic approach nicely complements those of other students of manakin courtship behaviour (e.g. Lill 1974, 1976; Foster, 1977, 1981; McDonald 1989*a,b*). Some of Prum's conclusions are of unusual interest. In several cases, derived male plumage traits evolved subsequent to the behavioural displays with which they are associated. For example, all four species of the Genus *Chiroxiphia* perform a coordinated cart-wheel display involving two or more males; this behaviour apparently evolved prior to the speciation events leading to the four contemporary species. However, in only one, the long-tailed manakin, has an extreme morphological trait—the long tail—evolved that clearly is associated with the cart-wheel display.

In a second study, Prum (1994) used a phylogenetic approach to investigate the evolutionary history of manakin social behaviour associated with lekking (see Chapter 15). Prum's analysis of the phylogenetic distribution of coordinated display behaviour indicates that it had five evolutionarily independent origins. In contrast, he concluded that lek breeding evolved only once in the common ancestor of all manakins, and that the few non-lekking contemporary species are derived from lekking ancestors. In addition, the spatial organization of the leks indicates that they have a strong phylogenetic, historical component. Lekking behaviour apparently evolved prior to the evolution of the morphological and behavioural displays that make manakin leks so striking both to female manakins and to human observers.

Prum suggests that the evolution of the suite of characters exhibited by males (e.g. leks, colourful plumage, exotic displays) placed constraints on the evolution of other reproductive strategies; i.e. an evolutionary transition from lek breeding to pair bonding is unlikely (see also van Rhijn 1990). One species, however, has secondarily lost lek-breeding behaviour. In the helmeted manakin, the male defends a nest territory, exhibits an extended pair bond, and may be monogamous. This departure from the usual pattern may be related to the kind of habitat this species occupies (Marini and Cavalcanti 1992; Prum 1994).

One of the most intriguing of Prum's (1994) conclusions is that cooperative 'team' display evolved five times in manakins. Thus, there appears to be some sort of genetically based predisposition within this lineage for the collaboration of males at leks.

7.2.2   Body size and sexual size dimorphism in blackbirds

Björklund (1991) investigated the evolutionary patterns of body size in grackles of the Genus *Quiscalus*. In particular, he studied tarsus length and tail length as they pertain to sexual selection and mating systems. Two species of *Quiscalus* are highly polygynous, four are 'regularly' polygynous, and two are socially monogamous. The question Björklund investigated is: has the great degree of sexual dimorphism in body size and tail length in the two highly polygynous species evolved as a result of their mating system? Björklund's phylogenetic analysis led him to conclude that there has been directional selection for long tails in males of these species, presumably due to female mate choice in a highly polygynous system. In contrast, Björklund suggests that the large body size of the two highly polygynous species is not the result of sexual selection and could even be the result of drift (see also Björklund 1990).

In a subsequent study of the relationships between sexual dimorphism, mating system, and body size in New World blackbirds, Webster (1992) evaluated a series of hypotheses developed to explain the relationship between body size and sexual dimorphism. Webster's analyses led him to reject Björklund's (1991) counter-intuitive suggestion that the correlation between body size and mating system was not due to sexual selection. Webster concluded that larger body size is causally correlated with mating system, either directly or indirectly (i.e. polygyny and/or coloniality lead to the evolution of large body size in both sexes), and that sexual dimorphism likewise is related to sexual selection, as well as to phylogeny and general body size.

7.2.3   Vocal repertoires in blackbirds

In a study of male song and female preferences in the common grackle, Searcy (1992) determined that although an individual male sings only one of the four male song types, female grackles injected with oestradiol to stimulate sexual responsiveness prefer an artificial song repertoire composed of all four types. Searcy suggested that this preference may be due to two general ethological phenomena—habituation and stimulus specificity (see Chapter 3). That is, females soon habituate to repeated playback of a single song type, which leads to a continuous decrease in response of females. In contrast, during repeated playback of repertoires containing four song types, decreases in response are reversed (responses increase) each time song types are switched. It is these recoveries that cause overall response to be greater for the repertoires than for the single song types. Searcy (1992, p. 71) concludes that the '... female preferences for repertoires may in general pre-date the evolution of song repertoires, which evolve to exploit the pre-existing female response bias.'

Irwin (1990) investigated the relationship between phylogeny, sexual selection, and song repertoires by assessing the pattern of mating system and repertoire size in six presumably monophyletic groups (clades) of blackbirds. Song repertoire size varies within each group, thus phylogenetic effects cannot explain this variation. Moreover, song repertoire is not associated with either mating system (e.g. polygyny) or with presence or absence of territoriality. Irwin's (1990) phylogenetic analysis suggests that Searcy's postulated sequence of evolutionary events may be backwards. The closest relatives of the monogamous common grackle are polygynous and have larger song repertoires. Thus, it appears that monogamy in this species may have evolved from polygyny and, concomitantly, that male song repertoire may have decreased to a single song type; i.e. both monogamy and the simple song type are apparently derived characters. If so, then current preference for larger repertoires by female common grackles may reflect an evolutionary history of more complex male song repertoires.

Additional factors should also be considered. First, to a sexually responsive female grackle, playback of all four male song types may represent, in effect, the presence of at least four singing males. Females often nest colonially in a single tree that contains a number of displaying males. It is possible that in nature the presence of several males may be required to attract a female; i.e. females may prefer to join groups of males that collectively sing all four song types. David Gray and Julie Hagelin (1996) also suggest that in common grackles, the song of several males may be more stimulating than a single male's song. They point out that pair formation takes place at singing assemblages and during group flights, which are initiated when a female departs and is followed by several males. These group flights are extensions of the activity of singing groups (Wiley 1976a,b). To help resolve this issue, the extent of multiple paternity within broods of common grackles, as well as the importance to the colonially nesting females of the presence of several or many males, needs to be determined.

7.2.4   Plumage dichromatism in blackbirds

It has been generally, if usually tacitly, assumed that the strong sexual plumage dichromatism that characterize many groups of birds (e.g. New World orioles, tanagers, wood warblers, etc.) have evolved as a result of sexual selection, with females preferring brighter males (Irwin 1994). An accompanying assumption is that the 'ancestral' condition was plumage monomorphism, with both sexes being dull in colour. However, the direction of evolutionary changes leading to the plumage characteristics of today's species has rarely been investigated. It is possible, for example, that, with regard to plumage colour and sexual dimorphism, the ancestral condition of some contemporary species with bright male plumage and conspicuous sexual dichromatism was not dull plumage in both sexes; that is, the bright plumage of males of contemporary sexually dimorphic species may not have evolved from monochromatic ancestors in which both sexes possessed dull plumage.

A phylogenetically based study of plumage dichromatism in the New World blackbirds provides some of the first evidence in support of the idea that sexual dichromatism may have evolved from sexual monochromatism (Irwin 1994). This is not surprising; what may be surprising, however, is that the ancestral condition in five of six clades of blackbirds appears to have been monomorphic bright plumage, with the subsequent evolution of dull, apparently cryptic, female plumage. What might the basis be for this counter-intuitive conclusion? Irwin suggests that bright female plumage, as in the oropendulas and caciques, might function in social interactions; however, this does not account for the loss of bright plumage in females of many other species, such as grackles and temperate zone blackbirds (e.g. red-winged blackbird). In these species, females are solely responsible for incubation, and, unlike female oropendulas and caciques, they incubate in open cup nests where they may be visible from above. Thus, it may be that natural selection has strongly favoured cryptic plumage of females in these latter species, while in the former, other selective agents have favoured plumage brightness in females.

Irwin emphasizes that her conclusions do not deny that sexual selection on male brightness may be occurring in these birds. (Plumage dichromatism is, in fact, greater in polygynous than in monogamous species.) What Irwin's conclusions do indicate is that plumage dichromatism and male brightness in icterine blackbirds is the result of a counter-intuitive evolutionary pathway that would not have been suspected in the absence of phylogenetic analyses.

Female red-winged blackbirds in Cuba, for example, are black like males, although they lack the red epaulettes (Whittingham *et al.* 1992; see also Section 4.7.1). This raises two possibilities. First, with regard to sexual dichromatism/monochromatism, Cuban red-wings may represent the condition from which North American red-wings are derived. Alternatively, the Cuban red-winged blackbird lineage may have evolved from monochromatic bright to strongly dichromatic (Irwin 1994), as seen in continental North America, and back to partially monochromatic bright.

7.2.5 Size dimorphism in lekking birds

Other studies of birds also provide phylogenetic approaches to mating systems. Because lek mating systems (Chapter 15) are thought to produce the most extreme variance in male mating success, and thus the strongest sexual selection, Payne (1984) postulated a relationship between lekking and sexual dimorphism. Höglund (1989) investigated this idea by examining the relationship between size and sexual dichromatism and lck mating systems. He concluded that there was no general relationship between either of these morphological traits and lekking, although the relationship was present in some specific groups of lekkers. Subsequently, using an alternate methodology, Oakes (1992) reported a significant correlation between lekking and size dimorphism, leading him to conclude that lekking does in fact promote the evolution of size dimorphism.

Höglund and Sillen-Tullberg (1994) re-analyzed the data on body size of lekking species and their non-lekking relatives. They concluded that while lekking species of some groups do show greater size dimorphism, overall, lekking species are no more size dimorphic than their non-lekking close relatives. Höglund and Sillen-Tullberg (1994) point out that sexual dimorphism is great in many kinds of birds, and may reflect strong sexual selection in both lekking and non-lekking social environments. All of these authors recognize that male birds have many means of display and that, while male–male competition often involves body size, there are species and situations in which larger size is apparently of no advantage to males. For example, Payne (1984) points out that small male size may provide competitive advantage to males of highly aerial species, such as hummingbirds.

7.3   Social parasitism and phylogenetic history

Phylogenetic studies have also contributed importantly to the subject of brood or social parasitism. Obligate brood parasitism is a rare phenomenon, occurring in one or more species of five traditional families of birds (black-headed duck, many cuckoos, honeyguides, cowbirds, whydahs, indigobirds, and the parasitic weaver). Within a family its occurrence ranges from a single species (e.g. black-headed duck) to every species making up the family (e.g. all 17 species of honeyguides are parasitic, insofar as known). Evolutionary histories of social parasitism in two of the best-studied groups of social parasites, the cuckoos and the cowbirds, are briefly considered here.

7.3.1   Cuckoos

It has long been thought that obligate parasitism had evolved independently in two different Subfamilies of cuckoos—the Cuculinae of the Old World (Fig. 7.2) and the Neomorphinae of the Neotropics—with facultative parasitism also occurring occasionally in the black-billed and yellow-billed cuckoos of North America (a third Subfamily, Phaenicophaeinae). Recently, however, Hughes (1996) presented a new phylogeny of cuckoos which drastically alters this picture. Hughes' analysis suggests that the neotropical

Fig. 7.2 The cuckoo of Europe is the world's most famous avian social parasite; most other Old World cuckoos, such as the great spotted cuckoo, are also parasitic. Reprinted with permission from Dover Publications, Inc.

parasitic species are actually part of the 'Old World' assemblage of parasitic cuckoos, as are the black-billed and yellow-billed cuckoos. Given the overall rarity of brood parasitism among birds, this scenario leads to the intuitively more appealing conclusion that brood parasitism evolved only once, rather than three times, in this family (Fig. 7.3). A single evolutionary origin of

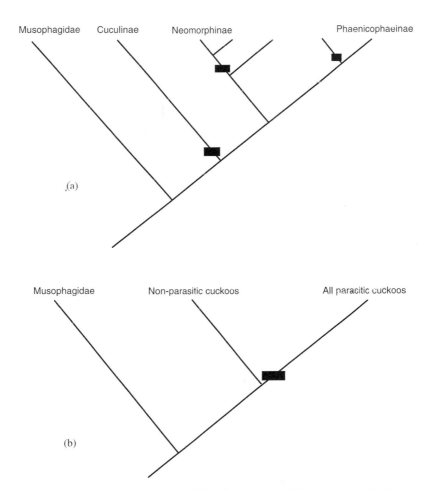

Fig. 7.3   Cladograms of brood parasitism in cuckoos. The outgroup is the touraco group, Family Musophagidae. (a) Earlier phylogenies of the cuckoo family indicated that obligate brood parasitism had evolved independently in two subfamilies, and occurs facultatively in a third: i.e., all 16 genera of the Cuculinae, two genera of the Neomorphinae, and *Coccyzus* of the Phaenicophaeninae. (b) A new phylogeny produced by Hughes (1996), based on combined osteological, behavioural and ecological data. Hughes' study suggests that the Old and New World parasitic species are actually most closely related to each other, and thus that brood parasitism evolved once, rather than three times, within the family.

parasitism in a single family certainly appears to be a more parsimonious conclusion than one in which this trait evolved independently three times.

Perhaps the most surprising aspect of Hughes' analysis of the evolutionary history of social parasitism in cuckoos is her conclusion that the black-billed and yellow-billed cuckoos are descended from an obligately parasitic ancestor and have, over evolutionary time, lost the parasitic habit. This interpretation contrasts strongly with the more usual assumption that certain traits exhibited by these species (e.g. occasional egg-laying before the nest is completed, extremely short incubation period, and short nestling period as compared to non-parasitic cuckoos) are 'preadaptations' to an obligately parasitic life style in the future (Hamilton and Orians 1965). Hughes' (1996) interpretation is that these traits reflect the parasitic reproductive strategy of the ancestor of the black- and yellow-billed cuckoos. This perspective seems less forced than attempting to explain the future adaptive significance of what clearly are maladaptive traits shown by these species (e.g. laying eggs before the nest is constructed).

7.3.2 Cowbirds

In the cowbirds, brood parasitism has long been assumed to be the result of a single evolutionary event (e.g. Lack 1968). Using the cytochrome-*b* gene, Scott Lanyon (1992) conducted a phylogenetically-based analysis of parasitism in this group and corroborated this assumption. Lanyon also determined that the primitive form of parasitism in cowbirds is host specificity, with host generality representing a derived condition. The brown-headed cowbird, the epitome of a generalist, with 216 species recorded as hosts, is also the most derived member of the assemblage. This result contradicts the view that as host species develop defence mechanisms against parasitism, the parasites should specialize on fewer hosts. While this may be correct for some other parasitic lineages, such as many Old World cuckoos, in which the defences of the hosts and the responses of the parasites to them seem to be more specialized, the opposite seems to have occurred in brown-headed cowbirds.

Lanyon (personal communication) has suggested that the differences between the cuckoos and cowbirds in the parasite–host relationship is based, in part, on the differences in the time elapsed since the evolution of social parasitism. Brood parasitism clearly is an extremely ancient trait in cuckoos. In contrast, it has evolved comparatively recently in cowbirds. Some cuckoos may have been parasitizing other species for millions of years before the emberizid lineage to which cowbirds belong had even evolved. Extending this line of reasoning, one might predict that in the fullness of evolutionary time, as hosts respond more effectively to the threat of parasitism, the brown-headed cowbird lineage eventually will, like many cuckoos, become more host-specific.

7.4   Cooperative breeding and phylogenetic history

It has become increasingly apparent that the taxonomic distribution of co-operative breeding among extant species has a strong phylogenetic component. This realization stands in stark contrast to the long-held view that cooperative breeding is a response to particular environmental conditions (see Chapter 14). Studies by Russell (1989) were the first to indicate that phylogenetic history has had a major impact on the taxonomic distribution of contemporary cooperative breeding systems. Cooperative breeding is relatively common in Australia. Russell (1989) points out that two major passerine lineages occur on that continent, the Parvorders Corvi and Muscicapae of Sibley and Ahlquist (1985). The evolutionary radiation of the 'old endemic' Corvi is thought to have taken place in Australia, while the species of Muscicapae present there are comparatively recent arrivals. Despite the frequent occurrence of cooperative breeding in Australia, not a single Australian member of the Muscicapae breeds cooperatively, whatever habitat type is occupied (Ford 1989, p. 147, Table 32). All of the 57 passerine cooperative breeders in Australia are part of the old endemic radiation. This separation of cooperative and non-cooperative species along phylogenetic lines is striking in view of the fact that environmental factors in present-day Australia have long been thought to promote the evolution of cooperative breeding (e.g. Rowley 1974; Ford *et al.* 1988).

Another analysis of the phylogenetic distribution of cooperative breeding by Edwards and Naeem (1993) provided additional new insights. Many of the best-known species of cooperative breeders belong to taxonomic groupings (e.g. genera or even subfamilies) where most or all species in the group breed cooperatively. Edwards and Naeem (1993) analysed the occurrence and distribution of cooperative breeding in 71 polytypic genera that contained at least one cooperative breeder, and compared its incidence in each genus with a random distribution among these genera. They did not deal with the various forms of cooperative breeding; therefore their analysis includes both species in which it is obligate and those in which it is incidental or opportunistic (Dow 1980).

Edwards and Naeem's results indicate that the most parsimonious assumption is that cooperative breeding in several lineages (e.g. genera) rose prior to many of the speciation events that occurred within those genera, rather than arising independently or *de novo* in each contemporary species (see also Peterson and Burt 1992). Thus, it would be unwarranted to make the assumption (which has been made repeatedly) that the ecology of any given contemporary cooperatively-breeding species will account fully for the evolutionary origin of cooperative breeding in that species.

In some ways, Edwards' and Naeem's study complements the views of Jamieson (1989*b*) and Jamieson and Craig (1987) concerning the origins of helping behaviour. The stimulus–response interaction of feeding nestlings must be ancient, having evolved early in the evolutionary history of birds, and possibly preceding the evolution of altricial nestlings. This implies that all species with altricial young possess the neural 'hard-wiring' necessary for development of either parental or helper feeding behaviour. Because of the ubiquitous nature of the feeding response and its presumed early origin in the evolution of birds, it is easy to see that the feeding of nestlings by non-breeders could have pre-dated speciation events (see also Peterson and Burt 1992), and could have been transmitted over evolutionary time as speciation occurred, just as if it were a morphological character.

Australian fairy-wrens can be used to illustrate the essential points of Russell (1989) and Edwards and Naeem (1993). Fairy-wrens belong to one of the 'old endemic' groups of cooperatively breeding Australian birds (Sibley and Ahlquist 1985; Russell 1989). All species in this widespread genus are thought to be cooperative breeders (Rowley *et al.* 1988). Despite the fact that contemporary species of fairy-wrens replace each other to a large extent geographically or by habitat, and despite striking differences in the plumage of males of some species, all the fairy-wrens appear to be fundamentally similar in both ecology and social behaviour. All are weak-fliers that forage extensively on the ground, traits also commonly found in many other group-living and cooperatively-breeding species (Stacey and Ligon 1987; Ford *et al.* 1988). Thus, it is reasonable to assume that the behavioural traits promoting or permitting group-living were present in the common ancestor of today's species. In short, the phylogenetic history of the old endemics of Australia provides a partial explanation for the existence of helpers in contemporary species of fairy-wrens. Additional examples of the role of phylogenetic history in the development of contemporary cooperative breeding systems are provided in Chapter 14.

7.5   *Mating systems and altricial and precocial young*

Another example of the usefulness of phylogenetic analyses in attempting to understand reproductive strategies of birds is a study by Temrin and Tullberg (1995). These authors investigated the influence of the condition of young at hatching—altricial or precocial—on length of pair bond and mating patterns. Although their categorization incorporates some fundamentally different social–reproductive phenomena (e.g. cooperative and classical polyandry, see Chapters 13 and 16), Temrin and Tullberg make several points of interest. (1) Although most avian species are altricial, precociality is the ancestral state in birds. (2) Transitions to short pair bonds are significantly more frequent in lineages with precocial young than in those with altricial young. (3) Transitions to female polygamy (polyandry, both cooperative and classical) is significantly more frequent in birds with precocial young than in birds with altricial young. (4) Transitions to male polygamy (polygyny) do not differ significantly between altricial and precocial species. (This last comparison is statistically significant only if passerines are excluded and if the maximum possible number of transitions are used. All passerines are altricial, and this group makes up almost 60 per cent of all living bird species.) Most of these points have been intuitively appreciated previously; however, this phylogenetic analysis provides an additional explanation for the evolution of certain mating patterns.

7.6   The origin and evolutionary history of bowers

The bowers of bowerbirds are a truly unique form of male display under the influence of strong sexual selection. From the ethological perspective, the evolutionary origin of bowers can provide insights into the origins of more typical behavioural displays (see Section 3.5). With regard to the origins of bower-building behaviour and bowers, there have been two schools of thought. One of these, the 'nest hypothesis' (discussed by Borgia *et al.* 1985), suggests that bowers represent courtship nests necessary to bring females into full reproductive readiness. Diamond (1982, 1986*b*), a modern-day proponent of the nest hypothesis, presents its essential point as follows:

> Males of many bird species court females and induce them to ovulate by constructing true nests and offering food. It is possible that bowers began as courtship nests that became free of size constraints as the function of egg incubation was transferred to a separate nest built by the female alone. (Diamond 1982, p. 101.)

Borgia *et al.* (1985) refute the nest hypothesis by raising four issues pertaining to the possible relationship between bowers and nests. (1) No bower resembles the nest of any species of bowerbird. (2) Bower sites and nest sites are completely different. Bowers are located on the ground and all bowerbird species nest in trees. Thus, bower sites do not resemble nest sites. (3) The decor of bowers does not resemble nests. No bowerbird decorates its nest. Nor is any species known to exhibit courtship feeding, thus there is no reason to assume that decorative items represent symbolic food. (4) Display at a nest by male bowerbirds is not required for stimulation of female reproduction. In the bower-building species of bowerbirds, nest-building is carried out by the female alone and it occurs before males and females associate.

Rather than being derived from nests, Gilliard (1963, 1969) and Borgia *et al.* (1985) suggest that bowers were preceded evolutionarily by cleared, undecorated courts, and that over time the courts were elaborated by the development of bowers. In addition, one species, the tooth-billed bowerbird, clears an arena or display ground on the forest floor and decorates it with fresh leaves turned upside down. In this species, males are promiscuous and females care for eggs and chicks without male assistance (Frith and Frith 1993). Three other members of the bowerbird family, the 'catbirds', are more like typical passerine birds in that they are thought to be socially monogamous with biparental care, and they do not display at arenas.

7.6.1   Possible origins of bowers

Evolutionary studies may shed additional light on the origins of bowers. Based on their DNA–DNA hybridization studies, Sibley and Ahlquist (1985) suggested that bowerbirds are most closely related to the lyrebirds (Fig. 7.4) and scrub-birds (traditional families Menuridae and Atrichornithidae) of Australia. With regard to the question of the evolutionary origin of bowers, this conclusion is tantalizing. Similarities between bowerbirds and lyrebirds in courtship and reproductive biology include the following. (1) Both lyrebirds and the non-bower-building tooth-billed bowerbird, as well as the bower-building species, clear display areas on the ground. (2) In both lyrebirds and nearly all bowerbirds, males are promiscuous and provide no parental care. (3) Lyrebirds and at least some bowerbirds have loud, prolonged song. More importantly, mimicry (which is generally an uncommon form of display in birds) is a major component of overall male display in both. (4) Plumage maturity requires several years both in male lyrebirds and males of those bowerbirds that have been studied (Lill 1979; Vellenga 1980; Pruett-Jones and Pruett-Jones 1983). (5) Sexual dimorphism in body size is slight in lyrebirds, as is also the case for bowerbirds. At first glance, this might appear to be surprising, because, among ground-dwelling species, a strong positive relationship generally exists between degree of size dimorphism and degree of polygyny/promiscuity (Payne 1984). Although the similarities between bowerbirds and lyrebirds listed here, which are unusual among passerine birds, could be due to convergence, according to the phylogenetic conclusions of Sibley and Ahlquist (1985), they could also reflect common ancestry.

Assuming that lyrebirds and bowerbirds are indeed related, both the similarities and differences in their display characters can be viewed as support for Gilliard's idea of the evolutionary transferral of structures attractive to females from the male itself to the bower. For example, the common ancestor of lyrebirds and bowerbirds may have displayed at a cleared arena on the ground, with the lyrebird lineage evolving extreme plumage ornamentation and complex song with mimicry. The subsequent evolution of most bowerbirds, on the other hand, led to a unique alternative form of 'ornamentation', the bower and its decorations, along with complex song and mimicry.

The suggestion that cleared courts preceded the development of bowers does not, however, respond to the question of why it is that bowers came to be used by males in courtship in this group of birds, rather than some other behavioural or morphological traits (e.g. as in lyrebirds). Once courts began to enhance male attractiveness to females, mechanisms for further increasing their attractiveness (e.g. placing items in the court) probably began to

Fig. 7.4   Lyrebirds may be closely related to bowerbirds (Sibley and Ahlquist 1985, 1990). Although bowerbirds and lyrebirds do not look alike, many aspects of their courtship, including prepared display sites on the ground, are similar (see text). Reprinted with permission from Dover Publications, Inc.

evolve in the ancestor of contemporary bower-builders, with the males most successful in developing and defending attractive bowers being favoured.

7.6.2   Phylogenetic patterns of bower construction

Kusmierski *et al.* (1993) used mitochrondial DNA cytochrome *b* to construct a phylogeny of bowerbirds, including those species that do not build bowers. These data were then compared with types of bower construction. Overall, there was impressive concordance between the cytochrome *b*-based phylogeny and type of bower construction (see Kusmierski *et al.* 1997 for a modification of this conclusion).

As mentioned above, males of the three species of catbirds are dull-plumaged, monogamous, and exhibit neither bower-building nor court-clearing behaviour. Kusmierski *et al.* (1993) suggest that the catbirds represent an ancestral condition. On the other hand, the cytochrome *b* data indicate that another species, the tooth-billed bowerbird, which does not build a bower, but which does clear a court and decorate it with leaves (see Section 3), diverged after the evolution of bower-building behaviour. Partly because the evolution of bower-building is such an unlikely event in the first place, Kusmierski *et al.* (1993) concluded that bower-building has been lost in the lineage leading to this species.

7.7   Conclusions and summary

In recent years it has become clear that a consideration of phylogenetic history should be included in any attempt to explain the evolution of morphological and behavioural traits associated with sexual selection and mating systems. As one example, phylogenetic analyses have provided a deeper understanding of both of these aspects of the reproductive biology of the neotropical manakins (Prum 1990, 1994) than could be done solely by use of comparative studies of contemporary species.

Phylogenetic studies may also stimulate new ways of looking at the evolution of behavioural characters. As one example of several summarized in this chapter, the long-standing view of brood parasitism in cuckoos is that it independently evolved twice, once in the Old World and once in the neotropics. In addition, it has been argued that many of the traits of two species that breed in the Nearctic, the black-billed and yellow-billed cuckoos, were precursors or 'pre-adaptations' for the eventual evolution of brood parasitism. However, this view has recently been turned upside down as a result of phylogenetic analyses which indicate that brood parasitism evolved only once in the cuckoos, and, moreover, that it has subsequently been lost in the black-billed and yellow-billed cuckoos (Hughes 1996).

Evolutionary trajectories in which traits of contemporary species are (apparently) less complex than those of their ancestors may prove to be rather common. In short, with regard to the direction of evolutionary change, it appears that phylogenetic analyses of avian breeding systems will continue to produce some fascinating surprises.

8 Sexual selection and speciation

8.1 Introduction

In this chapter, I consider the role of sexual selection in the process of speciation. This issue is an interesting one for some very different reasons. First, there are historical reasons, and second, the process of speciation is not so well understood or settled as many textbooks indicate. Indeed, Douglas Futyuma (1989) states that the subject of speciation is awash in unfounded and often contradictory speculation.

As mentioned briefly in Chapter 2, not long after Darwin elucidated the theory of sexual selection, and for many years thereafter, it was an ignored issue. West-Eberhard (1983) refers to this period as 'The forgotten era of sexual selection'. During this time, most of the leading students of evolutionary theory focused on issues related to speciation. Thus, this neglect of the topic of sexual selection occurred during the same period that the phenomenon of speciation—the key concern of neo-Darwinism—was receiving intensive study and thought. With the benefit of hindsight, it is a bit puzzling to realize that so many of the foremost students of speciation almost completely ignored sexual selection and apparently made no connection between the two phenomena.

During the 'forgotten era', however, two eminent ornithologists provided exceptions to this generalization. Sibley (1957) and Gilliard (1963, 1969) attributed the elaborate plumage in males of many groups of birds largely to sexual selection, rather than to the more popular interpretation referred to as 'species-isolating mechanisms' (see next section). These authors recognized that sexual selection was not only an important phenomenon in its own right, but that under some circumstances, it might also play an important role in the process of speciation. Sibley (1957) and Gilliard (1963) pointed out that decorative and presumably sexually-selected plumage and other ornaments are often very different in males of closely related species, and they suggested that sexual selection might be driving this kind of morphological divergence. Sibley (1957) also showed that extreme ornamentation can evolve in species long isolated geographically from any close relative and that, therefore, the evolved function of such plumage is unlikely to be for 'species isolation'. In considering the birds of paradise and bowerbirds, Gilliard (1969) suggested that two of the most conspicuous features of these groups—spectacular and bizarre courtship ornaments and displays—reflected the power of sexual selection, and he suggested that in 'arena' species (i.e. species in which males mate promiscuously and exhibit lekking behaviour and/or specialized male display sites; see Chapter 15), the pace of evolutionary change might be moving faster than in more typical socially monogamous species. Thus, both Sibley and Gilliard recognized that the bizarre courtship traits of groups of birds characterized by such mating systems were probably due to intense sexual selection.

Here I follow a classical biological definition of a species, i.e. 'Species are groups of interbreeding natural populations that are reproductively isolated from other such groups' (Mayr 1970, p. 12; see McKitrick and Zink 1988 for another view of species and speciation). The issue, however, is more complex than this straightforward definition might indicate. Among birds, there are many cases of interspecific hybrids. As one example, greater prairie chickens and sharp-tailed grouse are distinctive lineages that are sympatric (i.e. their ranges overlap), yet they interbreed in nature often enough that hybrids are not particularly rare; i.e. they are not completely reproductively isolated. Many similar examples exist, and the possible evolutionary significance of hybridization varies from case to case (see Section 8.5).

In addition to the questions raised by hybrids, when two similar populations exist in allopatry (i.e. in geographic isolation), there is the question of their specific status. For example, for many years the Florida scrub-jay and the scrub-jay of western North America were considered to belong to the same species (the implication being that if individuals from the two populations were to have the opportunity to interbreed, they would do so and would produce normal offspring. In fact, Florida and western scrub-jays have bred successfully in captivity (J. W. Hardy, personal communication). Recently, however, a review of the similarities and differences between individuals of three scrub-jay populations has led to the decision to recognize them as distinct species—the Florida scrub-jay, the western scrub-jay, and the island scrub-jay of Santa Cruz Island, California (American Ornithologists' Union 1995). This decision implies that if any two of these populations became sympatric, they would remain reproductively, and thus genetically, isolated.

In this chapter, I first review briefly the classical scenario for the multiplication of species that is generally thought to apply to birds. I then consider the possible role of sexual selection in the speciation process from three different aspects. These include genetic and phenotypic divergence of populations, the possible role of hybridization in speciation, and the possible interaction between geographic isolation, fluctuating asymmetry, and sexual selection in the process of speciation.

8.2   Speciation in birds: the classical scenario

Sibley (1957) provided a clear scenario for the classical neo-Darwinian model of speciation in animals. This model ignores the possible contributions of sexual selection to the speciation process. A brief outline of the most important factors and their temporal sequence is as follows.

1. So far as is known, divergence of a single, interbreeding avian population into two populations that may become distinct species always requires, first and foremost, geographic isolation, which is either complete (allopatry) or nearly complete (parapatry). Such isolation means that genetic changes appearing in one population, as by mutation, do not spread into any other population. Agents of natural selection will also inevitably vary between two isolated populations occupying different geographical areas, and this will, in turn, promote further differentiation between the populations. In short, isolation of two populations for a long period inevitably leads to genetic divergence and probably to separate co-adapted gene complexes.

2. If the two allopatric populations subsequently become sympatric ('secondary contact'), several possibilities exist. First, if individuals of each of the two populations have been isolated for enough time that they unanimously ignore members of the other populations as possible mates, and if they are ecologically compatible, then the speciation process has gone to completion: i.e. two sympatric species may co-exist where before there was one. Alternatively, if many members of each of the previously isolated populations are attracted to members of the other (i.e. if there is no assortative mating), and if successful breeding of individuals from the two 'daughter' populations occurs, with the resulting production of healthy, fertile offspring, then genetic exchange is widespread and the end result is one species. That is, during the period of geographic isolation, the two populations did not differentiate sufficiently to create reproductive costs to interbreeding.

3. Less clear-cut situations may also occur when two previously isolated populations come back into contact or overlap. One that has been critical to the interpretation of bright, distinctive ornamentation in birds, especially males, is as follows. When sympatry between two populations is re-established, some members of each sex of one or both populations mate with an individual from the other population, and such matings are unproductive. That is, either eggs fail to hatch, the young are physiologically inferior and die prior to attempting to reproduce, or the young do reach maturity and attempt to breed, but are either sterile or produce reproductively inferior offspring. Such unproductive matings, and the time, etc. invested in them, provide strong selection against those individuals that do not discriminate between prospective mates in their own population and those in the other.

The extreme costs of such mating 'errors' lead to the evolution of mechanisms that make possible clear recognition of the appropriate sort of mate. These mechanism are usually referred to as isolating mechanisms. If they serve to prevent inappropriate pairing in the first place, they are termed pre-mating isolating mechanisms and the speciation process has been completed. Again, ecological compatibility is necessary for the continued co-existence of the two new species.

Distinctive ornamentation of males of many avian species has often been uncritically labelled as an isolating mechanisms. Appearing, at least at first glance, to support this interpretation are observations of both naturally and artificially occurring hybrids. (1) In some cases, interspecific hybrids occur in the wild where males of the two parental species are conspicuously different in appearance and/or in associated behaviours (i.e. this indicates that mating 'mistakes' do occur). (2) Hybridization between certain species or groups of species (e.g. pheasants, ducks) occurs readily in captivity. This has been interpreted as indicating the potential for 'breakdown' of the putative isolating mechanisms under certain conditions. (3) Such hybrids or their offspring may be demonstrably 'inferior' in one way or another (e.g. Sharpe and Johnsgard 1966). In short, both the existence of hybrids and their reproductive inferiority have been used as evidence that the often striking difference in plumage of males, in particular, of two sympatric groups is causally related to avoidance of hybridization.

This, in brief form, is the classical model of speciation via geographic isolation and the evolution of isolating mechanisms. With regard to male ornaments and sexual selection, the important point in this scenario is that the bright feathers, distinctive displays, etc. of male birds, particularly those species whose ranges overlap the geographic ranges of closely related species, were long thought to have evolved in response to selection to perfect species recognition; i.e. the main evolved function of bright male plumage was thought to be unambiguous species identification. During the 'forgotten era', the majority of evolutionary biologists gave little thought to the roles of morphological ornaments, songs, or other displays in intrapopulation mate selection.

On the whole, a scenario incorporating a role of sexual selection, together with geographical isolation, in the speciation process appears to accord better with the empirical facts than does the isolating mechanism hypothesis. This view of the interaction between sexual selection, geographical isolation, and divergence of populations can account for differentiation in ornamentation that develops in complete allopatry, and, as discussed below, such differences may require relatively little genetic alteration or differentiation either in males, which exhibit the trait, or in females, which respond to it.

8.2.1   The question of isolating mechanisms: plumage patterns of male ducks

In previous sections I have suggested that ornamental plumage and other displays may often have an evolved function in the context of intrapopulation mate choice, rather than serving as a species-isolating mechanism. Although Sibley (1957) interpreted most cases of divergence of male plumage among closely related species in the former way, he concluded that in ducks of the Genus *Anas* an evolved function of bright, distinctive male plumage was to serve as a species isolating mechanism. He cited evidence of two sorts for this interpretation. First, hybridization in the wild with at least one other species has been recorded for every one of the nine North American members of the

Fig. 8.1   A pair of northern pintail ducks. Pintails hybridize relatively commonly with mallards both in captivity and in the wild. Reprinted with permission from Dover Publications, Inc.

genus; thus 'mating errors' do occur and natural selection would presumably favour the evolution of mechanisms to avoid such matings.

Second, the best evidence for an evolved function of male plumage as an isolating mechanism, at least at first glance, is the loss of such plumage in regions where interspecific mating 'mistakes' are not possible simply because only one species breeds there. This pattern is shown for insular derivatives of both mallards and pintails (Sibley 1957). In addition, in North America, males of more or less geographically isolated derivatives of the mallard—the black duck, Mexican duck, and mottled duck—exhibit female-like, cryptic plumage. The correlation between cryptic male plumage in these duck populations and the presence of only a single breeding species has been interpreted as evidence that the bright plumage that characterizes males of most North American species of *Anas* (e.g. mallard) has evolved in the context of species isolation. That is, in areas where only one species breeds and where mating mistakes thus could not be a problem, selection for cryptic coloration favoured the loss of conspicuous, distinctive plumage.

Although this is an appealing explanation for both the brightly plumaged and cryptically plumaged forms of *Anas* ducks, alternative explanations for plumage monomorphism also exist. Life history features of highly dimorphic species of the temperate region (McKinney 1986) suggest that the intensity of sexual selection may be greater than in the monomorphic, cryptic species. There is indirect evidence—specifically, the timing of the moults of males—that plumage is important in mate choice. Following breeding, males moult into a female-like, cryptic plumage that lasts for only a few weeks before another moult occurs that brings the male back into bright breeding plumage. The timing of moult into bright plumage corresponds to the unusual timing of pair formation, which may begin in early autumn. Thus, the prompt renewal of the conspicuous breeding plumage corresponds to the onset of the period of mate choice. This temporal coincidence of completion of moult and pair formation suggests that the bright plumage is important in mate choice (but see Chapter 6), whether or not an additional evolved function is species isolation.

Omland (1996a) suggests that all 13 of the monomorphic, cryptically-feathered taxa closely related to mallards (Livezey 1991) may be derived from a mallard-like ancestor, and that the weak effects of male plumage on female choice may have predisposed the evolutionary loss of the bright, 'greenhead' coloration of males. That is, because females do not rely heavily on male plumage in making mate choice decisions, selection on males to maintain bright plumage is weak.

Female black ducks often pair with male mallards, which has been interpreted as indicating that females of this species may prefer the bright, greenhead plumage of male mallards to the drab, cryptic plumage of males of their own species (e.g. Johnsgard 1960). In a study of mating patterns of free-living mallards and black ducks, Brodsky and Weatherhead (1984) found that male mallards preferred to mate with female mallards, but that, because of a male-biased sex ratio, many male mallards failed to obtain a mate and turned their attention to female black ducks. Mallard drakes were far more successful than black duck drakes in pairing with female black ducks. Subsequently, Brodsky *et al.* (1988) studied the development of female mate choice in these two species. These researchers reared ducklings in four different combinations—male and female mallards, male and female black ducks, male mallards and female black ducks, and male black ducks and female mallards. Upon maturity, females of all four groups preferred to mate with males of the type with which they had been reared. This result indicates that female mate preferences are based, at least in part, on prior experience, and that the bright plumage of male mallards is not innately particularly attractive to either female mallards or female black ducks (cf. Johnsgard 1960). In a free-swimming situation these same female mallards and black ducks were allowed access to males of both species, and all formed pair bonds with mallard drakes. This appeared to be due primarily to the fact that male mallards were always dominant to male black ducks, rather than to a female preference *per se*. Thus, in this case, male–male interactions appeared to override female preference in the process of pair-bonding (Brodsky *et al.* 1988). In contrast, in a study of dominance relationships in mallards and black ducks, Hoysak and Ankney (1996) found that male mallards were not generally dominant to male black ducks. Thus, the role of male–male competition in the hybridization of these two forms remains unclear.

With regard to the question of whether the bright plumage of male mallards serves as a species-isolating mechanism, these studies do not provide a definitive answer. First, the results of Brodsky *et al.* (1988) suggest that females do not have an innate preference for conspecific male plumage, which is a key assumption of the species-isolating interpretation of distinctive male plumage (Sibley 1957). Second, Omland's (1996*a*,*b*) work suggests that plumage is not very important to female mallards in intraspecific mate choice; whether it is used to identify males to species is unclear. Other male signals, including vocalizations, status displays, and bill colour are apparently more important to females than feather colours and patterns.

Finally, it should be remembered that in addition to the apparent importance of male dominance interactions in the process of pair-bonding in these ducks, another factor can dilute the importance of female mate choice—forced extra-pair copulations. Such copulations are known to occur relatively frequently in ducks, and they may actually be the basis for many or most of the records of wild, non-captive, interspecific hybrids (which have provided much of the support for the isolating mechanism concept).

8.3 *The role of sexual selection in speciation*

Over the past decade or so, a major change has developed in the way morphological and behavioural differences among males of closely related species are interpreted. As described above, for many years, the accepted, and almost sole, explanation for this kind of divergence of closely related forms was based almost exclusively on the geographic model of speciation. However, beginning with Darwin (1871), a number of authors (e.g. Sibley 1957; Gilliard 1963; see West-Eberhard 1983 and Barraclough *et al.* 1995 for additional references) have suggested that sexual selection might in some situations affect the speciation process or lead to unusual species diversity. The incorporation of sexual selection thinking into considerations of speciation and species diversity was a major step towards a fuller understanding of questions like: why is the red junglefowl red, the grey junglefowl grey, the green junglefowl green, and the Ceylonese junglefowl yet another colour? Among male birds, this kind of variation within a genus occurs frequently, and it usually cannot be satisfactorily explained either by variation in the ecologies of the related species or by invoking the concept of isolating mechanisms.

The proposed relationship between sexual selection and speciation is based on the notion that '. . . if divergence can lead to speciation (reproductive isolation), then anything that accelerates divergence should tend to accelerate speciation—especially if characters critical to survival or reproductive success are concerned' (West-Eberhard 1983, p. 175). Sexually selected characters are, by definition, critical to reproductive success. Therefore, it is no coincidence that some of the most spectacular examples of apparently rapid or recent speciation in birds occur in groups characterized by polygyny or lek–promiscuity (e.g. pheasants, hummingbirds, manakins, birds of paradise), and that in these birds it is the ornamental and/or behavioural display characters of males that show the greatest differences between closely related species (Sibley 1957; Prum 1990).

By comparing sister taxa of passerines, Barraclough *et al.* (1995) tested the idea that strong sexual selection will positively affect the rate of speciation. They found a significant positive correlation between the proportion of sexually dichromatic species and the number of species in those taxa, and presented a case that this sexual dichromatism evolved through female choice. This study provides some of the first and best 'broad-brush' evidence in support of the hypothesis that, within a taxonomic unit, sexual selection may have a direct effect on the rate or frequency of speciation. Similarly, Mitra *et al.* (1996) compared taxa with promiscuous mating systems to sister taxa with other types of mating systems and found that the former contained more species. Like Barraclough *et al.* (1995), Mitra *et al.* (1996) concluded that strong sexual selection is likely to be causally related to rates of speciation.

Coyne (1992, p. 515) summarizes the reasons that sexual selection affects speciation as follows: 'Sexual selection may be a major cause of speciation in animals because it is pervasive, leads directly to reproductive isolation, and is associated with many adaptive radiations.' It is also worth re-emphasizing that sexual selection accelerates the rate of divergence of isolated populations. An increase in the rate of divergence is, in itself, an important 'contribution' to the speciation process, because the amount of time that two isolated populations will remain in allopatry varies from case to case, and the more different the two become during the period of allopatry, whatever its duration, the more likely that the speciation process (reproductive isolation) will have gone to completion when or if the populations come into secondary contact.

8.3.1   Genetic control of ornamental traits

Birds are the most visible of terrestrial vertebrates, and no other group contains species exhibiting as much diversity in colour, colour patterns, and behaviour. Conspicuous differences in plumage colour patterns often occur between species that, on the basis of all other evidence, are extremely closely related. One of many possible examples is the golden pheasant–Lady Amherst pheasant species pair. Males of these two species exhibit strikingly different plumage colour patterns. Despite conspicuous differences in the appearance of males, the facts that (1) the two species readily hybridize in captivity, (2) any combination of hybrid can be produced (i.e. half and half, a quarter–three quarters, etc.), and (3) all such hybrids are completely interfertile (Johnsgard 1986) indicate that genetically the two forms are extremely similar. Several other congeneric pheasants also show conspicuous differences in male plumage, with little other evidence of genetic differentiation (Johnsgard 1986).

In his discussion of the evolution of sexual dimorphism over forty years ago, Sibley (1957, p. 173) made two very important points that have stood the test of time well. (1) The genetic basis for ornamental and display characters probably involves a very few genes that control only the relatively superficial characters of plumage and display movements. (2) In polygynous species, sexual selection produces high degrees of sexual dimorphism by action upon a few genetic factors in the males.

These points raise two issues relevant to the genetics of speciation. First, few genetic changes may be required for reproductive isolation; i.e. speciation often (perhaps usually) does not require a genetic revolution; change in only a few alleles may be sufficient. In fact, Coyne (1992) cites a few non-avian cases where as few as three or fewer loci have apparently been sufficient to produce genetic isolation between two populations. Thus, speciation sometimes has a relatively simple genetic basis.

Second, differences between two closely related species in ornamental traits (both morphological and behavioural) may be controlled by very few genes: some groups of related species, in which male plumage ornamentation and displays differ conspicuously, may not be very differentiated in other respects. This suggestion is supported by data indicating that hybrids with intermediate morphologies also exhibit intermediate behaviour. For example, in a study of plumage patterns and courtship behaviour of F2 (second generation) hybrid mallard and pintail ducks, a strong correlation $r = 0.76$) existed between plumage type and courtship display (Sharpe and Johnsgard 1966). That is, individuals that were more mallard-like in appearance also exhibited mallard-like courtship, individuals that more were pintail-like in appearance likewise were pintail-like in behaviour, and birds with intermediate plumage displayed in an intermediate fashion. Sharpe and Johnsgard (1966) concluded that the species-typical plumage and display traits of mallards and pintails have a very similar genetic basis and probably involve few genes. Studies such as this suggest that genes for phenotypic morphological characters, such as colour and pattern of display plumage and behaviour associated with display, are likely to be closely linked or that the same gene may have pleiotropic effects, controlling some aspects of both morphology and courtship behaviour in males. It is possible that the same genes also influence the preferences of females.

8.3.2   Fixation of newly evolved traits

In nearly all species of birds, all adults within each sex are very similar in overall appearance (e.g. colour patterns, structurally specialized feathers). That is, there is little intraspecific variation. There can be little doubt that the raw material—mutation—for variation in plumage colour, for example, occurs relatively frequently. In virtually all domesticated and semi-domesticated species, numerous mutant plumage colours occur and have been 'fixed' by artificial selection (see Sections 2.5.1 and 6.5). Yet in the wild forms of these species, all individuals of each sex possess plumage that is extremely similar; i.e. a single field-guide illustration of a male and female is sufficient to allow the identification of all adults of most avian species.

In reviewing the birds of paradise (BOP), Diamond (1986b, p. 21) considers the issue of uniformity of plumage within populations to be:

> . . . an unsolved problem: While adult males of many BOP species and even allospecies look drastically different from each other, individual variation within a population is no greater than in other bird species, and subspecific variation within most species is trivial. What forces promote such uniformity within each population or species but such divergence between species? Could the explanation involve sudden changes of female preference for male traits in founder populations?

As discussed above, female preferences for particular, and often unique, male display characters evolve in isolated populations, which may explain the divergence between closely related species.

The other, more rarely addressed issue raised by Diamond (1986b) is the uniformity of plumage traits within populations. At least three points in addition to strong directional selection may help to explain the intrapopulation uniformity of appearance that characterizes nearly all populations of wild birds.

1. As discussed above, it appears that very few genes may control conspicuous plumage characters, and that differences in just a few alleles may be sufficient to bring about speciation (Coyne 1992). Although the genetic bases

for individual traits are virtually unknown in wild birds, this is not the case for the domestic fowl and its ancestor, the red junglefowl. Domestic fowl show far more morphological variation in plumage colours, comb shapes, etc. than occurs across the four wild species of *Gallus*. Many important traits are known to be controlled by a single allele (e.g. single comb (wild type) and pea comb). The point is, fixation of a phenotypic trait controlled by one or very few genes should occur much more readily than one controlled by many genes.

2. A second factor may be involved in the fixation of critical phenotypic traits controlled by a few alleles. In domesticated animals, fixing a desired trait by artificial selection involves the inbreeding of a relatively small number of individuals. It is possible that small population size at some point in the species' history, producing a high level of inbreeding plus strong directional selection, may have been necessary to produce the uniformity of appearance that characterizes wild populations of birds and most other animals. In the United States, it has been well documented that a small initial population can rapidly expand numerically and geographically to occupy large areas (e.g. European starling, house sparrow, house finch); thus current range or numbers do not necessarily provide an accurate picture concerning either population size or geographical range throughout a lineage's history.

3. Mutations of genes controlling plumage and other morphological charac-ters occur frequently in captive birds, and this also probably occurs in wild populations. If so, natural and/or sexual selection removes most of such mutant phenotypes. The basis for selection against mutant plumage, for example, is unknown, but probably involves both interspecific (e.g. predation) and intraspecific (e.g. social interactions of one or more kinds) factors. New morphological traits that become species-typical ornaments are probably constrained in that they must be both attractive to females and not too deleterious to their male bearers. These factors, in addition to the nature of mutations, are probably major reasons why differences between closely related species (e.g. in plumage colour patterns) are often extensions or modifications of pre-existing traits, rather than being completely different.

In short, few genes may be involved in the control of many ornamental traits, and three lines of evidence suggest that fixing such traits should occur more readily when the genes responsible are few in number. First, in domestic fowl, where many mutations affecting the phenotype have been preserved by artificial selection, several plumage colours and colour patterns are controlled by different alleles at the same locus. The same is true for comb structure, tarsus colour, etc. Second, under artificial selection, traits are fixed by the inbreeding of a relatively small number of individuals. Small population size may have been involved in the fixation of the new ornaments. Third, most phenotypic traits that appear by mutation are removed via selection, both natural and sexual. This can help to explain why new traits are usually extensions or minor modifications of already existing traits.

8.4   Closely related species often differ conspicuously in male ornamentation: Australian fairy-wrens

The plumages of male birds, in particular, of many closely related species differ in one or more conspicuous ways. As discussed above, the answer to the question of why males of closely related species usually differ in details of plumage ornamentation has been related traditionally to the concept of species isolating mechanisms. However, for many and perhaps most species, a more complete answer might include a role for differing female preferences in isolated populations. New male morphological traits that prove to be attractive to females may arise by chance mutation in one of two allopatric populations, while other traits may arise independently in the other population.

Because of its climatic history, Australia has an unusually large number of allopatric populations of closely related forms (e.g. Keast 1958; Cracraft 1986; Ford 1989). This pattern is exhibited by populations accorded specific rank as well as by those regarded as subspecies. Plumage differences between closely related species are often striking. What makes the Australian avifauna unusual is the recency of isolation of so many populations. Differentiation in isolation has often not proceeded for a long enough time to produce major changes, yet congeners in separate populations often differ conspicuously in certain morphological traits, especially colours and colour patterns. That is, differences in aspects of plumage that probably have social (West-Eberhard 1983) and sexual significance often occur in these isolated counterparts (e.g. rosellas, Ford 1989; estrildid finches, Burley 1986*b*).

Australian fairy-wrens provide an especially good example of this pattern (e.g. see Pizzey 1980, plates 68 and 69). Eight species of fairy-wrens occur in Australia and, for the most part, they are geographically isolated. Whereas the males of all species and subspecies are brilliantly and usually distinctively coloured during the long breeding season, the females are, by comparison, uniformly dull in plumage. The morphology, social systems, and ecology of all fairy-wrens seem to be very similar, which suggests that the striking divergence of male plumage has progressed more rapidly than divergence of other features of their biology.

For many years, it was assumed that the cooperatively breeding fairy-wrens were monogamous, with a single breeding pair assisted by non-breeding helpers (e.g. Rowley 1965, 1981). Recently, however, it has become apparent that this is not at all the case, and that sexual selection is a powerful selective agent. In the splendid fairy-wren, for example, more than 65 per cent of a large number of young wrens sampled were sired by males outside their natal social group (Brooker *et al.* 1990). Similarly, in the superb fairy-wren, at least 76 per cent of genetically typed young were sired by males outside the social group, 95 per cent of broods contained at least one chick sired by a male from outside the group, and in 48 per cent of broods, all young were sired by extra-group males (Mulder *et al.* 1994). These data clearly suggest that sexual selection is strong in fairy-wrens.

Another line of evidence that sexual selection among male fairy-wrens may be unusually strong and that colours may play an important role in male mating success is the bizarre behaviour known as 'petal-carrying' (see Fig. 12.4). Petal-carrying behaviour has been recorded in at least six species (Rowley 1991). Early in the breeding season, male fairy-wrens of several, and perhaps all, species frequently trespass into neighbouring territories, where the intruder sometimes performs a display involving a flower petal, the colour of which is species-specific. There is a strong correlation between the frequency of petal-carrying and the initiation of clutches, which suggests that this behaviour is indeed an aspect of a male fairy-wren's reproductive strategy. In some species, the colour of the petal carried by males contrasts conspicuously with the male's plumage, while in others, the colour of the preferred petals are similar to the male's plumage. Rowley (1991) summarizes the possible significance of petal-carrying and suggests that competition for extra-pair or extra-group copulations may be the basis for this unique behaviour; i.e. petal-carrying may enhance either the attractiveness of males or the intensity of their performance, either of which may influence mate choice by the female.

The significance of petal-carrying is still unresolved and has many potential interpretations. For example, petal-carrying could (1) reflect a 'demand' by females for stimuli beyond the male's colourful plumage and exaggerated displays, i.e. a sort of super (ab)normal stimulus *per se*, (2) indicate that the intruding male comes with a handicap (e.g. Zahavi 1975), in that, for example, he is more visible to the male owner(s), or (3) signify that the intruder is dominant (and thus possibly genetically superior) to the male owner of territory. Separating these possibilities is a challenge for future workers.

In any case, the high level of male–male competition for matings, as suggested by reproductive anatomy (Mulder and Cockburn 1993) and documented by DNA fingerprinting (Mulder *et al.* 1994), together with the spectacular plumage differentiation and flower petal displays of male superb fairy-wrens, refute the earlier assumption of a monogamous mating system. A non-territorial, floating female may evaluate a prospective social mate by one set of criteria (e.g. ownership of an acceptable territory for nesting), while later focusing on different factors (bright striking plumage and display behaviour, including use of flower petals) for copulations outside the pair bond. In short, it appears likely that male–male competition for matings outside the 'pair bond' has driven the extreme development and differentiation of male plumage traits, bolstered by flower petals, well beyond what would have been the case had the birds been genetically monogamous. Such strong sexual selection may have promoted differentiation of male plumage and displays, and thus speciation, in the fairy-wrens.

8.5 The significance of hybridization

Production of hybrid offspring indicates a certain degree of genetic compatibility between individuals of two hybridizing populations. In ducks and pheasants, for example, hybrids between many species and even genera have been produced in captivity (e.g. Johnsgard 1986). The subsequent reproductive capability of the F_1 (first generation) hybrids provides an indication of the degree of genetic compatibility, and presumably the closeness of the phylogenetic relationships, of the parental populations. In some cases, both sexes of F_1 offspring are fertile, while in others, one or both sexes are sterile. However, even when F_1 birds are reproductively competent, their offspring, the F_2's, may, for one reason or another, be incapable of reproducing (e.g. mallard and pintail ducks, Sharpe and Johnsgard 1966).

The phenomenon of hybridization involving individuals from populations defined as species has played an important role in the development of thinking about the process of speciation. The traditional neo-Darwinian view has been that interspecific matings were invariably negative in their effects, and that hybrid offspring were one of the more conspicuous manifestations of such mating errors (e.g. the classic example is the sterile mule—offspring of a male donkey and a female horse). More recently, an alternative view has been suggested, along with some supporting data, that hybridization may not always produce inferior offspring and, moreover, in some cases, there may even be fitness benefits and thus selection for hybridizing individuals.

Whether one views hybridization as something clearly disfavoured by natural selection (e.g. Mayr 1963) or, conversely, as a potentially fitness-enhancing process, it clearly is not a rare phenomenon. Grant and Grant (1992) reviewed the literature on hybridization and determined that about 10 per cent of the extant avian species have bred in nature with another species and produced hybrid offspring. Here I consider briefly the phenomenon of hybridization, both from the traditional neo-Darwinian perspective and from a more recent and positive perspective.

8.5.1 Hybridization as a negative factor in the speciation process

The idea that matings between individuals of two species is deleterious to their own fitness and that, therefore, selection operates to prevent such mating is based in part on the observation that hybrid offspring produced in captivity are often reproductively inferior, if, in fact, viable hybrid offspring are produced at all. The costs of producing hybrid offspring have led to the view that an important final stage in the speciation process is the 'perfection' of pre-mating isolating mechanisms.

The most general genetic explanation for the frequent inferiority of the progeny of interspecific matings is that co-adapted gene complexes (groups of genes that have evolved positive interactions over the course of the separate evolutionary history of each species) are broken up by the combining of genes from heterospecific parents. An example of genetic incompatibility is sterility. Sterility of hybrids is one kind of evidence supporting a critical proposition of the neo-Darwinian scenario, namely that selection should disfavour inter-specific matings due to genetic incompatibility.

For example, under conditions of captivity interspecific hybridization, with sterility of offspring, is common. Hybrids often exhibit Haldane's rule: in the hybrid offspring of two species, when only one sex is sterile or non-viable, it is nearly always the heterogametic sex (Coyne 1992). Thus, in birds, female offspring of interspecific crosses are more often sterile than males (e.g. pheasants, Johnsgard 1986). Haldane's rule is apparently based on the sex chromosomes, which suggests that these chromosomes may play a major role in the genetic isolation of two hybridizing populations. (See Coyne 1992 for a discussion of possible proximate explanations for the manifestation of Haldane's rule.)

A study of hybridization in nature between pied and collared flycatchers (Alatalo *et al.* 1990*b*) found that there was no reduction in the breeding success of heterospecific pairs. However, backcrosses between hybrids and either of the two pure species showed greatly reduced fertility, which appeared to be associated with Haldane's rule, i.e. in many cases, none of the eggs of hybrid females hatched.

8.5.2   Hybridization as a positive factor in the speciation process

As described above, the traditional view of interspecific hybridization has been that such matings decrease the fitness of individuals engaging in interspecific mating. More recently, another view of the role of hybridization in avian evolution has emerged (Grant and Grant 1992; Pierotti and Annett 1993). This view suggests that in some cases interspecific hybrids have higher fitness than their parental species, and that hybridization can produce positive effects via introgression of new genes or alleles into the parental species.

In their comprehensive studies of the population biology of several species of Darwin's finches, Grant and Grant (1992) found that over a period of several years interspecific hybridization between certain species of finches was not uncommon, and that hybrid individuals were not inferior by any of several measures of reproductive success and fitness. That is, hybrids not only survived well, they also proved to be no less fertile, and in some cases they were apparently more fecund, than their parental species. Moreover, these hybrids showed no evidence of Haldane's rule nor was there any evidence for 'hybrid breakdown' in the F2 generation.

Why are these hybrid finches superior to individuals of their parental species in many important measures of fitness? Grant and Grant (1992) provide three possible answers to this question. First, they may be better at dealing with environmental variables; e.g. their intermediate beak size may, at certain times, make them better at exploiting certain types of foods. Second, the increased heterozygosity might provide increased vigour, especially if the parental forms were highly inbred as a result of reduction of population numbers. Finally, in other cases, hybrids have been shown to be superior in certain novel environmental situations associated with disturbed habitats. The Grants discount this last suggestion, while leaving open the possibility that either one or both of the other two may contribute to the success of hybrid finches. They also suggest that over the longer term, as climatic conditions change, the hybrids will come to be at an ecological disadvantage, and that the periodic events favouring hybrids, followed by periods where hybrids are disfavoured, will prevent the complete fusion of the hybridizing forms into a single panmictic population.

Drawing together several lines of evidence, Pierotti and Annett (1993) also suggest that hybridization may be favoured in some cases, and that it may be an important mechanism for promoting evolutionary change. These authors claimed that hybridization within genera occurs more frequently in avian families and subfamilies (e.g. gulls) where there is considerable male parental investment than in families with only moderate or no male parental care. (This conclusion is counter-intuitive; one might expect care-giving males to be more, rather than less, fussy in their mate choice decisions.) This may be due to the preference of females for certain traits of males that may be more strongly manifested in males of a related species than in conspecific males (e.g. a preference by females for large mates). Pierotti and Annett (1993) argue that adaptive hybridization may explain the pattern of large numbers of sub-species within certain groups (e.g. genera) that contain relatively few species. Recurrent hybridization, which may be favoured for any of a number of reasons, prevents the genetic isolation required for speciation, while it does not prevent the evolution of phenotypic traits adapted to local ecological/environmental conditions.

The arguments and evidence presented by Grant and Grant (1992) and Pierotti and Annett (1993) clearly indicate that much remains to be learned about the role of hybridization in the process of evolutionary change, and that the long-held view of hybridization as a maladaptive mating error inevitably disfavoured by natural selection requires modification, to some as yet undetermined degree.

8.5.3   Hybridization and plumage ornamentation

Interspecific hybrids have often been used as evidence that pre-mating isolating mechanisms have not yet been perfected, and, by extension, that ornamental plumage has evolved in response to selection against hybrids. However, the birds of paradise do not clearly support this interpretation. Among the birds of paradise, several hybrids, sometimes involving different genera (Gilliard 1969, p. 64) have been produced. Overall, such hybrids are rare (i.e. only 30–50 identified from the more than 100 000 skins sent to Europe between 1879 and 1924 for human adornment, Mayr 1963, p. 126). It could be argued that the existence of hybrids of bird-of-paradise species demonstrates that selection should favour the production of pre-mating isolating mechanisms, such as the distinctive ornamentation of males, while their rarity reflects strong selection against hybridization.

Alternatively, for several reasons it seems more likely that hybridization in the birds of paradise is incidental to species recognition, and thus does not imply selection for isolating mechanisms. First, although pre-European human alteration of the environment has been going on for a long time in New Guinea, and thereby possibly increasing contact and the potential for mating errors between species (e.g. the blue and raggiana birds of paradise, Gilliard 1969, p. 255), hybrids clearly are rare. Second, given the possibly important roles of supernormal stimuli and novelty in female choice in birds of paradise, which is suggested by the bizarre ornamentation and displays of males, the overall or underlying genetic similarity between congeneric species (Sibley and Ahlquist 1985), and possibly recent sympatry between some of them, the fact that matings occasionally occur between two species in which the males look very different should not be especially surprising. By this view, the extreme development of ornate plumage in male birds of paradise may have had little or nothing to do with evolved isolating mechanisms. Instead, male ornamentation may solely reflect the intensity of sexual selection, and possibly the weakness in the opposing forces of natural selection (see Section 8.7.2).

Third, if sexual selection speeds up the rate of divergence, then isolated populations assumed to be under strong sexual selection should be diverging rapidly compared to species not subjected to strong selection (see Section 8.7.2.1). Consistent with this view, the polygynous, ornate birds of paradise have apparently diverged quite rapidly on an evolutionary time scale (Sibley and Ahlquist 1985).

8.5.4   Sexual selection, hybridization, and introgression

It has been suggested that sexual selection accelerates changes in certain kinds of traits, thereby accelerating the process of differentiation of populations and ultimately contributing to speciation. It also appears that, among related species, the occurrence of strong sexual selection may increase the frequency of hybridization. Moreover, in some cases, the genes controlling sexually-selected traits apparently introgress more rapidly than other genes, which provides strong evidence for Sibley's (1957) suggestion that differences between populations (including species) in male ornamental characters are likely to be poor indicators of overall genetic differentiation.

These points are illustrated by a study in Panama of the white-collared and golden-collared manakins (Parsons *et al.* 1993). Most manakins, including these two species, have lek breeding systems with highly skewed male mating success, suggesting that sexual selection is especially strong in these birds. Parsons *et al.* (1993) surveyed phenotypes and genotypes of the manakins across a geographical hybrid zone and found that the golden-collar phenotype (plumage coloration) extends into areas where, based on mitochondrial and nuclear DNA, the birds were otherwise genetically white-collared. In contrast, distinguishing morphometric characters—tail length and beard (specialized throat feathers) length—shifted at the same site that the DNA changed. Thus, the ornamental colour that characterizes male golden-collared manakins has moved much farther into the range of white-collared manakins than have other plumage traits. Finally, in the 'hybrid zone', the birds are genetically and morphometrically indistinguishable from white-collared manakins, although their plumage characters had led to their previous designation as a subspecies of golden-collared manakin. These findings led Parsons *et al.* (1993) to suggest that the greater introgression of ornamental plumage traits may have been driven by intense sexual selection.

This study provides some of the best evidence obtained to date that 'superficial' ornamental traits can be selected apart from selection on other traits, either morphological or genetic, and it supports the proposition that sexual selection can accelerate change and thus contribute to the rapid differentiation of populations.

8.6 Ornaments, fluctuating asymmetry, and speciation

As described in Section 5.6.3, Møller (1993*c*) has suggested that the degree of fluctuating asymmetry of a particular trait is inversely related to the functional importance of that trait; i.e. the most important functional characters are under stringent genetic control, and are therefore not the characters likely to evolve ornamental functions; i.e. those morphological characters most likely to evolve into sex ornaments are less developmentally canalized than are more functionally important characters.

With regard to the issue of sexual selection and speciation, Møller (1993*c*) provides some intriguing ideas. He suggests that speciation events may be related to the low degree of developmental stability exhibited by secondary sexual characters subjected to intense directional selection. This might be especially true in small, marginal populations where levels of developmental stability may be lower than in more central populations. In support, Møller (1993*c*) offers the following suggestions: (1) functionally unimportant traits will be weakly canalized (controlled by few genes); (2) because of weak genetic control, such bilaterally paired traits will show more asymmetry than other, more functionally significant traits; and (3) these will be the traits on which females focus their attention. Finally, strong directional selection based on fluctuating asymmetry in ornamental traits will rapidly drive a small, marginal, or isolated population to differentiate from the founder population, potentially leading to the development of a new species.

8.7 Sexual selection and other factors promoting differentiation and speciation: sage grouse and birds of paradise

More than thirty years ago Gilliard (1963) suggested that in 'arena' species, i.e. promiscuous species exhibiting lekking behaviour and/or specialized male display sites, the pace of evolutionary change moves faster than in more typical socially monogamous species. This view is consistent with contemporary thinking.

In this final section, I describe two cases among birds in which it appears that sexual selection has promoted morphological and behavioural differentiation of ornamental traits. Many of the grouse and birds of paradise appear to be among the most strongly sexually-selected of birds. Thus, some species in these two groups should most clearly illustrate any relationship between intensity of sexual selection and rate of differentiation of isolated populations. In addition to sexual selection, differentiation of populations, including speciation, also requires other factors. Some of these are considered in the discussion of the birds of paradise presented below.

8.7.1 Sage grouse

In a study of the possible role of sexual selection in differentiation of allopatric populations, Young *et al.* (1994) analysed a variety of male display characters from a geographically isolated population of sage grouse in Gunnison County, in southern Colorado, and compared traits of these birds with the same traits of sage grouse from northern Colorado (Jackson County) and California (Mono County). Despite the great geographic distance between them, males from these latter two populations were very similar in size, and the acoustical components of their displays were identical (Young *et al.* 1994). Likewise the strut display rate of males from these two sites did not differ significantly. In short, comparisons of Mono and Jackson birds indicate that displays of male sage grouse are highly uniform over a vast geographic area.

In contrast, comparisons of displays of the Jackson–Mono birds with those from the isolated Gunnison County population revealed that the morphology and behavioural displays of the latter differed strongly from those of the other two populations. In addition to being about one-third smaller, acoustical displays of male sage grouse from Gunnison differed completely from Jackson and Mono birds in frequency and duration (Fig. 8.2). The same was true of both the visual component of display of Gunnison males, which included some elements not recorded in the other populations, and the strut display rate. The following differences were noted (Young *et al.* 1994, p. 1358):

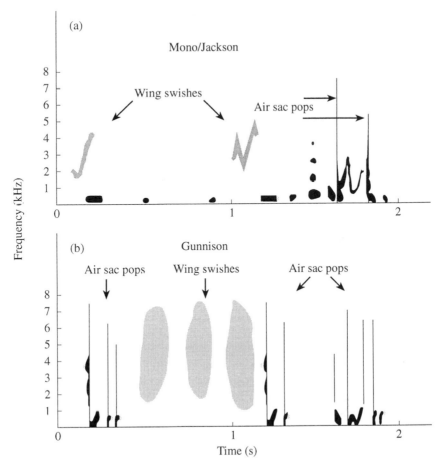

Fig. 8.2   Acoustical component of strut displays from two populations of sage grouse. Comparisons of display characteristics of the two populations suggests that sexually-selected traits may diverge rapidly in isolated populations. Reprinted with permission from Young, J. R. *et al.* (1994). Phenotypic divergence of secondary sexual traits among sage grouse, *Centrocercus urophasianus*, populations. *Animal Behaviour*, **47**, 1353–62.

Gunnison males have whiter rectrices and their filiplumes are thicker and used much more conspicuously: males tossed them above their heads throughout their strut display . . . . Their air sacs were compressed more often (twice in Mono/Jackson . . . versus eight times in Gunnison). In addition, the amount of wing movement was greatly reduced . . . . male sage grouse in the Gunnison Basin also often culminated their display by performing a 'tail wag' in which they vigorously shook their tails while remaining upright.

Finally, males of the Gunnison population displayed at a significantly slower rate than did those in Mono and Jackson counties.

Several of the differences between the displays of Gunnison males and Jackson–Mono males are in traits that have been shown to correlate with mating success in other populations (i.e. strut rate, acoustical component). Thus, it appears that sexual selection, specifically mate preferences of females, has played an important role in the divergence of the isolated Gunnison birds. The fact that Gunnison birds are about one-third smaller than sage grouse from other populations indicates that genetic divergence of ecologically important traits, in addition to reproductively important ones, has also occurred or is occurring. It would be extremely interesting to ascertain whether the differences in displays reported by Young *et al.* (1994) would prevent mating by individuals from the two populations. If so, this case would provide the most direct evidence to date for birds that sexual selection can accelerate morphological and behavioural differentiation in geographically isolated populations and thus contribute importantly to the process of speciation.

8.7.2 Birds of paradise

The birds of paradise are a moderately large passerine family (Paradisaeidae; 43 species). The group is confined to the Australasian region and 38 species are endemic to New Guinea and its satellite islands (Beehler *et al.* 1986). In terms of male ornamentation and display, these birds are best known as the group in which the process of sexual selection has gone to fantastic extremes. The goal of this final section is to examine a suite of factors in addition to female mate choice that may have contributed to the extraordinary development of ornaments and associated displays among males of a majority of the birds of paradise (Fig. 8.3).

According to Sibley and Ahlquist (1985, 1990), the sexually monomorphic birds of paradise known as manucodes branched from the dimorphic taxa 18–20 million years ago (long before the appearance of what is now New Guinea); thus they are not very closely related to the typical, ornately decorated birds of paradise. In contrast, the dimorphic taxa began to diverge only within the past 4–5 million years, and probably did so on New Guinea (see below). Despite the extreme diversity of male ornamentation and display behaviour exhibited by this group, they are in fact very closely related, leading Sibley and Ahlquist to conclude that the seven genera for which DNA was examined could all be contained in a single genus. Thus, these birds illustrate well the key point that apparently extremely divergent species—in terms of male plumage ornamentation and displays—may not in fact be very differentiated in other respects. Beneath the spectacular feathers of male birds of paradise is a homogeneous, very closed related group of species.

For purposes of this discussion I consider only the 'classic' bird-of-paradise species—those in which male plumage and display is extreme, and in which

Fig. 8.3 The birds of paradise exhibit almost unbelievable elaboration and diversity of male plumage characters designed to attract females. In one species, the twelve-wired bird of paradise, the 'wires' of the displaying male are drawn across the face of the female. This may be one of the few cases in which tactile (as opposed to visual or auditory) stimulation of the female is a key component of the male's display repertoire. Reprinted with permission from Dover Publications, Inc.

males are promiscuous and play no role in the rearing of young. In terms of body ornamentation and associated display, some of these birds are truly bizarre, as well as beautiful (see Cooper and Foreshaw 1977, Coates 1990, and Johnsgard 1994 for illustrations), and the often spectacular plumage development has been attributed generally to selection favouring pre-mating isolating mechanisms as well as runaway sexual selection (Gilliard 1969). Some birds of paradise produce an almost unbelievable courtship spectacle, which in several species is conducted in leks by several simultaneously displaying males. The concerted actions of several males presumably increases the overall stimulus to females. Thus, female birds of paradise are recipients of a 'bombardment' of stimuli from males displaying either singly or in groups.

The interspecific differences in ornamentation between males of closely related species (e.g. Genus *Paradisaea*) evidently evolved in geographic isolation, and apparently did so rapidly, as evidenced by the recency of divergence (Sibley and Ahlquist 1985). The combination of intense sexual selection, together with a number of environmental and historical factors, discussed in the following paragraphs, may have promoted rapid evolution of ornamental characters and displays of males. Certain of these factors may also have permitted the forces of sexual selection to go beyond what has occurred in most other groups of birds.

8.7.2.1 The geological history of New Guinea

In geological terms, New Guinea is quite young. During the Pliocene, 5–6 million years ago, a great mountain-building phase occurred that established much of the island. This geological activity was the result of the collision between the northward-moving Australian tectonic plate and the Pacific plate. New Guinea, located at the edge of these two plates, is a result of this collision. The northern coastal region is the youngest part of the island and continues to be geologically active, with ongoing rapid uplift. The topography of New Guinea is characteristic of geological youth, as indicated by steep valleys, cliffs, waterfalls, and frequent land slippage. The island is highly mountainous, with 66 per cent of the land area more than 300 m above sea level and 14 per cent higher than 1500 m. In addition to the Central Ranges, which form the spine of the island, there are 11 outlying ranges of varying heights, extents, and isolation. Thus, as a result of its geological and climatic history, mainland New Guinea (not to mention the numerous nearby islands) has provided numerous opportunities for geographical isolation of small populations. This is indicated by the high numbers of endemic species in New Guinea. All of the 11 separate minor mountain ranges contain distinct endemic subspecies of birds, and many support endemic species, including bird-of-paradise species, that are absent from the Central Ranges and are confined to only one or two of the isolated ranges (from Beehler *et al.* 1986).

8.7.2.2 The biogeographical history of New Guinea

The vertebrate fauna of New Guinea is primarily of Australian origin. Few placental mammals have ever reached New Guinea, and the same is true for several otherwise widespread families of birds. This is thought to be due to a deep water barrier between the islands of Bali and Celebes to the west and Lombak and the Moluccas to the east ('Wallace's Line', named for A. R. Wallace, who first recognized the biogeographical importance of this barrier to terrestrial animals). This deep channel apparently did not, however, prevent the dispersal to New Guinea of many kinds of plants. Most of the forest flora have their closest affinities with south-east Asia.

Of special relevance to the radiation of the birds of paradise is the fact that no fruit-eating squirrels or monkeys occur in New Guinea, although they are common west of Wallace's Line (Beehler 1989). The absence of such mammals means that some of the birds of paradise have almost exclusive access to certain highly nutritious capsular fruits. Possibly as a result of traits similar to those of their corvid relatives (e.g. large body size, strong bills and feet, omnivory), birds of paradise, in a sense, may have been pre-adapted to utilize difficult to exploit, high-quality fruits. Because these birds do not destroy the seeds by ingesting them along with the fruit, and because they range widely, they are effective seed dispersers. Thus, both the birds of paradise and the trees which bear the fruits upon which they feed may be more common than they would have been had mammalian seed predators been present.

8.7.2.3   Feeding ecology and mating systems

Beehler and Pruett-Jones (1983) and Beehler (1988, 1989) have determined that in the highly decorated, promiscuous birds of paradise, a complex relationship exists between frugivory and mating system. In New Guinea, several species of trees produce complex fruits that are large and rich in proteins or fats. These fruits are produced in relatively small quantities per tree, and over long periods of time. The fruits, usually protected by tough outer capsules, are used by birds of paradise, and, for some of these tree species, these birds are the only seed dispersers.

The birds of paradise use a dietary mix composed of highly nutritious fruits and arthropods. Because the fruits ripen asynchronously, the birds visit the same trees daily to take recently ripened fruits. This pattern of spatial and temporal dispersion of high-quality food has three features critical to the development and maintenance of the promiscuous mating systems, with males displaying either alone or in leks, that characterize the ornate birds of paradise. (1) It means that foraging ranges of individuals of these birds of paradise are large and that territoriality is absent, thus many females may move through a given area on a daily basis. This can promote a male strategy of remaining and displaying day after day at one or a few sites, known to the females in the area, rather than actively searching for females (see Chapter 15). (2) Use of high-quality food means that males can satisfy their own nutritional needs quickly and thus can afford to spend most of their time, month after month, at display sites. (3) Finally, dependence on nutritious fruits means that females can support themselves and their single (usually) nestling without male assistance, by eating fruit and feeding their offspring a mix of fruit and protein-rich arthropods.

In short, the economics of promiscuity—both male display effort and female-only care of young—appears in most cases to be directly related to these birds' dependence on high-quality fruits, which, in some cases, are harvested only by birds of paradise. (The sicklebill birds of paradise are exceptions to this pattern in that they are both polygynous and primarily insectivorous; Beehler 1987*a*).

8.7.2.4 *Ecological distribution of species*

Birds of paradise show markedly restricted altitudinal distributions (Pruett-Jones and Pruett-Jones 1986). Of special significance is the Pruett-Jones' finding that the elevational amplitudes (width of altitudinal occurrence) of bird-of-paradise species are significantly less than those of other New Guinea forest birds. It has been suggested that the width of the elevational amplitude occupied by a species is a measure of that species' relative adaptation to environmental variables across altitude (Terborgh 1971; Terborgh and Weske 1975). Thirty-two of 35 bird-of-paradise species show elevational amplitudes of only 500 to 1800 m. This could have implications for the development of strongly sexually-selected traits. The narrow elevational range of most species means that the diversity of factors making up natural selection may impinge less on these birds than would have been the case had they occupied a wider range of habitats. The narrow elevational range occupied may mean that a given bird-of-paradise species has to contend with (and thus evolved responses to) fewer environmental variables than most other types of New Guinea birds. Species occupying a larger diversity of habitats inevitably must contend with more environmental pressures, such as a greater number of potential predators or parasites, for example, than ones occupying a more narrow range. As a result of the relatively narrow range of habitat types occupied, plus the apparent ease of food gathering, the overall force of natural selection may be less in birds of paradise than in many other kinds of birds. This could affect the extent to which sexually-selected traits develop. For example, sexual selection favouring male ornamentation may have been less strongly resisted by natural selection, or not resisted as early in its evolutionary elaboration, as in other New Guinea species with wider altitudinal distributions.

8.7.2.5 *The drop-out phenomenon*

In addition to the geological and climatic factors that have promoted isolation of populations, Diamond (1972) has proposed an additional means by which geographical isolation of populations may occur in montane New Guinean birds; he terms this 'the drop-out phenomenon'. Diamond points out that much differentiation has occurred along the central spine of New Guinea's mountains and suggests that in sedentary tropical species, local extinctions within portions of the range create breaks in formerly continuous populations. The gaps in distribution so formed become barriers to gene flow, which allow isolation and differentiation over time. Evidence that this process occurs in New Guinea is available in some bird populations. Certainly the narrow altitudinal range occupied by most birds of paradise (Pruett-Jones and Pruett-Jones 1986) make the drop-out phenomenon more likely to occur than would be the case for species with more extensive elevational amplitudes.

The line of thought being followed here is that sexual selection in the polygynous species of birds of paradise—as manifested by spectacular male plumage elaboration and bizarre behavioural displays—would probably not have proceeded to the extremes present today in the absence of the historical and ecological factors discussed above. The plumage and display of male birds of paradise may have become so exaggerated in part as a result of relatively weak counteracting forces of natural selection (e.g. an absence of placental mammalian frugivores and predators). The narrow altitudinal ranges occupied might further reduce the number of factors comprising natural selection. (Strong natural selection would constrain the evolution of extreme plumage morphology and associated displays.) These factors, in a sense, have 'permitted' the components of sexual selection to drive the evolution of male ornaments to the marvellous extremes shown in this group.

8.8   Conclusions and summary

Despite Darwin's original insight that strong sexual selection might increase the frequency or rate of evolutionary divergence, this issue was ignored for many years. Differences in plumage and other display traits were thought to have evolved in the context of species recognition and, specifically, to serve as 'isolating mechanisms'. More recently, the possible importance of sexual selection in the speciation process has come to be more widely appreciated, although the proposed relationship between sexual selection and speciation remains rather speculative. Sexual selection is thought to increase the rate of divergence of allopatric or parapatric populations. Speciation (reproductive isolation) can occur with little genetic differentiation (e.g. mutations of as few as two or three loci), and this can account for the great overall genetic similarity sometimes seen in populations that behave as separate species. As a rule, sexually-selected traits differ more between populations of close relatives than do other characteristics. One study involving two species of manakins found that when two species came into geographical contact and hybridized, attractive male ornamental characters of one of them introgressed into the other population. This study also determined that the overall genetic mixing in the zone of hybridization was far less than the phenotypic appearance of males would suggest.

Although it is generally agreed that rapid differentiation of populations occurs as a result of strong sexual selection, the question of fixation of new display traits has received little detailed attention. In addition to directional selection, three factors may help to account for rapid fixation of newly evolved characters (e.g. feather colours): involvement of few genes or loci, small population size that promotes inbreeding and selection against most mutations.

In traditional speciation theory, hybridization has been viewed as a negative factor indicating that pre-mating isolating mechanisms have not been perfected. Evidence supporting this view is seen in the fact that hybrids are often sterile (e.g. Haldane's rule). Recently, however, it has been argued that hybridizing individuals may obtain positive fitness effects. There is now some supporting data to the effect that, under certain environmental circumstances, hybridization has positive effects for its practitioners; e.g. intermediate morphological traits may be superior to the parental condition, and increased heterozygosity may produce 'hybrid vigour.' It appears that in geographical areas of hybridization, genes controlling sexually-selected traits may introgress more rapidly and over a greater geographical distance than genes affecting other traits. This finding provides indirect support for the idea that sexual selection can accelerate differentiation.

9 Male mate choice and intrasexual competition

9.1 Introduction

This chapter deals briefly with two other issues that, together with female mate choice, make up the subject of sexual selection. The first is mate choice by male birds. In contrast to female mate choice, this subject has received little attention. The second topic, intrasexual competition for mates and matings, is widely recognized as extremely important in the lives of vertebrate animals. Nevertheless, it too has received far less attention than mate choice by females. Typically, such competition is among males; female–female competition, however, can also be intense in birds and is sometimes even fatally violent.

9.2 Mate choice by male birds

Most studies of mate choice in birds have focused on the interaction between female choice and male displays and/or ornamental traits. In lek-promiscuous species (see Chapter 15), males are notoriously non-discriminating. Males of some lekking grouse will attempt copulation with straw hats or other similarly inappropriate objects. For males of such species, inseminating females is the 'end-all' and characteristics of individual females are probably not important. In contrast, in monogamous species with biparental care (which includes the vast majority of bird species), males should be discriminating, simply because the fitness of males is strongly influenced both by their own parental efforts and by the traits of their mates (genetic or parental quality, fecundity and fidelity). Likewise, males of classical polyandrous species should evaluate prospective mates.

Thus, for a majority of species, mate choice should be important to males as well as to females. Circumstantial support for this suggestion is the fact that in many species of birds, males and females are similar in size and plumage. In fact, in many species representing a number of orders, females bear the same kinds of ornaments exhibited by males. However, to date, in only in a few cases have ornaments of females been shown to be important in mate choice (e.g. Jones and Hunter 1993; see Section 4.7).

What kinds of traits might males of socially monogamous species seek in prospective mates? First, we might ask, under what conditions should mate choice by male birds be important? Although few experimental studies have attempted to address this issue, the answer seems to be that males should exhibit choosiness in mate selection when they make a major parental contribution, and when their own reproductive success will be affected by qualities (genetic or phenotypic) possessed by their mates. These two points are true for the monogamous majority of avian species. Therefore, male mate choice is expected to occur primarily in socially monogamous species. Although experimental studies of mate choice by male birds are scarce, a few do exist. Four of these are briefly considered here.

Pinyon jay

In her studies of mate choice in captive pinyon jays, Johnson (1988*b*) found that males tended to choose dominant and larger-than-average females as mates. Subsequently, Johnson and Marzluff (1990) reported that the longest-lasting pairs of pinyon jays are composed of a relatively small male and a large female. In nature, dominance probably correlates with age and experience, thus males may prefer to mate with older, more experienced females. It is easy to imagine the benefits of mating with an experienced individual. Why might larger females also be preferred?

Pinyon jays often nest in late winter and early spring in regions where environmental temperatures are low, and a female must remain on the nest almost continuously, leaving it briefly only when its mate arrives with food (Marzluff and Balda 1992). Large females may be better incubators and brooders, in that they may be better able than smaller individuals to withstand the often cold and snowy environmental conditions typical of their nesting period, as a result of a more favourable surface area-to-volume ratio than that possessed by smaller females. Moreover, larger females may also be able to build up larger fat reserves prior to the onset of incubation than can smaller females.

House finch

Female house finches vary in plumage coloration. Some are a drab brown, while others exhibit a subdued version of male coloration. Moreover, unlike the usual situation in males of dichromatic species, first-year females are brighter than older females. To test for male preferences, Hill (1993a) conducted both laboratory experiments and field studies. In the former, male house finches clearly preferred the most brightly-plumaged females (just as females preferred the most brightly-coloured males). Perhaps not surprisingly, in view of the inverse relationship between age and feather brightness in females, coloration of females was not related to overwinter survival, reproductive success, or body condition. Thus, there is no indication that plumage brightness of females is related to quality. Among free-living birds, Hill determined that the 'highest quality' males preferred to mate with older females. When first-year females were removed from the analysis, males preferred brighter, as well as older, females. It appears that age of females is the primary criterion used by males, with colour being of secondary importance.

Hill (1993a) tentatively concludes that the attraction of males to more colourful females may be a correlated response (see also Muma and Weatherhead 1989); i.e. the genetic basis for female mate choice, which is adaptive (Hill 1991), also affects the mate choice behaviour of males, at least under laboratory conditions.

Barn swallow

In Møller's (1994) population of barn swallows, no females remained unmated for a breeding season. How promptly females obtained mates was correlated with tail length, which suggests that tail length in females is a sexually-selected trait. Long-tailed females tended to mate with the preferred long-tailed males. In addition, long-tailed females arrived on the breeding grounds earlier and were more likely to produce a second clutch than were shorter-tailed females. Møller concludes that the benefits to a male of mating with a long-tailed female can be accounted for by the greater reproductive potential of such females.

Zebra finch

In controlled mate choice trials, Burley et al. (1982) found that males preferred to mate with females wearing a black band over a naturally orange leg, and avoided females wearing blue bands. Although the functional significance of this preference was not known, Burley (1986a) subsequently showed that over a 15-month period in a laboratory situation, preferred (attractive) males produced twice as many offspring as unattractive ones, probably due to the greater parental investment of their mates. This study is discussed in more detail in Chapter 6, with regard to its mate choice implications for each sex.

9.3   *Male–male competition*

Probably nearly everyone reading this book has seen male robins (American or European) fighting on the lawn like tiny gamecocks. Male–male competition takes many forms in addition to overt aggression, and it occurs commonly in both monogamous and other mating systems. For birds, as well as for other groups of animals, intrasexual competition is generally more pronounced among males. Male–male competition is usually thought to be most intense in polygynous species, due to the much greater variation among males in obtaining matings (Payne 1979, 1984; Arnold 1983; see below). Classically polyandrous birds, in which females are the more competitive sex and provide no parental care, are the exceptions that prove this rule (Emlen *et al.* 1989).

Intense competition among males occurs for reasons general to almost all higher animals and for some reasons peculiar to birds. A competitively inferior male may leave no offspring, even though a single male can produce far more sperm than there are eggs potentially available to be fertilized. In contrast, female birds lay only a relatively few large and energetically costly eggs. As a result, the number of eggs (which represent the male's as well as the female's sole route to evolutionary fitness via offspring) and the number of females producing them are in limited supply. This great disparity in size and number of gametes produced by the two sexes means that females and the eggs they produce are a scarce and extremely valuable resource for males, in an evolutionary sense. (Even if an ejaculate is costly, which may well be the case, males can produce an ejaculate far more frequently and probably over a longer period than a female can produce eggs.) Thus, females, and more specifically their eggs, are the 'limiting factor' in the reproductive success of males, and this is thought to be the ultimate basis for the intense competition observed between males of many animal species.

Other factors, too, may contribute to the intensity of male–male competition. In many kinds of birds, for example, including socially monogamous ones, the adult sex ratio is not 50:50; rather, it favours males (e.g. Selander 1972; Breitwisch 1989; Hill *et al.* 1994*b*). Several studies have demonstrated a 'floating', non-breeding population of male birds in socially monogamous species (e.g. Hensley and Cope 1951; Stewart and Aldrich 1951). Regardless of the social mating system, numerical superiority of males automatically further increases the value of each female (and of each mating with a female), and thus the intensity of competition among males.

The fundamental difference in the number and size of the gametes produced by the two sexes and differences in parental contributions, together with a male-biased adult sex ratio, when it occurs, set the stage for many of the attrib-

utes related to the reproductive biology of birds. These range from conspicuous traits, such as the generally greater size and aggressiveness typical of
males (including monogamous ones), to the generally earlier return by males
of many migratory species to their breeding grounds in northern and southern
temperate zones. Time of arrival is an important component of male–male
competition for territories, and the early bird gets the territory and the female.

Male–male competition has been demonstrated by numerous manipulative
studies utilizing playback of recorded song. Typically, during the breeding
season, when song is most prevalent, territorial male passerines respond to
song (apparently interpreted by the bird as an intruder) so aggressively that
they are easily captured by human researchers. The functions of song and
other vocalizations in sexual selection were considered in Chapter 4. Experimental studies dealing with the role of morphological traits in male–male
competition are far less common. Studies considered here focus on the red
junglefowl, two other galliform birds (the ring-necked pheasant and the
turkey), and the pinyon jay. These three galliforms are strongly dimorphic in
morphology, with striking male plumage and facial ornaments, and are polygynous, while the pinyon jay is sexually monochromatic and socially
monogamous.

Red junglefowl
In most polygynous species, sexual differences in body size are pronounced,
and, in some, special morphological structures designed as weapons have
evolved, e.g. the metatarsal spurs of most pheasants and some other galliforms (see Section 4.8 for a discussion of spurs). One such species is the red
junglefowl, a member of the Asian pheasant radiation and the ancestor of the
domestic fowl or chicken (Fumihito *et al.* 1994). In this species, male–male
competition is extreme (Fig. 9.1). Adult male red junglefowl are highly aggressive; they fight by pecking and flogging with the wings and especially by
striking with the hard, sharp-pointed spurs. In these birds, the spurs do not
appear to be important in female mate choice (Zuk *et al.* 1990*a*), and probably
function solely in the context of male–male competition.

Ligon *et al.* (1990) studied the relationship between male–male competition
and morphological traits in red junglefowl to gain information about the nature
of the traits involved with this aspect of sexual selection. They analysed 40
morphological characters and selected 14 (most of which were not correlated
among themselves) that seemed to provide a description of all ornamental
characters. The staged male–male conflicts were of two sorts.

First, randomly drawn pairs of one-year-old roosters, referred to as cockerels, were allowed to fight. Fights were also staged between an older male
(cock) and a cockerel. The contestants had been spatially and visually isolated
from other males for several weeks prior to the contests, which leads to the
development of a high degree of 'cockiness', manifested as aggressiveness and
a readiness to engage any other male in combat. Fights were conducted on

Fig. 9.1   Two fighting male red junglefowl. This species is the ancestor of all domestic breeds of chickens. The ancient sport of cockfighting is based on the aggressiveness of the junglefowl, with artificial selection having produced their descendants, the gamefowl.

neutral ground away from the areas where the contestants had been penned. These contests were designed to ensure that the roosters could not injure each other (see Ligon *et al.* 1990 for details). The fights provided a biologically relevant means of assessing the relative 'condition' (endurance and perseverance) of each pair of contestants. The outcomes of the conflicts (determined by one bird's failure to maintain aggressiveness) were then correlated with the development of ornamental traits of each contestant. Comb size was the only morphological trait that proved to differ between winners and losers. This was true even in contests between the cocks and cockerels. Although the older birds were significantly heavier, as well as significantly more ornate in some plumage characters, they did not have larger combs and they lost a majority (non-significant) of the staged fights.

Second, dominance status for 86 haphazardly drawn pairs of free-ranging cockerels that had grown up together was also assessed. Because these roosters were individually known to each other and interacted frequently, fighting did not occur. Instead, one bird usually signalled its subordinate status simply by actively avoiding the other. In these measures of status, differences between winners and losers in comb size and body size were significant; i.e. bigger birds with larger combs tended to be dominant. This result is what one might expect among free-ranging young males that had grown up together.

Thus, the comb, which was correlated with winning fights and with general dominance status, appears to be something of an indicator trait for male red junglefowl. Neither of the two approaches described above provided any evidence that the highly ornate plumage of male red junglefowl is currently important in male–male competition (see Chapters 5 and 6).

Ring-necked pheasant
Mateos and Carranza (1997) studied the role of ornaments in male–male agonistic encounters in ring-necked pheasants. The most important characters

appeared to be the length of the ear tufts and wattle size. Depending on the social setting, males could either maximize or minimize wattle size and could likewise prominently exhibit or conceal the ear tufts; i.e. the wattles and ear tufts appear to be 'coverable badges' (Hansen and Rohwer 1986).

Mateos and Carranza (1997) propose that these head ornaments provide information about willingness to fight and fighting ability, and the ability to hold a territory. The total time spent displaying (which includes inflated wattles and erect ear tufts) is correlated with testosterone level, physical condition, and rank. Thus, these head ornaments are accurate indicators of male condition.

Turkey
Like red junglefowl and ring-necked pheasants, male turkeys possess a number of ornamental features. In particular, the length of the male turkey's snood, like the size of the junglefowl cock's comb and the male ring-neck pheasant's wattles, accurately reflects male condition. The turkey gobbler's snood is of primary importance to females as they make mate choice decisions (Buchholz 1995).

In addition to its importance as a signal to females, the length of the male's snood is predictive of the outcome of male–male competition. In addition, experimental studies using artificial males demonstrated that male turkeys assess snood length of potential competitors independent of other male characteristics (Buchholz 1997).

Pinyon jay
Johnson (1988*b*) determined dominance relations in male pinyon jays housed together in large aviaries by use of several indicators of the relative status of pairs of individuals, such as supplantings from perches or food, pecks, and stereotyped appeasement postures by subordinates. (See Balda and Bateman 1972 and Marzluff and Balda 1992 for illustrations of these postures.) Analysis of the relationship between male dominance and morphological traits indicated that heavier males with larger bills tended to dominate smaller, shorter-billed males. Thus, at least under conditions of captivity, large body and bill size are favoured in agonistic encounters among male pinyon jays. These traits, however, are probably not the main determinants of male mating success. Instead, testis size and plumage brightness were found to be the best predictors of male success with females (Johnson 1988*b*, see Chapter 6). Johnson's (1988*b*) study of captive pinyon jays suggests that the traits important in male–male dominance interactions and those important in female mate choice are not the same. This conclusion stands in contrast to the red junglefowl, ring-necked pheasant, and turkey, where a particular trait (comb size, wattle size, snood length) was related both to male–male competition and to female mate choice.

9.4   Female–female competition

In general, competition among females for mates is not nearly as widely appreciated as is such competition among males. For example, while acknowledging the existence of a few 'role-reversed' taxa, such as jacanas and phalaropes, Arnold (1994, p. 58) writes that '... our minds have become closed to the possibility of sexual competition between females.' If Arnold means that theorists have focused almost exclusively on sexually-selected competition among males, then this assertion is accurate. On the other hand, if he is referring to the scientific community as a whole, then it is incorrect. Considerable evidence indicates that many field ornithologists are aware of the existence of sexually-selected competition among female birds. For example, Arcese (1989) provides a number of references that document competition and territory defence among females. Females of the classically polyandrous buttonquail are extremely aggressive, so much so that they are used by humans in a spectator sport analogous to cock-fighting (Johnsgard 1991).

Depending on specific circumstances, conflict among females, regardless of its intensity, may or may not be classified as sexual selection. Females also compete for reasons having nothing to do with mate selection. When females clearly compete for a resource such as territorial space and the resources it holds, and are not particularly concerned with pairing with a particular male, this should not be viewed as a manifestation of sexual selection. In contrast, when females actively compete specifically for mates, as in the socially monogamous moorhen (Petrie 1983) and the classically polyandrous spotted sandpiper (Oring 1986), they are exhibiting the standard form of the intrasexual component of sexual selection.

Intrasexual competition among female birds, although typically less conspicuous than among males, possibly in part because it is also less common, is nevertheless widespread and can be fatally intense (e.g. Hill 1986). Extreme female–female competition appears to be related to the importance to the female of exclusive ownership of a territory and a mate. For example, in the socially monogamous black-headed grosbeak the male parent defends the territory, contributes significantly to incubation, and feeds the nestlings and fledglings. Hill (1986) described a case where two individually-marked female grosbeaks fought 'to the death'. The winning female owned a nest with eggs almost ready to hatch, while the loser, which had intruded into the nesting female's territory, sustained severe injuries and is known to have died between one and two days after the battle. So much for coy females. Competition doesn't get much more severe than this!

Why might female–female competition sometimes be so severe in a typical, sexually dimorphic, socially monogamous bird, such as the black-headed grosbeak? Several possible factors come to mind. (1) The breeding season of this species is brief. In central New Mexico, where Hill's study was conducted, nesting or renesting does not occur after 1 July, and, with time (i.e. the nesting season) running out, the intruding female probably did not have a territory or mate. (2) In this 'strictly monogamous' species (Hill 1986), territorial males are the most limited reproductive resource for females because, while few first-year males breed, most first-year females attempt to do so. On his study site Hill found that the ratio of territory-holding adult males (usually birds more than one year of age) to females was 0.71:1.00, which differed significantly from unity. Thus, these two females probably fought for access to a territory-holding, parenting male.

The patterns of female–female competition apparently follow the same rules that characterize male–male competition, with one important difference. Sexual competition among females probably is not usually causally related to a numerical shortage of males *per se*; instead, it is more likely to be based on a shortage of territory-holding, parental males. That is, female–female competition may be intense even in those species where there are more males (e.g. territory holders plus 'floaters') than females present in the population. (This appears to be true for black-headed grosbeaks.)

In woodpeckers, female–female competition can also be pronounced (personal observation). In addition to the issue of territorial space, this competition, like that of the black-headed grosbeak, may be due to the male's essential and continuing contributions to his—and his mate's—reproductive success over the entire reproductive cycle. Among woodpeckers, the critical importance of the male to the success of the breeding effort of the pair has been documented for four North American species with very different ecologies (Lawrence 1967). Male woodpeckers are typically highly territorial. They also assume the primary role in excavation of the nest cavity, contribute most of the incubation effort, including all nocturnal incubation (see Section 1.5.4), and provide most of the food to nestlings. Thus, the value to a female woodpecker of a territory-holding male goes far beyond the quality of the male's territory and the sperm he will contribute to the eggs. Even with the acquisition of territorial space and sperm, it would be impossible for a female woodpecker to reproduce without the ongoing contributions of the male to some or all aspects of parental care. Despite the full efforts of two parents, starvation of some nestlings occurs in broods of woodpeckers. Such 'brood reduction' appears to be of frequent occurrence, at least in temperate zone members of this group (e.g. Ligon 1970 and unpublished data), providing additional evidence for the importance of biparental care, and thus for the significance of female–female, as well as male–male, competition for mates.

Similarly, Arcese (1989) showed that competition among female song sparrows for mates and breeding sites was a major factor in determining their reproductive success. As in the grosbeaks and woodpeckers, female song sparrows fought for territorial space—nesting sites and food—and probably for paternal care of offspring (also see Liker and Szekely, 1997).

Dale *et al.* (1992) provide another example of female–female competition. Female pied flycatchers search for, and choose among, territorial males by moving rather rapidly from one male's territory to another. These mate searches by female flycatchers were often affected strongly by agonistic inter-actions with other females. The freedom of movement of searching females was constrained by the aggressiveness of other females, which decreased the number of males a female visited. This field study, which employed both observational and experimental techniques, provides strong evidence concern-ing the importance of competition among females as they attempt to choose mates and breeding territories.

Other than the many papers on agonistic interactions between females within groups of domestic fowl (e.g. Collias 1943), manipulative experimental studies of female–female competition are rare. One such study of a wild species was conducted by Johnson (1988a). Using captive pinyon jays, Johnson documented competition within groups of females. She recorded dominant–subordinate relationships for females housed together in large outdoor aviaries, in the manner described previously for males. However, unlike the results for males, no morphological trait or combination of traits clearly discriminated between dominant and subordinate females. Of the mor-phological traits measured, brightness of malar feathers and body weight seemed to be most closely related to social status.

Among female pinyon jays, dominance functions both in general social relationships and in mate acquisition. Johnson's (1988a) observations of captive jays suggest that, although dominant females may have more ready access to males, overt aggression is not the form of competition for mates usually taken by females. Instead, at least in large aviaries, female pinyon jays compete for the attention of a preferred male by calling to him and attempting to feed him, rather than by attempting to exclude other females via physical aggressiveness.

9.5   Conclusions and summary

Mate choice by male birds, which has to date rarely been studied, seems to be related to attributes of females that are associated with parenting. For example, male pinyon jays prefer to mate with females that are large and experienced. Both of these traits are likely to enhance the quality of parental care provided by the female. Similarly, male house finches appear to be most strongly attracted to older (more experienced) females that also possess relatively bright plumage.

Although female mate choice is the most studied aspect of sexual selection, to the human observer of typical songbirds, male–male competition, which takes many forms, is often much more conspicuous. In contrast, female–female competition is less frequently recorded, but, depending on the species, it can also be a major aspect of sexual selection. In birds, competition among females may be both rather common and of high intensity. This is because the vast majority of birds are socially monogamous with shared parental care. Under such a system, for individual reproductive success, a mate (especially one with resources such as a territory) is as important to a female as vice versa. A difference between the sexes is that males compete for access to females, some of which are their mates and some of which are not (extra-pair copulations), while females compete with each other primarily for male partners-in-parenting that possess territories.

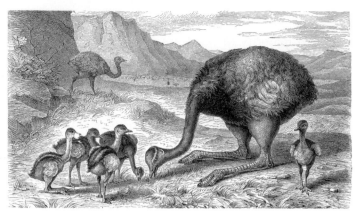

10 The benefits of oviparity and the evolution of parental care

10.1 Introduction

Why are there no viviparous birds? This question is critical to obtaining an understanding of parental care patterns and thus mating systems of birds. The benefits and costs of egg-laying have had effects on the reproductive biology of birds that cannot be overstated. The production of eggs, along with the evolution of physiological (e.g. endothermy) and behavioural (e.g. incubation) traits associated with the care of eggs and hatchlings, has set birds apart from all other animals with regard to parental care, and particularly with regard to paternal care.

All other classes of vertebrates contain both oviparous and viviparous species. Even in the Class Mammalia, a very few species, the monotremes or Prototheria, lay eggs. In the marsupials, or Metatheria, the first two-thirds of embryonic development takes place within a shell membrane in the uterus, and the young are born in a rudimentary condition. In our own group, the placental mammals, or Eutheria, the degree of development at birth is highly variable. In some species, the newborn is as helpless as the most altricial of birds, whereas in others, the young are born in a physically advanced state, and ready to run.

10.2 Avian oviparity

The issue of avian oviparity has been treated comprehensively by Blackburn and Evans (1986), who point out that biologists have widely assumed that viviparity, particularly of the placental mammal variety, is a superior mode of producing offspring. Blackburn and Evans (1986) refer to this as the eutherian bias—that birds are somehow constrained from this mode of embryonic development by one or more features of their morphology or physiology. Factors that have been proposed to constrain the evolution of viviparity in birds include flight and the anatomical and physiological factors associated with it, the cleidoic egg, the mode of sex determination, immunological obstacles, and special requirements of lung development (see Blackburn and Evans 1986, Table 1). Blackburn and Evans (1986) provide a detailed assessment of these factors and conclude that each of them is probably insufficient to constrain or prohibit the evolution of viviparity in birds. Blackburn and Evans (1986) suggest that the eutherian bias has led to a lack of appreciation for the extraordinarily successful reproductive strategies of birds, all of which ultimately are based on their universal oviparity.

10.2.1   The evolutionary transition from oviparity to viviparity

Any transition from egg-laying to birth of living young could not have occurred in one step. Each of the intermediate stages in the evolutionary transition from oviparity to viviparity must permit successful reproduction (Blackburn and Evans 1986). Studies of reptiles, in which viviparity has repeatedly evolved from oviparity, demonstrate that such a transition must result from an evolutionary increase in the proportion of the developmental period that precedes oviposition. That is, there must be a gradual lengthening of the period of egg retention and intrauterine development. Viviparity is achieved when the egg is retained until term.

During this evolutionary transition, natural selection leads to a reduction in the thickness and degree of calcification of the shell membrane. Following the loss of the shell and the attainment of viviparity, matrotrophy (maternal investment of nutrients, etc. during embryonic development) can evolve. Something like this scenario can be seen in the development of marsupials, in which the shell membrane is retained for more than half of the intrauterine period of development; newborn marsupials, however, are not as fully developed as a newly-hatched altricial bird. Thus, it appears that, given strong selection for the intermediate stages between typical oviparity and viviparity, there is nothing that would prohibit the evolution of viviparity in birds, as has occurred repeatedly in reptiles, not to mention other major taxa of vertebrates. This leads to the next point.

10.2.2   Why are there no egg-retaining birds?

Even if viviparity proper in birds has not been strongly favoured by natural selection, the question still arises: why do birds not retain eggs for some period of time? Intrauterine egg retention would decrease the period of egg vulnerability and parental incubation responsibilities. The only birds known to retain their eggs during early development are certain parasitic cuckoos, in which the embryos develop through the 25-hour primitive streak stage before the egg is laid (Payne 1973; Blackburn and Evans 1986). This appears to be a specific adaptation associated with the reproductive strategy of social parasitism, in which selection has favoured early hatching of the parasitic chicks compared to the chicks of their hosts.

Blackburn and Evans suggest that if there were strong selection for it, egg retention should be achievable in birds. Thus, the question is, could a period of prolonged egg retention be beneficial to birds as we know them? Perhaps the best way to view this issue is by considering some of the costs of egg retention. There are only two ways by which retention could be accomplished, via either (1) the simultaneous retention of the entire clutch of eggs, followed by oviposition of the entire clutch, as occurs in oviparous reptiles, or (2) short-term retention of each individual egg, followed by its oviposition and then ovulation of the next egg.

With regard to the first scenario, given the size of a bird's egg relative to the space available in the female's abdomen, it is unlikely that for most species the number of fully-formed eggs of a typical clutch could be retained within the female. (Thus, the limited space can be viewed as a constraint that inhibits the evolution of retention of an entire clutch of eggs; this would not apply to species that produce a one-egg clutch.) If this is so, then birds can produce more eggs per clutch by evicting each egg from the abdomen soon after it is produced rather than by retaining eggs. The second possibility also has costs associated with it. Short-term retention would probably decrease clutch size and/or the number of clutches produced per year because of time and resource limitations imposed by seasonality. In fact, it would take considerably longer to produce a clutch, and thus hatchlings, if each egg were retained for a time in the female's reproductive tract. For the majority of species that lay an egg a day (more or less), retention of each egg for only a single day would almost double the length of time required to produce a clutch of partly-incubated eggs. For example, a four-egg clutch would require seven days to produce.

Moreover, if embryonic development had begun in each egg prior to its being laid, then either incubation would have to get underway with the laying of the first egg or incubation would have to be withheld from the eggs until the clutch was complete. Fresh eggs can remain viable for many days before beginning to develop, but for contemporaneous species, this is probably not true of eggs that receive a period of incubation followed by a longer period of no incubation. Thus, if each egg was retained in the uterus one by one, the number of eggs in a clutch that remained viable and hatched would almost surely be fewer than what is seen in a species-typical clutch.

In short, one of the great benefits of the avian mode of reproduction, the ability to produce a nest full of simultaneously hatched young, would probably be decreased or lost if each egg were retained for a period of time within the female's reproductive tract. Finally, Blackburn and Evans (1986) point out that under the conditions of consecutive retention of eggs, the development of viviparity would entail a clutch size of one, by definition. Presumably, for those species producing a clutch size of one (e.g. most penguins, procellariforms), the benefits of two care-givers under certain ecological conditions have overridden any selection for viviparity.

10.2.3 Benefits of sequential egg production

If the benefits of laying an egg promptly are greater than the benefits of egg retention, then the latter should not evolve. It appears that there are significant benefits associated with producing a clutch, one egg at a time, as promptly as possible. This is probably associated both with flight and with a high metabolic rate.

1. By sequentially concentrating resources on the production of each egg, a clutch can probably be completed as rapidly as physiologically possible. That is, given the relatively large size of bird eggs, it is likely that a clutch can be produced more quickly in this way than by any alternative mechanism. For the majority of bird species, the rate of production is about one egg every one– to two days (Gill 1995). Prompt production of a clutch, followed by incubation, often by both parents, allows simultaneous development of all eggs. This pattern of egg production also means that if the clutch is lost to predation, either before or after it is completed (which probably happens more often than not in most species), rapid initiation of a new clutch is possible.

2. By rapid clutch production and simultaneous hatching of young, the benefits of parental care, particularly biparental care, can come into play. The ability of both parents to contribute to the development of offspring is less available in viviparous mammals, and must be a major benefit of oviparity as practiced by birds. In contrast to mammals, both male and female birds can contribute fully and equally to the development of the embryos via incubation and to nourishment of the nestlings as soon as the eggs hatch, simply because, in nearly all cases, the adults gather food from the environment to feed their chicks. In contrast, paternal care is rare in mammals, which must be due, in part, to the inability of males to provide these kinds of care to their offspring.

Recognizing these benefits of the avian mode of reproduction illustrates the weaknesses of the eutherian bias. Rather than egg retention being a potentially better, but unattainable, strategy due to some sort of constraint, the absence of egg retention in birds may instead be because, along with the complementary and critical parental care adaptations, non-retention is actually the more productive strategy; i.e. for birds, non-retention of eggs may be the most profitable way to produce offspring. In short, the benefits of oviparity, rather than constraints on the ability to evolve viviparity, may best explain why all birds are oviparous. Blackburn and Evans (1986) point out that the suite of avian reproductive strategies provides an alternative means of achieving the main benefits of egg retention and viviparity. First, incubation provides an endothermic environment for embryonic development. Second, post-hatching parental care provides both a continuation of a suitable thermal situation (brooding) and an abundance of high-quality nutrients. Third, embryonic wastes are managed via the non-toxic uricotelic system. Fourth, the thick calcareous, but porous, shell and membranes protect the egg from desiccation and from attack by microbes, and provide a mechanism for oxygen exchange from the shell to the allantois, the organ of respiration. Fifth, cryptically-coloured eggshells and construction of elaborate nests provides protection (though far from perfect) to the eggs and hatchlings. Other lineages that contain both oviparous and viviparous species, most notably reptiles, do not possess all of these traits, which, in concert, have made the avian mode of reproduction so successful.

In short, although the coadapted suite of reproductive specializations exhibited by extant avian species probably evolved early in the history of birds, there is no good reason to consider avian oviparity as inferior to the mode of reproduction practised by placental mammals.

10.3   The origin and evolution of parental care

Parental care is one of the most conspicuous characteristics of birds. The term 'parental care' of course includes many components, such as nest building, incubation, feeding and protection of nestlings and fledglings, and so forth; thus, parental care is composed of a highly complex system of behaviour. With the exceptions of the social parasites and some poorly studied megapodes that bury eggs singly and desert them (extremely minimal parental care), one or both sexes of all species of birds exhibit parental care. Beyond this generalization, the diversity of birds ensures that the details of parental care will also be diverse. This section considers the origins of parental care in the Class Aves.

Monogamy is by far the most common mating system in birds (Lack 1968) and, associated with this fact, biparental care is the norm. Gilliard (1963) estimates that 99 per cent of the world's birds exhibit biparental care. Most writers on the subject of avian mating systems have assumed that most or all recent systems have been derived from a monogamous system with shared parental care (see references in van Rhijn 1984 and Wesolowski 1994). van Rhijn (1984) points out that this hypothesis is based on the physiology of bird reproduction and on the fact that, in recent birds, the monogamous biparental care system prevails.

Although biparental care probably appeared early in the evolution of neognathous birds (all modern birds except the paleognathous ratites and tinamous), monogamy/biparental cannot be the answer to the question of the *origin* of parental care. At some point parental care must have been preceded by an absence of care. With this in mind, a key question arises: what factors led from no parental care to any form of parental care? This, then, leads naturally to the next critical question: what was the initial form of parental care? Finally, if uniparental care was the first parenting system to evolve, what was the evolutionary sequence of events that led from uniparental care to biparental care? Addressing these issues should help to expand our thinking about the various mating systems found in contemporary birds.

10.3.1   The ancestral state of non-care in birds and the origins of parental care

Because parental care is so rare in living reptiles, and so comparatively rudimentary in those reptiles that do exhibit it, it is reasonable to assume that the absence of care was the situation in the ancestors of birds (but see Section 10.4.2 below). According to Wesolowski (1994), the limited maternal care seen in crocodilians probably evolved independently of the evolution of parental care in the avian lineage. Moreover, in birds, exclusive maternal care probably evolved independently several times from shared parental care.

In considering the origins and subsequent evolution of parental care, the possibility should be kept in mind that avian patterns of parental care may have evolved even before birds were birds! (See Wesolowski 1994, p. 41 for a refutation of this idea.) Whether parental care of some form preceded the evolution of feathers—and thus preceded the evolution of birds—is not known. If the initial form of parental care was simply protection of a cluster of eggs, then it is possible that the origin of parental care in birds did indeed appear before feathers. However, if endothermy evolved after the appearance of feathers, as seems likely, then we might expect the evolution of more critical parental care, such as the transfer of heat from the parent to the egg, i.e. incubation, very early in the history of true birds.

Wesolowski (1994, p. 44) provides a detailed treatment of the evolution of the coadapted suite of features that must be involved with the evolution of parental care:

> … in the process of parental care evolution, not only do parental responses change but properties of eggs and young change too. The eggs and young become more and more dependent on parental activities, up to the point at which they totally lose the ability to develop normally without care. Beyond this limit, the care becomes obligatory, regression to the state without care is no longer possible, and the question, Why care at all? becomes immaterial. One has only to predict which sex should provide care.

This quote emphasizes a critically important point to keep in mind. Not only did parental care evolve, but patterns of embryonic development and characteristics of hatchlings also changed over time. As one obvious example, at some point in the past the eggs of the ancestors of birds did not require the heat provided by parental incubation in modern birds. The transition from eggs capable of developing without incubation to eggs utterly incapable of development without it must surely have been a gradual one, even if it occurred within a relatively short period of geological time, e.g. a few million years. Thus, as parenting behaviour was evolving, the development patterns of embryos (eggs) and hatchlings (e.g. transitions from precocial to semi-precocial to altricial nestlings) were also developing, and these two factors became so thoroughly intertwined that parental care became an almost inescapable task for birds.

In keeping with his point that embryonic and chick development evolve in parallel with parental behaviour, Wesolowski (1994) also argues that egg size of early birds increased greatly in mass and quality prior to the evolution of parental care, due to selection favouring superprecocial young (related to the evolution of flight), and that parental care evolved of necessity to enhance the survival of large (and thus valuable) eggs. Once intensive parental care evolved, selection then favoured production of smaller eggs (Fig.10.1).

With regard to the relationship between parental care and the parenting sex, for a variety of reasons it is unlikely that biparental care arose directly from a state of no care. In the first place, the reproductive roles of non-parenting vertebrates are very different (e.g. the reproductive behaviour of male and female lizards shows little or no overlap). In contrast, one of the most striking things

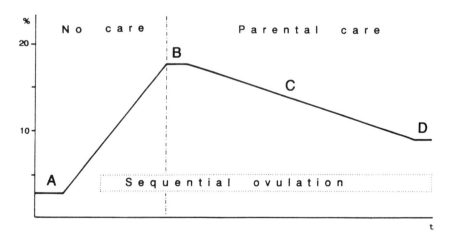

Fig. 10.1   Change in avian egg size over evolutionary time, during the reptilian–avian transition. (A) Reptile (superprecocial); (B) early bird (superprecocial); (C) contemporary bird (precocial); (D) contemporary bird (altricial). The position of events along the time axis marks the relative sequence of appearance, but not the absolute amount of time elapsing between the events. Reprinted with permission from Wesolowski, T. (1994). On the origin of parental care and the early evolution of male and female parental roles in birds. *American Naturalist*, **143**, 39–58.

about most birds is the similarity of the roles of males and females in the business of rearing young. It is improbable that the numerous and complex evolutionary changes—physiological, endocrinological, and behavioural—required to turn non-parental animals into parenting individuals would occur simultaneously in both males and females. In general support of the idea that, in birds, uniparental care evolved first, biparental care in fish has evolved independently several times, and it is thought to have always been derived from uniparental care of one sort or another (Gittleman 1981; Gross and Sargent 1985).

10.4   Paternal care: the original form of parental care?

One of the most unusual aspects of the reproductive biology of birds is the prevalence of male parental care. From our perspective as mammals, both empirical facts and theory make the evolution of maternal care easy to appreciate. The development of sole paternal care in birds (and some fish and amphibians) is less intuitively obvious. Nevertheless, there is accumulating support for the proposition that exclusive paternal care may have been the original form of parental care in birds (van Rhijn 1984, 1990; Ezlanowski 1985; Handford and Mares 1985; Wesolowski 1994). Wesolowski (1994) suggests (1) that uniparental, rather than biparental, care probably evolutionarily succeeded the absence of parental care, and (2) that sole paternal care was the first parenting system to appear in birds. Wesolowski assumes that feathers and powered flight arose independently of, and possibly prior to, the evolution of the reproductive strategies considered here. Also, like van Rhijn (1984) and Handford and Mares (1985), he assumes that territoriality of males, based on habitat suitable for breeding, preceded any parental care. The case for paternal care as the original form of parental care is based, in large part, on two points, one hypothetical and the other empirical (for an alternative view, see McKitrick 1992).

10.4.1   Hypothetical scenario for the evolution of paternal care

van Rhijn (1984) presented a model for the evolution of parental care systems in birds. According to this model, in the earliest phase, a male should remain with a female for as long as the female continues to lay eggs (mate-guarding), and a female might remain in the proximity of a particular male if that male possessed a territory that held resources of importance to the female (see also Handford and Mares 1985; Wesolowski 1994). Based largely on the concept of evolutionarily stable strategies, van Rhijn suggests that the original avian parenting system was most likely to have been paternal care, and he notes that biparental care with similar roles occupies a central position among the various parental care systems. van Rhijn (1984, p. 103) concludes that '...

since monogamous paternal care systems can easily evolve towards all recent mating systems in birds, it is advocated that the evolution of parental care in birds was primarily based on a monogamous paternal care system.'

In the hypothetical early situation, before either sex engaged in parental care, why should a male remain with a single female, and vice versa? Using the idea that evolutionary change may proceed rather easily in one direction, but that a reversal is either very difficult or impossible (as seems to be true for many morphological and physiological traits), van Rhijn (1990) analyses the great diversity of mating systems in the Order Charadriiformes. As an example of the irreversibility of certain behavioural traits, he argues that the lekking system of the ruff, with exclusive maternal care, is unlikely to lead to major evolutionary changes in mating and parenting systems in the future; it is a dead-end system (but see Prum 1994 and Section 7.2.1). In contrast, certain other systems can more readily be modified by natural selection. For example, monogamous paternal care can lead to biparental care systems with similar male and female roles, which, in turn, may evolve in various directions. (Presumably because of the several preconditions required for its evolution, polyandry is never readily evolved (Ligon 1993); circumstantial evidence for the truth of this assertion comes from the rarity of polyandry in birds, despite the fact that males of the vast majority of avian species exhibit well-developed parental care.)

In their review of parental care and mating systems of ratites and tinamous, Handford and Mares (1985) independently develop an evolutionary scenario that leads to a conclusion similar to van Rhijn's. Like van Rhijn, they do not make the usual assumption that monogamy with shared parental care was the ancestral condition in birds. Instead, the starting point of their model is species with precocial young that require care from only one parent. Following Trivers (1972), Handford and Mares argue that if males are territorial, and if uni-parental care is sufficient for offspring survival, then females should desert and male parental care will ensue. Whether the male has only one mate or several is a secondary issue based on varying ecological conditions that primarily affect females. From the territorial male's perspective, once it has accepted parental duties—incubation—and has some confidence of paternity, it should prefer to acquire eggs as rapidly as possible. This is done by encouraging several females to deposit eggs in its nest, as seen in rheas, the ostrich, and some (possibly most) tinamous. This differs somewhat from van Rhijn's (1984, 1990) scenario, which couples paternal care with monogamy as the original mating–parenting system.

Wesolowski (1994) also presents arguments to the effect that sequential ovulation occurred very early in the history of birds, and that this could have hindered the initial evolution of parental behaviour in females. This is because, in the absence of any parental care or sophisticated hiding of the eggs, placing all of the eggs in one site risks loss of all of them. (All eggs are in one basket.) Megapodes (discussed below) provide an example. In those megapode species without parental (paternal) care of a nest site, the female lays each egg in the soil in isolation and the egg is not visited again (Jones *et al.* 1995). Such dispersion of the eggs would decrease the potential benefits and thus the likelihood of the evolution or maintenance of parental care, either maternal or paternal, in these or other birds. The fact that many non-parenting reptiles do lay clutches of eggs in one site does not invalidate this point; except for megapodes, most of which inhabit islands with few natural egg predators, birds typically do not bury their eggs. In general, burying of the eggs in reptilian fashion would negate many of the benefits of specialized parental care, e.g. incubation and associated rapid embryonic development.

Probably the initial form of avian parental care was protection of the eggs from predators at a central site, the proto-nest, on the surface of the ground (Fig. 10.2). For territorial males, defending the nest site should present relatively small costs, as compared to the potential costs of nest defence to females, in terms of reduced egg production. Moreover, assuming a monogamous relationship and a high confidence of paternity (van Rhijn 1984, 1991), it would be directly to the male's benefit to have the female continuing to produce eggs that he fertilized, rather than terminating laying to assume some incubation duties. (Again, an analogy is provided by the malleefowl, a monogamous megapode, in which the male tends the nest while the female produces very large, costly eggs.)

Alternatively, the male may have received the eggs of several females, as occurs today in the ostrich, the rheas, and a number of tinamous, as well as in the Australian brush turkey (Jones *et al.* 1995). The system of egg care at a single site shown by some megapodes—burying eggs and monitoring the mound (e.g. the monogamous mallee fowl and the promiscuous brush turkey; Jones *et al.* 1995)—is thought to be derived from more typical patterns of avian parental care (Clark 1964). It may be analogous, however, to the hypothetical ancestral condition outlined here in some important ways (see Section 10.5.4 below).

In the initial stages of parental care, females or males may have simply scraped a depression for reception of the eggs. Males could thus guard the eggs from small egg predators by periodically covering the eggs with their body, which would at the same time provide energy to the developing eggs in the form of body heat. Once this form of protection became routine, together with the transfer of heat to the eggs, embryonic development would be greatly accelerated. And once egg development had evolved to the point that incubation was essential, parental care became obligatory. With this stage reached, additional parenting and egg specializations could develop in a coevolutionary manner (Fig. 10.2).

An analysis of the phylogeny of parental care in fish (Gittleman 1981) parallels the scenarios presented here to a surprising degree. The most common transition among parenting fish is from no parental care to male parental care,

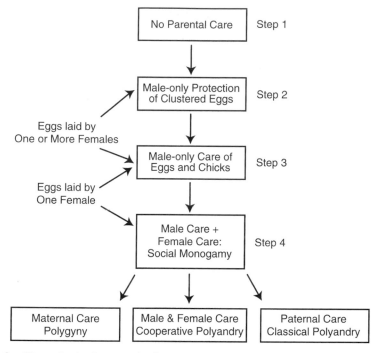

Fig. 10.2   Hypothetical scenario for evolution of various forms of parental care. Step 1. No parental care, beyond choosing a safe site for egg deposition, had yet evolved. Step 2. Male birds (or, conceivably, the reptile ancestor of birds) may have provided protection at a single site, the proto-nest. Step 3. Males alone provided care to the eggs and chicks of one or more females. Among living species, this system exists in the most primitive of living birds, the ratites and tinamous, and in a few charadriiform birds. Step 4. Once parental care became indispensable to the welfare of young, biparental care evolved; this led to male–female cooperation and monogamy, which is by far the predominant parenting–mating system among living species. Other mating systems, such as polygyny with sole female care, cooperative polyandry, and some cases of classical polyandry with sole male care, appear to be derived from monogamy with biparental care. See text for additional discussion.

and the next most common transition is from male care to biparental care. In fish, the route from no care to maternal care is thought to have occurred no more than once.

In brief, according to the scenarios presented by van Rhijn, Handford and Mares, and Wesolowski, incubation was initially provided solely by males, as is still the case in the paleognathous ratites and tinamous. Exclusive male care also occurs rarely in the main group of living birds, the Neognathae. These groups are discussed below and in Chapter 16.

10.4.2   Parental care and the fossil record

The subject of the early evolution of birds has entered a very exciting period. In recent years, discoveries of many new fossil birds have revolutionized thinking about the evolutionary history of the Class Aves (Chiappe 1995; Feduccia 1995, 1996). Currently, two scenarios for the early evolution of birds are on the table. One popular, but debatable, notion is that birds are present-day, direct descendants of dinosaurs (see Feduccia 1996 for an opposing view and for references). Some dinosaurs apparently laid eggs in nests and possibly provided care for them (Norell *et al.* 1995). However, in view of the existence of parental care in the form of egg protection in various living fish, amphibians, and reptiles, this does not indicate that the parental care of birds is directly descended from, or homologous to, that of dinosaurs. This is not to deny, however, the possibility that the origins of avian parental care may be located in the evolutionary history of the reptilian ancestors of birds, rather than in the history of birds proper.

A view of bird evolution, prominently championed by Alan Feduccia, is that birds are derived from a reptilian lineage that preceded the dinosaurs, the thecodonts, or basal archosaurs (Feduccia 1996). Feduccia (1995, 1996) summarizes the evidence that a major radiation of birds—the enantornithines—occurred during the late Mesozoic, and that it was completely wiped out at the end of the Cretaceous, along with the dinosaurs and most of the Mesozoic mammals, as well as the better known avian lineages of the Cretaceous, the Hesperornithiformes and the Ichthyornithiformes (see Fig. 1.3).

According to Feduccia's (1995, 1996) review of the fossil evidence, of which there is a surprising amount, all living groups of birds are derived from a group labelled 'transitional shorebirds' that survived the Cretaceous cataclysm. Feduccia suggests that the lineage leading to the living ratites and the tinamous, the Paleognathae, separated from the Neognathae by no later than the early Paleocene. The neognathous transitional shorebirds gave rise to all other post-Cretaceous birds, which evolved relatively rapidly into most of their present diversity within a 5- to 10-million-year period during the Paleocene and Eocene. This idea is revolutionary. If correct, it means that storks, owls, hummingbirds, falcons, etc. are descended from early 'shorebirds' and evolved at a rate far faster than anything previously proposed. Indeed, some fossils from this period are apparently truly mosaics in their characteristics (e.g. the Eocene shorebird-duck *Presbyornis*, the shorebird-ibis *Rhynchaeites*, and others; see Feduccia 1996).

With regard to the evolution of parental care, Feduccia's evolutionary scenario suggests some intriguing points. First, if all living neognathous birds are descended from shorebirds similar to some of those occurring today, then some of the most primitive living shorebirds may resemble the founders of most of the world's current avifauna. Second, exclusive paternal care is the reproductive pattern of nearly all of the most 'primitive' of living birds, the paleognathous ratites and tinamous, plus buttonquail and certain distinctive

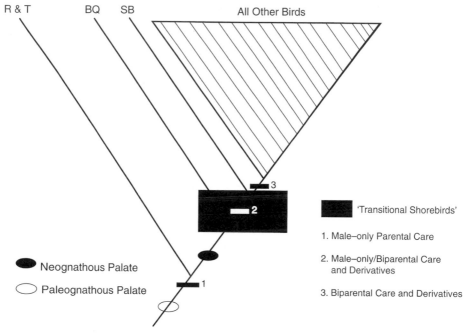

Fig. 10.3   Cladogram showing the hypothetical evolution of biparental care from male-only care, based on living lineages of birds: R & T, ratites and timamous; BQ, buttonquail; SB, shorebirds. 1. Male-only care is virtually the rule in ratites and tinamous, as well as in the neognathous buttonquail. 2. All neognathous birds may be derived from 'transitional shorebirds'. The 'black box' represents the transitional shorebirds (Feduccia 1995, 1996). In some living groups of shorebirds, sole paternal care may be primitive, while in others it may be derived from biparental care, which is associated with monogamy. 3. Biparental care and monogamy are the rule in most other living groups of birds; other parenting–mating systems, such as polygyny and cooperative polyandry, are thought to be derived from monogamy. See text for elaboration.

shorebird lineages, and this rare, shared characteristic may have come from an early common ancestor (Fig. 10.3).

10.5   Paternal care in certain groups of living birds

To the extent that living birds can tell us about the reproductive behaviour of their ancestors, the empirical evidence that exclusive paternal care is primitive in birds comes from some of the most distinctive and probably most ancient living lineages of birds. The four groups considered here are (1) the Paleognathae—ratites and tinamous, (2) certain of the shorebirds, traditional

Order Charadriiformes—jacanas, painted-snipes, plains wanderer, (3) the taxonomically enigmatic buttonquail, Family Turnicidae, and (4) the megapodes or mound-builders, Family Megapodiidae.

Several points suggest that consideration of these groups may be useful. (1) As described above, several authors have independently come to the conclusion that exclusive paternal care was likely to have been the original form of parental care in birds and, by extension, that other systems evolved from it. (2) The groups considered here show this form of parenting. (3) The ratites and tinamous diverged first from the lineage that gave rise to all other modern birds. Small, fully volant paleoganths known as lithornithids are known from the Paleocene and Eocene of North America and Europe (Houde 1988). Lithornithids are considered to be 'prototinamous', the ancestral types of all living paleognaths (Feduccia 1996). (4) All other modern birds are derived from generalized 'transitional shorebirds' and had diverged by the Eocene (Feduccia 1995, 1996). (5) The relationships of the buttonquail to other birds are more than uncertain; according to Sibley and Ahlquist (1990), they are apparently completely unknown, which suggests that buttonquail separated very early from other neognathous avian lineages. In view of the fact that they were long allied taxonomically with the plains-wanderer (a shorebird, see below), they could be an early offshoot of the shorebird lineage. (6) Although the unique form of male parenting exhibited by some megapodes is thought to be derived from the more typical mode of egg care and incubation (Clark 1964; Jones *et al.* 1995), the extensive parenting role of the male of some species, along with the absence of care by the female (Australian brush turkey, malleefowl), closely resembles the hypothesized paternal care of the first parenting birds. For this reason, rather than because of their phylogenetic affinities, they are considered here.

10.5.1   Ratites and tinamous

In the paleognathous ratites and tinamous, exclusive male parental care is the rule (Handford and Mares 1985). There are 10 living species of ratites, including the ostrich, two species of rheas, three species of cassowaries, the emu, and three species of kiwis. The kiwis, cassowaries, and emu are primarily monogamous (Fig. 10.4). Males of the ostrich and rheas are polygynous, accumulating the eggs of several females in their nests, while the mating pattern of females is sequential polyandry. (One female ostrich, the 'major hen', remains with the nest to aid the male in incubation; see below.) Living ratites vary greatly in size, are found in a diverse array of habitats, and differ greatly in their ecologies, e.g. forest dwelling, slow-moving nocturnal kiwis weighing as little as 1 kg at one extreme, and open country, fast running, diurnal ostriches weighing up to 130 kg at the other. Thus, there are no striking similarities in the ecologies of the ratites that might explain the universal importance of paternal care.

Fig. 10.4   A pair of cassowaries. With their casques and brightly coloured faces and wattles, cassowaries are the most ornamented of the ratites. As in most other ratites and tinamous, male cassowaries provide all incubation and care of chicks. Reprinted with permission from Dover Publications, Inc.

Similarly, all 47 species of tinamous exhibit exclusive paternal care, insofar as known. Beyond this fundamental similarity in parental roles of males and non-parental roles of females, the types of sexual bonds recorded in this group are variable, e.g. monogamy, simultaneous polygyny with all females laying in one nest, and sequential polyandry or promiscuity (see Ridley 1978 and Handford and Mares 1985 for reviews). Compared with the ratites, the greater variation in mating systems of tinamous probably reflects (1) the larger number of species that occupy a variety of habitats, and (2) the much greater predator pressures with which tinamous must contend (as a result of their small size, terrestrial habits, and continental distribution). Given the varying selective pressures on a relatively large number of species over a long evolutionary history, the exclusive male parental care found in tinamous indicates a remarkable conservatism in their breeding biology. It is also worth noting that the galliforms, which are convergent with tinamous in many ways, have no species exhibiting this form of parental care.

Why do almost all species in these two sister groups exhibit the same, extremely rare, parental care system: exclusive paternal care? Clearly, current ecological factors alone cannot account for the rare parental care system of the ratites and tinamous. Both van Rhijn (1984) and Handford and Mares (1985) make a connection between the ancestral condition of male parental care that they espouse and the prevalence of that system in the ratites and tinamous. van Rhijn (1984, p. 118) writes:

> ... the various parental care systems in recent birds seems to have evolved from a monogamous paternal care system, which originated after a stage of postcopulatory mate-guarding by the male. It is possible (and, building on that starting point, certainly the most parsimonious solution) that polygynous paternal care systems, such as in many ratite birds, evolved directly from this primitive state.

Handford and Mares (1985, p. 98) suggest that, in kiwis, '... male parental care may have been their ancestral condition.' Similarly, Wesolowski (1994, p. 43) concludes that '... unaided male care in Eoaves [ratites and tinamous] seems to reflect the primitive state preserved from the early stages of bird evolution.'

The suggestion made by these writers, that male parental care in ratites and tinamous reflects a common evolutionary history, just as certain basic morphological or DNA similarities do (see also Ligon 1993), are supported by two points.

1. Because of the early divergence of the lineage leading directly to the living ratites and tinamous from the one producing all other birds, the former two groups might be expected to retain a number of primitive characters. For morphological characters, this appears to be the case (Cracraft 1974). The argument suggested here is that, as with morphological characters, 'primitive' behavioural–physiological characters related to reproduction—specifically exclusive paternal care, with or without monogamy—can also be retained over evolutionary time.

2. A paternal care system characterizes all ratites and tinamous (the initial form of parenting, according to van Rhijn 1984, 1990), and most ratites are thought to be primarily or exclusively monogamous, but with no maternal care (Handford and Mares 1985). For those that are not—ostriches, rheas, and some tinamous—the polygamous systems are easily derived from monogamy with male parental care.

Why is maternal care largely absent in this group? The taxonomic distribution of maternal care—present in the ostrich, absent in the other ratites—is interesting. In addition to the historical explanation of paternal care discussed earlier, the biogeographical history of ratites may be relevant to this question. The ancestors of most of the extant and recently extinct ratites came to occupy areas that lacked an array of placental mammalian predators: moas and kiwis in New Zealand, the emu, cassowaries, and an extinct group, the mihirungs, in Australia and New Guinea, the extinct elephantbirds of Madagascar, and rheas in South America. In contrast, the fossil history of the ostrich lineage indicates that it evolved on the Eurasian–African land masses. The ostrich and its ancestors have had to contend with a diverse and numerous array of large terrestrial mammalian predators, scavengers, and large ungulates, and, uniquely among the ratites, the female ostrich, along with the male, provides a full quota of parental care. Several female ostriches lay in the nest of a single male (primitive trait), and the females compete for the opportunity to provide parental care, with the dominant female (the 'major hen') assuming a major role in incubation after having driven the other females away (Bertram 1992). After the eggs hatch, the major female joins the male in shepherding the young. Thus, maternal care in the ostrich—both incubation of eggs and escorting of chicks—may be 'derived' traits (Harvey and Pagel 1991).

To conclude, the ratites and tinamous are thought to be ancient avian groups that are most closely related to each other. An explicit consideration of phylogenetic affinities and conservatism in mating system evolution will be essential for development of a more nearly complete explanation of the mating and parental care systems of these two groups, which share a rare parental care system: exclusive paternal care. In view of the diversity in ecology and morphology of the living ratites and in the ecology of tinamous, the nearly universal pattern of male-only parental care in these groups, insofar as known, is remarkable, perhaps especially so in view of its rarity in the other 99.9 per cent of all living birds (Ligon 1993).

10.5.2   Shorebirds: jacanas, painted-snipes, plains-wanderer

Although the case for the primitiveness of male-only parental care in some shorebird lineages is more speculative than for the paleognaths, several interesting points about paternal care in shorebirds merit a brief consideration. (1) Among neognathous birds, obligate male-only care occurs almost exclusively in this one taxonomic group. (2) As discussed above, all modern neognathous birds may be derived from 'transitional shorebirds' (e.g. Feduccia 1995, 1996), thus distinctive shorebird lineages (i.e. lineages that have been long separated from other charadriiforms) should be a good place to look for behavioural and physiological characters that are ancient in the neognathae. (3) The parental roles of three such groups of the order Charadriiformes—jacanas, painted-snipes, and plains-wanderer—are characterized behaviourally by paternal care; i.e. females of most of the species in these families provide little or no parental care (e.g. Jenni and Betts 1978; Ridley 1978).

How might these three points be interpreted? First, the fact that exclusive paternal care, so rare in birds overall, appears in three of the most distinctive groups of this ancient order suggests that it may have evolved very early in the history of the group, and that other mating systems may have been derived from it, as suggested by van Rhijn (1990). Second, two of these three families, the jacanas and the plains-wanderer, are sufficiently different from most of the other shorebird groups that various authors have placed them in the Order Gruiformes (see Sibley and Ahlquist 1990 for a review of their taxonomic histories). Their morphological distinctiveness and the DNA–hybridization evidence suggest that these three groups (all within the Parvorder Scolopacida of Sibley and Ahlquist 1990) diverged from other shorebirds early in the radiation of this group, which makes their shared obligate paternal care all the more striking. It may also support the notion that male-only care is ancestral to other parental care systems in this group. If indeed this eventually proves to be the case, then the diversity of mating–parental care systems seen in the shorebirds (e.g. Pitelka *et al.* 1974), such as simple monogamy, double-clutching monogamy, polygyny, and lek-promiscuity, are derived from it, as suggested by van Rhijn (1990). A hypothetical phylogenetic history of this sort is illustrated in Fig. 10.2.

Recently, Szekely and Reynolds (1995) conducted a phylogenetic analysis of parental care in shorebirds (comprised of the Parvorders Scolopacida and Charadriida of Sibley and Ahlquist 1990). Their results suggest that biparental care was ancestral in this group as a whole (I will return to this point), but that male-only care was ancestral in the Scolopacida (which includes the plains-wanderer, seedsnipe, jacanas, painted-snipes, and scolopacids—woodcock, snipe, sandpipers, curlews, etc.). Thus, the conclusions of Szekely and Reynolds partially support the suggestion that male-only care is primitive in this sub-set of charadriiform birds.

With regard to the question about the primitiveness of male-only parental care, Szekely and Reynolds (1995) state that their results indicate that although the most likely ancestral form of care in shorebirds was biparental, male care as the ancestral condition in shorebirds would increase the length of the tree by only one step. Szekely and Reynolds (1995, p. 59) argue that such an ancestry is unlikely, however, for three reasons.

> First, biparental ancestry is supported by the choice of either sandgrouse or auks as the outgroup: both have biparental care, as do most of the next nearest relatives, such as the clade which includes diurnal raptors, herons, flamingos, storks and divers. ... Second, the inclusion of Laroidea superfamily of the Charadriida ... will not change the ancestral type of care or transitional patterns. This is because 25 species of pratincoles, coursers, gulls, terns, auks and allies (Laroidea) for which we have readily accessible data all have biparental care until the chicks fledge. ... Third, biparental ancestry is supported by the alternative phylogeny of Strauch (1978).

Despite these points, there may be some reasons to question whether biparental care is ancestral in shorebirds. First, as noted above, the alternate conclusion that paternal care is primitive would increase the length of the tree by only one step. Second, the conclusion that biparental care is ancestral is based on Szekely and Reynolds' (1995) choice of outgroups, and the two out-groups they chose may not be appropriate for two reasons. 1. One of the out-groups, the alcids, are also charadriiform birds, and, according to Sibley and Ahlquist (1990), they are relatively recently derived (i.e. their ancestors were shorebirds). 2. Szekely and Reynolds (1995) also used sandgrouse as an out-group, yet they, like the alcids, belong to the shorebird order Charadiiformes (see Gill 1995, p. 648). Similarly, Feduccia (1996, p. 225) refers to sand-grouse as 'shorebird derivatives'. Because biparental care is the predominant parenting system among all living neognathous birds, including the charadrii-forms, almost any outgroup chosen will exhibit biparental care. As discussed in previous sections, although biparental care clearly is the prevalent form of parenting in modern birds, it does not necessarily follow that its ubiquity indicates that it is the ancestral form of parenting. If the scenario of avian evolution presented by Feduccia (1995, 1996) is correct, *all* major groups of neognathous birds, including the biparental sandgrouse, are derived from 'transitional shorebirds'. This means that if the outgroup chosen is a member of the Neognathae, it would almost inevitably represent a taxonomic unit descended from shorebird-like ancestors. Among neognathous birds, the phylogenetic relationships of buttonquail are unknown (Sibley and Ahlquist 1990). Thus, buttonquail may be at least as logical an outgroup as sandgrouse. If buttonquail had been chosen as the outgroup, the analysis by Szekely and Reynolds (1995) would have led to the conclusion that paternal care is primitive in the Charadriiformes. Moreover, if early, but recognizable, shorebirds are truly the foundation stock—the ancestors, as it were—of all surviving neognathous birds (Feduccia 1995, 1996), it could be argued that the only appropriate outgroup for the shorebirds and their descendants is the Paleognathae. As with the buttonquail, this would lead to the conclusion that exclusive male care, rather than biparental care, is primitive in shorebirds (see Fig. 10.2).

Although I consider Szekely and Reynolds' paper to be one of the most important contributions to date with regard to the issue of the evolution of parental care patterns in charadriiform birds, I suggest here that the issue of the origins of exclusive paternal care, along with Feduccia's suggestion that transitional shorebirds gave rise to the neognathous avifauna, merits additional study. Because of the proposed basal position of transitional shorebirds in the evolutionary history of neognathous birds, choice of an appropriate outgroup for shorebirds may be especially difficult.

The analyses of parenting patterns in shorebirds by Szekely and Reynolds (1995) yielded some additional important conclusions:

1. Male parental care is significantly more common overall than female care (males incubate and remain with their brood in 87 per cent of the species, whereas females do so in only 56 per cent of the species). The frequency of exclusive male parental care in shorebirds stands out in comparison with all other major groups of neognathous birds. In the vast majority of avian orders, biparental care is the rule; among other neognathous birds, exclusive, or almost exclusive, paternal care is otherwise known to occur only in the buttonquail, some megapodes (see below), and in coucals, a distinctive group of cuckoos (see Section 16.10).

2. Over the entire phylogenetic tree, transitions from one parenting system to another were frequent, with most being from male care to either biparental or female care. There were no transitions from female to male care. Szekely and Reynolds (1995, pp. 60–1) write:

Overall, our results do support the 'male → biparental' hypothesis (van Rhijn 1985, 1990) more strongly than the 'biparental → uniparental' hypothesis (Jenni 1974; Pitelka *et al.* 1974) with two caveats. First, an ancestral state of predominantly male care applies to only one clade, the Scolopacida. The parental care of jacanas, painted-snipes and plains-wanderer in all but two species is provided by the males. These species diverged long ago from the rest of the scolopacids. Second, there were as many shifts directly to predominantly female care (6–8) as there were toward full biparental care (5 11, Fig. 2b). If there were intermediate 'stepping stones' in these transitions such as biparental care or multiple clutches per pair, these must have been very transient. The general trend was toward more reductions in male care than in female care ...

3. The numerous evolutionary shifts from one form of parental strategy to another in this group (see Szekely and Reynolds 1995, Fig. 2b), plus the ancestral nature of paternal care in the Scolopacida, lead to two points. First, the primitive nature of male care can help to account for the unusual diversity of parenting–mating systems in shorebirds, and second, it can help to explain why classical polyandry exists in shorebirds, along with all other systems, and is absent from nearly all other avian taxa.

The logic being followed here is that the evolutionary appearance of exclusive paternal care from no parental care may have occurred very rarely, possibly only once, in the common ancestor of the paleognathous and neognathous birds (Fig. 10.3). Alternatively, it conceivably evolved twice, once in the common ancestor of the ratites and tinamous, and again in the transitional shorebirds, the putative ancestors of all neognathous birds. Either one or the other of these two suggestions must be correct, or something thus far unidentified about the biology of charadriiform birds, in addition to, or other than, their evolutionary history has promoted the repeated evolution of parental care primarily or exclusively by males. Both possibilities could be correct or partially so. That is, paternal care and polyandry may be primitive in some extant lineages of shorebirds, while in other cases they may be derived from another mating–parenting system. For example, various lines of evidence indicate that the polyandrous–paternal care system of the spotted sandpiper is probably derived from monogamy with shared parental care. In this species, females have the capacity to assume parental duties and, depending on circumstances, regularly do so (Oring 1982; see Section 16.6). The possibility that polyandry–paternal care is ancestral in some shorebirds, while in others it is derived from biparental care, is supported by the analyses of Szekely and Reynolds, who found many transitions in this group from one mode of parenting to another, with the general trend being more reductions in male care than in female care.

10.5.3   Buttonquail

The buttonquail are a small, Old World family consisting of 15 species, 14 of which are members of the Genus *Turnix*. Although they have usually been placed in the Order Gruiformes, according to Sibley and Ahlquist (1990) their relationships to other birds are completely obscure. Individual species in this group are thought to be either monogamous or polyandrous. In any case, the male typically incubates the eggs and rears the young without female participation. Clutch size is most commonly four eggs (Bruning 1985). Thus, clutch sizes of buttonquail appear to be similar to those of jacanas, painted-snipes, the plains-wanderer, and most other shorebirds (i.e. typically four eggs, or, rarely, fewer).

Is it possible that the male-only parental care system of these birds, together with their small truncated clutch size, may be telling us that the buttonquail are early offshoots of the basal 'transitional' shorebirds that have retained, with little modification, the parenting system of the early Neognathous birds? The one non-*Turnix* species of buttonquail, the quail plover or lark buttonquail of Africa, moves and postures in a manner reminiscent of another distinctive charadriiform group, the coursers (Family Glareolidae). Taylor (1986, p. 78) writes: 'On ground, moves rapidly through grass in crouch, but also runs fast over open ground like miniature courser; stands, courser-like and watches intruder.'

In their synthesis of the evidence concerning the charadriiform affinities of the plains-wanderer, formerly placed with the buttonquail in the Order Gruiformes, Olson and Steadman (1981) comment on the courser-like movements and postures of the plains-wanderer. Thus, interestingly, both the plains-wanderer and the quail plover have reminded some field observers of coursers. Subtle behavioural traits or 'mannerisms' such as these, may appear to be trivial, but experienced field ornithologists, as well as classical ethologists with an interest in phylogeny, know that seemingly superficial characters such as postures and distinctive ways of moving often parallel phylogenetic relationships. (They also know that it is usually difficult to make a convincing case for phylogenetic relationship by use of such observations.)

The important suggestion here is that the buttonquail, plus the shorebird lineages discussed in the previous section, may have retained exclusive paternal care, which has been identified by several authors as most likely to have been the initial pattern of parental care in birds. If transitional shorebirds are indeed the basal stock of all neognathous birds (Feduccia 1995, 1996), buttonquail, like all other neognathous birds, are derived from transitional shorebirds.

10.5.4   Megapodes or moundbuilders

The megapodes are a distinctive, monophyletic family of galliform birds of the Australian and western Pacific regions, with 19 species in six genera (Jones *et al.* 1995). The group is unique and famous for its basic mode of parental care of eggs. Eggs are either laid in mounds especially constructed to serve as incubators or are buried in soil or sand. Heat is produced either by solar radiation, geothermal activity, or organic decomposition. This does not mean, however, that parental care is lacking. In some species, the construction and maintenance of the mound demands extensive and prolonged investment by males. Newly hatched chicks are highly developed, with functional flight feathers and the ability to use them. At hatching the chicks leave the mound and immediately take up an independent and solitary life.

Most megapode species are thought to be monogamous (Jones *et al.* 1995), and in the few well-studied mounding species, it is primarily the male that deals with the construction and maintenance of the mound. This is true in the two best-studied species, the monogamous malleefowl and the promiscuous Australian brush-turkey. Male malleefowl may be occupied with tending the mound for up to 11 months out of the year. Wesolowski (1994) discusses briefly the possibility that the major parental role played by male megapodes is primitive, rather than derived.

Compared to some other galliform birds of similar size, the eggs of megapodes are unusually large (Jones *et al.* 1995). Eggs of the Melanesian megapode weigh up to 18 per cent of adult body weight, and eggs of the brush-turkey are 12 per cent of adult weight; in contrast, eggs of pheasants often weigh about five per cent of adult hen weight (Clark 1964). Similarly, megapode eggs have a very high ratio of yolk weight to albumen weight. Such eggs require a large investment in their production, and these characteristics of the egg are correlated with several aspects of megapode reproductive biology. (1) Production of eggs may require the female to spend most of her waking hours foraging. (2) Eggs are produced slowly, one every several days. This places a premium on paternal care of the growing 'clutch' of eggs. (3) Large, rich eggs are probably required to produce such uniquely precocious offspring. (4) Due to the variability in mound conditions, the incubation period of some megapode eggs is also variable, and it is long. Frith (in Clark 1964) reports natural incubation periods for the malleefowl that range from 50 to 96 days! Variation in the incubation period of this magnitude is probably unique to megapodes, and it indicates that embryonic viability is far less dependent on a constant incubation temperature than is the case for most, and perhaps all, other kinds of living birds. In brief, the general reproductive strategy of the well-studied, care-giving species of megapodes is as follows. In both the monogamous malleefowl and the promiscuous mound-building Australian brush-turkey, the male constructs the incubation mound and tends it and the eggs for months on end. The female's primary reproductive task is to produce large, costly eggs, and to lay a lot of them. The overall mean number for the malleefowl is 18.6 eggs per season; however, there is a great deal of inter-season variability (Jones *et al.* 1995).

Although burying eggs in the soil may suggest a reptilian approach to reproduction, this system is not usually thought to be primitive in birds (Clark 1964; Jones *et al.* 1995; see also Wesolowski 1994 (Interestingly, however, the pre-hatching position of megapode embryos differs from that in all other birds and is said to resemble that of reptiles; Oppenheim 1972.). Instead, the burying of eggs is a derived, specialized form of parenting that probably evolved from more typical nesting behaviour; i.e. the mounds may have evolved from more typical nests. The megapode system of reproduction could have evolved and been maintained only in a geographic region that held relatively few, or no, placental mammalian predators (see Jones *et al.* 1995).

To conclude, although we will probably never have unequivocal evidence on the point, it appears possible that the male-only parenting systems of the ratites and tinamous is a direct legacy from the earliest of modern (i.e. post-Cretaceous) birds. Moreover, this may also be true of some of the relict-ual shorebirds such as jacanas, painted-snipes, and the plains-wanderer, and possibly the taxonomically enigmatic buttonquail. This speculation is consistent with analyses suggesting that sole paternal care was the earliest form of parental care in the lineage leading to today's avifauna (van Rhijn 1984, 1990; Handford and Mares 1985; Wesolowski 1994). Megapode reproduction has also been discussed here because parental care in the two best-studied species, the malleefowl and Australian brush-turkey, appears to be similar to the hypothesized original form of parenting. Specifically, males produce and defend a valuable resource—the 'nest' mound—which attracts one (mallee-fowl) or several (brush-turkey) females to lay their eggs, and the male then provides all of the subsequent parental investment. Thus, the parental invest-ment patterns of males of these megapode species parallel those of the paleog-nathous birds: in both groups males provide sole care of the eggs, albeit in different ways.

10.6   The evolution of biparental care

It is reasonable to assume that biparental care would have evolved only if par-enting by females as well as males increased the production of mature off-spring, which would provide net fitness gains to both sexes. If, over evolutionary time, environmental conditions, biotic and abiotic, made it more difficult for a single parent to rear young as successfully as could be done by two parents, then selection should favour increasing post-laying parental care by both the male and the female. Factors that could favour the evolution of biparental care might include, for example, the appearance of more, or more effective, types of nest predators, such as an increasingly diverse set of preda-tory mammals, combined with decreasing environmental temperatures from the Mesozoic into the Cenozoic.

In early birds, as in some ratites and tinamous today, the male may have routinely accepted eggs from several females, which would produce a nest full of eggs in short order. Other things being equal, in terms of potential pro-ductivity, this appears to be an effective system for both males and females, simply because males may receive a full clutch of eggs in a very short time (and a quick replacement clutch, if needed) and because females may continue to lay eggs in the nests of available males throughout the breeding season, rather than producing a few eggs and then ceasing laying to care for them.

Once reproductive success became tied to parental care, however, biparental care could in many cases be superior to uniparental care. This would be especially true with the evolutionary development of altricial young. When it became in the female's best interest not only to contribute care along with the male, but also to exclude the eggs of other females from the nest, the stage was set for a transition to biparental care and then to monogamy. An illustrative example of this point is seen in the ostrich (the only paleognathous species exhibiting biparental care), where, along with the male, a single 'major' female incubates the eggs and later shepherds the chicks. Parental care by the female ostrich makes it possible for her to manipulate eggs, by excluding those of other females, in order to increase her own reproductive success (Bertram 1979).

Although sole maternal care is not uncommon in birds, there is no evidence to suggest that exclusive maternal care preceded biparental care; i.e. no group of living birds provides a hint that sole maternal care has given rise to biparental care. The opposite is generally assumed to be the case; i.e. most, if not all, exclusive maternal care systems (usually associated with polygyny or promiscuity) are thought to have arisen from biparental (usually monogamous) systems. For example, Wesolowski (1994, p. 41) writes: 'The taxonomic distribution of groups with uniparental female care ... clearly demonstrates that maternal care has evolved several times independently from the shared biparental care typical of the majority of birds.'

10.7   Conclusions and summary

The oviparous mode of reproduction as exhibited by birds provides many clear benefits and should not be viewed as a reproductive programme inferior to that of eutherian mammals. Benefits of oviparity include the following. Several or many eggs can be rapidly produced. Incubation permits embryonic development under an optimal temperature regime. By placing the eggs in a nest and initiating incubation at the end of egg-laying, all of them will hatch at the same time. This maximizes the benefits of parental care, particularly biparental care. The ability of two parents to contribute in a major way to the development of their offspring, in effect doubles this component of reproductive effort as compared to most mammals.

Once males held territories with sites suitable for placing eggs and attracting females, it would have been to the male's advantage to guard the eggs. This could best be accomplished by accumulating them at a single location. The first form of parental care in birds was probably exclusive male care. One or more females laid eggs for the male at a specific site—the proto-nest. Initially, paternal care was probably restricted to egg guarding. One way to protect the eggs was to cover them with the body, and what eventually evolved into incubation behaviour may first have been most important as a means of protecting eggs. Covering the eggs with the body led to the evolution of behavioural and morphological characteristics, such as the incubation or brood patch in one or both sexes, which provides an almost constant and optimal temperature for embryonic development, thus accelerating the rate of development.

Flight probably evolved prior to, or simultaneously with, the evolution of costly parental care. Early in the history of birds the demands of flight, together with little or no parental care, may have led to selection for super-precocial young. This, in turn, led to larger and thus more expensive eggs. Male care of the nest and eggs allowed females the freedom to forage full-time and thus to produce more and possibly larger eggs than would have been possible if they had guarded the nest. Once eggs acquired their typical avian characteristics (e.g. relatively large size, proportionately large yolk), and once males began to apply body heat to the eggs, which greatly increased the rate of embryonic development, the selective benefits of providing parental care greatly increased.

When parental care, together with increased dependence of the egg and hatchlings on such care, became fully established, selection could favour the evolution of maternal, as well as paternal, care. With biparental care came selection for smaller eggs (associated with the benefits of producing more young per brood) and more specialized forms of care, such as parental role specialization by sex. These changes probably occurred along with the evolution of semi-precocial and altricial young. Altricial species predominate in today's avifauna.

When the eggs and hatchlings absolutely required parental care, as is true of nearly all living birds (the only exceptions are the few megapodes that simply bury their eggs singly in the soil and abandon them—minimal parental care)—then selection could not favour the loss of parental care. The only exception is social parasitism—*somebody* has to incubate those eggs and look after those chicks.

This sequence of steps can explain the evolution of the parenting and mating systems of nearly all birds, because once biparental care evolved and became essential, monogamy prevailed. Most other mating systems, such as polygyny with no male parenting (e.g. turkey) or polyandry of the sort seen in the spotted sandpiper, are almost certainly derived from monogamous systems. The following chapters consider some of the benefits and costs of each of the major mating systems and the factors that promote them.

11 Social monogamy

11.1 Introduction

Social monogamy, by far the most common mating system in the Class Aves (Lack 1968), can be recognized by the apparently exclusive association of one female and one male as mates over some definable period of time, such as a nesting cycle, a breeding season, or even a lifetime. In birds, monogamy is usually closely associated with biparental care. (Although biparental care and monogamy actually refer to two distinct suites of behaviour—parental care and mating system, respectively—the two terms are so closely linked that they are often used interchangeably.) As noted previously, unlike any other group of vertebrates, social monogamy is the prevalent mating system in birds, with well over 90 per cent of all species estimated to exhibit this form of male–female association (Lack 1968). In view of its rarity in other classes of vertebrates, the frequency of monogamy in birds is striking and is a key behavioural–reproductive characteristic that sets this group apart from all other vertebrate taxa.

This prevalence of social monogamy and associated biparental care throughout the Class Aves has often been used as the basis for the assumption that these are the primitive or original mating and parenting systems of birds, and that other, less common mating–parenting systems are derived from them. In the previous chapter, it was suggested that although biparental care and monogamy may have appeared early in the evolutionary history of avian parental care, this was not the original form of avian parenting. The prevalence of monogamy among living birds indicates that it has stood the test of time very well; that is, for most species, a 'better' system has not evolved. This statement does not deny the possibility of the loss of monogamy within a lineage and its subsequent evolutionary resurrection. Many such transitions surely have occurred in the major avian lineages (e.g. Szeleky and Reynolds 1995). Moreover, most other mating systems, probably including all forms of polygyny, may be derived from monogamy with biparental care (see Fig. 10.2).

What factors set the stage for the evolution and maintenance of biparental care and monogamy in birds? Oring (1982) suggests that the combination of oviparity and endothermy promoted biparental monogamy in this group because these two factors increased the importance of intensive parental care to the eggs and young; i.e. care by two parents may often have been required for the successful production of offspring. As pointed out in Chapter 1, in almost all avian species, oviparity and endothermy do demand some measure of parental care, and this has probably been true for many millions of years. Thus, at this stage in the evolution of all birds, parental care of some sort is almost always required for successful reproduction.

Among contemporary species, however, biparental care is not always necessary by any means, as demonstrated by the many polygynous species in which only the female parent provides care for eggs and nestlings, as well as the few classically polyandrous species in which the male has sole parental responsibility. Moreover, it has also often been demonstrated that in some socially monogamous species, one parent can rear at least some young to fledging. Thus, even in some monogamous species, biparental care is not always essential for successful reproduction. There can be little doubt, however, that for the vast majority of birds (from penguins to petrels to partridges to potoos to passerines), two parents are far more productive of young than one. For example, in the great majority of male removal studies, fatherless chicks had reduced survival rates (e.g. see reviews by Wolf *et al.* 1988 and Bart and Tornes 1989; see also Section 11.4).

11.2   Monogamy: the neglected mating system

Although monogamy is by far the prevalent mating system in contemporary birds, according to Douglas Mock, it is also the 'neglected mating system'. Mock (1985) points out that many of the most influential writings on mating systems of birds have focused primarily on how best to explain the evolution, or at least the maintenance, of mating systems other than monogamy. He suggests that this is because monogamy as a '... package seems bland' (Mock 1985, p. 1). Prior to Mock's assertion that monogamy has been the neglected mating system, Oring (1982, p. 10) wrote:

> In most instances, avian oviparity and endothermy together have resulted in advantages to both sexes in remaining together for some or all phases of the reproductive cycle. The greater the proportion of the reproductive cycle during which pair members remain together, the less opportunity there is for additional pair bonds to form or for extra-pair relationships to occur. Clearly, constraints are sufficiently great in birds that relatively few species have become polygamous or promiscuous. The relative rarity of polygamy has directed attention to environmental determinants of avian polygamy–promiscuity, at the expense of understanding factors that cause variations in duration and nature of monogamous relationships.

Mock's assertion that monogamy has been neglected was first presented at a symposium, 'Avian monogamy' (Gowaty and Mock 1985), held at the 1982 Annual Meeting of the American Ornithologists' Union (AOU). In 1995, another AOU symposium, 'Male and female reproductive strategies within socially monogamous mating systems', focused primarily on extra-pair copulations and fertilizations. The nature of the topics discussed in this latter symposium appears to offer support for Mock's contention that monogamy *per se* is viewed as bland.

Although extra-pair matings within socially monogamous systems (see Chapter 12) continue to receive more attention than systems characterized by social and genetic monogamy, the prevalence of extensive paternal care in contemporary birds suggests that a high confidence of paternity, of at least some portion of the brood, must have been a common condition during the evolution of parental care and mating systems. Once both females and males could 'count on' extensive parental investments by their social mates, the door was opened for the evolution of strategies involving extra-pair copulations and fertilizations (see Chapter 12) and, less commonly, even for desertion of the young by one parent or the other.

11.3   Factors promoting the evolution and maintenance of monogamy

In addition to the perception of blandness, monogamy is sometimes presented as a secondary or non-preferred reproductive strategy. For example, in their influential paper, Emlen and Oring (1977, p. 217) state: 'The prevalence of monogamy in birds is due primarily to the inability of most species to take advantage of any environmental "polygamy potential".' Similarly, Davies (1991, p. 283) reviewed the data and concluded '... that the predominance of monogamy in many birds arises not, as Lack proposed, because each sex has the greatest success with monogamy, but because of the limited opportunities for polygyny.' This conclusion may be true for some small temperate-zone passerines in which males neither incubate the eggs nor feed the incubating female; it is probably not correct, however, for the diverse array of monogamous–biparental non-passerine species or for some passerine groups (e.g. Corvidae, see Dunn and Hannon 1989).

Wittenberger and Tilson (1980, p. 199) provide three preconditions for the evolution and maintenance of monogamy. (1) females must obtain benefits from monogamous pair bonding that are not otherwise obtainable; (2) females must be able to ascertain the true mated status of potential mates; and (3) mates do not desert. Although these points may have been necessary for the evolution of monogamy, all three are probably not required for its maintenance. Certainly, desertion of the mate and brood occurs regularly in some socially monogamous species (e.g. Beissinger 1990; see also Persson and Öhrström 1989).

These authors also present five hypotheses for the development of monogamy (see Wittenberger and Tilson 1980 for references). Monogamy should evolve: (1) when male parental care is both non-shareable and indispensable to female reproductive success; (2) in territorial species, if pairing with an available, unmated male is always better than pairing with an already mated male; (3) in non-territorial species, when the majority of males can reproduce most successfully by defending exclusive access to a single female; (4) even though the polygyny threshold is exceeded, if aggression by mated females prevents males from acquiring additional mates; and (5) when males are less successful with two mates than with one (Wittenberger and Tilson 1980, pp. 199–200).

The five hypotheses of Wittenberger and Tilson (1980) can be reduced to fewer, more inclusive categories. For example, Bertram Murray (1984, p. 109) makes the important distinction between facultative and obligate monogamy:

> Facultative monogamy refers to situations in which individuals fail to acquire additional mates because of a shortage of members of the opposite sex resulting from an unbalanced sex ratio. In obligate monogamy, selection is for monogamy *per se*, that is, both male and female gain from a monogamous relationship. Facultative monogamy, then, has a non-adaptive origin because many of the males are monogamous only because females are unavailable, whereas obligate monogamy has an adaptive origin because it is selected for.

The first hypothesis listed by Wittenberger and Tilson (1980) is close to Murray's obligate monogamy, while the last four can be viewed as variations of facultative monogamy. Making a distinction between obligate monogamy and facultative monogamy is also useful because it emphasizes that unqualified mating system labels such as 'monogamy' are actually catch-all terms that include a multitude of phenomena (see also e.g. Oring 1982; Gowaty 1985, 1996*a,b*; Mock 1985; Birkhead and Møller 1996; Johnson and Burley 1997). Here I follow Murray's approach: monogamy is obligate in species where biparental care is essential for the production of independent offspring. Facultative monogamy, on the other hand, occurs due to some other sort of constraint on non-monogamous mateships.

Mock (1985) points out that the hypotheses for monogamy presented by Wittenberger and Tilson (1980) simply reverse the logic of polygyny models (see Section 13.2). Although polygyny in modern birds usually, and perhaps always, has been derived from monogamy, it is not satisfactory to simply flip the logic of polygyny as an explanation for monogamy. Monogamy probably preceded evolutionarily the development of the polygynous systems seen today. Thus, monogamy needs to be accounted for in its own right. As Murray states, in obligate monogamy, selection is for monogamy *per se* (or, more precisely, selection is for biparental care), thus it should be explained on the basis of the factors favouring it. The position advocated here is that even facultative monogamy probably cannot be adequately explained simply by modifying or reversing the ecological explanations that may largely account for the appearance of polygyny in primarily monogamous systems. Keeping in mind that 'monogamy' is something of an umbrella term for a variety of mating strategies, explanations for its evolutionary origins and maintenance should address the selective factors promoting obligate monogamy, as well as those factors constraining males to monogamous mateships (facultative monogamy).

11.4   Obligate monogamy and biparental care

In Chapter 10, I reviewed the case for the proposition that exclusive paternal care was the first parental care system to evolve in birds, and that biparental care and monogamy developed from a paternal care system. Here, I address the question, why are most birds monogamous? Some answers can be found in assessing the costs and benefits of monogamy. It appears certain—indeed a truism—that for most birds the net benefits of monogamy outweigh its net costs, and that, since the Mesozoic, in the great majority of birds selection has continued to refine the benefits to be obtained for both males and females by this mating system and associated biparental care. This does not argue against the view that the benefits vary greatly in kind and in degree, or that monogamy is always the superior mating strategy for both sexes; clearly, that is not the case. Instead, the point is simply that for the vast majority of avian species, biparental care and monogamy are more productive for both males and females, on average, than any alternative reproductive strategy. This generalization appears to hold for many precocial species (e.g. Schneider and Lamprecht 1990), as well as for most altricial ones.

To account for the prevalence of monogamy, I follow Lack (1968) and begin with the assumption that over their evolutionary development, monogamy and biparental care have been, and usually have remained, causally associated. Lack (1968, pp. 4–5) suggested that, 'The main advantage of monogamy is that ... both male and female leave, on average, most offspring if both help to raise the brood.'

Subsequently, the essential role of the male parent in monogamous species has been questioned. Although this issue has not been studied at all in most orders and families of birds, the available evidence suggests that for most types of birds, biparental care at one or more stages of the nesting cycle is indeed essential for success (e.g. Bart and Tornes 1989). For many species, shared incubation may be the most critical aspect of biparental care. Shared incubation is the rule for all species in 62 of the 92 (67 per cent) neognathous families of non-passerine birds recognized by Van Tyne and Berger (1959); in most (perhaps all) species in which shared incubation is the norm, two individuals are probably required for successful hatching of the brood; i.e. biparental care is critical. In the western sandpiper, a species with shared incubation, removal of either parent during incubation led to 100 per cent nest failure (Erckmann 1983). Provisioning of the incubating females is also critical in some other non-passerine families in which male incubation is rare or non-existent (e.g. Accipitridae, Strigidae). Similarly, seven of eight female American kestrels (Falconidae) abandoned their nests after experimental removal of their mates (Bowman and Bird 1987). The one female kestrel that reared young alone had lost her mate after 24 days of incubation. (The incubation period is 27 days; it is doubtful that this female could have completed successful incubation if her mate had been removed two or three weeks earlier.)

The counter-evidence, based mostly on monogamous, temperate zone passerines, has indicated that when experimental removals occur after the brood has hatched, the female parent can often rear part or all of the brood. Even in the majority of these studies, however, loss of the male parent leads to a significant decrease in productivity (Table 11.1). The percentage decrease in productivity was often associated with the stage of the nest cycle when the male parent was removed, as well as with current environmental conditions (i.e. male contributions were more important in some years than in others). Three studies of small songbirds, in which the fates of the chicks were followed to independence after removal of the male parent (Smith *et al.* 1982, Greenlaw and Post 1985, Wolf *et al.* 1988), indicated that production of young decreased from about 37 to 66 per cent. This suggests that for some species paternal care may be most critical after the young have left the nest, a phase of the nesting cycle not commonly studied.

In short, under most realistic circumstances, would a male bird actually be better off to desert and attempt to obtain a second mate than to aid one female in the rearing of their young? Although the answer will vary from species to species and, in some cases, from year to year, overall the available data (e.g. Bart and Tornes 1989) do not provide an affirmative answer.

Constraints that limit a male's chances of obtaining a second mate undoubtedly do play an important role in the maintenance of monogamy in many cases (see next section). However, if all birds are considered (rather than just a relatively few readily accessible and commonly studied temperate zone passerines,

Table 11.1   Effects of male-removal on nesting productivity in some altricial birds

Species	Effect on survival from incubation to independence	Source[2]
Tree swallow	*; *[1] (2 studies)	1
Blue tit	*	1, 2
Great tit	*	1, 2
Eastern bluebird	*; NS[1] (2 studies)	1, 2
Pied flycatcher	*	1, 2
Savannah sparrow	*	1, 2
Yellow-headed blackbird	*	1
Snow bunting	*	1, 2
Northern cardinal	NS	1, 2
Seaside sparrow	*	1, 2
Dark-eyed junco	*	1
Song sparrow	NS	1, 2
White-throated sparrow	NS	1
House wren	*	2
Rock dove	*	2
American kestrel	*	2

[1]Significant negative effect; NS, not significant.
[2]See (1) Wolf *et al.*, 1988, Table VI, and (2) Bart and Tornes, 1989, Table 5, for original references.

which are not representative of avian diversity), I suggest that the evidence does support Lack's perspective of the relationship between biparental care and monogamy. That is, for most types of birds, monogamy has been, and continues to be, the most productive mating system for both males and females, because of special characteristics that make biparental care so effective in this group (e.g. see Section 11.7). Moreover, the evolution of the cooperative, precisely synchronized biparental care seen in most socially monogamous species suggests a high degree of genetic monogamy in its evolutionary background despite the frequent occurrence of extra-pair fertilization (see Chapter 12).

11.5   Facultative monogamy

Many factors apparently tend to limit or constrain the development of non-monogamous relationships. In most cases these should not be viewed as alternative explanations to obligate monogamy, but rather as factors that complement biparental care of offspring in promoting monogamy. Facultative monogamy is illustrated by Freed's (1987) paper entitled, 'The long-term pair bond of tropical house wrens: advantage or constraint?' Tropical house wrens are permanently monogamous, with very low rates of separation. Although long-established pair bonds produce no net advantage, in terms of more offspring, pairs maintain their mateship from year to year, usually for life, because of certain constraints. There is a limited number of territories and a large, non-breeding population of floaters, thus breeding vacancies are extremely rare and therefore are extremely valuable. Freed (1987, pp. 520–1) suggests that:

> Tropical house wrens may be monogamous simply because each sex denies opportunities for its partner to associate freely with individuals of the opposite sex, and their monogamy may be permanent because neither sex has opportunities to go elsewhere.

In the tropical house wren, and in many other facultatively monogamous passerines, only females incubate the eggs and brood the chicks. In such species, there is 'polygyny potential' either because paternal care is not essential or because it is shareable. In the case of the tropical house wren, the polygyny potential is not realized for the reasons discussed by Freed. Other factors can also constrain the development of non-monogamous mating systems.

11.5.1   Synchrony in the onset of breeding within a local population

Breeding activity in a local population of a given species typically begins rather synchronously and involves all or nearly all of the territory-holding and reproductively competent individuals in the population. This is often as true of tropical and desert regions as it is of highly seasonal, temperate parts of the Earth, and demonstrates that members of a local population possess similar physiological responses to environmental cues. Synchrony in the onset of breeding alone will reduce opportunities for extra matings for males, because all, or nearly all, of the females will be mated to the territory-holding males, leaving few females available and undefended (e.g. Emlen and Oring 1977; Weatherhead 1979). Although breeding synchrony alone is insufficient for the development of monogamous mating systems, it should make departures from monogamy more difficult for males than would be the case if nesting activity could be initiated at any time.

11.5.2   An unbalanced sex ratio can promote monopolization of females by males

The available evidence suggests that for most avian species, there are more adult males than females present in the population. Randall Breitwisch (1989, Table 1) reviewed the bases and evidence for differential mortality of males and females. For many non-migratory species, females, on average, disperse farther than males, which is thought to lead to greater mortality of that sex (see Breitwisch 1989 for references). Similarly, in some migratory species, females travel farther than males and, as a result, may suffer heavier migration-related mortality. Other sources of sex differences in mortality include sexual differences in body size (e.g. Downhower 1976; Boag and Grant 1981), secondary sex characters (e.g. Selander 1965, 1972), or reproductive roles.

For example, incubating females may be particularly vulnerable to predators. This is thought to be a major source of mortality in ground-nesting species, such as most temperate zone ducks (McKinney 1986), and it may lead to the male-biased sex ratios reported for this group. It may also be important where ecological or environmental factors make females especially reluctant to leave the nest. For example, pinyon jays often nest in late winter when air temperatures are low, and incubating females sit so tightly on the nest that they can sometimes be touched by the human investigator (personal observation). Probably not coincidentally, the females appear to be highly vulnerable to predators at this time (Ligon 1978; Marzluff and Balda 1992). In short, in many species or populations, and for many different reasons, there are more males than females.

While there is little good evidence that a population sex ratio skewed towards males in itself promotes polyandry, or, for that matter, that a majority of females in a population promotes polygyny (but see Persson and Öhrström 1989), it does appear that a male-biased sex ratio can be one of the factors that encourages monogamy. This is because in such species even a single female is a very valuable resource to each male; i.e. a male with a single mate will be far better off than the average male. A male-biased sex ratio also means that male–male competition for a mate will probably be more intense than would be the case with an evenly balanced sex ratio; moreover, it would often be extremely difficult for a male with one female to acquire a second mate, while at the same time retaining the first one. Thus, it appears likely that monogamy in a number of species is promoted by the great value of even a single female mate to a male, as a result of a male-biased sex ratio. All of this suggests that the demographic environment, as well as the physical one, can serve as an important agent promoting monogamy. A group that illustrates this point well is the north temperate zone ducks. In most species, there are considerably more males than females, thus a male that sires most or all of the ducklings in a single brood will fare well above the average for the male population as a whole.

Other possible examples of monogamy based primarily on male defence of a female, rather than on the importance of paternal care, occur within the strikingly sexually dimorphic pheasants. Although this might suggest a male-biased sex ratio, as often occurs in ducks, good data on natural sex ratios are probably non-existent for most species of pheasants. The very few observations of free-living species, other than the well-studied ring-necked pheasant, introduced and widespread in western Europe and North America, suggest that monogamy may be more common in this group than is generally assumed (Fig. 11.1).

The field studies of Beebe (1990), one of the few western ornithologists (probably the only) ever to make behavioural observations in the wild on nearly all species of pheasants, together with the literature survey of Johnsgard (1986), suggest that of the 16 currently recognized genera (containing 1–10 species) six are uniformly monogamous and five are uniformly polygynous. In

Fig. 11.1 A number of highly ornate pheasants, including the grey peacock pheasant, are thought to be monogamous (Johnsgard 1986). Reprinted with permission from Dover Publications, Inc.

the other five genera, there appears to be both interspecific and intraspecific variation in mating systems, with some species being primarily monogamous, while congeners are polygynous to varying degrees.

The widely held, but poorly documented, assumption that most pheasants are polygynous is probably based on three factors: (1) the one well-known species, the ring-necked pheasant, is clearly polygynous; (2) pheasants are favourites of aviculturists, and, under typical conditions of captivity, one male is usually confined with two or more females; and (3) the spectacularly beautiful plumage of males of many species simply looks as though it should be correlated with strong sexual selection via polygyny.

If we knew as little about the natural history of temperate zone ducks in the wild as we do about Asiatic pheasants, and if we based a guess about the mating systems of ducks purely on the degree of sexual dimorphism and brightness of male plumage, many biologists would probably assume that most ducks are not monogamous—and they would be wrong. The available data suggest that many pheasants, like the ducks, are actually primarily monogamous. If so, a male-biased sex ratio may prove to be one factor promoting this mating system in this group, as in ducks.

11.5.3   Aggressive intra-sexual territorial defence by females of monogamous
         species

Another factor promoting facultative monogamy is the intrasexual aggressiveness of females of many monogamous species. In assessing the importance of
female–female competition for territorial space/resources and parental care by
the male, Arcese (1989, p. 109) concludes that '... models that do not include
the influence of female spacing behaviour on mating systems can be rejected.'
(See Davies 1991 for models that consider female spacing behaviour.)

In a variety of socially monogamous species, territory-holding females
aggressively exclude intruders of their sex. This behaviour is common in
woodpeckers and in those passerines (e.g. Hill 1986; Arcese 1989; Veiga 1992)
in which sharing either the resources of the territory or the male's parental
effort with a second female would be costly to the primary female's reproductive effort. Because polygyny in some species, especially passerines, often (but
not invariably; see Section 13.4.5) does decrease the reproductive output of
primary females, the intolerance of females towards others of their sex can be
viewed as reflecting a conflict of interest between the members of a monogamously mated pair.

11.6   Other factors related to avian monogamy

The subject of monogamy involves consideration of factors in addition to
those that promote it in the first place.

11.6.1   Physiological synchronization between the sexes: an effect of selection
         for monogamy

In birds, the process of parenting is extremely intricate and complex. Pair
bonding, nest site selection, nest construction, egg laying and incubation, care
of newly-hatched, delicate nestlings (brooding, providing the appropriate
food—often different than the food of adults, fecal sac removal in woodpeckers and most passerines), feeding and protection of nestlings and fledglings, etc. demand a repertoire of highly refined forms of parental behaviour.
Adding to this complexity is the fact that, for many species, both the male and
female must be prepared to play their roles in an appropriate manner and the
roles are often sex-specific (see Section 11.7). Thus, in the vast majority of
avian species, the male and female must be physiologically synchronized, both
with the stage of the breeding cycle and with each other (Tinbergen 1953).
What this means, in turn, is that in species in which paternal care is absolutely
essential for reproductive success, males probably cannot be precisely synchronized simultaneously with the endocrine state and nesting cycles of two
females at different nests, or simply that a single male cannot be in two places
at one time. This latter is especially likely to be a constraining factor when the

male provides nocturnal incubation, as in non-parasitic cuckoos and the woodpeckers.

11.6.2   Intra-pair cooperation

In birds, more than any other major animal group, the reproductive strategies and goals of the two sexes are similar. (This does not, however, mean that the goals are identical or that conflicts of interest between the sexes do not occur; e.g. conflict over the amount of care invested by each partner; Davies 1991.) This is reflected by the prevalence of monogamy and the associated high frequency of more-or-less equivalent parental investment. In those species where each sex is dependent on the contributions of its mate for their mutual reproductive success, a high level of intra-pair cooperation and reciprocal aid-giving may occur (Ligon 1983). Oring (1982) and Rowley (1983) discuss the importance to an individual of a monogamous species of maintaining its mate in good physical condition. One sex provides aid to the other in the 'expectation' that it (the donor) will receive compensation via their shared offspring, i.e. by mutually increased reproductive success.

Beginning with the simplest form of male parental care (possibly initially only the guarding of eggs), we see parental strategies in today's birds that include specific male and female parental roles and contributions that range from very similar to very different. The potential for major and often equal contributions by each sex, together with the fitness rewards associated with biparental care, are what have made monogamy the prevalent mating system in birds. Once biparental care evolved, the specific and diverse forms it has taken have been modified by ecological pressures as the various avian lineages diversified.

Male birds of many species not only feed their mates, they also may protect them from predators, sometimes at real risk to themselves. Thus, the level of cooperation between members of a pair is often conspicuous, and is readily understandable when the short- and long-term goals of each individual are considered—maximization of life-time reproductive success.

Despite large overlap in the evolutionary interests of the two sexes, their interests are not identical. In many cases, males will be alert to, and may seek, mating opportunities outside the pair bond (see Chapter 12); in a number of usually-monogamous species, males may occasionally mate polygynously (see Chapter 13). However, by no means do males have all the advantages in this regard. In birds, females of socially monogamous species, as a rule, have as much or more opportunity to selfishly harm the interests of their mates as vice versa. There is now abundant evidence that females of many monogamous species regularly seek extra-pair copulations and lay eggs not sired by their social mates (see Chapter 12). This means that males which provide care for chicks that are not their genetic offspring are being exploited.

11.6.3   Benefits of long-term pair bonds

A significant aspect of monogamous mateships pertains to reproductive or
fitness benefits gained by retention of the same mate from one breeding period
to the next. Rowley (1983) reviewed the patterns and incidence of re-mating
(pairing with the same individual over successive breeding periods or seasons)
in monogamous birds and concluded that, other things being equal, mating
with the same individual from one year to the next can have definite benefits in
terms of lifetime reproductive success. Reasons for this are several. (1) In
many long-lived birds, the process of pair-bond formation is a time-consuming
affair. When the same mate is retained, this process occurs only once. (2) In
many species, egg size increases with the female's age, and egg size may be
related to the nutrition (i.e. quality and quantity of yolk) available for the
developing embryo and thus its quality or condition at hatching. (3) Clutch
size tends to increase with age. (4) Laying date tends to come earlier with
increasing age or experience, especially with the persistence of the pair bond.
(5) Hatching success may vary with the age of the parents. (6) The quality of
parental care or chick-rearing improves with age and experience. In addition,
as pairs of birds become older, they tend to rise in the local social hierarchy,
which may eventually lead to acquisition of the best breeding sites or
territories (Rowley 1983 provides references supporting these points). It
should be noted that these factors are also relevant to explaining why an individual of either sex generally prefers to obtain an older, rather than a younger,
mate (see Section 11.8.3).

11.6.4   Factors preventing remating

Given the difficulties of acquiring more than one mate, and given that many
benefits of retaining the same mate exist, why is long-term monogamy not
more common than it is? According to Rowley (1983), the single most important factor preventing more frequent re-mating is the death of one or both
members of the pair between one breeding season and the next. Rowley points
out that if annual mortality exceeds 30 per cent, there is automatically less
than a 50–50 chance that both members of a pair will be alive in the following
year. This alone means that, for most small, temperate-zone birds, natural
selection will not favour strong re-bonding behaviour; i.e. waiting for a previous mate to arrive (as in a migratory species) will not usually be selectively
advantageous if the odds are high that the absent mate is dead (Rowley
1983).

As one might expect, there is also a relationship between the kind of movements a species or population employs (i.e. dispersal patterns and migration)
and the probability of reuniting to breed with a prior mate. Permanent
residents often occupy the same territory year-round. In many such species, the
male and female remain together and pair bonds are extended, often for as
long as both members remain alive. Excluding studies of mate retention in

some seabirds (see Rowley 1983 for references), stability of pair bonds over extended periods of time is probably best documented in cooperative breeders, many of which are extremely philopatric. Several thorough, long-term studies (10 or more years) involving large numbers of pairs of individually marked birds have indicated that established breeders usually remain together on the same territory for as long as both remain alive (see Stacey and Koenig 1990).

Migratory species, which comprise a large fraction of the avifauna of the southern as well as the northern hemisphere, seem to be least likely to maintain pair bonds between one breeding season and the next. This is thought to be due to high mortality as a result of the hazards of migration and to the fact that in some species the sexes overwinter in different areas and follow different migratory schedules. There are exceptions to this generalization, however; in some species, such as geese and cranes (Fig. 11.2), mated pairs migrate together with their young of the year over great distances, and show long-term pair fidelity.

MIKE RAMOS
1993

Fig. 11.2   In the world's only natural population of whooping cranes, adults are monogamous and the young birds follow their parents from the breeding grounds in northern Canada to the Gulf Coast of Texas.

11.7 Role of the male parent in socially monogamous systems

In the great majority of bird species, the male contributes directly to the welfare of its offspring in one or more important and conspicuous ways; nevertheless, it often has been assumed that males make lesser total contributions than females do. Among contemporary monogamous birds, variation in the kinds and extent of male parental investment is striking. At one end of the spectrum are those species where the male conducts virtually all of the parental duties once the eggs are laid (e.g. tinamous and ratites, some shorebirds). Then there are those species in which the two sexes appear to contribute equally and in virtually identical ways. This is probably true in species such as procellariiforms that lay a one-egg clutch and that forage far from the nest (Lack 1968), and it may also be the prevalent pattern in many other orders of birds (e.g. loons, grebes, penguins).

In other monogamous species, males make major contributions that differ in large or small ways from those of females. For example, male woodpeckers conduct most of the nest-cavity excavation and contribute all nocturnal, as well as about half of the diurnal, incubation (Lawrence 1967). In other birds, males make major contributions that differ qualitatively from those of females. Some falconiforms, most and perhaps all strigiforms, and geese and swans, among others, fall into this category. For example, female owls do all of the incubating, while their mates provision them and later also provide most or all of the food for the young (e.g. Gehlbach 1994). Male razorbills escort their single chick to sea when it is still small (Fig. 11.3), while the female plays no further role in parenting (Wagner 1992a). Hornbills provide some of the best-known and most spectacular examples of role differentiation, with the female sealed inside the nest chamber for many weeks while her mate provides food for her and the nestlings (Kemp 1995).

There are relatively few monogamous species in which the female conducts all of the parental care. The temperate-zone dabbling ducks are especially

Fig. 11.3 A male razorbill becomes a single parent when it leads its single chick out to the open sea.

interesting in this context, in that the males leave their mates at the onset of incubation and thus long before their offspring hatch. Thus, it appears that male ducks make no contributions to the welfare of their young. Appearances, however, can be deceptive. The drake, in fact, does contribute to the future well-being of his offspring-to-be in a subtle, but highly important, way. By guarding his mate from the sexual advances of other males during the time she is accumulating the nutritional reserves necessary to produce a clutch of eggs, the male allows the female to concentrate on foraging (e.g. McKinney 1985). This appears to have a direct effect on the quality and number of eggs the female produces, and thus on the parental success of the male (see McKinney 1985 for elaboration and references).

Finally, excluding the social parasites, there are a very few monogamous species in which neither sex contributes parental care. Most megapodes are apparently monogamous (Jones *et al.* 1995), and many of these species provide no care of the eggs after they have been placed in an appropriate site.

Thus, in the great majority of birds, the male contributes directly to the welfare of its offspring in one or more important ways, although it is often assumed that males make a lesser total contribution than females. In most cases of biparental care, the impression that the male contributes less overall has not been documented, and it would be hard to do so, if, in fact, this is the case. In the following sections I mention briefly some of the many ways that male birds provide critical parental investment.

11.7.1   Courtship feeding

One of the most common forms of parental investment by males, apart from care of offspring, is what has long been termed 'courtship' feeding. In an earlier era of ornithology, courtship feeding was thought to be purely symbolic in nature (e.g. Lack 1940), rather than of major functional significance. However, it has been recognized for some time that courtship feeding is often critically important to egg formation (see Breitwisch 1989, p. 26, and Gill 1995, pp. 408–9, for references). Consider the great weight gain females of many species undergo just prior to and during egg laying. The female reproductive tract increases in weight several hundredfold, fat reserves develop, and the eggs themselves also represent an impressive gain in mass in a very short period of time. Thus, the functional significance of large amounts of food provided by males becomes apparent.

As an example, in the green woodhoopoe, a female that weighed 65 g in non-breeding condition (mean mass of 60 female woodhoopoes from the same site was 64.3 g; Ligon and Ligon 1978) weighed 91 g just before egg laying commenced, an increase in mass of 40 per cent! In this cooperatively breeding species, the male breeder and helpers normally feed the female extensively prior to, during, and following egg laying, thus to a large degree 'covering' the cost to the female breeder of egg production (J. D. Ligon and S. H. Ligon, unpublished data).

In several, probably many, species males provide nearly all the food ingested by females during the period of egg production. If the female incubates and broods alone, she may also be fed extensively throughout incubation and into the nestling period (e.g. Ligon 1968*b*, 1978; Newton 1986; Gehlbach 1994). Although translating the calories used by males to procure and deliver the food items, plus the calories produced by those foods, into 'egg equivalents' will not be easy, it appears that male feedings during egg-laying, as well as during incubation, are often great and may make production of a clutch of eggs possible.

The importance of courtship feeding has been quantified in the rifleman, a member of an endemic New Zealand family (Sherley 1989). The rifleman is a very small (5.0–7.0 g), hole-nesting species that usually rears two broods per year. The total mass of the first clutch of eggs is about 84 per cent of the female's body weight, and the second about 72 per cent. Males feed their mates only before and while the first clutches are being laid, and during that time contribute almost half (42 per cent) of the food they gather to the females. Males preferentially feed females large food items and their contributions amount to 35 per cent of the females' total food intake. Sherley (1989) calculated that food provided by the male was the difference between the female's maintenance requirement and the extra energy required for oogenesis. Thus, by extensively feeding his mate, the male rifleman provides the raw materials that the female converts into eggs.

It is clear that in this case male parental investment is considerable and that overall it may equal or exceed that of the female. The male gains from this effort in that nesting begins earlier, which permits a second brood. Although the male does not feed the female as she produces the second clutch, he is solely responsible for the still-dependent first brood. Thus, the food provided by males first for the laying female and later for their offspring apparently makes a doubling of the annual reproductive output possible, is of large fitness benefit to both parents, and especially makes clear the benefits that can be gained by role-specific cooperative teamwork of each member of the pair.

11.7.2   Other contributions to reproduction by males

Male birds of most kinds participate in the majority of activities involved in rearing young. This is more conspicuous in altricial than precocial species, simply because in the former parental care requires more complex behaviour.

Nest construction
With regard to participation of both sexes in nest building, Skutch (1976, p. 99) writes: 'In non-passerine families, building by both sexes is the

prevailing mode. This is true of herons, pigeons, gulls and terns, cormorants, hawks, kingfishers, motmots, trogons, non-parasitic cuckoos, barbets, jacamars, puffbirds, and woodpeckers, to mention only a few.' Among passerines, there is tremendous variation in contributions by males to nest construction.

In at least some of these groups, males provide major parental investment in nest preparation in ways not matched by their mates. For example, in most woodpeckers, males contribute far more effort to excavation of the nest cavity than do females. Lawrence (1967) determined that two male hairy woodpeckers conducted about 95 per cent of the cavity excavation, and she recorded a similar pattern for yellow-bellied sapsuckers. Similarly, Crockett (1975) reported that, in Williamson's sapsuckers, males were responsible for excavating nests (Fig. 1.5), a process which required several hours each day for three or four weeks. Skutch (1976) provides many additional examples of paternal contributions to nest construction.

In short, the role of the male in nest building varies from species to species, with males of some doing little, while in others, such as the woodpeckers, males probably contribute more parental effort overall than do their mates.

Incubation

At one extreme are those species, which include many passerines, in which males neither develop a brood patch nor exhibit any incubation behaviour. At the other are those species in which the male provides most or all of the incubation. Shared incubation is one of the most conspicuous and, in some species, most critical aspects of monogamy and biparental care. In many socially monogamous birds, males contribute extensive incubation of the eggs. In such species, which make up the majority of avian orders (see Section 11.4), monogamy is almost bound to be obligatory. Among birds as a whole, the specific incubation role taken by each sex varies widely. As Skutch (1976, p. 153) writes:

> Although all the members of a species rather closely follow the same pattern of incubation, which in a number of instances prevails throughout the family, in the whole class of birds these patterns have proved to be amazingly diverse. Almost every scheme that one can imagine for keeping a set of eggs intermittently or constantly covered has been adopted by one species or another.

Feeding of chicks

Males of most monogamous altricial or semi-altricial species feed their young. As is also true of other aspects of parental care, there may or may not be differences in the role played by each sex. In species that produce food internally, such as pigeon's milk or the oils secreted in the proventriculus for their young by procellariiforms, the male parent is as capable as the female of feeding the young. Excluding raptorial birds, more or less equal feeding contributions by both parents is probably typical of many species. For example, Lawrence (1967) presented data showing that feeding of nestlings by the male and female of four species of North American woodpeckers was close to equal.

In other species, however, the male clearly has a larger role in the actual provisioning effort. Typically, this is true when the female remains at or near the nest to brood or guard the chicks. Examples include corvids (e.g. male pinyon jays provide 70 per cent of the feeds; Marzluff and Balda 1992) and both diurnal and nocturnal predators (falconiforms and strigiforms). Among small owls, it appears that, at least in some cases, the male provides virtually all of the food brought to nestlings (e.g. Gehlbach 1994), and a similar pattern is seen in some hawks and falcons. Moreover, in these species, before the chicks hatch, the male captures and delivers most of the food ingested by the incubating female. Thus, in raptors, where prey may be especially costly to capture and deliver to the nest, it certainly appears that, overall, the male more than 'pulls his weight'.

A particularly interesting morphological difference between the sexes related to provisioning young is found in sandgrouse (Fig. 11.4). In these Old World, arid-country birds, water is brought to the chicks by means of specialized belly feathers that hold moisture, and the abdominal feathers of males are more structurally specialized for water-carrying than are those of females (Cade and Maclean 1967).

Finally, in addition to incubation and feeding young, males of many monogamous avian species may contribute more than females to brood defence,

Fig. 11.4 In the monogamous sandgrouse, the belly feathers of males are specialized to hold water that is flown to the chicks. Reprinted with permission from Dover Publications, Inc.

which is potentially dangerous. Regelmann and Curio (1986) and Breitwisch (1988) provide numerous references documenting this point.

These few examples illustrate that extreme parental specialization—role differentiation—occurs frequently within monogamous systems and that reproductive efficiency and effectiveness are increased by the sex-specific role of each parent. Once again, maximizing the number of young reared by sharing the costs of reproduction, whether the specific tasks are the same or different, must be the single most important factor behind the prevalence of monogamy in birds.

11.8   Costs of monogamous mateships

In the previous section, I stressed the positive aspects of monogamous mateships in birds. The prevalence of monogamy is testimony to the benefits of this mating system. Even so, monogamy also has costs, which should be viewed from the perspective of each sex. Despite large overlap in the evolutionary interests of the two sexes, their interests are not identical. This is illustrated well by those systems in which males of usually monogamous species attempt to obtain more than one mate, and by those systems in which females copulate with one or more additional males.

11.8.1   Sexual conflict: additional mates

In many species, males will be alert to, and may seek, extra-pair matings, to the extent that they do not disrupt the pair bond (see Chapter 12). Similarly, females of a variety of monogamous species actively seek copulations, and thus presumably fertilizations of their eggs, outside the pair bond. Fitness benefits to females of such liaisons are considered in the next chapter. The most spectacular examples of the percentage of fertilizations from outside the social/mating unit are two species of cooperatively breeding Australian fairy-wrens, where very few of the young hatched in a nest are the offspring of the apparent mate of the resident female.

In addition, in some species that are typically monogamous, males occasionally acquire a second mate. Davies' (1990, 1992) studies of the dunnock provide an outstanding example of sexual conflict of this sort (see Section 13.5.4). This point is also illustrated by the pied flycatcher, in which a male with a territory and mate acquires a second, separate territory and attempts to attract a female to it. It appears that neither the first nor the second female is aware that she is 'sharing' the same male (Alalato *et al.* 1987). In species where males provide extensive parental care, this conflict of interest is potentially, and actually, damaging to one sex—here, one or both of the females. In the house sparrow, the second female of polygynous males has lower reproductive success than the primary female because the male provides no care at her nest (Veiga 1990).

11.8.2 Desertion as a selfish reproductive strategy

Desertion of young by a parent occurs in birds, as well as in humans. This conflict of interest may occur in monogamous pairs of birds and is manifested by desertion of the nest and chicks by one sex or the other, leaving the remaining parent to rear the young unaided. Desertion of mate and offspring has been studied in snail kites (Beissinger and Snyder 1987; see Beissinger 1990 for additional references) and penduline tits (Persson and Öhrström 1989).

Male snail kites provide a lot of early parental investment—apparently more than the female (Beissinger 1987a)—by carrying out most of the nest construction, feeding the female prior to clutch completion, and incubating the eggs. Females also incubate, and both male and female provision the young in a typical biparental fashion. Subsequently, however, either sex may desert the nestlings, with females deserting twice as frequently as males. Snail kites typically desert after brood reduction has taken place and when the remaining chicks no longer require brooding, either late in the nestling period or in the post-fledging period when they are still dependent on parental feeding. The number of chicks in the nest is related to the probability of desertion. In experimental manipulations of brood size conducted in Venezuela, desertion occurred at all nests fledging one chick, at half of the nests fledging two, and at none of the few nests with experimentally enlarged broods of three young (Beissinger 1990). The deserted parent routinely was able to successfully complete the task of brood-rearing. The chief benefit for the deserter is thought to be increased opportunity to re-mate and produce another brood. The remaining parent, the 'tender', may incur substantial costs. First, tenders lose considerable time that could potentially otherwise be devoted to re-nesting. Second, they may have increased energy expenditures, which could add up to a significant portion of the bird's lifetime reproductive effort. Despite these costs, Beissinger and Snyder (1987, p. 486) conclude that:

> ... there is no viable alternative strategy to tending their offspring because desertion occurs relatively late in the nesting cycle (Trivers' cruel bind), when the probability of successfully completing brood rearing is high, while the likelihood of being successful in a new nesting attempt is low.

An even more striking case of nest desertion by members of either sex occurs in the penduline tit of Eurasia. In a small Swedish population, uniparental incubation and nestling care was the rule, and either the female (18 per cent), the male (48 per cent), or both individuals (34 per cent) of every pair deserted. Persson and Öhrström (1989) suggest that two critical factors make this kind of mating system possible: uniparental care and a long breeding season. In addition, the covered nest is well insulated, which reduces the necessity of brooding the chicks (i.e. foraging time for the single parent is increased, as compared to cup-nesting species), and the nest is located in a very safe place, at the end of small twigs and often over water.

It appears that this system is driven primarily by the female, who decides whether to remain as the sole parent or to desert. Most females remain with their first clutches, leaving the male free to seek a second mate. With females occupied with parental duties, the operational sex ratio (Emlen and Oring 1977) becomes increasingly male-biased. This, in turn, has two important effects. First, with relatively and absolutely fewer females available in the population, it may pay males to assume parental care. This occurs at the same time that females have the option of exploiting the 'surplus' of males by becoming polyandrous. That is, as the season progresses, a female's chances of obtaining a new mate are much greater than they are for a male.

11.8.3   Dissolution of the pair bond

Although long-term and even lifetime pair bonds are apparently important in many kinds of birds, mateships are occasionally terminated while both pair members are alive and in good physical condition. Rowley (1983, Table 15.6) found that the annual 'divorce rate' among migratory species ranged from 9 to 100 per cent. A cost of monogamy is that pair bonds sometimes form between high-quality and low-quality individuals. Divorce can cancel future costs of this sort to the higher quality bird. Thus, the deserting bird may usually be following an adaptive, and thus potentially profitable, strategy, while at the same time the deserted, presumably lower-quality, individual suffers a cost when its mate leaves. Choudhury (1995) reviews the hypotheses that seek to account for divorce in birds, and concludes that most of the hypotheses boil down to the 'better option' model, which proposes that an individual should divorce when it has the opportunity to obtain a better mate and/or a better territory (see also Ens *et al.* 1996 for a comprehensive review of divorce in monogamous birds). Two studies will serve to illustrate this point.

A study of divorce in the European oystercatcher by Ens *et al.* (1993) indicated that individuals desert a mate primarily to improve their own prospects for future reproductive success, rather than deserting due to some sort of incompatibility with the partner. In this species, a major consideration is the relative quality of territories, which has a large effect on reproductive success. Similarly, divorce in willow tits seems to be primarily a female strategy to increase future reproductive success (Orell *et al.* 1994). This is accomplished primarily when the female can desert her current partner in order to pair with a higher-ranking male. In the oystercatcher and willow tit, a change of mate did not lead to a reduced reproductive output in the breeding session following the divorce. Thus, for both of these very different types of birds, it appears that females that leave their mates to join another male improve their reproductive success.

In contrast, after obtaining a new mate, the reproductive success of the deserted male decreased. This suggests that in each case females chose a 'better option' (a higher ranking male or a better territory, or both), and that their former mates suffered reduced reproductive success when they paired with a yearling female. Similarly, in the willow tit, males sometimes deserted young mates with whom they had no prior breeding experience, when an older female became available. This further suggests that for both sexes future expectations rather than previous breeding history is likely to be the usual basis for divorce in birds.

These two studies, plus several others (e.g. Harvey *et al.* 1984; Marzluff and Balda 1992), have also determined that, after the effects of female age were standardized, new pairs did not have lower reproductive success than old pairs, and they contradict the many studies which have concluded that new pairs did less well than old pairs (see Rowley 1983 and Orell *et al.* 1994 for references). Finally, in a relevant study, Pärt (1995) experimentally controlled the age of first breeding in female collared flycatchers. Pärt concluded that high-quality females first breed at a younger age and that female quality may best explain the apparent relationship between prior breeding experience and reproductive performance.

11.9   Sources of variance in reproductive success of monogamous species

It is frequently and convincingly argued that within-sex variance in mating success is a major factor contributing to the intensity or strength of sexual selection (see Section 2.5). This is the presumed basis for a relationship between degree of polygamy and degree of sexual dimorphism (Payne 1984). Although high variance in reproductive success among males is thought to be the rule for polygynous species, variance can also be high even in species considered to be both socially and genetically monogamous. Among such species, high variance in reproductive success can arise in several ways. Clutton-Brock (1988) and Newton (1989) provide a number of excellent case studies of factors affecting variation in lifetime reproductive success.

1. Across a diverse array of species (e.g. sparrow hawk, Newton 1986; screech owl, Gehlbach 1994; kingfisher (Fig. 11.5), Bunzel and Drüke 1989; great tit, McCleery and Perrins 1989), an important factor affecting lifetime reproduction is simply the number of years the individual lives. Variation in longevity may not be due simply to the vagaries of chance, but in some cases may reflect genetic differences among individuals in quality (e.g. Birkhead and Goodburn 1989).

2. In territorial species, a large, non-breeding (e.g. 'floating') population of one or both sexes often exists. Many mature but unmated individuals die before breeding, thus increasing the overall variance in reproductive success (Mock 1985). This source of variance has gone almost completely unrecognized, in large part, no doubt, because it is so difficult to study non-breeding birds and to ascertain their fate. However, in some species,

Fig. 11.5 In the monogamous kingfisher, as in many other species, a critical factor affecting lifetime reproductive success is simply how long an individual lives. Reprinted with permission from Dover Publications, Inc.

non-breeding individuals of both sexes are extremely philopatric, allowing their lifetime reproductive success (or lack of success) to be determined. For example, in the monogamous and cooperatively breeding green woodhoopoe, about 80 per cent of all individuals of both sexes that attain maturity never breed successfully (Ligon and Ligon 1989). A similar pattern has been described for the meadow pipit (Hötker 1989; see Clutton-Brock 1988 and Newton 1989 for additional examples).

3. If reproductive success is strongly affected by territory quality, as it often is, variance in reproductive success among breeders can be increased greatly as a result of differential productivity on different territories. For green wood-hoopoes, variation in territory quality leads directly to great variance in life-

time reproductive success among the minority of birds that ultimately attain breeding status (Ligon and Ligon 1988). About 90 per cent of all wood-hoopoes that became breeders left no breeding offspring and about 20 per cent of the successful male and female breeders produced 80 per cent of the young woodhoopoes that eventually attained breeding status (Ligon and Ligon 1989). This pattern of reproductive success among breeders is similar to that seen in the cooperatively breeding Florida scrub-jay (Fitzpatrick and Woolfenden 1988).

4. If extra-pair copulation (see Chapter 12) is relatively common in a socially monogamous species, with some males being especially successful in this pursuit, variation in overall reproductive success among males will be increased relative to genetically monogamous species. Although Wittenberger (1981, p. 439) states '... that extrapair copulations do not generate much inter-sexual selection for male secondary sexual characteristics', there is now ample evidence that this conclusion was erroneous. For example, in the splendid fairy-wren of Australia, a cooperatively breeding species thought until recently to be monogamous (socially and genetically), at least 65 per cent of more than 90 chicks sampled were not sired by any male in their social group (Brooker *et al.* 1990; Rowley and Russell 1990). A similar pattern recently has been described by Mulder *et al.* (1994) for the superb fairy-wren (see Section 11.4.3). In addition to this documentation of extensive extra-pair fertilizations, the spectacularly beautiful plumage of males of this and other species of fairy-wrens, together with their bizarre flower-carrying behaviour (see Fig. 12.4), also indicate that male fairy-wrens may invest considerably in extra-pair mating effort.

5. Hill *et al.* (1994*b*) found that although extra-pair fertilizations were low in the monogamous house finch, other factors could generate considerable variance in male reproductive success. In addition to a strongly male-biased sex ratio, where up to 50 per cent of the male house finches fail to obtain mates, brighter males begin nesting significantly earlier than do duller males. This leads to a potential difference of up to two broods of young per year, and should exert a powerful selective agent favouring more brightly coloured males.

These few examples illustrate that, for a variety of reasons, considerable potential for high variance in lifetime reproductive success exists in socially, and even genetically, monogamous species (e.g. house finch).

11.10   Phylogeny and monogamy

It has usually been assumed that monogamy/biparental care was the ancestral mating–parenting system in birds. However, as discussed in Chapter 10, although monogamy and biparental care must have occurred early in the history of neognathous birds, this was probably not the first mating–parental care system to evolve in this group; i.e. its evolution may have been preceded by exclusive paternal care. Biparental care was so well suited to the basic breeding pattern of birds—rapid production of rapidly developing eggs placed in a single location, the nest—that most of the evolutionary refinements that have evolved since the Eocene have served to make monogamy and biparental care more effective. (Recall that all modern lineages of non-passerine birds may have appeared by the Eocene (Feduccia 1995, 1996), and that nearly all orders and most families and species of birds are characterized by monogamy and biparental care.) Among most non-passerine groups, exceptions to this generalizations are scarce (e.g. Altenberg *et al.* 1982), although, of course, some do exist (e.g. grouse, hummingbirds).

Many passerines have deviated from monogamy and most of this deviation has been from monogamy to polygyny. Correlated with shifts away from monogamy is the ability of a single parent to provide all of the incubation and, often, all of the provisioning of nestlings. As Searcy and Yasukawa (1995) point out, reduction of paternal care is closely correlated with an increase in polygyny. The ability of some passerine species and groups of species to shift from monogamy–biparental care to polygyny–uniparental care is especially impressive when they are compared to 'passerine-like' non-passerines (typically small arboreal or aerial birds: swifts, pigeons, and doves, most coracii-forms, piciforms, colies, non-parasitic cuckoos), nearly all of which exhibit what appears to be obligate biparental care. In short, mating systems other than social monogamy are rare among most non-passerine birds and are less common than monogamy even among passerines.

Sequential polyandry has been claimed for a few passerines, but these cases could also be viewed as 'serial monogamy' (Ford 1983). Middleton (1988) described what he viewed as polyandry in the American goldfinch. Similarly, Seutin *et al.* (1991) described a case in the common redpoll that they labelled 'sequential polyandry'. In both the goldfinch and redpoll studies, a relationship was described that involved, at different times, a single female and two males. In cases such as these, it is a fine point to distinguish between serial mono-gamy and sequential polyandry, if indeed such a distinction is valid and mean-ingful. These cases illustrate the problem of terminology in the area of avian mating systems (see Johnson and Burley 1997).

11.11   Conclusions and summary

In Chapter 10, a case was presented in support of the proposition that the first form of parenting in birds was exclusive paternal care. Biparental care and apparent or social monogamy probably evolved relatively soon thereafter. Thus, although monogamy and biparental care may not have been the original parenting and mating systems in birds, this behaviour and associated physiological/endocrinological characteristics probably evolved early in the history of this group and have since been, by all available measures, the overwhelmingly predominant reproductive strategies in this group. Despite its prevalence in birds, monogamy has not received the level of attention by avian behavioural ecologists that has been accorded to other, far less common, mating systems. This has led to the assertion that monogamy is a 'neglected' mating system.

The key factor influencing both the evolution and maintenance of monogamy is parental care by males; this is primarily because of two features, oviparity and endothermy. With the appearance of these traits, parental care became essential to the survival of young birds. Except for producing eggs, the male is potentially as capable as the female as a parent. This means that male birds, unlike male mammals, can contribute critically to the well-being of their offspring at all stages of the reproductive cycle; when biparental care is essential, monogamy is 'obligate'. The importance of biparental care also can account for the evolution of the uniquely complex cooperation in reproduction that occurs among mated males and females of most avian species.

Other factors have also helped to promote and maintain 'facultative' monogamy and biparental care in most avian lineages. Some of these include (1) synchrony at the population level in the onset of breeding, (2) male-biased sex ratio with monopolization of females by males, and (3) aggressive defence by females of the male and the territory. In addition, in many monogamous species, there are apparently some measurable benefits of maintaining long-term pair bonds.

One of the most striking things about monogamy and biparental care is the level of male contribution to the overall reproductive effort. Although the specific kinds of contributions by males vary widely, as one might expect in a group as large and as diverse as birds, the most common pattern is that males contribute a great deal. In fact, for some, and possibly for a good many, species, a case can be made that the overall contribution of males at least equals that of females, even when the cost of producing a clutch of eggs is included. Depending on the species, males may construct the nest, deliver a great deal of food to the female in the form of 'courtship feeding', incubate the eggs, feed the nestlings and fledglings, and defend the young from dangerous predators. Except for laying the eggs (which, by definition, is the exclusive purview of females), males can, and frequently do, participate extensively in all aspects of parenting.

The social monogamy and biparental care practised by the great majority of birds is by no means always paralleled by genetic mate fidelity (see Chapter 12). In addition to extra-pair copulations/fertilizations, which form the subject of the next chapter, males may attempt to attract a second mate, often at a cost to the first. Alternatively, either sex may desert, leaving the remaining parent with the full responsibility of rearing their (presumably) shared offspring.

In some species characterized by long-term mateships, dissolution of the pair bond, or 'divorce', occasionally occurs. This appears to be a strategy by the initiator of the break to improve its overall prospects for future reproduction, either by acquiring a mate perceived to be superior to its current one, or by gaining access to a superior territory on which to breed, or by obtaining both.

Although high variance in reproductive success of males is usually associated with polygynous and promiscuous systems (e.g. see Chapter 13 and Searcy and Yasukawa 1995), several factors can promote variance in reproductive success among males of monogamous species. These include (1) the unmated, non-breeding portion of the overall population—many individuals either never breed at all or they are never successful breeders, (2) differences in territory quality, (3) the frequency of extra-pair fertilizations, and (4) in genetically monogamous species, earlier and more frequent nesting by more attractive males.

Although social monogamy–biparental care is by far the most important mating–parenting system in birds, as measured by the number and diversity of species practising it, secondary strategies designed to increase fitness have evolved repeatedly in both female and male birds. The next chapter deals with what is apparently the prevalent means of increasing individual fitness within monogamous social systems, the seeking of extra-pair matings by individuals of both sexes.

12 Extra-pair copulations and their evolutionary significance

12.1 Introduction

As discussed in Chapter 11, monogamy is the social mating system practised by the vast majority of birds. This chapter deals with the subject of copulations and fertilizations carried out by individual males and females that are not members of a social mateship. These phenomena are widely referred to as extra-pair copulation (EPC) and extra-pair fertilization (EPF), respectively. Extra-pair paternity (EPP) refers to chicks sired by a male other than the social mate. EPCs may occur with or without EPF, due, for example, to sperm competition. Because EPC is usually recognized by the detection of chicks in a brood that had not been sired by their apparent father, this term is commonly used to refer to extra-pair copulations that lead to fertilization of eggs, and this is the convention followed here.

It is most appropriate to use the term 'extra-pair copulation' for species in which there is a recognizable mating relationship between one male and one female (Westneat *et al.* 1990); thus, it seems to be most applicable to monogamous species. Although the label 'extra-pair' is not quite right for social-mating systems involving more than two birds, the term 'extra-pair copulation' is often applied to polygynous systems when a female copulates with a male other than her social mate. It is clearly inappropriate for promiscuous-lekking mating systems, where a female has no established social–sexual relationship with a male, beyond a brief visit that possibly culminates in copulation. Nor are matings of polygynandrous species (multiple males per female, and vice versa, within a stable social group; see Section 13.5.4), such as the acorn woodpecker or dunnock, examples of EPCs. In such species, a female may have a long-term, stable social relationship with two males, both of which sire her offspring. Thus, EPC is not defined simply by multiple paternity of a brood of young birds. Instead, EPC occurs when two individuals that are not members of the same social unit copulate.

For many years it was assumed that, in monogamous birds, the social mating bonds between males and females accurately reflected their genetic mating patterns; e.g. males of monogamous species fertilized all of the eggs of their mates. Since the 1960s, avian mating systems have been an active area of research, particularly from an expressly ecological perspective. During this period a general and critical assumption was that the social or apparent mating system accurately reflected the genetic mating system. Thus, discovery of the phenomenon of EPC has had major ramifications for the semi-classical ecological classifications of mating systems, such as those of Verner and Willson (1966), Orians (1969), Emlen and Oring (1977), Oring (1982), and Wittenburger (1981). All of these authors assumed (along with other ornithologists) that the apparent mating system was an accurate indicator of the parentage of broods of young birds.

During the past decade, however, a scientific breakthrough has captured the attention of those avian behavioural ecologists interested in mating systems. The discovery of widespread occurrence of EPCs has probably done more to revolutionize thinking about mating systems than any other single empirical discovery in at least the past 30 years (e.g. Westneat 1987*a,b*; Burke *et al.* 1989). This subject has been reviewed by Westneat *et al.* (1990) and Birkhead and Møller (1992), and the interested reader is referred to those sources for more extensive coverage.

A good deal of evidence has now accumulated which indicates that 'apparent' or social monogamy and gametic or genetic monogamy do not necessarily closely coincide (e.g. Ford 1983; McKinney *et al.* 1984; Gowaty 1985, 1996*a*; Westneat *et al.* 1990; Birkhead and Møller 1992, 1996). The same has been shown to be true in some polygynous systems, where more than one male may fertilize the brood of a given female (e.g. Bray *et al.* 1975; Gibbs *et al.* 1990; Westneat 1987*a,b*, 1990, 1993, 1995). The obvious benefits to males initially led to a focus on the benefits of EPCs to members of that sex. Subsequently, it has become widely appreciated that inclusion of the issues of EPC into considerations of mating systems requires attention to the subject of mating strategies of female, as well as male, birds (e.g. Gowaty 1996*a,b*; Birkhead and Møller 1996).

In contrast to many aspects of parental care, which often appear to be tightly programmed, strategies of EPC can be extremely flexible and opportunistic. Among males of the monogamous majority of species, both mating effort and parental effort are often critical to the production of young birds, with the roles of each sex being well defined. In contrast, for those few species in which EPC has been studied, its frequency among members of each sex has varied, with some males and females obtaining many, while others obtain none.

Numerous studies of EPC have now been conducted via biochemical techniques (protein electrophoresis and especially DNA fingerprinting), which document that a brood of young birds may be sired by more than one male, i.e. by a male or males other than their 'social' or apparent parent (see Birkhead and Møller 1992; Gowaty 1996*b*, Table 1; Table 12.1). Because EPC appears to be a common phenomenon in birds, the studies documenting it might lead one to conclude that genetic monogamy is rare. DNA fingerprinting, however, has demonstrated that genetic monogamy (all chicks sired by the social mate) appears to be the rule in a number of species (see Section 12.5); i.e. in many cases, social and genetic monogamy apparently correspond closely, even in birds possessing different kinds of social-mating systems. In view of the interspecific variation in the frequency of EPC, it is interesting that, to date, no ecological correlates of extra-pair paternity have been detected (Birkhead and Møller 1996).

Mate 'fidelity' may also be high in polygynandrous birds. In the best-studied of such species, the dunnock, only one nestling out of 133 sampled was sired by a male from outside its social unit (Davies 1990). Thus, it is clear that students of avian mating systems must be alert to the possibility that the observed social–reproductive relationship between males and females of a given species or population may or may not reflect the actual genetic relationship of the putative parents to their presumed offspring. In this chapter, I review briefly some of the benefits and costs of EPC for both male and female birds; most of these points, plus others, are treated by Westneat *et al.* (1990), Birkhead and Møller (1996), and Gowaty (1996*a,b*). In addition, I consider some of the possible effects of EPC on the evolution of mating systems and coloniality.

12.2   Benefits and costs of mating with multiple partners

Westneat *et al.* (1990) and Birkhead and Møller (1992) have recently provided comprehensive syntheses of the ecological and social conditions that might promote EPC. Although the major significance of EPC is fertilization of eggs, it is important to note that this behaviour may also serve social functions unrelated to the more obvious direct genetic fitness benefits of egg fertilization. In this section, I discuss briefly some of what are currently thought to be the major benefits and costs of EPC. More comprehensive lists that include additional possible benefits and costs can be found in Westneat *et al.* (1990, Table 1) and Birkhead and Møller (1992, Tables 10.1 and 11.1).

12.2.1   Benefits of EPC for male birds

For males of promiscuous species, maximizing the number of copulation partners is an obvious strategy to promote fitness. The potential benefits of

EPC should be similar for males of monogamous species. In both cases, all other things being equal, those males fertilizing the most eggs will sire the most offspring. According to current evolutionary theory (e.g. Trivers 1972; Davies 1991), a male bird should accept opportunities to obtain genetic paternity of additional young, regardless of its species' social or apparent mating system, so long as any associated costs are acceptable. That is, a male of a monogamous species should seek and accept extra-pair matings to the extent that its primary reproductive strategy—i.e. acquiring a mate, mate-guarding, and caring for eggs and/or chicks—is not harmed (i.e. as long as the net benefits outweigh the costs). As noted above, in many monogamous birds, but apparently not all, that is exactly what males do. By producing essentially cost-free genetic offspring (i.e. offspring cared for by another male), an individual male bird can greatly increase its reproductive success, both per year and over its lifetime.

Westneat *et al.* (1990) list five factors that should affect the likelihood that males of a given species or population will benefit by pursuing a strategy of EPC.

1. *Value of male parental care.* The more time and effort a male is required to expend in parental duties, the less likely it is to engage in major EPC efforts. Thus, for species in which paternal care is limited or non-critical, we might predict a higher level of EPC activity than in species in which the male is fully occupied with caring for nestlings.

2. *Density of breeding individuals.* Many cases of EPC come from studies of colonial species, and it has commonly been assumed that high bird densities result in more opportunities for EPC than exist for non-colonial species. This appears to be supported by observations which suggest that males attempt EPC at higher rates in denser colonies. Alternatively, it may be that colonial species are easier to observe than non-colonial ones, and that rates of EPC corrected for observability are similar in colonial and non-colonial species. Genetic studies are required to resolve this point.

3. *Breeding synchrony by females.* Other things being equal, the greater the degree of breeding synchrony in a population, the less likely it is that EPC will be widespread. This is because if all of the females in a local population are fertilizable at about the same time, the value of mate-guarding by the typical mated male will be increased, thereby simultaneously inhibiting search for EPC opportunities. (See Stutchbury and Morton 1995 for an alternative view.)

4. *Ability of males to guard their mates.* One of the usual assumptions about EPC has been that it is primarily a male strategy (Gowaty 1996*a*). Based on this view, mate-guarding by males has usually been interpreted as an effort by the guarding male to defend its mate from copulation-seeking males from elsewhere. Alternatively, mate-guarding may often actually represent an attempt by males to deter their mates from seeking EPC (Wagner 1993). Factors which

demand that males leave their mates for periods of time might well lead to an increase in the frequency of EPC for either or both of these reasons (i.e. unguarded females are more likely to be approached by an EPC-seeking male, and, conversely, unguarded females are also more likely to approach a male for copulation). The importance of mate guarding by males is indicated by the fact that in some species within-pair parentage is positively related to male size (e.g. Kempenaers *et al.* 1992; Weatherhead and Boag 1995; Wagner *et al.* 1996*a*; Yezerinac and Weatherhead 1997).

5. *Behaviour of females with regard to EPC attempts by males.* A key assumption of the points mentioned above is that EPC is largely a male reproductive strategy. However, this assumption may frequently be partially incorrect. Females may not only tolerate EPC by males, they may search for and solicit them. Moreover, females may seek males that exhibit certain kinds of traits. In the yellow warbler, for example, males have reddish streaking on the breast, and extra-pair mating success is positively related to the amount of streaking (Yezerinac and Weatherhead 1997). Possible benefits to females of such behaviour are discussed in the following section.

12.2.2   Benefits of EPC to female birds

Until fairly recently, the importance of EPC to females was not fully appreciated (see Gowaty 1996*a,b* and Birkhead and Møller 1996). Thus, it is worth emphasizing that seeking EPCs may be just as important a reproductive/ fitness strategy for females as for males (see Wagner 1993), although the specific benefits probably differ between the sexes. As discussed above, males gain in number of offspring sired by successful EPC. Females, on the other hand, typically do not realize the benefits of increased number of offspring. Benefits of EPC accruing to females are often more subtle, but, based on the apparent frequency that females initiate them, they are probably highly significant. Potential benefits to females fall into three general categories: material, genetic, and social.

Material benefits
For females, potentially important material benefits of EPC have been suggested to be of two basic types: (1) food via courtship feeding and (2) paternal care. There is little observational evidence, however, to support either of these suggestions, and Birkhead and Møller (1992) conclude that it is unlikely that female birds in socially monogamous species engage in EPC either to gain food via courtship feeding or to gain additional care with rearing their offspring. Their conclusion is probably correct for the typical pair-only monogamous breeding unit.

However, among at least some cooperatively breeding species that appear to be both socially and genetically monogamous, females may interact sexually with male helpers. For example, in the green woodhoopoe, male helpers sometimes attempt to mount the breeding female of the group. Copulation was occasionally observed between a male helper and the breeding female, but only after the eggs were laid (Ligon and Ligon 1978). This sexual interaction appears to prompt the male helper involved to increase the rate of delivery of food to the female and possibly to the nestlings, which suggests that the male helpers have '… been duped into assuming they had some paternity … .' (Birkhead and Møller 1992, p. 203). Alternatively, mixed paternity may occur in broods of young woodhoopoes, as has proved to be the case in another cooperative breeder, the stripe-backed wren, also thought, prior to studies using DNA fingerprinting techniques, to be both socially and genetically monogamous (Rabenold *et al.* 1990).

Dunnocks often breed in trios composed of one female and an alpha and a beta male (Davies 1990). In these trios, the amount of parental care provided by the beta male depends on his frequency of copulation; if the alpha male completely excludes the beta male from mating, beta contributes no care to the nestlings. Because frequency of copulation, genetic parentage, and paternal care patterns are all well known, the dunnock provides the best evidence currently available concerning a relationship between copulations (prospective fertilizations) and providing care to young birds.

Genetic benefits
This category includes a variety of possible benefits. For example, the first category considered, fertility insurance, is a genetic benefit only with regard to obtaining or not obtaining genes carried by spermatozoa.

1. Fertility insurance. A female bird might mate with more than one male in order to ensure that viable sperm will be available as the ova are released from the ovary. Wetton and Parkin (1991) found that in house sparrows about 14 per cent of nestlings were unrelated to the attendant male, and that the occurrence of cuckoldry was not affected by male age or experience, pair bond duration, or time within the breeding season. However, extra-pair young were significantly more common in broods that included some infertile eggs. These facts led Wetton and Parkin (1991) to conclude that sperm competition can be strongly influenced by the fertility of the cuckolded male, and that females may benefit from extra-pair copulation as insurance against their mate's infertility.

Similarly, in a study of the benefits of EPC to female razorbills, an alcid of the North Atlantic, Richard Wagner (1992*b*) tested predictions from three hypotheses. He found that only one, insurance against male infertility, was not refuted by any of his results. Wagner suggests that the pattern of within-pair versus extra-pair copulations implies that females might be storing sperm of extra-pair males as insurance against infertility of their own mate. The fact that female razorbills lay only one egg per year supports the assumption that fertility insurance may be extremely important to them. Birkhead and Møller

(1992) argue that a more likely explanation is that poor-quality females that cannot produce fertile eggs are also mated to poor-quality males, and as a result they seek extra-pair copulations. These authors conclude that '... the idea that female birds copulate with more than one male to ensure fertilization of their eggs seems unlikely' (Birkhead and Møller 1992, p. 201).

As a broad generalization, this conclusion may be premature. Sheldon (1994) suggests that fertility insurance may be as likely an explanation for EPCs as the hypothesis that EPC-seeking females are obtaining superior genes from the EPC partner. Sheldon (1994) describes tests that could distinguish between the two hypotheses.

2. Inducing sperm competition. Another possible adaptive explanation for why female birds might choose to copulate with more than one male is to induce sperm competition. Such competition could benefit females in either of two ways. First, if production of competitive sperm is a heritable trait, females could benefit by producing sons with such sperm (e.g. Lifjeld et al. 1993). Second, if the competitiveness of a particular male's sperm is related to his vigour and health, then, by mating with more than one male, a female could have a majority of her offspring sired by the healthiest available male.

This could be especially important in species, such as, for example, the polygynous red junglefowl, in which females often cannot prevent forced copulations, nearly all of which are carried out by subordinate males (Thornhill 1988). That is, one way for a female to respond to a forced copulation is to also obtain sperm from one or more other males of her choosing, which would dilute the sperm of the forced copulator, thereby reducing the chances that his sperm would fertilize eggs. The intense male–male competition characteristic of red junglefowl, frequently manifested as extreme aggressiveness (Ligon et al. 1990), may also favour a strategy by females to induce sperm competition. In a free-living situation, female choice of mating partners may be severely constrained by the long-term dominance of one male over others. Thus, much female 'choice' may be conducted, in part, by inducing sperm competition via occasional surreptitious copulations (Collias and Collias 1985; Thornhill 1988), combined with the capacity for storage of sperm.

3. Genetic quality. The good genes hypothesis can account for EPC by females of socially monogamous or polygynous species that are mated to a male of lesser quality than other, neighbouring males, (Westneat et al. 1990; Birkhead and Møller 1992). Females may obtain genes from a morphologically (e.g. larger, more attractive) or physiologically superior (e.g. disease-resistant) male for fertilizing their eggs, thereby increasing the attractiveness and/or viability of at least some of their offspring.

That females of species without paternal care apparently prefer to mate with certain, apparently high-quality, males is documented in Chapter 6. Møller (1992a) provides discussion and evidence to the effect that female birds engage in extra-pair copulations in relation to the phenotypic quality of their mates relative to the quality of other males in their neighbourhood: mates of top-quality males are expected to avoid EPC, while mates of below-average males should seek EPC. Thus, females of socially monogamous or polygynous species in which paternal care of young is important will also seek to mate with males exhibiting signals of higher quality than those of the female's mate (e.g. Burley and Price 1991; Kempenaers *et al.* 1992; Hasselquist *et al.* 1996). Relative dominance can also affect the pattern and frequency of EPC. High-ranking male white ibis, which also tended to be the paired males, were more likely to perform successful EPC than were low-ranking males; at least in part this was due to female solicitations (Frederick 1987*a,b*).

It has often been suggested that good genes arguments are fallacious because fitness is not heritable (see Section 2.3). In birds such as the red junglefowl, however, many components of fitness, such as resistance to disease and certain parasites, are highly heritable (e.g. Hutt 1949; Hartmann 1985). Like many other species, red junglefowl are extremely philopatric (Collias *et al.* 1966), thus the parasites or diseases to which individuals of one generation are exposed are likely to be present in the environment when their offspring are produced. Under these conditions, selection of mates likely to possess heritable resistance to locally and temporally important pathogens could have major fitness consequences for females.

4. Genetic diversity. This hypothesis suggests that females could enhance their overall reproductive success by increasing the genetic diversity among a brood of offspring beyond the level possible with a single sire. For example, in an unpredictable environment, especially one with numerous diseases or parasites, genetic diversity of offspring might increase the chances that at least some young will be successful (Westneat *et al.* 1990). This suggestion is an extension of one of the general arguments made for sexual reproduction (Williams 1975; Sherman *et al.* 1988).

5. Good genes plus genetic diversity. Obtaining both good genes and genetic diversity may be important to female birds. For example, if females cannot determine precisely the diseases to which prospective mates possess heritable resistance, and if disease, in the broad sense, is an important selective agent, then mating with more than one male may provide insurance that at least some offspring within a brood will be resistant to whatever pathogens occur locally and currently. Thus, as pointed out by Westneat *et al.* (1990), it is difficult to distinguish between the good genes and genetic diversity scenarios. Birkhead and Møller (1992, p. 206) suggest that these two hypotheses might be distinguished by manipulating a sexual ornament, such as tail length, and looking at female copulation preferences: 'The genetic-diversity hypothesis predicts that females should copulate with males with both shortened and elongated tails. The genetic-quality hypothesis on the other hand predicts that females will prefer to copulate with males with elongated tails.'

Female red junglefowl behave in a manner consistent with both the good genes and genetic diversity hypotheses (Ligon and Zwartjes 1995*b*). Using male red junglefowl with either large or small combs, these authors found that females mated preferentially with the large-combed individual, especially upon first exposure. This supports a good genes interpretation (see also Zuk *et al.* 1995); however, nearly all females also chose to mate at least once out of five times with the small-combed male. Following the reasoning of Birkhead and Møller (1992), these mate choice tests suggest that, in addition to seeking to mate with males exhibiting traits indicative of health and vigour (i.e. large comb), sperm from more than one male is also sought. If this interpretation of the females' mate choice behaviour is correct, the study of Ligon and Zwartjes (1995*b*) indicates that one reproductive strategy of female red junglefowl is procurement of multiple paternity within a single brood of offspring, and it supports the proposition that both good genes and genetic diversity may form the basis for female mate choice in this species.

Social benefits
In his study of razorbills, Wagner (1991, 1992b) provided the first evidence that monogamous female birds pursue mountings by males outside their fertilizable periods (although most did not permit cloacal contact). Female razorbills visited traditional mating arenas located on flat boulders *after* laying, when their mates were absent from the arena, and were almost always mounted by an extra-pair male. Even though females often resisted mounting, the opportunity to interact sexually with males seems to be the only plausible reason for visiting the mating arena at this time, and it did appear as though females sought interactions with a number of males. The function of this behaviour for the female may be to assess males, either as prospective future mates or as EPC partners during the next breeding season. This appears to be true for European oystercatchers. EPCs between male and female oystercatchers in neighbouring territories usually occurred in the male's territory. Heg *et al.* (1993) surmised that the EPCs they recorded were related to possible future mate changes. In any case, sexual encounters in a non-reproductive context provide some benefit to females (and presumably also to males).

12.2.3   Costs of EPC for males

Birkhead and Møller (1992, Table 10.1) list six potential costs to males of performing EPC. These include (1) sperm depletion and ejaculate production costs, (2) increased risk of cuckoldry, (3) increased likelihood of divorce, (4) risk of parasite or disease transmission, (5) risk of injury from female's partner, and (6) increased risk of predation. Of these, Birkhead and Møller consider only one of them, increased risk of cuckoldry, as likely to be important. If a mated male bird seeks EPC some distance from its mate during the time that the mate is vulnerable to fertilization, the would-be cuckolder may be cuckolded. This is thought to be the chief reason that males often begin to prospect for EPC after their mate has begun incubation (Ford 1983; see Birkhead and Møller 1992, Table 10.2). Mate-guarding by males of many species can be interpreted as support for the proposition that being cuckolded when their mates are susceptible to fertilization has been an important selective agent.

12.2.4 Costs of EPC for females

Although it is clear that EPC can be an important reproductive strategy for females, it is also likely that the interests of the female and a potential extra-pair copulatory partner will often differ. That is, females may not desire to copulate with a particular male, or at a particular place or time. Thus, EPC has potential costs for females, and, accordingly, they sometimes resist the efforts of males to mate with them. Westneat *et al.* (1990) list several potential costs to females in the form of responses by their mates. These include both retaliation or punishment by the male mate and the withholding of parental care (see next section). In addition, by mating with additional males, females may increase chances of exposure to parasites or diseases carried by the males.

Birkhead and Møller (1992) also provide a list of possible costs to females of EPC. These include (1) fertilization by a male of low or unknown quality, (2) physical retaliation by the male mate, (3) risk of injury by extra-pair males, and (4) foraging costs associated with the proximity of their guarding mate. Although all of these costs are plausible to a greater or lesser degree, to date there is little evidence documenting their actual importance to female birds (see Westneat *et al.* 1990 and Birkhead and Møller 1992 for additional discussion of costs).

12.3 The relationship between EPC and paternal care

As previously discussed, the great majority of avian species exhibit monogamy and extensive paternal care. It is clear that in many such species the male's contribution is essential to the successful production of a brood of young. The question raised here is whether EPCs are more common in species in which the male's parental effort is not required than in species in which parental care is absolutely critical. According to current theory, such a trade-off is to be expected for two rather different reasons. First, a male bird cannot be doing two things at one time; e.g. diligent feeding of nestlings means that the time invested in this endeavour probably cannot be used to seek EPCs. Second, when extra-pair paternity is high within a population or species, due to EPCs by their mates, males might be expected to respond facultatively by contributing less parental care. Phrased in another way, might the frequency of EPC reflect a female mating strategy that is related to the extent that females can rear young without male assistance?

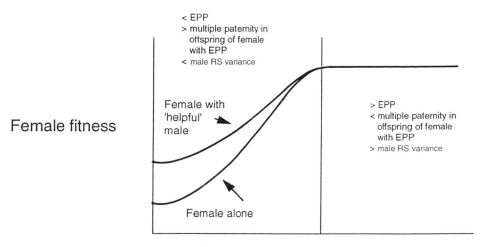

Fig. 12.1   The constrained female hypothesis predicts genetic mating patterns of socially monogamous females. (EPP = extra-pair paternity; RS = reproductive success.) Reprinted with permission from Gowaty, P. A. (1996*a*). Battles of the sexes and origins of monogamy. In *Partnerships in birds. The study of monogamy* (ed. J. M. Black), pp. 21–52. Oxford University Press.

This suggestion is similar to Gowaty's (1996*a,b*) 'Constrained Female' hypothesis, which suggests that the mating strategy of females is based on the degree to which they rely on male parental care in the rearing of their offspring. Specifically, because most females do not have the opportunity to form a pair bond with the highest quality male(s) in the area, their optimal reproductive strategy may be to seek sperm from such males, while forming social mateships with an available lower-quality male who will provide parental care. Gowaty suggests that, within a species, females of intrinsically high quality or females in especially productive habitats might be better able to rear young without parental assistance. As a result, such females would be expected to show higher levels of EPC than females more dependent on parental contributions by their mate (Fig. 12.1).

Within species, a relationship between EPC and subsequent paternal care has been documented in the dunnock (Burke *et al.* 1989; Davies 1990, 1992). In polyandrous male–male–female trios, beta males usually provide care only if they have had opportunities to sire offspring. In addition, the amount of care provided by the beta male dunnock appears to be related to the number of copulations obtained. (However, males are apparently unable to recognize which chicks in the nest are more likely to be their genetic offspring, as they do not selectively feed certain offspring.)

For obvious reasons, genetic parentage should be important to males, and this has been empirically demonstrated in a few cases. In addition to the dunnock, this is true for the reed bunting (Dixon *et al.* 1994), a species in which extra-pair paternity accounts for 55 per cent (118/216) of young and 86 per cent (50/58) of nests. Male reed buntings can apparently gauge the EPC activities of their mates on a brood-by-brood basis, and they provide more paternal care at nests containing a lower proportion of extra-pair chicks.

In other species or situations, however, there appears to be no relationship between parentage of a brood and the level of paternal care (e.g. Gavin and Bollinger 1985; Frederick 1987*b*; Westneat 1988*a*, 1995; Whittingham *et al.* 1993; Wagner *et al.* 1996*b*). In monogamous pairs of dunnocks, males provision nestlings even if their actual paternity is low. Similarly, in the polygynous red-winged blackbird, males fed chicks in the nests of the females nesting in their territories equally, and males with low paternity of several broods also provisioned at the same level as males with high average paternity (Westneat 1995).

One of the unresolved mysteries related to the subject of EPC is why males of a species characterized by frequent cuckoldry should provide a high level of parental care, in view of the fact that some of the chicks in their nests are unlikely to be their genetic offspring. Under such circumstances we might expect males to attempt to counteract the problem of potential cuckoldry (see Møller and Birkhead 1993 for discussion of this issue).

There are also other possible explanations. First-year male purple martins, some of which are frequently cuckolded, fed nestlings at a rate comparable to males with complete paternity of their broods (Wagner *et al.* 1996*b*). Wagner *et al.* address and reject four theoretical explanations proposed to account for the absence of a relationship between paternity and feeding effort by the male. These include: (1) males increase their fitness more by attempting EPCs than by feeding offspring; (2) a male's paternity is the same from one breeding attempt to the next; (3) insufficient cues are available to allow males to appraise their paternity; and (4) provisioning does not reduce survival (see Wagner *et al.* 1996*b* for references). Wagner *et al.* (1996*b*) offer another explanation—that in a colonial species like the purple martin, poor parenting performance by a young male in the presence of future mates or EPC partners can have an effect on the male's future social status and thus its fitness. In short, if current parental performance can have long-term effects on the male's fitness, it would pay the male to expend effort in the feeding of chicks in its nest, whether or not those chicks are its genetic offspring.

As in the red-wings and purple martins, polygynous male indigo buntings do not show a relationship between paternity and paternal care (Westneat 1988*a,b*; Morton *et al.* 1990), nor did caged male tree swallows provision nestlings less after having seen their mates engage in EPC (Whittingham *et al.* 1993). Westneat (1995) suggests that male birds have few reliable cues about the paternity of chicks in their nests, and that, apparently, no mechanism has evolved for assessing paternity. He contrasts the responses of male dunnocks with those of male red-wings, pointing out that a male dunnock of a male–male–female trio has the responses of the other male, including its provisioning effort, to make assessments; in contrast, monogamous male dunnocks and red-wings lack this source of information. Moreover, a beta male dunnock can reduce his provisioning coincident with low probability of paternity, secure in the knowledge that the alpha male is feeding the young. In contrast, a male dunnock of a monogamous pair, or a male redwing, may have poor or unreliable information on his paternity prospects, and is likely to lose the chicks in the nest if he fails to provide parental care.

In most species for which EPC has proved to be of regular occurrence, many broods of nestlings are of mixed paternity, with the social mate being the actual sire of some of the young birds. This pattern should reinforce parenting behaviour by the male. Despite the negative results between paternity and provisioning for red-winged blackbirds summarized above, Westneat (1995) suggests that probability of paternity could still be influencing paternal care of nestlings. The frequency of mixed (not to mention complete) paternity of a brood means that, on average, a male bird should obtain fitness benefits as a result of parenting effort, regardless of the proportion of young that are its offspring.

12.4   Evolutionary effects of EPC on mating strategies and social structures

The importance of EPCs in sexual selection is evidenced by the relationship between this mating strategy and ornamental traits. For example, Møller and Birkhead (1994) investigated the relationship between brightness of male plumage and sexual dimorphism in brightness and reported that both were correlated with high levels of extra-pair paternity.

This section explores the idea that EPCs may not only influence the evolution of traits such as plumage colouration, they may also destabilize mating systems, especially monogamous ones, which can lead to the evolution of alternative mating systems. If environmental conditions permit females of monogamous, biparental species to rear young without male participation, this frees the female from mating constraints; i.e. the male mate has lost the 'leverage' provided by his parenting contributions. That is, females may be freed, to varying degrees, to seek copulations with preferred, presumably higher quality, males. This kind of shift would, at the same time, increase the intensity of sexual selection.

In a study involving 52 species, Møller and Birkhead (1993) found a significant relationship between probability of paternity and extent of parental care, specifically the feeding of nestlings and fledglings, leading them (Birkhead and Møller 1996) to conclude that interspecific comparisons support the idea that in species in which biparental care is essential for reproductive success, EPCs are rare or non-existent. At the other extreme, in species in which the female is completely unconstrained by the need for male parental care, multi-male fertilization of clutches is also rare; here females are free to mate solely with the single 'best' male available. Many species, however, lie between these two extremes. In these cases, females practice both social monogamy and genetic polyandry. For this strategy to work, the female must apparently 'consider' her social mate's genetic interests, as well as her own. Commonly, the answer appears to be a compromise, with the brood containing some offspring sired by an extra-pair male and some sired by the social mate.

12.4.1 EPC and the evolution of polygyny and lekking

The question raised in this section is whether a chronically high frequency of EPC (due to a change in ecological conditions, for example, that decreases the necessity of extensive parental care by both sexes) in a socially monogamous species could become an effective selective agent for a fundamental change in male reproductive strategies. In addition to mate-guarding (possibly the most common response of males), which may be ineffectual, reaction to predictably frequent cuckoldry could take one of two forms, or a mix of both: (1) a decrease in parental care, and (2) an increase in EPC effort. This latter could have a snowball effect, in that the 'problem' of EPC would be increased for nearly all mated males. Based on his studies of genetic paternity, paternal care, and mating system in the indigo bunting, Westneat (1988b) suggests that a cuckolded male loses less from reducing parental effort than a male that is not cuckolded, and that a cuckolded male is more likely than a non-cuckolded male to pursue EPCs.

In addition, might a high level of EPC in a population increase a male's mating effort via attempts to acquire second mates? Males could attempt to become polygynous, at the expense of parenting efforts. That is, if females become increasingly free to seek EPCs for any reason, thus causing the social mate to lose confidence in paternity, this could lead to a change in the mating system (Fig. 12.2). For example, a change in environmental conditions might allow females to rear their young successfully with little or no male assistance. That is, under some circumstances, polygynous mating behaviour could *originate*, at least in part, directly in response to a chronically high level of female-driven EPC. The frequency of polygyny (the number of males actually successful in attracting a second mate) may then depend on habitat features, such as those discussed by Emlen and Oring (1977), and these will often vary from site to site.

The well-studied indigo bunting provides a possible example of a causal relationship between frequency of EPC, mating system, and male parental care. In this species, the rate of EPC is high and many mated males (including all those in their first breeding season) provide no parental care. Some male

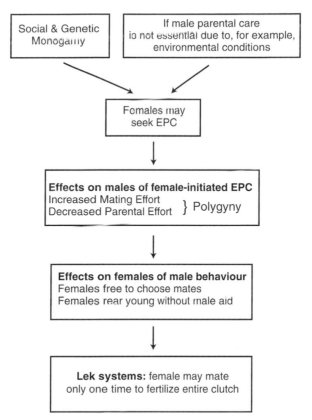

Fig. 12.2.   A hypothetical role of EPC in the evolutionary shift of mating systems from monogamy to polygyny, and possibly even to lek-promiscuity. This scenario is similar to the 'constrained female hypothesis' of Gowaty (1996*a*) and the arguments of Birkhead and Møller (1996).

buntings are polygynous, and such males provide even less parental care than monogamous males (Westneat 1987*a*,*b*, 1988*a*,*b*). In addition to monogamous and polygynous mateships, male and female indigo buntings employ EPC as a secondary reproductive tactic. Westneat (1990) found that although the total per cent of EPC at two sites (Michigan and North Carolina) were similar, levels of polygyny were very different (14 per cent versus 3 per cent, respectively). It is possible that environmental factors affecting mating strategies differ between Michigan and North Carolina. For example, more food in Michigan could decrease the amount of male parental care required, while at the same time increasing the frequency of polygyny.

While data are not available to provide strong support for the idea that high levels of female-initiated EPC should favour a shift by males from monogamy to polygyny, available information is compatible with this suggestion.

Although distinguishing between cause and effect in this case would be difficult, if not impossible, it is conceivable that parental care by male indigo buntings initially decreased over evolutionary time to the low levels seen today as a result of high frequencies of EPC. Correspondingly, levels of EPC-solicitation by females may have further increased as females became increasingly able to rear their broods with little aid from their mates. If so, this case provides an example of the kind of scenario outlined in Fig. 12.2.

In a more general context, Westneat *et al.* (1990) also consider the possible relationship between polygyny, EPC, and environmental homogeneity. These authors refine the operational sex ratio of Emlen and Oring (1977) by dividing it into two components: (1) 'the ratio of sexually active males to unpaired females (a measure of the potential for polygyny; Emlen and Oring [1977]), and (2) the ratio of the number of sexually active males to fertilizable, but paired females (a measure of the potential of EPC)' (Westneat *et al.* 1990, p. 343). These ratios are based on the ecological conditions affecting the settlement of males and females, which limit the potential for polygyny. Ecological conditions may also make it possible for females to rear young with little or no male assistance. This allows females to increase their EPC activity (see also Gowaty 1996*a,b*). In turn, this selects for males to (1) further reduce their parental effort, and (2) increase their effort to mate polygynously and to obtain EPCs. Following this line of reasoning to its limits, one possible outcome of the release of females from the constraint of parental contributions by males is the development not only of polygyny, but also of leks. As the parental roles of males disappear altogether, there should be progressively less selection even for polygyny (Fig. 12.2). Viewed in this way, the concentration of males that define leks may reflect the mating strategies of females (see Chapter 15). For example, in many or all lekking grouse, females typically copulate only one time. This suggests that when females are free to choose the 'best' male from among many, mixed paternity, as seen in many species, is not a strategy favoured by females (cf. Section 12.2.2).

As an example of the scenario outlined in Fig. 12.2, consider the marsh-breeding red-winged blackbird, a polygynous species with frequent EPC and, depending on the population, little or no paternal care (e.g. Searcy and Yasukawa 1995), and the lek-promiscuous sage grouse, a species with no paternal care and with no pair bond (i.e. every copulation can be viewed as an EPC). In both of these species, males and females congregate either at a marsh (redwings) or at a traditional mating site, the lek (sage grouse). A critical factor that distinguishes the breeding system of redwings from that of sage grouse is the fact that female redwings use the marsh for nesting, whereas female sage grouse do not nest at the lek site. In the former species, the marsh habitat concentrates females, and thus males, with the latter competing for desirable sites (territories)—those sites most attractive to females—on the marsh. Male sage grouse may congregate on leks specifically to attract females, and, within the lek, the males, like male redwings on a marsh, are ter-

Origins of social monogamy

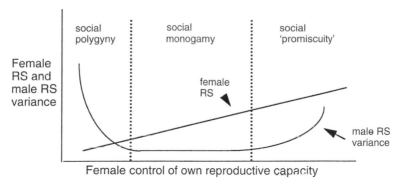

Fig. 12.3   A graphical model of the relationship between the female control of mating decisions and variance in male reproductive success (RS). Reprinted with permission from Gowaty, P. A. (1996). Battles of the sexes and origins of monogamy. In *Partnerships in birds. The study of monogamy* (ed. J. M. Black), pp. 21–52. Oxford University Press.

ritorial. Reproductive success (RS) of both male redwings and male sage grouse is skewed (Fig. 12.3), and in both cases this skew may be based, in part, on their locations within the marsh or on the lek.

In brief, the existence and frequency of EPC in birds may have ramifications well beyond its importance in monogamous mating systems. EPC may be one major causal factor that has shifted avian mating systems from monogamy to polygyny and conceivably even to lekking systems (see also Section 15.11).

12.4.2   EPC and divorce

Another line of evidence supporting the notion that EPC may influence mating systems is the positive relationship between divorce and EPC: 'Thus, we find that species with higher frequencies of cuckoldry also show higher divorce rates' (Cezilly and Nager 1995, p. 10). Cezilly and Nager conclude that the 'better option' hypothesis (Ens *et al.* 1993) best accounts for this relationship. The better option hypothesis suggests that divorce in birds is usually based on a decision by one of the pair members that is seeking to improve its breeding status or productivity by mating with an individual of higher quality than its current mate. Thus, both EPC and divorce, particularly those initiated by females, are driven by the same goal. Both strategies also rely on extensive variation in mate quality within populations and on intense intrasexual competition for high-quality mates (Cezilly and Nager 1995).

12.4.3   EPC as a possible initiator of colony formation

In addition to the possibility that EPC may have played a role in driving the evolution of mating systems from monogamy to polygyny and lekking, it has also been suggested that EPC may have been a critical agent in the evolutionary development of coloniality. Eugene Morton and colleagues (1990) proposed that attempts by male purple martins to obtain EPCs might select for clustering of breeding birds, and that this could favour the evolution of coloniality. Countering this suggestion, Richard Wagner (1993) argues that there are at least two reasons why the effects of EPC on male fitness should not select for coloniality. First, for females there are potential costs to truly forced copulations, and females should therefore avoid them (i.e. a high risk of forced copulations should make coloniality an unattractive option for females). Second, males that suffer increased risks of cuckoldry in colonies should avoid breeding colonially if they can breed solitarily.

Wagner (1993) has modified the idea relating EPC to coloniality as follows: if EPC is an important reproductive strategy of females, they should prefer to be located in areas where males are most likely to be. Wagner's hypothesis of colony formation predicts that males cluster in response to the presence of females that seek extra-pair copulations, and he draws an analogy between the development of leks, where females go specifically to obtain sperm for their eggs, and colonies, where females go to obtain mates and also, according to his scenario, to obtain sperm from one or more males in addition to their social mate.

Studies on purple martins in Maryland provide some support for this hypothesis. Adult female martins and their mates arrive back at their nest sites (nearly all of which in the eastern United States are artificial, man-made structures, 'martin houses') in the early spring, take up residence, defend the holes from other adults, construct a nest, lay eggs, and begin incubating by the time other, typically younger males and females arrive. Morton *et al.* (1990) suggested that the adult male 'owner' of a martin house then may behave in a way designed to attract younger males, which then soon attract additional females as their mates. The older male may obtain many EPCs from the mates of young males, while the converse is not true.

The idea that adult males actually recruit younger males was not corroborated by Stutchbury (1991*a*), who studied a purple martin population in Oklahoma. Instead, she found that first-year males, whether or not they were dyed to mimic adult plumage, typically obtained a nest cavity by persistent intrusion and eventual eviction of a bird that had previously controlled several cavities. The previous owner was often an adult male with its own nest in a nearby cavity (Stutchbury 1991*b*). Once a young male gained control of a cavity, he also typically gained a mate almost immediately. The presence of the mates of these young males apparently provides the original territory owner, a fully adult male, with a grand opportunity to increase his reproductive output for the year. The older male proceeds to copulate with these secondary females and in this way greatly increases his paternity (Morton *et al.* 1990; Wagner *et al.* 1996*b*).

As mentioned, Morton *et al.* (1990) interpreted this situation as evidence in support of the hypothesis that increased opportunity for males to engage in extra-pair fertilizations led to the evolution of colonial breeding in purple martins. In contrast to this view, Wagner *et al.* (1996*b*) suggest that it is females seeking EPCs that has driven the evolution of colony formation in this species. Female martins either (1) pair with old males and avoid EPC, or (2) pair with young males and obtain EPC from old males. Young males attempt to counter their mates' tendency to mate with old males by mate-guarding. Two factors significantly affected the degree of paternity of young males. Those young males that were larger than their mates and that mate-guarded most diligently sired more chicks than did smaller or less diligent males. Thus, it appears that in purple martins EPCs are not forced, but rather that the opportunity for females mated to young males to obtain EPCs is affected by the behaviour of their mates. Wagner *et al.* (1996*a*, p. 379) suggest that the evidence favours '... the female-driven 'hidden-lek hypothesis' of colony formation which predicts that males are drawn to colonies when females seek extra-pair copulations.' Although they differ in a critical detail, namely which sex actually favours group living specifically in order to obtain EPC, both of these studies are novel in suggesting that increased opportunities for EPC have driven the evolution of coloniality.

Close study of the martin system has brought some interesting issues to light. In Oklahoma, old males do not encourage young males to set up house-keeping; instead, young males fight, persevere, etc. to gain a nest hole and then a mate. In contrast, in Maryland, adult males allow younger males access to cavities in their martin houses. Subsequently, the old males may obtain many EPCs from the mates of the young males, while the latter obtain none. From an ultimate or evolutionary perspective, why would the prospect of a high level of female infidelity attract young male martins? That is, why do young males fight persistently to obtain a nest cavity at a site where there is a high probability that they will then be cuckolded? For the idea of Wagner *et al.* (1996*a*) to work, females must be in the vicinity of an established—and already mated—old male, prior to obtaining a social mate and nest site. Apparently, under these conditions, the only option for young males is to fight to obtain a nest site near the old male, and to accept the high risk of cuckoldry.

Consider the age and social status of the individual members of a martin colony and their reproductive options. (1) The major beneficiary of an EPC strategy—as measured by number of offspring—is the old male; this male sires all of the nestlings in his nest, plus chicks in several other nests. (2) The mate of the adult male benefits by breeding early and with a successful adult. (3) Young males may benefit primarily by obtaining partial paternity of their brood—a few chicks are better than none. Without obtaining access to nest sites, they could not attract a mate and breed at all. This point is significant in that it provides a plausible explanation for why young males attempt both to deter EPC by their mates and to 'accept' what appear to be high levels of cuckoldry. (4) Finally, young females may be pursuing the most complex strategy, in that they must retain their young male mates, which provide care for their offspring, while at the same time obtaining sperm from the older, dominant male. A male-biased sex ratio may give females some leverage; i.e. young males do not have the option of divorce and re-mating. For an immature male purple martin, partial paternity is apparently better than nothing.

In short, when several aspects of this interesting and complex system are considered, it appears that the benefits to each age and sex category are sufficient to provide net gains for all participants, compared to the most likely alternative, which, for the partially cuckolded young males, would be not breeding at all.

12.4.4 Fairy-wrens: the ridiculous extreme in EPC

A suggestion offered above, that a high level of EPC within an initially monogamous species could lead to the evolution of an alternative approach to mating may receive support from studies of two Australian species, the splendid and superb fairy-wrens. These birds are cooperative breeders, with many social units containing helpers-at-the-nest. Within a cooperatively breeding group, there is a single breeding female (usually), a primary male, typically older and brighter than other group members, and a few helpers, most of which are males. For many years, since Rowley's (1965) ground-breaking study on the superb fairy-wren, it was assumed that the primary male and the female were the breeders in the group and that the helpers were usually the grown offspring of the apparent breeding pair.

Based on the frequent presence in the same group of a closely related, putatively mated, pair, Rowley *et al.* (1986) reported a uniquely high level of inbreeding in the splendid fairy-wren. Such inbreeding seemed unlikely in view of the facts that (1) no comparable levels had been reported for any other species of bird, cooperative breeder or otherwise, and (2) inbreeding avoidance seems to be a common phenomenon in other highly social avian species. However, what turned out to be the case was even more surprising than the original report of extreme inbreeding. In the splendid fairy-wren, the great majority of young birds hatched in a given territory are not sired by the resident top male, and, moreover, a minimum of 65 per cent of the young birds were not sired by any male within the social unit (Brooker *et al.* 1990). Subsequently, this pattern was also documented for the superb fairy-wren (Fig. 12.4), where at least 76 per cent of the young birds were sired by extra-group males (Mulder *et al.* 1994). In terms of percentage of EPC, the superb fairy-wren is apparently the reigning champion among birds, with the splendid fairy-wren not far behind.

Fig. 12.4   Male superb fairy-wren carrying flower petal. 'Petal-carrying' (Rowley 1991) has been recorded in at least six species of fairy-wrens. It appears to be an accessory form of ornamentation used by males seeking extra-pair copulations. With regard to extra-pair copulations, this species is the reigning world champion

Mulder *et al.* (1994) argue that helpers provide females with an alternative source of paternal care, thus obviating the need to provide their social 'mate' with copulations to maintain his confidence of paternity at a sufficiently high level to assure his assistance with brood care. This interpretation is supported by the fact that in breeding/social units consisting of only two individuals, the male obtained significantly more paternity (but never exclusive paternity of a brood) than the apparent breeder in groups containing male helpers. It appears, then, that the more helpers a female fairy-wren has for rearing her brood, the more actively she seeks EPC; i.e. as a result of the presence of helpers, females are largely freed of the constraint of exchanging EPC for male parental care. This is analogous to the situation in lekking species: when females do not rely on parental care by males, their choice of copulation partners may become almost unrestricted.

12.5   Why are females of some monogamous species 'true' to their mates?

The biochemical documentation of frequent EPCs in many species, together with numerous additional behavioural observations of such copulations, has led to the conclusion that, for both male and female birds, seeking and obtaining matings outside the pair bond can be an important aspect of an individual's overall reproductive strategy. Within socially monogamous mating systems, EPCs increase the intensity of sexual selection and, for some males, it

Table 12.1   Some socially monogamous species for which a high degree of genetic monogamy[1] has been documented by DNA analysis

Species	Percent chicks sired by extra-pair male	Chicks	Broods	Reference
common loon	0	34	20	Piper *et al.* 1997
Northern fulmar	0	—	28	Hunter *et al.* 1992
blue duck	0	14	10	Triggs *et al.* 1991
black vulture	0	36	16	Decker *et al..* 1993
American kestrel	8.3	—	—	See Gowaty 1996*b*, Table 1
purple sandpiper	1.2	82	26	Pierce and Lifjeld 1998
Wilson's phalarope[2]	0	51	17	Delehanty *et al.* 1998
oystercatcher	1.5	65	26	Heg *et al.* 1993
Eurasian dotterel	4.5	44	22	Owens *et al.* 1995
Eastern screech-owl	0	80	23	Lawless *et al.* 1997
European bee-eater	2	100	65	Jones *et al.* 1991
red-cockaded woodpecker	1.3	80	44	Haig *et al.* 1994
dusky antbird	0	15	9	Fleischer *et al.* 1997
Florida scrub-jay	0	ca. 130	50	See Gowaty 1996*b*, Table 1
blue tit	11	314	34	Kempenaers *et al..* 1992
stripe-backed wren	10	69	22	Rabenold *et al.* 1990
bicoloured wren	8.6	222	99	Haydock *et al.* 1996
great reed warbler[3]	3	678	162	Hasselquist *et al..* 1995
willow warbler	0	120	19	Gyllensten *et al.* 1990
wood warbler[2]	0	56	13	Gyllensten *et al.* 1990
pied flycatecher	4; 7	135; 98	27; 17	Lifjeld *et al..*, 1991; also see Gowaty 1996*b*, Table 1
Eastern bluebird	8.4	83	21	Meek *et al.* 1994
bearded tit	0	—	—	See Gowaty 1996*b*, Table 1
common bushtit	0	59	9	Bruce *et al.* 1996
bull-headed shrike	10.1	99	24	Yamagishi *et al.* 1992
European starling	8.7	62	14	Pinxten *et al.* 1993
common myna	0	—	—	See Fleischer *et al.* 1997
palila	0	20	12	Fleischer *et al.* 1994
white-throated sparrow	11	62	21	See Gowaty 1996*b*; Table 1
zebra finch (wild)	2.4	82	25	Birkhead *et al.* 1990
silver-eye	0	—	—	Robertson & Kikkawa 1994
house finch	8.3	119	35	Hill *et al.* 1994*b*

[1]About 90 per cent or more
[2]Rarely polyandrous
[3]Polygynous

is probably a major means of enhancing lifetime reproductive success. For females, it is the way to obtain genes from high quality males when such males are not their social mates. Although EPC appears to be a common reproductive strategy in many birds, it has also been documented that in a number of species, it is rare (Table 12.1).

This leads to the question, why extra-pair fertilizations are apparently virtually non-existent in other species. Those species in which EPCs apparently do not occur are characterized by contributions by males of extensive parental care, which are critical to successful production of young. In a review based on 18 species with extensive parental contributions by each sex, Gladstone (1979) reported that females actively resist forced EPCs, and he may have been the first to suggest a relationship between male parental effort and probability of paternity. The most probable explanation for genetic monogamy in such species is that (1) males contribute their vital parenting effort only when prospects of genetic paternity are high, and that (2) females are unwilling to risk loss of the male's parent contributions; i.e. the possible genetic benefits to be gained by females via procurement of EPC are outweighed by the risk of loss of paternal care (see also Piper *et al.* 1997 and Delehanty *et al.* 1998).

A number of species representing a variety of taxonomic affiliations and lifestyles are apparently genetically monogamous (Table 12.1). One such species is the black vulture. Decker *et al.* (1993) determined paternity for 16 complete families of black vultures by use of DNA analysis. All parents could be assigned to their apparent offspring and all other adults could be excluded as parents, which indicates that black vultures are genetically strictly monogamous, as their social mating system indicates. Male, as well as female, black vultures are required for successful reproduction because mated pairs share equally in incubation and care of young. This mutual dependence of each sex on the full participation of its mate in care of eggs and nestlings may be the major factor promoting monogamy. For example, a male black vulture that 'suspects' that it has been cuckolded might fail to participate in incubation; this would provide strong selection against females with a tendency to mate outside the pair bond.

A recent study of the genetic mating system of the cooperatively breeding red-cockaded woodpecker detected almost no evidence of EPC (Haig *et al.* 1994). Woodpeckers are among the long list of birds in which shared parental care is crucial to the successful production of a brood of young. (As in the black vulture, shared incubation may be especially critical.) Thus we might expect that EPC is not sought by females of this group of birds because of their dependence on a high level of paternal care, as discussed above.

EPCs are also low in two socially monogamous cooperatively breeding birds of Venezuela, the stripe-backed and bicoloured wrens (Rabenold *et al.* 1990; Haydock *et al.* 1996). In the former species, 90 per cent (62/69) of the young birds were the offspring of the principal male and the principal female, with six of the remaining seven being sired by another male group member.

The numbers are similar for the bicoloured wren; 92 per cent of the juvenile birds are the offspring of the principal pair (Haydock *et al.* 1996). Rabenold *et al.* (1990) suggest that the prospect of obtaining some paternity may help to explain why the secondary males tolerate what may be a long tenure as an auxiliary, or helper. However, their data, like those of Haydock *et al.*, suggest that the reward of parentage is rarely obtained by such males.

Finally, Fleischer *et al.* (1997) report that four studies of tropical, socially monogamous species of passerine birds have failed to detect any evidence of extra-pair fertilizations (see Table 12.1). These authors predict that most tropical, socially monogamous birds will also exhibit genetic monogamy because most such species are sedentary year-round on territories, maintain long-term pair bonds, and nest asynchronously with other pairs. Fleischer *et al.* (1997, p. 513) suggest '... that genetic monogamy may still be the dominant form of mating system in socially monogamous songbirds simply because more songbird species live in the tropics than in the Temperate Zones.' Thus, it is possible that, across socially monogamous avian species as a whole, the frequency of EPC leading to fertilizations is less prevalent than the studies conducted in the north temperate zone appear to have suggested.

12.6   Conclusions and summary

The term extra-pair copulation (EPC) is most appropriate for socially monogamous species, since the breeding unit is a pair—a single male and a single female. This term is also applied to polygynous species, where the female, but not the male, is socially monogamous. The discovery that EPC is a widespread reproductive strategy has had a major effect on our understanding of avian mating systems. It is now widely appreciated that for many species extra-pair copulations that lead to fertilizations are a significant portion of the overall mating strategies of both male and female birds. It is also important to note, however, that in some other species this is not the case. It appears likely that as the genetic parentage of more obligately bi-parental species are studied by use of biochemical analyses, the degree of genetic monogamy will prove to be correlated with the relative importance of male parental care.

The main benefit of EPCs to males is clear: it is a means of siring additional offspring over and above the number of eggs laid by the social mate. EPC also appears to be of major importance for females. Although they do not gain additional offspring, they may profit by obtaining material benefits from the male non-mate or by obtaining sperm from a second male to ensure against possible infertility of their social mate. Probably most significantly, they stand to gain genetic benefits of two sorts. First, it appears that in a number of species, females seek matings from high-quality males, i.e. males that exhibit phenotypic signals of their condition or health. A female of a monogamous species can obtain parental care from her low-quality social mate, while at the same time she obtains genes for her offspring from a high-quality male. Second, females of some species may seek matings with more than one male to increase the genetic diversity of offspring. This could be particularly important for species faced with many unpredictable environmental challenges in the form of diseases and parasites. Heritable resistance to many pathogens is known to occur in domestic fowl, and this is probably also true for wild species.

In addition to its importance as an individual reproductive strategy, the phenomenon of extra-pair copulations has the potential to exert effects on population- or species-level mating systems. EPC as a female strategy may play a causal role in the evolution of polygynous systems from monogamous ones; this possibly extends even to the evolution of lek behaviour. Such an effect on mating systems is related to the fact that one response of a mono gamous, parenting male to decreasing certainty of paternity should be an increase in mating effort. For example, over evolutionary time cuckolded males might be expected either to seek additional mates—polygyny—or to devote all of their reproductive effort to mating. The latter strategy could result in the evolution of promiscuous-lek systems.

It has also been suggested that EPC could facilitate the initiation of colony formation. The idea is that males cluster in response to the presence of females that are seeking extra-pair copulations. By this view, there is an analogy between the development of leks, where females of certain species go specifically to obtain sperm for their eggs, and colonies, where females go to obtain mates, and also possibly to obtain sperm from one or more males in addition to their social mate.

Finally, in most species for which EPC has been documented, mated males do have partial paternity of the brood. Females may seek either to mix paternity of broods or to 'reward' males for their parental effort via some genetic parentage. This scenario is supported by studies on two species of Australian fairy-wrens. In these species, the female controls matings and typically has nest helpers. The provisioning efforts of the helpers decrease the value of the 'breeding' male's parental care, and, presumably as a result of this, females mate primarily with males outside their social units. Supporting this interpretation is the fact that in units consisting of simple one male–one female pairs (no helpers), the male has greater paternity than when helpers are present.

13 Multiple mates: polygyny and cooperative polyandry

13.1 Introduction

The subject of polygyny has been a key topic in avian behavioural ecology for more than thirty years. Polygyny is identified by a mating association between one male and two or more females, with each female possessing its own nest and young. Why females would choose to mate with an already-paired male, when unmated males appear to be available, is a question that has fascinated many students of avian mating systems.

In contrast to polygyny, cooperative polyandry is recognized by the presence of two or more males and one female at a single nest; all males may sire offspring in a single brood and later provide parental care to eggs and chicks in the typical avian fashion. Cooperative polyandry is apparently much rarer than polygyny and has not received nearly as much attention from ornithologists. Like polygyny, however, it presents a very interesting question, namely, what conditions would lead males to share matings with a single female and the resulting paternity of chicks within a single brood?

It should be re-emphasized that mating patterns are not as discrete as labels such as polygyny or polyandry might suggest. In fact, the social mating systems of some species tend to grade from one type to another, just as genetic mating systems do. For example, within a single species, most notably the dunnock of England and Europe (Davies 1990), one might find a single male mated with two or more female mates (polygyny), one female laying in a single nest and mated with two or more males (cooperative polyandry), and two or more breeders of both sexes (polygynandry), in addition to simple pairs (monogamy). Fortunately perhaps, for those avian behavioural ecologists who prefer clean, unambiguous terminology, it appears that the mating strategies of few other species are as labile as those of the dunnock.

Although both polygyny and cooperative polyandry are probably derived from monogamy with biparental care, their evolutionary development has been very different. In polygyny, male parental care may be either present or absent, while in cooperative polyandry, parental care by males always appears to be present. In many highly polygynous species, for example, the yellow-rumped cacique (Robinson 1986), paternal care of young is completely absent. Thus, for this system to develop fully, the female must be able to rear a brood of young with no assistance from the male. In contrast, in cooperatively polyandrous systems, both sexes are involved in the rearing of young in the manner typical of monogamous–biparental species.

13.2   Polygyny

All typical polygynous mating systems among contemporary neognathous birds are probably derived from monogamous systems. In some polygynous species, males may provide care for the chicks of at least one of their mates' nests; in others, however, no paternal care at all is provided. In contrast, polygyny in the paleognathous ratites and tinamous is associated with exclusive or almost exclusive paternal care. In some members of the latter groups, several females may mate with one male, with all of them laying their eggs in the male's nest. The male alone then provides all of the subsequent parental care (Bruning 1974; Fernández and Reboreda 1998). Thus, the relationship between polygynous mating systems and parental care is fundamentally different in paleognathous and neognathous birds. This chapter focuses exclusively on the better known and more typical polygyny of neognathous birds where paternal care is limited or absent. Paternal care in paleognathous species was discussed in Section 10.5.1.

Polygyny has been extensively studied because it is relatively common, but not too common, and particularly because it has received a lot of stimulating theoretical treatment (see Section 13.3), compared to either monogamy or polyandry. Most of what is known about factors promoting polygyny comes from extensive study of a few species, most notably the pied flycatcher of Europe and the red-winged blackbird of North America. According to Searcy and Yasukawa (1995), more than a thousand papers have been published on redwings, most of which relate to the issue of polygyny.

As mentioned previously, the emphasis on polygyny has had the effect of downplaying the significance of monogamy, the overwhelmingly most prevalent and important mating system in the Class Aves. In contrast to the perceived blandness and resulting neglect of the subject of avian monogamy (Mock 1985), for the past thirty years polygyny has been a favourite topic of avian behavioural ecologists (e.g. Verner 1964; Verner and Willson 1966, 1969; Orians 1969; von Hartmann 1969). The attention directed in the late 1960s by some North American workers to polygynous species raised an interesting point: species were classified as polygynous if as few as five per cent of the males in a study population had more than one mate (Verner and Willson 1969). As pointed out by Ford (1983), Murray (1984), and Mock (1985), this has led to the peculiar situation where up to 95 per cent of the males and 100 per cent of the females in a population could be socially monogamous, yet that population would be classified as polygynous. Ford (1983) suggested that it makes more sense to view very low rates of polygyny (e.g. 5 per cent) as deviations from monogamy, rather than to view high rates of monogamy (e.g. 95 per cent) as deviations from polygyny.

In the 1960s polygyny was thought to be a rare mating system in passerine birds (Verner and Willson 1969; von Hartmann 1969). Subsequently, as more North American passerines became better studied, the number of species known to exhibit polygyny, at least occasionally, increased. Verner and Willson (1969) identified 38 species of North American passerines as polygynous. Fourteen years later, in 1983, that number had risen to 55 (Ford 1983), an increase of 45 per cent (for European species, see Møller 1986). This rise was due to the discovery of occasional polygyny in many species previously thought to be monogamous. Even so, the attention that has been given to the phenomenon of polygyny by avian behavioural ecologists does not parallel its frequency in nature.

The shift from a monogamous mating system to a polygynous one is clearly readily accomplished by many kinds of birds, perhaps especially temperate-zone passerines (although among non-passerines, it also occurs regularly in some owls and hawks, e.g. Altenburg *et al.* 1982; Carlsson *et al.* 1987; Simmons 1988; Korpimaki 1991; Gehlbach 1994). This is illustrated by the large number of North American species categorized as opportunistically polygynous: 'Polygyny rare, often only one case' (Ford 1983, Table II). In some cases, polygynous males provide care for one or more broods of young, while in others, male care is either non-existent or is limited to defence of the area in which the nest is placed. Thus, particularly in passerines, polygyny usually occurs in species in which part or all of a typical brood of young can be reared by a single parent—the female. As one example of the flexibility of mating systems and parental roles that occur in some species, male European starlings may be either monogamous or polygynous, and, if polygynous, males may provide incubation to the eggs of either one or both females (Smith *et al.* 1995).

As already emphasized, as a 'system', polygyny can be highly labile, appearing rarely or occasionally in a variety of primarily socially monogamous species, and in a variety of ecological conditions (Emlen and Oring 1977; Searcy and Yasukawa 1989, 1995). A major contribution of the ecological approach to interpreting mating systems lies in its ability to explain shifts within species from monogamy to polygyny, or vice versa, and to explain the existence of both monogamy and polygyny in two or more closely related species. In recent years it has become more fully appreciated, however, that current ecological conditions alone are insufficient to account fully for all polygynous systems in the absence of consideration of other factors, such as evolutionary history (e.g. Searcy and Yasukawa 1995, pp. 149–53).

Polygyny, like other mating systems, can be subdivided in a variety of ways. Emlen and Oring (1977) and Oring (1982) recognize three major types of polygyny (Table 13.1), while Wittenberger (1981, Table 11–1) lists five types of polygyny, plus four of promiscuity. The latter are considered by Emlen and Oring (1977) to represent forms of polygyny. This variation in classification illustrates both the flexible nature of this mating system and the inadequacy of a single word like 'polygyny' to indicate the complexity and diversity of what has been considered to be a single category of avian mating systems (see also Johnson and Burley 1997).

Among the recognized mating systems, the study of polygyny is especially relevant to the topic of sexual selection, for one additional major reason: the high variance in male mating success resulting from polygyny is thought to have driven the evolution of the extreme traits—ornamental and behavioural—that characterize males of many highly polygynous species. The relationship between polygyny, the evolution of extreme display characters, and variance in reproductive success was discussed in Section 2.5.

Most cases of polygyny, excluding promiscuous–lek systems, which are sometimes classified as polygynous, involve male territoriality. (The yellow-rumped cacique provides an exception to this generalization; Robinson 1986). Males of many species usually occupy territories that differ in one or more

Table 13.1   Basic categories of avian polygyny[1]

Resource defence polygyny Males control access to females *indirectly*, by monopolizing critical resources.

Female (or harem) defence polygyny Males control access to females *directly*, usually by virtue of female gregariousness.

Male dominance polygyny Mates or critical resources are not *economically monopolizable*. Males aggregate during the breeding season and females select mates from these aggregations.[2]

---

[1]From Emlen and Oring (1977).
[2]Emlen and Oring (1977) and Oring (1982) considered leks to be a particular form of polygyny. In this book, I refer to the mating system of lekking species, such as the grouse, as lek promiscuity, because males are clearly promiscuous (see text).

ways (e.g. in vegetation structure), and this variation may correlate with the number of females occupying a given territory. The best known and most enduring model, the polygyny threshold model, discussed below, has attempted to ascertain the relationship between territory quality and/or male quality and the number of females mated to a particular male.

13.3   Factors promoting polygyny

The fact that polygynous mateships occur regularly in a variety of species that are basically socially monogamous (Ford 1983) illustrates the ease that individuals of some species, depending on their ecology and patterns of male and female parental care, can shift from a monogamous to a polygynous relationship (see preceding section). Apparently, a basic requirement for the occurrence of polygyny, in both territorial and non-territorial species, is either male parental care that is shareable by two or more broods of young or no necessity for paternal care. In the former case, when males provision young at two nests, there may be a cost to one or both of the female parents in terms of a reduced number of young produced. Polygynous passerines, in which males provide absolutely no parental care, such as the yellow-rumped cacique (Robinson 1986), provide examples of the latter situation.

It is a truism that polygyny should be favoured by a male, so long as its net production of offspring is increased. For most polygynous species, this is often the case. A female bird, on the other hand, may view a polygynous relationship very differently. For those species in which paternal care of chicks is essential to rearing the entire brood, a mated female should attempt to thwart her male's efforts to obtain a second female. In these species, several of which are discussed below, the interests of a mated pair are potentially in conflict and a polygynous arrangement signals that the male has 'won'. However, as discussed in Chapter 11, for the majority of birds, the male's parental role is so important, and so non-shareable, that polygyny is not an option.

13.3.1   Polygyny and habitat type

A major factor contributing to the emphasis on ecological factors in the study of mating systems was the apparent relationship between polygyny and certain habitats. In North America, studies of blackbirds (see Orians 1980 and Searcy and Yasukawa 1995 for references) and marsh wrens (Verner 1964, 1965; Leonard 1990) focused on the ecological factors favouring polygyny in these marsh-dwelling species, and a causal relationship was drawn between this mating system and the occupancy of marsh habitat. Several species of polygynous birds in North America do inhabit such an environment; however, most (9/14) of the species listed by Verner and Willson (1966, Table III) belong to a single taxonomic group, the family Icteridae, thus the phylogenetic aspects of this habitat-mating system relationship also need to be considered (Lack 1968, p. 30; Searcy and Yasukawa 1995, pp. 148–53). With many additional species being added to the list of passerine polygynists, confirmation of a causal relationship between a particular type of habitat and this mating system has become increasingly unlikely.

Moreover, although two of the wren species listed by Verner and Willson as polygynous are marsh-dwellers, two others (house wren and winter wren) also classified as polygynous live in woodlands (Verner and Willson 1969, Table 1). The case for a general causal tie between polygyny and marsh-grassland habitat was further weakened by the fact that polygyny in European birds is not related to such habitats (von Hartmann 1969; Møller 1986). On the contrary, in this region, avian polygyny is related most often to availability of nest cavities. For example, in pied and collared flycatchers about 10 per cent of the males have two mates. This occurs when a single male defends two or more potential nest holes in separate, isolated territories (Lundberg and Alatalo 1992); secondary female flycatchers may not know about the male's first mate.

13.3.2   Adult sex ratio and mating system

Some workers (e.g. Murray 1984) have suggested that a female-biased sex ratio might promote polygyny. This perspective, however, had earlier been labelled as discredited (Carey and Nolan 1979). In contrast to the apparent importance of absolute number of males and females as a factor promoting monogamy in certain cases (e.g. ducks), sex ratio is less likely to be a causal factor promoting polygyny. For example, more than thirty years ago Verner and Willson (1966, p. 143) wrote:

> ... we can no longer regard a population's mating system as a necessary result of its sex ratio, however true this may be in some cases. Polygyny has been recorded in several populations in which the sex ratio has not been shown to deviate significantly from 1:1, and in some polygynous populations a slight imbalance favours males.

Instead, it is the resources held by a male—e.g. a territory with suitable nesting conditions—rather than the total number of males that generally will be more important (see below).

Emlen and Oring (1977) emphasize the possible role of sex ratio in determining mating systems by describing a concept they label as the operational sex ratio (OSR). The OSR is defined '... as the average sex ratio of fertilizable females to sexually active males at any given time' (Emlen and Oring 1977, p. 216). This distinction between the overall population sex ratio and the sex ratio based on available individuals is compatible with Wittenberger's (1981) and Murray's (1984) arguments, namely that it is the ratio of territory-holding males to females that promotes the development of social polygyny, rather than differences in quality among territories. In the red-winged blackbird, for example, only males holding territorial space within a marsh are potential mates or neighbours of females. No matter how many of them there are, for those males without territories in an appropriate nesting habitat, females are unavailable.

There are, however, certain situations in which the overall sex ratio *per se* can influence the manifestation of polygyny. In some non-territorial species, the prevailing mating system may be either social monogamy or polygyny (Murray's facultative monogamy). For example, in free-ranging domesticated helmeted guineafowl (a species considered to be monogamous in the wild, Crowe *et al.* 1986), when the sex ratio is unity or when males are more numerous than females, social monogamy is usual (Elbin *et al.* 1986). On the other hand, when the sex ratio is manipulated so that females outnumber males, facultative social polygyny is seen, in that some males attempt to 'herd' and control access to two or more females (personal observation). When male guineafowl are able to sequester more than one mate, this is largely due to the gregariousness of females, which makes possible 'female defence' polygyny (Emlen and Oring 1977). The key point, again, is that under certain conditions the population sex ratio can influence the prevalent mating system. In this case, male guineafowl have the potential to benefit from a female-biased sex ratio.

Reproduction in the guineafowl appears to parallel that of the monogamous dabbling ducks of the Northern Hemisphere (McKinney 1986) in some interesting ways: (1) in nature the sex ratio of both ducks and guineafowl is probably skewed in favour of males, and males are primarily socially monogamous until the clutch is completed and the female commences incubation, (2) after their mates begin to incubate, male ducks and guineafowl actively attempt to obtain matings elsewhere, and (3) female ducks and guineafowl incubate alone for several weeks. However, in the guineafowl, but not in the ducks, males return to the nest of their primary mate when the eggs hatch, and contribute essential parental care to the chicks in the form of extensive brooding, herding, and feeding (Elbin *et al.* 1986; personal observation). Although both the guineafowl and ducks have precocial hatchlings, the demands of parental care differ greatly. Although newly hatched ducklings require relatively little care (either brooding or feeding), newly hatched guineafowl require extensive biparental care. Moreover, in nature guineafowl occur in an exceedingly predator-rich environment (Africa), which might further increase the benefits of biparental care.

Although the ducks (discussed in Chapter 11) and guineafowl appear to be socially monogamous, it is clear that in both, an important additional reproductive strategy of males is to obtain matings with additional females. Social monogamy is probably the rule in these species because more or less exclusive possession of one mate is far preferable to having none, and because retention of even a single female, plus some acceptable level of assurance of paternity, requires ongoing and continuous mate-guarding. In addition, although they do not incubate the eggs, male guineafowl, but not male ducks, play a major parental role after the chicks hatch.

13.3.3   Active female choice in polygynous systems

The question of whether females in polygynous systems actively choose among males is an important one. In some cases, females choose certain environmental features for nesting, such as vegetation structure, and thus incidentally become the social, and perhaps genetic, mate of the male controlling the chosen area. In other cases, however, it appears that females choose to nest within the territories of particular males (e.g. Andersson 1982*b*).

In an imaginative study of mate choice in a polygynous species, Bensch and Hasselquist (1992) captured female great reed warblers from a marsh, placed radio transmitters on them, and then released the birds at a new site. These newly-arrived females visited several territorial males before settling with one of them, and females chose to mate with males singing the long song (used to attract females). Females also chose to mate with early-arriving males, which suggests either that those males were of high quality or that they held superior territories, or both. These authors concluded that a female great reed warbler actively chooses to mate with the male offering the best breeding option. Thus, in this case, secondary females appear to be making a considered mate choice decision, possibly with regard to male quality and/or territory quality, as well as the mating status of males (either already mated or unmated).

Searcy and Yasukawa (1995) review a number of studies that attempted to determine the basis of choice of nesting sites by female red-winged blackbirds. They concluded that there is little evidence to indicate that females choose to mate with particular males; i.e. male quality is not a determining factor. On the other hand, females do appear to prefer territories possessing certain attributes, such as habitat type, vegetation type and density, food abundance, and presence of elevated perches.

13.3.4   Polygyny and paternal care

Although males of most regularly polygynous species do not incubate the eggs, often they do provide food for at least some of their nestlings. Not infrequently, polygynous males provision only one brood of nestlings at a time; typically, this is the older brood. There can also be behavioural flexibility by females to accommodate this behaviour of males. For example, in marsh wrens a secondary female times its nesting cycle so that the male will be available—will have completed its parental duties to the offspring of its first mate—to provision her nestlings (Leonard 1990).

Males also often show flexibility in their responses to more than one brood of nestlings. Male bobolinks occasionally feed nestlings of second broods, particularly if those broods are large (Martin 1974; Wittenberger 1980). Male pied flycatchers may respond to experimentally manipulated primary and secondary broods in ways that optimize their reproductive interests in the two broods (Lifjeld and Slagsvold 1991). Male flycatchers invested more heavily per young in larger broods, suggesting that they are attempting to optimize the combined effort of their two mates. (At a proximate level, a larger, and possibly hungrier, brood also provides a stronger auditory stimulus than does a smaller brood.) Finally, male yellow-headed and red-winged blackbirds switch their provisioning efforts to the secondary brood when primary broods are experimentally reduced in size (Patterson *et al.* 1980). Thus, in these and other passerine species, polygynous males have the capacity to respond in an adaptive fashion to their putative offspring in two or more nests.

13.4   Models of polygyny

As discussed briefly in Chapter 1, over the years polygyny has received the lion's share of attention from behavioural ecologists interested in avian mating systems. This is particularly true with regard to the development and modifications of theoretical models that have been proposed to explain the benefits of polygyny to females or to males. An alternative approach asks the question: does polygyny always involve costs to females? Here I provide a brief review of the major models of avian polygyny.

13.4.1   The polygyny threshold model

As mentioned, the phenomenon of avian polygyny has stimulated a lot of theorizing compared to other mating systems. One theory, in particular, has been responsible for much of the empirical work on polygyny. For the past thirty years, the polygyny threshold model (PTM) of Verner and Willson (1966) and Orians (1969) has provided the major theoretical framework for the many field studies of avian polygyny. The PTM argues that an uneven spatial distribution of important resources makes it possible for some males to control access to those resources by virtue of their territorial behaviour. On the basis of resource availability per female, females will decide whether to mate with an already mated male on a higher quality territory or with an unmated male on a lower quality one (thereby avoiding the problem of sharing resources with other females). The polygyny threshold is that point on a continuum of territories ranked by quality at which a female will do as well or better by mating with an already-mated male on a higher quality territory than with an unmated one on a lower quality territory. In addition, it has often been assumed that the quality of a particular territory and the intrinsic quality of the male holding the territory were causally correlated. As stated by Searcy and Yasukawa (1989, p. 323): 'Essentially, the PTM assumes that females pay a fitness cost for polygynous mating, so that polygyny occurs only if females are sufficiently compensated for choosing already-mated males by obtaining better territories or mates.'

Why did the relatively rare mating system of polygyny, and especially the PTM, stimulate so much interest among students of avian mating systems? It is worthwhile to consider this question, because it provides some insights regarding the way our science works. Several points come to mind. (1) The PTM appeared at about the same time as the publication of George Williams' (1966) classic book, *Adaptation and natural selection,* which emphasized these two phenomena—adaptation and natural selection—at the level of the individual. (2) This emphasis on individuals provided a clear perspective on the way natural selection might be operating on male and female birds. A strong focus on individual selection provided an explanation of polygyny not based on differential sex ratios, for example, to which, up to that time, this mating system had been causally linked (see Section 13.2.1 above). (3) The period from the 1960s to the early 1970s was also a very stimulating time for students of evolutionary ecology and sociobiology as a result of the appearance of many landmark theoretical works (e.g. Crook 1962, 1965; Hamilton 1964, 1971; Maynard- Smith 1964; Trivers 1971, 1972, 1974; Selander 1972; Alexander 1974; Wilson 1975). Thus, the time was ripe for the appearance of explanations of mating systems clearly based on maximization of fitness by individuals of each sex. (4) Like the studies of Crook (1962, 1965), the PTM illustrated the adaptive, dynamic aspect of mating systems. That is, mating systems were not 'merely' fixed species traits; on the contrary, they represented adaptive responses to the environment. The ecological classification of mating systems by Emlen and Oring (1977) followed this general conceptual framework. (5) Some species exhibiting polygyny are geographically widespread, abundant, and breed colonially, and thus are readily available for researchers to study. In North America, the red-winged blackbird has been the prime avian species for study of the ecology of polygyny. Because redwings are so common and usually breed in restricted, two-dimensional marsh habitats, large amounts of data can be obtained relatively easily and comparative measurements of factors related to territory quality, for example, readily obtained (e.g. see Searcy and Yasukawa 1995).

An assessment of the polygyny threshold model
For the past thirty years, the Verner–Willson–Orians PTM has been the primary paradigm for the evolution of polygyny in territorial altricial birds. It is hard to overstate the effect this concept has had on empirical study of avian mating systems. However, despite the interest and research generated, convincing documentation that the model, in its original form, has general validity has not appeared. This may be the result of several factors. First, the model may be inappropriate for explaining the evolutionary origin of polygyny in marsh-nesting birds (Wittenberger 1976, see below), which were the original

sources of the inspiration, nor is it thought to be appropriate for forest-dwelling cavity-nesters (Alatalo *et al.* 1981). On the basis of these points alone, one might question its general applicability (see also Davies 1989).

Second, over the years the proper predictions of the PTM have remained unclear (Davies 1989). For example, in a revision of the model, Garson *et al.* (1981) predicted the following: (1) secondary females should rear as many off-spring as primary females mating at the same time; (2) quality of territory and/or of the male should correlate positively with harem size; (3) males on whose territories females settle first should obtain the largest number of mates; and (4) the earliest settling females, because they have the most choice, should be the most successful. Altmann *et al.* (1977) and Vehrencamp and Bradbury (1984) consider points 1 and 4 as incorrect or unnecessary predictions of the model. Thus, part of the reason that confirmatory tests of the model are so few is that, apart from the difficulties of measuring either male quality or territory quality, the model itself has been easy to interpret in a number of different ways.

Although several studies have shown a correlation between certain environmental features and 'harem' size, others have not (see Oring 1982 for references). Oring (1982) describes several elaborations and exceptions to the PTM: e.g. the original model did not take into account the possibility of temporal change in quality within a breeding season (Garson *et al.* 1981), or the steepness of the slope of the habitat quality curve (Wittenberger 1976, 1981). In other cases, the model probably is not applicable when one male holds two discrete, separate territories (e.g. pied flycatcher, Alatalo *et al.* 1981, 1982). Searcy and Yasukawa (1989, 1995) discuss some of the criticisms levelled at the PTM, and stress that no single model will account for all cases of territorial polygyny. In addition, Slagsvold and Lifjeld (1994) have argued that two of the models outlined below, the asynchronous settlement model (Leonard 1990) and their own defence of the male parental investment model, are modifications of the PTM. If this view prevails, the question arises: how much modification can the PTM undergo and still actually be the same model?

13.4.2 Deception hypothesis

An alternative explanation of polygyny, the 'deception hypothesis' (von Haartman 1969; Alatalo *et al.* 1981), suggests that females pay the cost of polygyny because the male conceals its bigamous relationships from the females. This is seen most typically when a male defends two spatially separated territories and attracts a female to each of them (e.g. pied flycatcher). The second female to arrive may be unaware, at least initially, of the existence of the male's first or primary mate, thus the deception (but see Stenmark *et al.* 1988).

Polygyny occurs regularly in Tengmalm's owl (Korpimaki 1991), and it also occurs in some other small owls (e.g. eastern screech owl, Gehlbach 1994).

For male owls, this appears to be a profitable strategy as their net reproductive success is enhanced as a result of siring two broods. In Tengmalm's owl, as in many polygynous passerines (e.g. Johnson and Kermott 1993; Johnson *et al.* 1993), secondary females fledge fewer young than monogamously-mated or primary females. This is because the male preferentially provisions young of the primary female, leading to the starvation of many of the secondary female's offspring.

Fig. 13.1 Male Tengmalm's owl with two mates at their nest cavities. All three birds are closely watching a hunting pine martin. (See text.)

Given the critical nature of the male's parental contributions, why do secondary females mate with an already-paired male, even when nearby unmated males are available? Korpimaki (1991) suggests that early in the season, before their eggs are laid, males generously provision secondary females, thereby 'deceiving' them into pairing with them. Costs to the secondary female of their apparently maladaptive mating decision appear later, after their chicks have hatched.

Sonerud (1992) provides an attractive alternative to the deception hypothesis to explain polygyny in Tengmalm's owl, based on the possible historical importance of nest predation in primeval forest. Sonerud's thesis is that in natural old forest—where the owls have spent most of their recent evolutionary history—pine martins (*Martes martes*) are common predators of the owls (Fig. 13.1). In contrast, in most managed forests today (where the majority of studies of these owls have been conducted), pine martins are scarce or absent. The upshot of Sonerud's argument is that in an area where the risk of nest predation is 50 per cent or more, a secondary female can do as well, on average, as a primary or monogamously-mated female.

This argument depends on two assumptions, both of which are reasonable: (1) the nests of the primary and secondary females are equally vulnerable to pine martins, and (2) if the primary female's nest is lost to a predator, the male will turn his full provisioning efforts to the nestlings of the secondary female. Sonerud suggests that if nests suffer high predation rates, there is no need to invoke 'deception' to explain why females mate with an already-mated male. Given that the male provides adequately for the secondary female prior to and during egg-laying, and given that there is an even chance that either she or the primary female will lose their nest to a predator, then '... it may simply not have mattered whether a female was aware of a male's mating status or not' (Sonerud 1992, p. 873).

13.4.3   Asynchronous settlement model

A study by Leonard (1990) of polygyny in marsh wrens, the very species originally giving rise to the PTM (Verner 1964; Verner and Engelsen 1970), failed to support this model. Leonard's findings led her to present an alternative, the 'asynchronous settlement model' (ASM), which is based on the facts that (1) as long as bachelors were available, newly arriving females settled with them rather than with already-mated males, (2) secondary females (those settling with an already-mated male) fledged as many young as did monogamous females settling at the same time, and (3) males feed nestlings at only one nest at a time. The first point suggests that early nesting may be more important to females than other considerations, such as territory or male quality. Secondary females settling in a territory did so in a temporal pattern that assured them of male assistance in the feeding of their nestlings; specifically, their young hatched after the male had completed parental duties at the nest of the primary female. In contrast to the more complex decisions required of females by the PTM, in Leonard's (1990) model, prospecting females need only to know a male's mating status (mated or unmated) and nesting stage of the first female.

13.4.4   Defence of male parental investment model

The 'defence of male parental investment model' (DMPIM), presented by Slagsvold and Lifjeld (1994), emphasizes the role of female–female aggression in the development of polygynous or monogamous mating systems. The models discussed previously deal largely with the costs and benefits of mating decisions by secondary females. In contrast, the DMPIM focuses explicitly on the effects of polygyny on the primary female. Costs to the primary female of sharing a mate can also be great, and such females may exhibit a variety of mechanisms to discourage (1) their mate from bonding with a second female and (2) a second female from settling with their mate.

A clear difference between the DMPIM and the asynchronous settlement model is that the former predicts that the secondary female will breed early relative to the primary female because they would then receive more male help (true for pied flycatchers). In contrast, the ASM suggests that secondary females will attempt to begin breeding considerably later than primary females, with the expectation of greater male involvement with the secondary brood (true for marsh wrens).

13.4.5   Cost and no-cost models

Searcy and Yasukawa (1989, 1995) provide a good review of the various ideas that have been developed to explain avian polygyny. First, these authors conclude that, although a lot of evidence exists that males benefit as a result of polygynous relationships, there is no evidence for male coercion. That is, males apparently cannot force females to breed within their territories. This means, necessarily, that polygyny is a female choice phenomenon. The general question then becomes, why do some females choose to mate polygynously? Or, if they do not actually choose a polygynous situation, why do they accept it, when territory-holding, unmated males are available?

Searcy and Yasukawa (1989) present a hierarchical classification of models designed to account for this mating system (Table 13.2). These authors describe two basically different classes of female choice models: no-cost and cost. In no-net-cost models of polygyny, all females mating polygynously (both primary and secondary) should do as well as females mating monogamously, and they may or may not obtain benefits of group nesting. In net-cost models, on the other hand, some or all of the polygynously-mated females do less well than if they were mated monogamously. This approach is a good one, it seems to me, because it cuts right to the essential issue: do female birds pay a price for polygyny? The answer, as might be expected, is that in some cases

Table 13.2   A hierarchical classification of models of the occurrence of territorial polygyny

---

I. Male-coercion model: males force females to mate polygynously.
II. Female-choice models: males cannot force females to mate polygynously.
   A. No-cost models: female fitness does not decrease with increasing harem size; that is, there is no cost of polygyny.
      1. Benefit models: female fitness increases with increasing harem size.
         a. Directed-choice model: females choose mates according to breeding-situation quality (BSQ)[†].
         b. Random-choice model: females choose mates randomly.
      2. No-benefit models: female fitness does not change with increasing harem size.
         a. Directed-choice model: females choose mates according to BSQ[†].
         b. Random-choice model: females choose mates randomly.
   B. Cost models: female fitness decreases with increasing harem size.
      1. Skewed-sex-ratio model: females are forced to pay the cost of polygyny because of a lack of unmated males.
      2. Balanced-sex-ratio model: unmated males are available when females choose mated ones.
         a. Compensation model (polygyny-threshold model): females are compensated for the cost of polygyny by acquiring a male of high BSQ[†].
         b. No-compensation models: females are not compensated for the cost of polygyny.
            i. Search-cost model: females accept already-mated males because searching for an unmated male is costly.
            ii. Deception model: mated males conceal their mating status from females.
            iii. Maladapted-female model: females knowingly choose males even though more-adaptive strategies are open to them.

[†]Breeding-situation quality.
Reprinted with permission from Searcy, W. A. and Yasukawa, K. (1989). Alternative models of territorial polygyny in birds. *American Naturalist*, **134**, 323–43.

they do, while in others they do not. In both cases, we then want to know more and the hierarchy of issues provided by Searcy and Yasukawa (1989, 1995) leads to successive levels of analysis. An important conclusion emerging from this approach is that all of the models considered above are cost models. (Why this is important will become apparent.)

Some examples of no-cost polygyny
Searcy and Yasukawa (1989, 1995) evaluated polygyny in red-winged black-birds and concluded that for females of this species there is no reproductive cost to polygyny (e.g. starvation of nestlings did not increase as level of polygyny increased). In view of the importance of red-winged blackbirds in the study of the best-known cost model, the PTM, this conclusion may make the generations of avian behavioural ecologists who have been seduced by the PTM uncomfortable.

What this means, in brief, is that the first and most critical assumption about polygyny in this species is erroneous, and that if Searcy and Yasukawa are correct, the PTM does not even apply to redwings! Similarly, their review of polygyny in yellow-headed blackbirds suggested that '... females act as if polygyny has no cost or benefit' (Searcy and Yasukawa 1989, p. 338). Another example of no-cost polygyny has been described for the corn bunting. In this species, polygyny appears to be best explained by a random pattern of settlement by females (Hartley and Shepherd 1995).

Some examples of cost polygyny
For several species, costs of polygyny to females have been identified as reduced parental contributions by the male, which reduces the reproductive success of at least some of the polygynously mated females. Slagsvold and Lifjeld (1994) present a thorough review of the costs of polygyny to females. Significantly, they focus on costs to the primary female (first female to settle and nest) of the presence of secondary females. In general, costs to the primary female have not received much explicit consideration, and have even been regarded as of little significance (see Slagsvold and Lifjeld 1994 for references). In many polygynous species, however, the primary or resident female is aggressive to other females, which are potential second mates and thus competitors for the male's parental contributions; i.e. male parental care is usually the 'resource' defended by the primary female.

In one of the two most-studied polygynous passerine species, the pied flycatcher of Europe (see Lundburg and Alatalo 1992 and Slagsvold and Lifjeld 1994 for references), some males control two spatially separated territories, and may attract a female to each of them. When this occurs, the primary female suffers a cost in terms of young produced, because of reduced male provisioning at its nest. The cost of shared paternal care is even greater for the second female.

Similarly, in some other polygynous species, such as the blue tit, both primary and secondary females suffer costs associated with reduced paternal feeding of their offspring (Kempenaers 1995). Primary females attempt to counter this effect of secondary females by aggressively excluding them from the territory, if they detect their presence before they (the primary females) begin to incubate their eggs.

Another well-studied species for which female costs have been tentatively identified is the marsh wren. Secondary females time their broods to hatch when the male is not occupied with feeding chicks in the nest of the primary female. As a result, the number and weights of chicks fledged by primary and secondary females do not differ (Leonard 1990). Although the evidence for costs to secondary females appears to be slim, Searcy and Yasukawa (1989, p. 337) tentatively conclude that female marsh wrens may incur a cost of polygyny, because they '... act as if there is a net cost of polygyny by preferring unmated males to mated males, staggering settlement times in the same territories, and segregating spatially within territories.'

In the house wren, severe costs associated with mating polygynously have been identified. Secondary females produce significantly fewer offspring than do either first females in polygynous units or monogamously mated females (Johnson *et al.* 1993). This cost to secondary females is due, in part, to reduced male aid in feeding young. It is also partly due to the necessity of brooding the nestlings, which means that the foraging time of the female is restricted. In addition, it appears that remaining on the nest with small chicks is necessary to protect them from intruding house wrens (Johnson and Kermott 1993).

13.4.6   Status of the major models of polygyny

Despite major research investment over about three decades, the original version of the PTM has not been convincingly confirmed for any species. Might this indicate that the premise of this model is basically flawed? Criticisms of certain procedures and assumptions associated with testing the PTM have already been mentioned (Altmann *et al.* 1977; Vehrencamp and Bradbury 1984, pp. 262–3; Searcy and Yasukawa 1989; Leonard 1990). Although recent authors continue to support it by adding modifications and qualifications (e.g. Slagsvoeld and Lifjeld 1994), the fact remains that thirty years after its appearance, there is still is no compelling empirical support for the original model.

Searcy and Yasukawa (1989) convincingly argue that no single hypothesis can explain territorial polygyny in all species exhibiting this mating system. They suggest that different models might apply to different populations within a species, and that even within populations different models may be necessary to explain polygynous matings by different females. If these suggestions are correct, then it should not be too surprising that attempts to verify a particular model as the sole correct explanation of avian territorial polygyny have not been successful.

Another factor that casts additional doubt on the validity of the PTM is the question of matings between females and males other than their 'apparent' or social mates. A critical assumption of the original model, and of all studies attempting to test it, is that males sired (were the genetic parents of) all the young birds hatched on their territories. Extra-pair fertilizations, however, are known to be common in both red-winged and yellow-headed blackbirds (Bray *et al.* 1975; E. Davis, in Wittenberger 1976; Gibbs *et al.* 1990; Westneat 1995). The discovery that many nestling redwings are sired by males other than the individual in whose territory the female nests further supports the conclusion of no-cost polygyny to females (i.e. choice of nest site does not dictate a female's mating choices).

This discussion leads to two explanations for territoriality in male red-winged blackbirds: (1) males compete for territorial space on marshes where the females will nest, and (2) a territory provides the male with a secure base from which to assess and approach/attract females nesting both inside and outside its territory. These points are supported by the fact that most display behaviour of males is directed towards other males rather than towards females (Peek 1972; Smith 1979).

The above comments bring us back to the original question. Has the PTM received sufficient empirical support to justify its retention as the theoretical underpinning for future studies of the ecology of polygynous mating systems? Alternatively, should we recognize that although it has had extraordinary success as an idea, its chief contribution has been as a stimulant for many empirical field studies, rather than providing the answer to why polygyny occurs in certain birds?

My own view is that the PTM (in the broad sense) has been more than fairly tested and that the exceptions and qualifications to it, along with the considerable negative evidence, cast sufficient doubt on the theory to lead to the conclusion that it is not an adequate general model for the evolution of polygyny (see Davies 1989, Searcy and Yasukawa 1989, 1995, and Slagsvold and Lifjeld 1994 for detailed discussion and review of numerous empirical studies). At this point in time, students of territorial polygyny would probably be well-advised to approach the study of polygyny along the lines suggested by Searcy and Yasukawa (1989, 1995).

This conclusion does not deny the heuristic value of the PTM over a period of many years. In discussing another hypothesis, Wilson (1971, p. 334) asked, '... is the theory both good and true, or is it just good', in that it has stimulated a great deal of interest and research. By this standard, for almost thirty years the polygyny threshold model has clearly been a good theory.

13.5   Cooperative polyandry

Polyandry is the general term used for mating systems where (1) a single female mates either sequentially or simultaneously with two or more males during a single breeding season, and (2) two or more males mate with only one female per nesting effort or per season. Mating systems that fall under the heading of polyandry are of two, fundamentally different, sorts. These are generally referred to as *classical* polyandry and *cooperative* polyandry (Faaborg and Patterson 1981; Gowaty 1981; Oring 1986). Although both classical and cooperative polyandry are characterized by a mating bond between one female and two or more males, they are otherwise completely dissimilar and almost certainly have had very different selective backgrounds.

To emphasize that classical and cooperative polyandry are distinctly different phenomena, they are considered in separate chapters. The key distinguishing characteristic of classical polyandry, the subject of Chapter 16, is sex-role reversal, with each male possessing its own nest where it incubates the eggs and later provides care for its (usually) precocial young, typically with no help from either the female or any other individuals. In contrast, cooperative polyandry is recognized by the presence of a stable social unit consisting of two or more breeding males and one breeding female. (This clearly distinguishes cooperative polyandry from monogamy with extra-pair copulations.) Some or all males may sire offspring in a single brood and, along with the female parent, provide care to eggs and chicks in a single nest. Like polygyny, cooperative polyandrous systems are almost certainly derived from monogamy–biparental care (see Chapters 10 and 11).

13.5.1   Theoretical issues related to cooperative polyandry

Cooperative polyandry provides an unusually interesting evolutionary puzzle, namely, why should a dominant male bird, 'Alpha,', ever voluntarily share paternity with another male, 'Beta', especially when Alpha will then provide extensive parental care to young birds that may not be its own offspring? On the face of it, the sharing of paternity appears to contradict classical evolutionary theory. Yet it is now amply documented for a few species that often peaceable 'mate-sharing' by males does occur frequently.

To the extent that generalizations can be made about cooperatively polyandrous systems, it appears that, in most cases, the basis for multi-male occupancy of a territory is related to ecological factors. Specifically, the common theme appears to be that, under certain conditions, a coalition of males can defend an important resource better than a single male can, and that the defended resource is so valuable that the benefits of a male alliance outweigh the costs of sharing paternity (Fig. 13.2). This interpretation has been offered for the Galápagos hawk (Faaborg and Bednarz 1990), white-winged trumpeter (Sherman 1995*b*), pukeko (Jamieson *et al.* 1994), and acorn woodpecker (Stacey and Koenig 1984; Koenig and Stacey 1990). It is not the correct interpretation for the polyandrous mating patterns of the dunnock (Davies 1990); in this species, males are clearly competitors for copulations. Each of these species is discussed below.

To date, cooperative polyandry has not received either the empirical attention or the theoretical development accorded polygyny. Reasons for this are several. First, cooperative polyandry is a rare mating system compared to polygyny. Second, polygyny can be so clearly facultative, and it often appears to be so clearly associated with ecological factors, that its theoretical development is tractable (see previous section). In contrast, several authors have expressed the view that production of a general model for polyandry seems unlikely (e.g. Erckmann 1983; Oring 1986; Clutton-Brock 1991). Probably the major reason that no single explanation has proved satisfactory is that classical and cooperative polyandry are only very superficially similar; i.e. they are fundamentally different phenomena. Thus, we should not expect a single explanation to suffice.

Cooperative polyandry also cannot be explained satisfactorily by recasting polygyny models (e.g. Gowaty 1981). However, cooperative polyandry is sometimes explicable, at least in large part, by use of an ecological approach. For two species that exhibit cooperative polyandry, the dunnock and the alpine accentor (Davies 1990, 1992; Davies *et al.* 1995; Hartley *et al.* 1995), it appears that the kinds of ecological factors often considered to be responsible for the evolution of polygyny, such as habitat structure, defendability of

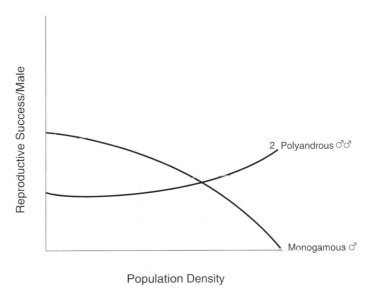

Fig. 13.2   Under certain ecological conditions, two (or more) males can achieve greater lifetime reproductive success by engaging in cooperative polyandry than by having exclusive access to a female, even if paternity is shared to some degree. This is thought to be due to the ability of coalitions of males to defend territorial space and the resources it holds. In this hypothetical figure, habitat occupied by monogamous and polyandrous social units is of uniform quality. (In many systems, however, territories vary greatly in intrinsic quality and larger groups typically hold the more valuable territories.) At low densities each polyandrous male is about half as successful as a monogamous one. However, as population density increases, the ability of a single male to defend resources adequate for breeding declines. In this hypothetical system, at very high densities, monogamous males are incapable of successful breeding as a result of intergroup territorial conflict. In contrast, individual males in polyandrous groups may become increasingly successful, approaching the level achieved by monogamous breeders at low population densities, as they evict the latter from their territories.

resources, variation in territory quality, etc. have also influenced the evolution of cooperative polyandry.

A fundamental distinction associated with the patterns of parenting remains, however, between cooperative polyandry and polygyny. In polygyny, each female has a nest and may or may not receive assistance in parenting from the male (i.e. each nest has either one or two care-givers). In contrast, in co operative polyandry, two or more males provide parental assistance at the nest of a single female (i.e. each nest has three or more care-givers). Thus, in polygynous systems, males would appear to gain more than females, whereas in cooperative polyandry, females may benefit at the expense of males (Davies 1991).

13.5.2   Frequency of genetic polyandry

Before considering cooperative polyandry further, it is important to point out that, although this kind of social organization has been claimed for a number of species, data on the single most critical issue—genetic parentage—has been gathered for far fewer of these. Specifically, the question is, to what extent do males actually share paternity? Extra-pair fertilization in socially monogamous species appears to be of common occurrence in many species (see Chapter 11). However, to date, few data exist concerning the frequency of shared paternity among males within a cooperatively polyandrous group. This kind of information was unobtainable prior to the development of biochemical means of determining paternity. The limited evidence, based on starch–gel electrophoresis or DNA fingerprinting, indicates that although multiple paternity does occur in a number of species, its frequency generally is poorly known (Table 13.3).

Before reviewing a few studies of cooperative polyandry, it should be mentioned that in some species, including the pukeko, acorn woodpecker, the dunnock and its fellow prunellid, the alpine accentor (Davies *et al.* 1995; Hartley *et al.* 1995), social-mating relationships may be polygynandrous, involving both two or more males and two or more females. Smith's longspurs

Table 13.3   Some documented cases of cooperative polyandry and/or polygynandry

Species	Nature of evidence	Frequency of multiple Paternity of broods	Reference
Galápagos hawk	Observational, DNA analysis	Common	Faaborg *et al.* 1995
Pukeko	Obs., DNA analysis	Common	Jamieson *et al.* 1994
Dusky moorhen	Obs.	Possibly common	Garnett 1980
Tasmanian native hen	Obs., DNA analysis	Rare	Gibbs *et al.* 1994
White-winged trumpeter	Obs.	Possibly common	Eason and Sherman 1995; Sherman 1995*a,b*
Acorn woodpecker	Obs., protein electrophoresis	Unknown	Joste *et al.* 1985; Mumme *et al.* 1985
Alpine accentor	Obs., DNA analysis	Common	Davies *et al.* 1995; Hartley *et el.* 1995
Dunnock[1]	Obs., DNA analysis	Common	Burke *et al.* 1989; Davies 1990
Smith's longspur	Obs., DNA analysis	Common	Briske 1992, 1993

[1]Not truly 'cooperative' polyandry; see text.

are also considered to be polygynandrous (Briskie 1992, 1993), but particular sets of males and/or females do not form discrete mating units. A female longspur, however, may obtain help with the feeding of her offspring from two or more males.

13.5.3   Social copulations

In addition to the usual difficulty of observing copulations, there is another critical problem with relying exclusively on behavioural observations in the study of mating strategies: in at least some putatively cooperative polyandrous species, some copulations or copulation-like behaviour appear to be of a social nature apart from the usual function of transfer of sperm to the female's reproductive tract. Making the assumption that all such copulation-like behaviours may actually fertilize eggs will tend to obscure any relationship between frequency of (apparent) copulations and genetic paternity. For example, groups of the putatively cooperatively polyandrous Harris' hawk engage in a type of behaviour labelled *backstanding* (Faaborg and Bednarz 1990), where male group members stand on the back of the female in a copulation-like arrangement. Backstanding hawks three deep have been observed! The frequency of genetic polyandry, if it occurs at all, is unknown for the Harris' hawk (see below).

Similarly, in the pukeko (see below), descriptions of copulation-like behaviour engaged in by all group members (Craig and Jamieson 1990) suggest a social function of some sort. Intra-sexual mountings by both sexes are common, and noisy calls that attract other group members are given prior to copulation.

Keeping in mind the fact that the extent of genetic polyandry is unknown for most of the species that have been classified as cooperatively polyandrous or polygynandrous, let us assume, for the sake of discussion, that it does occur regularly and frequently in those species. This allows a return to the question: why, or under what conditions, do males voluntarily share paternity of one brood of young? One idea is that the dominant or alpha male obtains a critical net benefit, as measured in terms of lifetime fitness, from the presence and activities of the beta male, and that to obtain beta's alliance, alpha must share copulations, and thus paternity, with beta. In addition, if beta has the option to breed elsewhere, a cost to alpha of 'enticing' beta to remain on the territory and contributing to the welfare of the other group members might include shared paternity. For example, additional males might be required to successfully defend a high-quality territory. This kind of interdependence of males is probably the single most important reason for cooperative polyandry in some species, such as the Galápagos hawk and possibly the pukeko (see below). Stacey (1982) provides a more complete discussion of factors leading to male–male cooperation and cooperative polyandry.

13.5.4   Some studies of cooperative polyandry

Cooperative polyandry has been reported to occur regularly in one or more species of at least four families of neognathous birds. Here I present a summary of what is known concerning the existence, frequency, and adaptive significance of cooperative polyandry or polygynandry in two hawks, two gallinules, a woodpecker, and a passerine. In all of these examples, genetic polyandry—multiple paternity of a single brood—has been confirmed by use of molecular techniques. However, it is apparently quite rare in at least one of the best-known examples, the Tasmanian native hen.

Galápagos hawk
The Galápagos hawk provides an especially interesting case of cooperative polyandry (Faaborg *et al.* 1995). Breeding units of these birds contain a single female and from one to eight unrelated males; typically, such a group consists of a female and two or three males. Groups are territorial, and larger units are probably better able to defend, and possibly to expand, their territories than are smaller groups. It is likely that in this species, male alliances, probably in relation to territorial defence, are critical to the interests of each male. Typically, new groups are initially composed of a female and several males; thus, a group is thought to be at its largest when it is newly formed. Over the years, changes in the composition of groups are affected primarily by mortality; as founding males die, group size dwindles (Faaborg and Bednarz 1990).

What makes this species especially interesting is not only the recent confirmation of regularly shared paternity among unrelated males (Faaborg *et al.* 1995), but also the fact that there seems to be absolutely no competition for paternity among the males of a social unit (DeLay *et al.* 1996). As mentioned above, selection theory tells us that males should complete for paternity of offspring, and this has been confirmed time and time again. In the Galápagos hawk, however, there is no evidence that males compete for matings; on the contrary, good observational data indicate that they do not. It does remain possible, however, that sperm from different males compete within the female's reproductive tract or that the female has some ability to control which sperm fertilize her eggs. That said, the apparent absence of any, even subtle, dominance interactions among males related to egg fertilization is striking.

Reproductive output per year is low in Galápagos hawks, with broods generally numbering one or two chicks. Thus, there are often fewer nestlings in a brood than the number of adult male group members. This means that, at a given nesting, most of the males are not related to the chicks for whom they provide care. Although the contributions of a given non-father to the welfare of the chicks has no immediate genetic benefit for him, each male may be 'paid back' over the longer term by having an opportunity to sire a chick or chicks in the future with the other group males providing food, etc. for his offspring.

Because of the egalitarian sharing of copulations—and paternity—and because these birds are long-lived, it appears that a male's genetic fitness is maximized by giving aid to all chicks, which may or may not be his offspring, with the longer-term 'expectation' that his net production of offspring will at least equal that of the other males in his group.

Harris' hawk

The Harris' hawk (Fig. 13.2) has been considered to be polyandrous since Mader (1975) published observations to the effect that about half of the breeding units he observed in Arizona consisted of more than two birds and that, in one trio, both males copulated extensively with the female (Mader 1979). However, the backstanding behaviour that is common in Harris' hawks (described above) may complicate interpretations of copulatory behaviour, especially if it is observed from a distance (Faaborg and Bednarz 1990). Limited studies in New Mexico provide no evidence that genetic polyandry is of regular occurrence in these hawks. Electrophoretic analysis from one multi-male group indicated that (1) the two adult-plumaged males could not have been father and son, (2) they probably were not brothers, and (3) one male sired both nestlings (Faaborg and Bednarz 1990).

These data are too few to draw any conclusions about the question of genetic polyandry in Harris' hawks and, at this point, whether genetic poly-andry even occurs in this species remains unknown. However, the data on group development, based on marked birds, are extensive, and indicate that in New Mexico most groups seen during the prolonged breeding season consist of a mated pair, an adult-plumaged male offspring of the breeding pair hatched in the previous year, and one to four young hawks hatched that year (Faaborg and Bednarz 1990). This kind of group composition, where a mature offspring remains with its parents, is standard in cooperative breeders and in itself offers no support to an assumption of genetic polyandry in this species. In the case of the Harris' hawk, benefits to all individuals of sociality, namely cooperative hunting, can account for their group-living behaviour (Bednarz 1988).

White-winged trumpeter

Cooperative polyandry in this species, a member of a small and little-known family of South American gruiform birds, has been described by Eason and Sherman (1995) and Sherman (1995*a*,*b*). Trumpeters live in polyandrous groups, with two or three adult males and from one to three adult females, plus young produced by the group (Fig. 13.3). The adult males, which are often unrelated, compete to mate with the dominant female. Only this individual lays eggs. Although the dominant male attempts to mate guard, one-third of the copulations with the breeding female are obtained by subordinate males.

The adaptive benefit of such groups appears to be related to territorial defence: two or more adult males can more successfully defend a territory than

Fig. 13.3 Family group of cooperatively polyandrous white-winged trumpeters. Two or more unrelated males may copulate with the sole breeding female.

can a lone male (see also pukeko, below). This is thought to be the causal factor promoting alliances of unrelated males.

Tasmanian native hen

Tasmanian native hens were one of the first species assumed to be polyandrous, with two or three males thought to be brothers mating with a single female and sharing parentage (Ridpath 1972). This was one of the first putative examples of 'wife-sharing' (Maynard-Smith and Ridpath 1972). In this species, group sizes range from two to seven, with mating units typically consisting of a single female and one or two males. Maynard-Smith and Ridpath (1972) assumed that, in this species, a male-biased sex ratio preceded the development of polyandry (but see Goldizen *et al.* 1998).

A recent study, however, casts doubt on the conclusion that polyandry is a common mating system in Tasmanian native hens. Gibbs *et al.* (1994) found that in five of six groups that contained either two or three putative male parents, only one male sired chicks; in the sixth group, there was evidence of shared paternity. Their data also indicated that in groups containing two females, only one of them laid eggs. Gibbs *et al.* (1994, p. 369) concluded that genetic monogamy was the predominant mating system of the groups of Tasmanian native hens that they studied. These authors also noted that '… in multi-male groups of native hens more than one male is often observed to copulate with the female' and that '… behavioral observations of copulatory behavior may not provide true estimates of male, or in some cases, female reproductive success in these birds' (Gibbs *et al.* 1994, p. 369).

Thus, as indicated earlier, with regard to mating systems in general, appearances can be deceiving. Although behavioural polyandry is indeed common in the Tasmanian native hen, it turns out that genetic polyandry is rare, contradicting the conclusions and interpretations of Maynard-Smith and Ridpath (1972). Mating systems may vary geographically, however, and it remains possible that genetic polyandry will prove to be more common in populations of Tasmanian native hens other than the one studied by Gibbs *et al.* (1994).

Pukeko
The pukeko, the New Zealand representative of a widespread species of gallinule, exhibits a behaviourally and genetically polygynandrous mating system (Craig and Jamieson 1990; Jamieson *et al.* 1994). It is one of three gallinules on the short list of cooperatively polyandrous/polygynandrous species (Table 13.1). The breeding biology of the pukeko is unusual in several respects. Concerning the issue of cooperative polyandry, courtship is a communal social event. Although there is a well-developed dominance hierarchy, the dominant or alpha male does not attempt to guard fertilizable females. On occasion the alpha male does interrupt mating attempts by subordinate males, but usually he quietly observes or ignores mating by other males. During the egg-laying period, females may copulate with up to three males within a few minutes.

A study population on New Zealand's North Island was composed of closely related individuals and contained non-breeding helpers. Groups in a second population on South Island contained from one to three unrelated males and one or two unrelated females. In this population, there were no non-breeding helpers (Jamieson *et al.* 1994). DNA analyses determined that shared paternity occurred in nearly all groups of pukekos in both populations. This difference between two populations of the same species in basic group structure illustrates both the basic similarities and the differences between cooperative polyandry and cooperative breeding.

Thus, with regard to the sharing of parentage by unrelated males, the reproductive system of South Island pukekos is very similar to that of the Galápagos hawk. It is likely that in both species, male alliances related to territory defence are extremely important to the overall well-being of each individual male.

Acorn woodpecker
The breeding system of the acorn woodpecker is variable and complex. At the Hastings Reservation in California, two, or rarely three, females sometimes contribute eggs to a single nest (Mumme and Koenig 1983). In contrast, two breeding females within one group have not been recorded at Water Canyon in New Mexico (Koenig and Stacey 1990). At both sites, however, molecular data have demonstrated the existence of multiple paternity within a brood

(Koenig and Stacey 1990). Because both multiple males and multiple females contribute genetically to a single brood of young, this species is polygynandrous in California, while in New Mexico, it is apparently cooperatively polyandrous.

To date, the frequency of multiple paternity has not been determined for either population. Koenig and Stacey (1990) discuss the difficulties encountered in attempts to work out the genetic mating system of these woodpeckers and list four types of indirect evidence suggesting the occurrence of multiple paternity. These authors speculate that by making mating cryptic and difficult to detect, females increase the parental uncertainty of each male, which, in turn, leads to a greater contribution at the nest by each male than would be the case if a male was certain that it was not the parent of some offspring (see also Stacey 1979).

Although these interpretations concerning the adaptive significance of multiple paternity may well be correct, especially in light of the similarities between acorn woodpecker and dunnock mating systems (see below), confirmation or modification of the views concerning the acorn woodpecker's mating system awaits additional information on genetic parentage.

Dunnock

The dunnock, studied by N. B. Davies and colleagues (see Davies 1990, 1992), provides an especially clear example of a flexible, ecologically-based mating system. This variable and, at least at first glance, complex mating system is well understood, thanks in part to detailed genetic data on parentage obtained via DNA fingerprinting (Burke *et al.* 1989). Breeding units of dunnock may be monogamous, polyandrous, polygynous, and polygynandrous (see Davies 1990, Table 15.1). These combinations of male and female breeding units develop as extensions of individual male and female territoriality (see Davies 1990, Fig. 15.2).

The dunnock studies are especially relevant to two of the general questions raised earlier concerning cooperative polyandry. First, in polyandrous trios, the secondary or beta male often obtains a large fraction of the paternity of broods, about 45 per cent on average. Thus, genetic polyandry is common in this species. Second, unlike Galápagos hawks or pukekos, male dunnocks do not cooperate in sharing paternity. The alpha male does not voluntarily 'allow' the secondary, beta male to copulate with the females as a strategy designed to obtain benefits via the presence and contributions of beta. Instead, the female behaves in a manner designed to thwart the alpha male's attempts to mate-guard, and to encourage and obtain copulations from the beta male, by sneaking away from alpha and associating furtively with beta. Beta males that have copulated with the female, and that have a high probability of fathering chicks, provide considerable paternal effort. In contrast, beta males that have been excluded from mating with the female provide little or no parental care.

A female dunnock's reproductive interests are best served in polyandrous situations where two males contribute genetically to the brood and then help to rear it. The alpha male, on the other hand, has less reproductive success than would be the case if he were able to enforce genetic monogamy. Thus, in polyandrous arrangements, there is a conflict of interest between the alpha male and the female, with the female often winning. This is not always the case, however; the tables can be turned. When polygyny occurs, the male is far more successful than either female, as measured by the number of young produced per season, and each female is less successful than she would be in a socially monogamous situation. In contrast, in 'selfish polyandry', where only the alpha male fertilizes eggs (genetic monogamy), he is as successful as the female. In cases such as this, where the beta male is excluded from paternity, only the female and the alpha male contribute care to the nestlings.

The existence of a similar polygynandrous system in a close relative of the dunnock, the alpine accentor (Davies *et al.* 1995; Hartley *et al.* 1995), suggests that even in these two species, which seem at first glance to exhibit the epitome of behavioural flexibility and opportunism, there may be an important phylogenetic component to their mating systems (Davies *et al.* 1995).

13.6   Conclusions and summary

Polygyny is a social mating system that involves a single male and two or more females, with each female having its own nest. Cooperative polyandry, on the other hand, is a mating system composed of a female and two or more males that care for the eggs and young at a single nest. Both systems are probably derived from monogamy with biparental care. Otherwise, however, their evolutionary development has been very different. In polygyny, male parental care may be either present or absent, while in cooperative polyandry, parental care by males appears to be always present. In many highly polygynous species, paternal care of young is completely absent. Thus, for this system to develop fully, the female must be able to rear a brood of young with no assistance from the male. In contrast, in cooperatively polyandrous systems, both sexes are involved in the rearing of young in the manner typical of monogamous– biparental species.

Polygyny is the most-studied avian mating system. Several models have been offered to account for the significance of polygyny, the best known of which is the polygyny threshold model (PTM). The PTM and its offshoots assume that it is costly for female birds to mate with an already-mated male. This leads to the question, what are the benefits that compensate polygynous females? Despite a great deal of research effort designed to demonstrate the costs and compensating benefits to polygynous females, the basic assumptions of the PTM have never been agreed upon and, overall, the model not been convincingly supported. The PTM, however, has been extraordinarily successful as a stimulant for empirical research, and its merits have been fairly tested over a period of many years. It is time for researchers in this area to develop a new paradigm.

An alternate approach, suggested by Searcy and Yasukawa (1989, 1995) is to classify models of polygynous systems as either 'cost' or 'no-cost' in terms of their effects on females. In no-cost systems, all females mated polygynously should do as well as females mated monogamously. In cost systems, on the other hand, some or all of the females should do less well than females mated monogamously. Two exhaustively-studied polygynous species, the red-winged blackbird of North America and the pied flycatcher of Europe, illustrate these categories. Female redwings suffer no costs of polygynous mating, while some female pied flycatchers do incur a cost.

Cooperative polyandry is almost surely derived from monogamous systems. The dunnock is especially illustrative in showing how polyandrous situations can develop from monogamous ones. In this species, the interests of the two males are different enough that 'sharing' of the female does not occur. Moreover, if a male has little or no prospect of actually siring offspring, it declines to provision the chicks in the female's nest. In contrast, other species appear to contradict theoretical evolutionary dogma. In a few species, the absence of intense male–male competition for mates and matings, which occurs in most birds, is striking. Unrelated males of the Galápagos hawk and the pukeko of New Zealand do not compete for matings, and dominant males do not usually interfere in the copulatory activities of their subordinate fellow group members. In addition, in these species, as their behaviour would suggest, the male members of a coalition share paternity. It appears that male–male alliances—probably based on mutual defence of the common territory—have strongly favoured the evolution of conflict-reducing behaviours, even at the cost of reduced parentage, at least as measured over a single breeding season.

In four studies of species that are apparently cooperatively polyandrous, electrophoresis or DNA fingerprinting has determined the existence of genetic polyandry. Thus far, regular, frequent occurrence of genetic polyandry has been documented for the Galápagos hawk, pukeko, and dunnock and its fellow prunellid, the alpine accentor, and its existence has also been documented in the acorn woodpecker. The behaviour of male Galápagos hawks and pukekos indicates that fellow group members may 'voluntarily' share paternity, while in the dunnock, females often pursue a multiple-male strategy that is contrary to the reproductive goals of the males of the group. Perhaps surprisingly, recent studies of one of the most famous putative cases of cooperative polyandry, the Tasmanian native hen, indicate that genetic monogamy actually is the rule in this species. Thus, as is also true for other mating systems, within the social system labelled cooperative polyandry, a diverse array of individual strategies occur.

14 Cooperative breeding

14.1 Introduction

Cooperative breeding, sometimes referred to as communal breeding, is a rare and specialized phenomenon that occurs in fewer than three per cent of the world's approximately 9700 species of extant birds. Cooperative breeding is defined by the presence in a stable social unit of non-breeding 'helpers', which assist in the care of young birds, in addition to breeding individuals (from a single pair to several or even most group members). It is these non-breeding helpers that distinguish cooperative breeding from both cooperative polyandry and polygynandry (see Chapter 13), and from true 'communal' breeding systems, where two or more females may contribute eggs to a single nest (see Section 14.6). The distinctions between these kinds of mating systems are not always sharp, however, because in a number of species more than one pair of breeders plus non-breeding individuals occur in a single social unit.

For several years (around the late 1970s and 1980s) cooperative breeding was a popular topic in behavioural ecology because it appeared at that time that this phenomenon might pose a serious problem for Darwin's theory of evolution. For almost two decades, workers focused on the question, 'Why do helpers help?' More specifically, why do individual animals—i.e. the non-breeding helpers—forego breeding on their own and help rear other individuals' offspring? Interest in this issue was fuelled by the appearance of an exciting idea, often labelled 'kin selection' (Hamilton 1963, 1964; Maynard-Smith 1964), that could account for the seemingly altruistic behaviour of helpers.

With the appearance of a number of detailed studies (most of which are summarized in Stacey and Koenig 1990; see also Brown 1987), it became apparent that the phenomenon labelled cooperative breeding does not refer to a particular mating system. Instead, cooperative breeding is an umbrella term that includes a diverse array of reproductive–mating and social tactics. Within the category of cooperative breeding occur monogamous (genetic and/or social), polygynous, cooperatively polyandrous, and polygynandrous mating systems (e.g. see Stacey and Koenig 1990). Moreover, although many authors have sought generalizations about ecological factors promoting cooperative breeding, they have been only partially successful in producing a consensus. In large part, this lack of success is due to the diversity of social systems that have been placed together under the banner of cooperative breeding, solely as a result of the existence of non-breeding group members, the helpers.

An additional complicating ecological factor in the study of cooperative breeding is the issue of territoriality, which is a conspicuous and major aspect of the lives of most cooperative breeders. Variation in territory quality (e.g. Ligon and Ligon 1988; Stacey and Ligon 1991) is intertwined with the possible effects of helpers *per se* on production of young birds (e.g. more young are often produced on high-quality territories and, as a result, such territories often have more helpers than low-quality ones; e.g. see Ligon *et al.* 1991); in contrast, territoriality is absent in some other cooperative breeders. Thus, as has been emphasized repeatedly in the discussions of other mating systems, a single term like 'cooperative breeding' does not reflect the diversity of the socio-behavioural and ecological phenomena included in it.

Most, if not all, of the important issues relevant to cooperative breeding systems are basically related to the costs and benefits of sociality. In this respect, cooperative breeding is similar to other forms of group-living and is not as distinctly different from other kinds of sociality in birds and mammals as has sometimes been claimed. To the extent that cooperative breeding can be viewed as a unitary phenomenon, it is part of a continuum of social patterns. For example, Koenig *et al.* (1992) point out that some of the earliest and most important insights into the issue of delayed dispersal in cooperatively breeding birds (see below) were first presented by Verbeek (1973), who developed his ideas through study of the social, but non-cooperatively breeding, yellow-billed magpie.

There are some interesting parallels between the development of the study of avian polygyny and avian cooperative breeding. Growth of both disciplines was based on keystone sociobiological writings of the 1960s and 1970s (e.g. Hamilton 1963, 1964; Maynard-Smith 1964; Williams 1966; Trivers 1971, 1972; Alexander 1974; West-Eberhard 1975; Wilson 1975), and interest in both was further stimulated by a number of influential papers (polygyny—e.g. Crook 1965; Verner and Willson 1966; Orians 1969; Emlen and Oring 1977; cooperative breeding—e.g. Brown 1974, 1978; Emlen 1978, 1982*a,b*).

Like the subject of polygyny, and unlike cooperative polyandry, there has been a great deal of theorizing, together with many empirical studies, about the phenomenon of cooperative breeding. These general approaches led to two major reviews, Brown's (1987) *Helping and communal breeding in birds*, and *Cooperative breeding in birds*, edited by Stacey and Koenig (1990). Since publication of these volumes, ideas and perspectives concerning various aspects of cooperative breeding have continued to develop. First, since the appearance of these two books, there has been a good deal of discussion about the issue of delayed dispersal (e.g. Emlen 1991, 1994; Stacey and Ligon 1991; Koenig *et al.* 1992; Mumme 1992*a*; Walters *et al.*, 1992*a,b*), which for almost two decades was an uncontroversial aspect of cooperative breeding, and a few new empirical studies on this subject have appeared (e.g. Komdeur 1992; Walters *et al.* 1992*b*; Komdeur *et al.* 1995). Second, the question of helping behaviour has been revisited from an 'unselected' perspective (Jamieson and Craig 1987; Jamieson 1989*b*; see also Ligon and Stacey 1989, 1991; and Emlen *et al.* 1991). The third, and probably most important, new development is explicit recognition of the role of phylogenetic history as a critical piece in the puzzle of cooperative breeding (e.g. Russell 1989; Peterson and Burt 1992; Edwards and Naeem 1993; Ligon 1993; Zack 1995), which has led to a refinement of some of the major questions and current thinking about them.

In brief, most early considerations of cooperative breeding focused on the possible genetic benefits to helpers to be obtained via their aid to relatives. More recently, delayed dispersal of non-breeders and direct (personal) benefits to all group members, as a result of group living and intra-group cooperation, have received more attention. In addition, it is now appreciated that the role of phylogenetic history in the evolution and maintenance of cooperative breeding systems must be given due consideration.

14.2   Helpers-at-the-nest

Cooperative breeding has been of special interest to sociobiologists because of the presence and behaviours of non-breeding auxiliaries or helpers (e.g. Brown 1987). Helping behaviour has almost universally been viewed as adaptive in either or both of two ways. First, it has often been assumed that the helpers gain indirect, genetic benefits to their own fitness by promoting the welfare and thus potential survival and reproductive success of relatives. This requires that helping behaviour itself increases the production of young birds related to the helper ('kin selection'). Second, helpers may obtain direct benefits by gaining something of importance to their own future reproductive effort (e.g. parenting experience, subordinate allies, space in which to breed in the future). Of course, this is not actually an 'either–or' dichotomy. Helpers could benefit simultaneously in both ways. The relative importance of these two kinds of potential benefits was a rather contentious issue a few years ago, but it has become more clearly appreciated that, as with other aspects of cooperative breeding, a single answer to the question 'Why do helpers help?' that is applicable to every cooperatively breeding species, will probably never be forthcoming. This is simply because the diversity of cooperative breeders and the different selective pressures on them ensure that the significance of helping behaviour will vary from case to case (Ligon and Stacey 1989, 1991).

14.2.1   Indirect and direct benefits of helping

Indirect benefits

Much has been written about the possible significance of helping behaviour by non-breeders, particularly with regard to the issue of indirect or kin-selected benefits. For indirect benefits to be a selective factor, the first thing to demonstrate is that helpers *per se* (and not some correlated factor, such as territory quality) do, in fact, increase the fitness of related individuals. In some instances this critical point has received affirmative support (Brown *et al.* 1982; Leonard *et al.* 1989; Mumme 1992*b*). In a few species the issue of territory quality does not complicate the existence of a relationship between the presence of helpers and production of young. In some of these cases helping *per se* appears to have a strong effect on production of fledglings. Two such species are the white-fronted bee-eater (Emlen 1990) and the pied kingfisher (Reyer 1990).

In contrast, in some other species there is no relationship between the presence of helpers or number of helpers and the production of young birds, either per nest or per annual reproductive cycle. Examples include the Harris' hawk (Fig. 14.1) and the green woodhoopoe (Fig. 14.2) (Bednarz 1987; du Plessis 1989, 1991; Ligon and Ligon 1990; see Ligon and Stacey 1989, 1991 for additional examples and discussion).

A recent study of the effect of helpers in the Seychelles warbler illustrates the difficulty in demonstrating an unqualified benefit of helpers. Komdeur (1994) considered both territory quality and number of helpers and found that the presence of a single helper was significantly correlated with increased reproductive output of its parents on high-, medium-, and low-quality territories. In contrast, on low- and medium-quality territories, the presence of two helpers significantly decreased reproductive success compared to groups with a single helper. Three helpers had a negative effect on reproductive

MIKE RAMOS
1993

Fig. 14.1 In the cooperatively breeding Harris' hawk, the presence of helpers does not affect production of fledgling hawks on either a per nest or a per annum basis (Bednarz 1987).

success, regardless of territorial quality. Thus, in this case, as in others, the answer to the question, 'Do helpers help?', is 'sometimes'. Across the board, and even within a single species, the overall effects of helpers may vary from a significant positive effect, to no detectable effect, to a significant negative effect, on the production of young birds.

Direct benefits
As discussed by Alexander (1974), for individuals of some species there are clear direct benefits to group-living that override its costs. In all social species of birds and mammals, individual group members, including helpers, probably always obtain one or more of a variety of benefits by virtue of group-living. However, it does not follow from this truism that more young are necessarily produced directly as a result of the assistance rendered by helpers. Even if more young are not produced, there are many potential direct benefits to be derived from helping behaviour *per se*. For example, in the Seychelles warbler,

individuals with experience as helpers before becoming breeders have higher lifetime reproductive success than do birds without such experience (Komdeur 1996).

In most species, the majority of helpers are related to the recipients of their assistance. Given the way in which groups usually grow in cooperatively breeding species—by retention of young birds in their natal group—this is not surprising. In several species, unrelated helpers also occur commonly (e.g. secondary helpers in the pied kingfisher, Reyer 1990). Typically, even when unrelated helpers occur regularly, they are less common than related helpers (e.g. green woodhoopoe, Ligon and Ligon 1990).

In the superb fairy-wren, however, helpers are frequently unrelated to the young birds they help (Dunn *et al.* 1995). Fairy-wren helpers regularly aid in the rearing of unrelated chicks; this is because the father of the chicks is from outside the group and is unrelated to the helpers, and the helpers' mothers have frequently either died or moved elsewhere. In this species, helpers do not increase the number of young produced, which, as noted above, is the first requisite for the assumption that helpers help. Dunn *et al.* (1995) tested several possible benefits of delayed dispersal and helping and concluded that helping behaviour serves as a payment to breeders, which allows helpers to stay in the group. A corollary of this line of thinking is that territories with several helpers are probably better (e.g. safer) places to live than are territories with few or no helpers (Ligon *et al.* 1991).

Other kinds of direct benefits to helpers are numerous, including acquisition of territorial space for breeding, as seen in Florida scrub-jays. An older male helper may acquire a portion of the parental territory by 'budding.' The male then attracts a mate to his new territory (Woolfenden and Fitzpatrick 1984; see also Mumme 1992*b*).

14.2.2 The stimulus–response nature of provisioning behaviour by helpers

Williams (1966, p. 208), Jamieson and Craig (1987), and Jamieson (1989*b*) have argued that helping behaviour *per se*, i.e. the feeding of nestlings by helpers, should not be considered as an adaptation, because '... the feeding of nestlings in communal breeders is maintained by the same stimulus–response mechanism that results in parents feeding their own young or host species feeding parasitic young ...' (Jamieson and Craig 1987, p. 80). According to this view, the feeding of nestlings by helpers is basically no more than a manifestation of a general trait among altricial birds, namely an 'automatic' response (placing food) to a stimulus (a gaping, vocalizing nestling's mouth). Jamieson and Craig point out that non-breeding birds of a variety of species, including individuals too young to be 'hormonally primed', will respond to this stimulus. Other lines of evidence, such as many cases of interspecific feeding of nestlings and the phenomenon of avian brood parasitism, also support a stimulus–response interpretation of feeding behaviour. This stimulus–response interaction will almost always be maintained by natural selection because of its overwhelming importance in parental care (Jamieson 1989b). This is because (1) all birds hatch with the possibility of becoming parents (i.e. the cost of responding to a nearby gaping mouth will rarely exceed its benefits), and (2) for most avian species, production of offspring is the sole means of maximizing individual fitness.

Schoech *et al.* (1996) provide hormonal evidence that appears to refute Jamieson's suggestion. In the cooperatively breeding Florida scrub-jay, prolactin, which mediates parental behaviours in vertebrates, is produced in helpers. These authors found that prolactin levels were elevated in non-breeders prior to their exposure to nests or nestlings, contrary to what Jamieson's hypothesis suggests. Moreover, prolactin levels of individual helpers were significantly correlated with the number of visits to the nest and the amount of food delivered to jay chicks. Finally, non-breeders that fed nestlings had higher levels of prolactin than non-breeders that did not provide food to nestlings. A similar pattern has been found for Harris' hawks (Vleck *et al.* 1991).

In short, the evidence obtained by Schoech *et al.* (1996) strongly implicates prolactin as a proximate stimulus for helping behaviour. However, several lines of evidence suggest that, to some extent, helping behaviour—the feeding of nestlings by non-breeders—is programmed or 'hard-wired' into altricial birds, and that it is not necessarily an adaptive behaviour.

First, with regard to 'hard-wiring', in an aviary, hand-reared juvenile western scrub-jays capable of self-feeding responded to the begging of juvenile pinyon jays by feeding them (Ligon 1985). Because these birds were so young and because a helper has never been recorded in western scrub-jays, it is doubtful that these juvenile jays had elevated prolactin levels. (Western scrub-jays, however, are probably descended from cooperatively breeding ancestors; see Section 14.4.1.1.)

Second, with regard to helping as an adaptation, the presence of non-breeding auxiliaries that are actively prevented from feeding nestlings provides additional support for the suggestion that helping—i.e. feeding of nestlings—is not an adaptation *per se* and that the feeding response is programmed (Rabenold 1985) into altricial birds independent of any the adaptive effects of helping. Scrub-jays in Oaxaca, Mexico, live in groups that apparently contain a single breeding pair. However, unlike their well-known cooperatively breeding relatives in Florida (Woolfenden and Fitzpatrick 1984), breeders in this population do not allow other group members to feed nestlings, although they attempt to do so (Burt and Peterson 1993). This provides particularly compelling evidence that, in this case, helping behaviour (the attempts to feed nestlings) is not being promoted currently by selection. Somewhat similarly, in green jays of South Texas, young birds remain with their parents throughout their first year, but do not behave as helpers. Following the fledgling of the newly produced juveniles, the one-year-old birds are evicted from the natal territory by their male parent (Gayou 1986); in contrast, green jays in Colombia do act as helpers in a typical fashion (Alvarez 1975).

Third, while the elevated prolactin levels characteristic of non-breeding helpers in Florida scrub-jays and Harris' hawks can be viewed as an adaptive response to nestlings, this proximate mechanism in itself provides no insight into the adaptive significance of helping (i.e. the specific nature of the benefits obtained by the helpers).

Although Jamieson and Craig (1987) and Jamieson (1989*b*) are probably correct about the evolutionary origin of the key behaviour that identifies cooperative breeding, their view is not sufficient to account for the subsequent variations in its evolutionary significance (Ligon and Stacey 1989, 1991; Emlen *et al.* 1991). Even if it is not an adaptation proper, in some species helping behaviour appears to have reliable benefits for the helper at least as often as for the recipients; i.e. it is adaptive. If so, then the non-adaptive interpretation of feeding by helpers proposed by Jamieson and Craig is useful for understanding the evolutionary background of helping at the nest, but it is insufficient to account for the current significance or effects of the phenomenon.

One implication of the view of Jamieson and Craig is that, for some species, helping behaviour may have neither detectable costs nor detectable benefits. (In some cases this seems to be true; see Ligon and Stacey 1989, 1991.) In such species, helping behaviour may be manifested simply because non-dispersal of some matured offspring provides opportunity for them to respond to the stimulus of begging nestlings. In many of the 'opportunistic' cooperative breeders of Australia (Dow 1980; Ford 1989), helpers may occur for no more than this reason. For few of these species is there, at present, good evidence that helpers *per se* increase production of young birds. Ford *et al.* (1988) describe the suite of environmental factors that may contribute to the frequency of natal philopatry in that region.

14.3 Delayed dispersal

It is widely recognized that certain ecological conditions tend to be associated with cooperative breeding. There has been less agreement, however, as to just what those conditions are. The lack of unanimity in accounting for delayed dispersal is due primarily to the fact that the diverse array of cooperatively breeding birds differ in (1) the relative importance of various constraints to dispersal and (2) the significance and kinds of benefits to be obtained by remaining in the natal territory. These are important points, and the thinking of all writers on the subject of cooperative breeding has been coloured by the peculiarities of their own study species. This tendency is understandable, but it should be explicitly recognized by both specialists and non-specialists that the biases or emphases of various students of cooperative breeding have been affected, to some extent, by their own empirical studies. Three cooperatively breeding coraciiform birds in Kenya—the colonial white-fronted bee-eater (Emlen 1990), the colonial and non-territorial pied kingfisher (Reyer 1990), and the non-colonial, highly territorial green woodhoopoe (Ligon and Ligon 1990)—illustrate this point. Data concerning the basic questions of delayed dispersal and the significance of helpers, along with the authors' interpretations of these phenomena, differ conspicuously for these three species.

14.3.1   Constraints to dispersal

Recent reviews of factors proposed to account for delayed dispersal include Brown (1987), Stacey and Ligon (1987, 1991), Walters *et al.* (1990, 1992*a,b*), Emlen (1991, 1994), and Koenig *et al.* (1992) and Mumme (1992*a*). Over the past twenty-plus years, the factor most often said to prevent dispersal of young birds was 'habitat saturation,' a concept introduced in the context of cooperative breeding by Selander (1964). Habitat saturation refers to a shortage of territorial vacancies or openings in breeding habitat.

In their review, Koenig *et al.* (1992) consider several issues related to delayed dispersal. These include relative population density, the fitness differential between early dispersal/breeding and delayed dispersal, the fitness of floaters, the distribution of territory quality, and spatiotemporal environmental variability. Their general conclusion is '... that no one factor by itself causes delayed dispersal and cooperative breeding' (Koenig *et al.* 1992, p. 112). They go on to present a model called the 'delayed-dispersal threshold model,' which they characterize as analogous to the polygyny-threshold model of Verner and Willson (1966) and Orians (1969). Koenig *et al.* focus on the cost and benefit factors that cause a would-be helper either to remain in its natal territory or to leave it. This approach can aid our understanding of what goes into dispersal decisions for individuals of many cooperatively breeding species. It may also be important in helping to understand the intraspecific variation in cooperative breeding once it has evolved (e.g. helping to explain why one population of acorn woodpeckers breeds cooperatively, while another does not; Stacey and Bock 1978). The approach of Koenig *et al.* (1992) in most cases, however, will probably not give us insight into the evolutionary *origins* of cooperative breeding, including delayed dispersal (e.g. see Peterson and Burt 1992).

14.3.2   The benefits-of-philopatry hypothesis

Stacey and Ligon (1987) presented a model and arguments to the effect that ecological constraints, and specifically habitat saturation, might not be the best way to approach the question of the origins of delayed dispersal. These authors suggested that the initial development of cooperative breeding should be viewed as an adaptive response to particular, and sometimes specialized, benefits either of philopatry *per se* or of social living. This perspective has been labelled the 'benefits-of-philopatry' hypothesis. Stacey and Ligon (1987) suggested that the fundamental basis of cooperative breeding for many species is the overriding importance of philopatry, either because it enhances present or future access to an important but limited resource owned by the group or because group living *per se* provides some important benefit. That is, the direct benefit of staying at home, rather than the cost of dispersing, may first have been responsible for natal philopatry. For example, in a number of species individuals may remain in groups even when suitable habitat is clearly available (e.g. Zack and Ligon 1985*a*; Ligon *et al.* 1991; Williams *et al.* 1994). Conversely, young birds of most species disperse irrespective of the presence or absence of unoccupied habitat.

In short, under certain conditions, it is usually specific kinds of benefits of philopatry that promote cooperative breeding, and these vary from species to species (e.g. see red-cockaded woodpecker below). Typically, these critical benefits can be obtained most readily or only in the natal territory. This perspective emphasizes the biological requisites and peculiarities of individual species.

Stacey and Ligon (1991, p. 833) further developed their arguments concerning the benefits-of-philopatry hypothesis and re-emphasized that this model has two critical components: '... (1) crucial benefits obtainable in the home territory and (2) variation among territories in the resource(s) of special importance, such that territory identity is the primary variable affecting individual lifetime reproductive success.' Their model predicts high variance in territory quality among cooperative breeders and lower variance in non-cooperative breeders. Stacey and Ligon illustrated these points by use of studies of two cooperative and one non-cooperative breeding species. Several other studies have also investigated intra-population variation in territory quality and its effects on dispersal patterns (e.g. Zack and Rabenold 1989; Zack 1990; Walters 1990; Walters *et al.* 1992*a,b*; Ligon *et al.* 1991; Koenig *et al.* 1992; Komdeur 1992; Komdeur *et al.* 1995).

A major difference between the explanation of delayed dispersal of Koenig *et al.* (1992) and that of Stacey and Ligon (1987, 1991) is based the levels of analysis issue (Sherman 1988). Although they are useful for helping to understand the factors behind the decision by young birds of a cooperatively breeding species either to disperse or to stay at home, the points and model put forth by Koenig *et al.* (1992) fail to provide an explanation for the *origins* of delayed dispersal. Stacey and Ligon (1987, 1991), on the other hand, do provide an explanation for the origins of delayed dispersal based on particular benefits related to the biology of some species (e.g. red-cockaded woodpecker); for others, however, their explanation, too, is incomplete. Neither Stacey and Ligon (1987, 1991), Koenig *et al.* (1992), nor Emlen (1994) sufficiently considered the phylogenetic component of cooperative breeding. Some specific examples of this are discussed in Section 7.4 and below.

In short, the essence of the differences in the perspectives of Stacey and Ligon (1987, 1991) and those of Koenig *et al.* (1992) is that the former authors suggest that a critical benefit related to the biology of the species in question can be obtained by remaining indefinitely in the natal territory (especially a high-quality one) and this benefit is critical to the evolution and maintenance of delayed dispersal. In contrast, Koenig *et al.* (1992) argue that extrinsic factors constrain dispersal. These two perspectives address different steps in the evolution and maintenance of cooperative breeding, and therefore are not really opposing explanations of the same issues. For example, cavities in living pines, a costly self-constructed and therefore extremely valuable resource that is passed from generation to generation, provides the benefit of philopatry for the red-cockaded woodpecker (discussed in the following section). The woodpeckers are not constrained from emigrating, and, in fact, some young males and almost all young females usually do so. For young males, however, it is access to a cavity roost that disposes some of them to remain in their parents' territory (Walters 1990). In contrast, the congeneric downy and hairy woodpeckers, which do not breed cooperatively, excavate cavities relatively quickly and easily in dead wood, and natal philopatry is unknown. Dead wood is relatively common in the habitats they primarily occupy, thus for these species there are no major or unusual benefits to philopatry.

It has been argued that the ecological constraints models and the benefits-of-philopatry model are simply two sides of the same coin (Emlen 1994), and that whether one wishes to emphasize costs or benefits of natal philopatry is a matter of taste. It is correct, of course, that dispersal decisions must be based on both costs and benefits. This is true for each and every 'decision' any animal makes. However, in their original forms the two models make very different predictions (see Table 1 in Stacey and Ligon 1987), thus the models themselves must also be somewhat different.

14.3.3   Some empirical studies of delayed dispersal

Red-cockaded woodpecker
The red-cockaded woodpecker provides an unusually clear example of a specific benefit of natal philopatry. These birds depend on a single, self-constructed, critical resource, namely cavities excavated in living pine trees and used for roosting and nesting (Ligon 1970; Walters 1990). Cavities require much time to excavate and, in good habitat, each occupied territory typically contains from one to several cavities. The critical variation in territory quality, which is related to factors associated with the presence, number, and quality of cavities, appears to be the basis for delayed dispersal by many of the young males produced in a given territory. In contrast, nearly all young females disperse from their natal territories. With regard to the benefits-of-philopatry model, Walters *et al.* (1992*a*) interpret the philopatry of these birds in the same way, but use the term 'critical resource model'. Walters *et al.* (1992*a*)

artificially created suitable cavities that attracted immigrant woodpeckers to settle in areas where previously there had been none. The birds promptly established territories centered on these cavity trees, experimentally confirming the importance of cavities to this species. The founders came from some distance, and they demonstrate that individual red-cockades are not constrained from dispersing. A safe roost site is the initial benefit of philopatry for young male red-cockaded woodpeckers.

Red-cockaded woodpeckers are especially instructive with regard to the question of the origin of the delayed dispersal aspect of cooperative breeding, for the following reason. Because this species is the only member of its large genus to exhibit delayed dispersal, and because the benefits associated with natal philopatry are well understood, we can confidently assume that in this case delayed dispersal and cooperative breeding are derived traits. The environmental factor that set the stage for delayed dispersal/cooperative breeding by ancestral red-cockadeds was occupancy of open, fire-maintained pine forest, where, prior to fire suppression by humans, dead trees were rare. Occupying this habitat required the excavation of cavities in living trees, which are fire-resistant. Such cavities are extremely costly to construct and are passed from one generation to the next for as long as the tree remains alive. The benefits of access to a roost cavity in a living tree made delayed dispersal an attractive option to young male red-cockadeds. Apparently this is not an option available to young females, nearly all of which disperse at a few months of age and become floaters. The fact that a large proportion of one sex tends to be philopatric, while the other disperses, demonstrates that, for red-cockaded woodpeckers, benefits of philopatry (for males), rather than costs of dispersing (accepted by females), provide a better explanation for natal philopatry in this species than constraints to dispersal.

Seychelles warbler

In considering the merits of the two major categories of explanations for delayed dispersal (constraints versus benefits and variation in territory quality), the studies by Komdeur and colleagues on the cooperatively breeding Seychelles warbler (Komdeur 1992; Komdeur *et al.* 1995) are illuminating. Komdeur (1992) found that both habitat saturation and variation in territory quality were important in explaining dispersal patterns of these warblers. When a number of Seychelles warblers were first released on an island that had none of the birds, all of them bred; i.e. none served as helpers (Komdeur 1992; Komdeur *et al.* 1995). This seemed to support the habitat saturation or extrinsic-factor interpretation of delayed dispersal on the 'home' island. As space for high-quality territories filled, however, birds often opted to remain on them as helpers rather than to occupy available lower-quality territories as breeders; i.e. variation in territory quality, rather than an absence of occupiable territories, seemed to influence their dispersal decisions, in agreement with the benefits-of-philopatry model.

Komdeur *et al.* (1995) suggested that the benefits-of-philopatry and the marginal habitat models might not really be different hypotheses. They write:

> Our study demonstrates clearly that the causes of variation in dispersal tendencies at one level may not account for variation at other levels; the primary cause of within-population variation in delayed dispersal is variation in territory quality, but the primary cause of between-population variation in delayed dispersal is the degree of habitat saturation. (Komdeur *et al.* 1995, p. 707).

Thus, while it appears that the two models should not be viewed as mutually exclusive alternatives, they are different hypotheses. They are not, however, directly competing ones. Instead, they address different aspects of the maintenance of cooperative breeding; i.e. they focus on different questions or issues.

The studies of Komdeur *et al.* (1995) demonstrate conditions under which Seychelles warblers exhibit cooperative breeding. As illustrative as they are, these studies do not, however, shed light on the issue of the *origin* of cooperative breeding in this species. As Komdeur *et al.* (1995) point out, this species' close relatives in similar environments show no tendency to breed cooperatively under any circumstances. Why the Seychelles warbler, alone among the members of its genus, has evolved this inclination is unknown. However, a congener, the Henderson reed-warbler, which is also confined to a small island, often occurs in trios composed of either two unrelated adult males and a female or two unrelated females and a male. Parentage and parental duties are shared (Brooke and Hartley 1995). This situation represents an alternative pathway to group-living, and may suggest a mechanism for the evolution of cooperative breeding proper, as seen in the Seychelles warbler.

Superb fairy-wren
Thanks to Ian Rowley (1965), the superb fairy-wren has a special place in the history of the field of avian cooperative breeding. Two more recent studies attempted to elucidate the basis of delayed dispersal in this species. To test for habitat limitation in superb fairy-wrens, Pruett-Jones and Lewis (1990) conducted removal experiments at a densely-populated study site at Canberra, in southern Australia. These authors found that male helpers almost invariably emigrated when a vacant territory with a female occupant was experimentally created nearby. This led them to conclude that habitat saturation, together with a shortage of females, accounted for the philopatry of male fairy-wrens. Both of these results generally agreed with those of Rowley (1965).

In another study, at a low-density site near Armidale, NSW, where critical habitat was patchily rather than uniformly and continuously distributed, Ligon *et al.* (1991) tested for a relationship between territory quality and number of helpers. All helpers from 14 territories holding more than two birds were removed. Later, at the end of the long breeding season, the productivity of those territories was compared with 20 territories originally holding only a single male and female. These authors also removed females from a number of territories, early in the breeding season, to test for the shortage of females, a potential constraint to male dispersal first proposed by Rowley (1965) and supported by the study of Pruett-Jones and Lewis (1990). Finally, because the territories of fairy-wrens at their study site were centered on thickets, usually blackberry brambles, Ligon *et al.* were able to create artificial, high-quality territories in order to investigate the issue of saturation of suitable habitat.

With regard to delayed dispersal, Ligon *et al.* (1991) obtained three interesting results. First, the territories where all but two individuals were removed produced significantly more offspring over the course of the breeding season than did the control territories, where neither of the two original birds was removed. Second, for every female experimentally removed, a replacement appeared. This usually occurred quickly, demonstrating that additional, floating females are present, at least early in the breeding season, when male helpers might be expected to be most interested in setting up their own territories. Third, unoccupied but acceptable sites for territories were available. For example, specific territories might become occupied, then unoccupied, and then reoccupied during the course of the extended breeding season. This, plus the fact that experimentally created territories remained unoccupied, indicated that at this site habitat acceptable to fairy-wrens was not limited. This result suggests that territories holding helpers were superior in quality. In short, despite the availability of unoccupied territories of acceptable quality and the availability of non-breeding females, male helpers often opted to remain in territories that appeared on the basis of vegetational structure to be of high quality. Thus, in one study (Pruett-Jones and Lewis 1990) the habitat saturation and female constraint interpretations of delayed dispersal were supported, while another (Ligon *et al.* 1991) led to different conclusions. These differences and the resulting interpretations are probably due to conspicuous habitat differences at the two study sites and the responses of fairy-wrens to those differences.

A key point is that at both sites—one in which the habitat apparently was saturated and one in which it was not—helpers were present. Thus, although habitat saturation and a shortage of females promote delayed dispersal at the Canberra site, the absence of habitat saturation and the availability of non-breeding females at the Armidale site, along with delayed dispersal, argues that these factors can provide neither an explanation for the origin of co-operative breeding in superb fairy-wrens nor even a general explanation for the maintenance of delayed dispersal (i.e. one that is appropriate for all populations of this species). This does not deny, however, that both constraints—habitat saturation and a shortage of females—exist in some populations. The models of Koenig *et al.* (1992) should help us to understand the inter-population differences in dispersal strategies and thus the maintenance of cooperative breeding in populations of superb fairy-wrens with very different spatial and demographic parameters. However, explanations for the origins of cooperative breeding in the fairy-wrens probably lie in the evolutionary history of this group, rather than in current ecological conditions (see Section 7.4).

Green woodhoopoe

In the green woodhoopoe (Fig. 14.2), the critical factor determining the quality of a given territory is the number of roost cavities and their degree of security from predators. In Kenya, at high elevations, where night-time temperatures are low, woodhoopoes require cavities in which to roost (Ligon and Ligon 1978, 1990), apparently in order to maintain a positive energy balance (see Ligon *et al.* 1988; du Plessis and Williams 1994). Occupied territories vary greatly in productivity, with a minority of territories being net producers of woodhoopoes, many of which eventually disperse, while a majority of occupied territories are sinks, with more birds dying on them than are produced. Such territories continue to exist (to be occupied) only as a result of immigrants from the more productive territories (Ligon and Ligon 1988).

Fig. 14.2 In the socially monogamous and cooperatively breeding green wood-hoopoe the female is fed extensively by her mate and helpers prior to egg-laying. Reprinted with permission from Dover Publications, Inc.

Because cavity roosting is imperative to green woodhoopoes, and because suitable cavities are scarce, dispersal by both sexes is conservative (Ligon and Ligon 1988). For this species, there is thus an internal constraint to dispersal based on physiological factors. Stated in another way, woodhoopoes obtain an intrinsic benefit, a warm and more-or-less safe place to roost, by remaining in their natal territories until a vacancy occurs in a nearby territory with roost sites. In contrast, at some sites in South Africa, roost cavities are not scarce and dispersal is frequent. As a result, in these areas most social units consist of only a simple pair (du Plessis 1989).

These data illustrate that delayed dispersal in green woodhoopoes is due, at least in part, to the presence of a critical resource (roost cavities) in the natal territory and the rarity of that resource outside the territory. There is a critical benefit to natal philopatry as a result of the importance of a physiological limitation (constraint) that requires roost cavities. With regard to the issue of delayed dispersal, the important point is that an essential benefit is located within the natal territory, and it is variation among territories in the amount and quality of this benefit that promotes variation in dispersal strategies.

Finally, as in the fairy-wrens, there is the issue of phylogenetic history. At least four of the five members of the woodhoopoe genus *Phoeniculus* live in groups, and probably all are cooperative breeders.

14.3.4   Delayed dispersal: general conclusions

The apparent disagreements between the different models of delayed dispersal are based largely on what might be viewed as differences in scale or 'level of analysis' (Sherman 1988). The approach of Koenig *et al.* (1992), which emphasizes constraints to dispersal, is useful for understanding intraspecific differences and differences among closely related species of cooperative breeders in dispersal patterns. It best addresses what Zack (1995) has termed the 'tactics of dispersal' for a young non-breeder of a typical territorial cooperatively breeding species. It does not really address the *origin* of cooperative breeding from a preceding state of non-cooperative breeding for any species.

14.4   Cooperative breeding and cooperative polyandry

In terms of similarities to other mating systems, cooperative breeding appears to be most like cooperative polyandry (Oring 1986). To distinguish between these two systems requires knowledge of the parentage of chicks; accurate information of this sort became available with the advent of DNA fingerprinting techniques. It is now known that in some instances both phenomena occur within the same social unit of birds. That is, within one group some individuals may be non-breeding helpers, while others may be cryptic breeders. Oring (1986) lists six species thought at the time of his review of polyandry to be cooperatively polyandrous. Cooperative polyandry has since been confirmed, via DNA fingerprinting, for three of them. However, co-operative polyandry may in fact not be an important aspect of the social systems of two others. Instead, these may be typical cooperative breeders. The book *Cooperative breeding in birds* (Stacey and Koenig 1990) provides summary chapters of 17 species of avian cooperative breeders. Only a few years ago nearly all of these species were assumed to be both socially and genetically monogamous. Recently, this view has changed dramatically; in several cases male helpers are now known to actually 'help' in more ways than one—by siring of some of the offspring of the single female breeder. Thus, it is now recognized that cooperative polyandry is an important part of the story for a number of cooperative breeders.

Hartley and Davies (1994) determined that in cooperatively breeding birds (here including cooperatively polyandrous species), the number of helpers is greater, on average, when the helpers are close relatives of the brood than when the helpers are males that may be the parents of some of the chicks. They suggested a causal relationship between mating/social system and group size, namely that group sizes are smaller in cooperatively polyandrous species because the helper's relatedness to the brood decreases with an increasing number of helpers; i.e. with more potential fathers in the group, the probability for a given individual of siring any offspring decreases. In addition, with more potential sires, the proportion of young sired by any individual will decrease. This explanation ignores that fact that in many cooperatively breeding species, most helpers are the offspring of one or both of the mated pair. Inbreeding at the level of parent and offspring is not known to occur in any cooperatively breeding species, thus group size *per se* in typical cooperative breeders prob-ably is unrelated to likelihood of extra-pair matings. If we view retention of young within their natal group as a form of extended parental care, irrespective of their significance as helpers, then mean group size for a given species may have implications unrelated to the issue of helping. In typical cooperatively breeding species, group size often corresponds closely to the productivity of the previous breeding season; i.e. the more young birds produced in year x, the more helpers will be present in year x+1. Thus, for a given species, year-to-year variation in mean group size may be great.

Another obstacle to the argument of Hartley and Davies (1994) is the data they used. Several of the cooperatively breeding species that they categorized as monogamous are now known to exhibit cooperative polyandry or other non-monogamous mating systems. The diversity of strategies being followed by individuals of 'cooperatively breeding' species is far more complex than indi-cated by Hartley and Davies (1994). The suggestions of Hartley and Davies (1994) were probably influenced by their studies of the dunnock, which is not a cooperative breeder by any usual definition (i.e. there are no non-breeding helpers-at-the-nest). These authors showed that female dunnocks usually chose not to mate with more than two males, even when additional ones were available, leading them to conclude that females would not gain any additional parental help with a third male, and that, moreover, three males would provide extra problems for females in the form of sexual harassment. Thus, for this species it does appear that females set a low upper limit (two) on the number of males they will encourage to become mates, and that a cost–benefit framework from the female's perspective can probably account for the number of males in typical dunnock breeding unit. As described earlier, male dunnocks do not openly share copulations with the female; mixed paternity is brought about by female manipulation of males. In some other species, however, two or more males do openly copulate with the female, and in these the obvious key evolutionary question is, 'Why do two or more males share reproductive opportunities?' This is not predicted by the long-standing view of intense competition among males for mates and matings (Trivers 1972). Because males of single (monogamous) pairs are almost invariably more successful than males in polyandrous groups, as measured by the number of young per male, increased reproductive success for each individual male is probably not a complete explanation for the mate-sharing that occurs among males in cooperatively polyandrous situations (see also Section 13.5).

14.5   Phylogenetic history

In addition to delayed dispersal and the helper phenomenon, there is another major aspect of cooperative breeding that until fairly recently was not appreciated. For most of the period of the modern study of cooperative breeding (*c.* 1970–present), the role of phylogenetic history in the overall picture was largely unrecognized by most workers, who focused instead on adaptive explanations of delayed dispersal and helping behaviour (see Section 7.4). Current conditions and the animals' responses to them were used to explain, at least implicitly, both the maintenance and the evolutionary origins of cooperative breeding, which were treated typically as a single issue.

Because avian cooperative breeding is so widespread both geographically and taxonomically, it has long been assumed to reflect adaptive responses to current ecological conditions (e.g. Dow 1980). The existence of cooperative and non-cooperative populations of the same species (e.g. scrub-jays, green jay), together with populations of many species that contain social units consisting of both simple pairs and pairs plus helpers (e.g. Florida scrub-jay), does appear to support the assumption that this is a flexible social system that can readily appear under certain (albeit restricted and uncommon) ecological conditions. However, Russell (1989) and Edwards and Naeem (1993) demonstrate that in many cases cooperative breeding has a strong phylogenetic component. With regard to the phylogenetic distribution of cooperative breeding and the sequence of evolutionary events culminating in the species present today, an extremely interesting pattern appears to be emerging. In a variety of avian lineages in which cooperative breeding occurs along with close relatives that do not breed cooperatively, it appears that cooperative breeding is the ancestral, rather than the 'derived', condition (Table 14.1). That is, it appears that the sequence of evolutionary events usually assumed—contemporary cooperatively breeding species having evolved from a non-cooperative predecessor—may, in a number of cases, be backwards.

What might be the significance of this recurring pattern? First, several of these groups (Table 14.1) may be derived from the 'old endemic' lineage that originated in Australia (Sibley and Ahlquist 1990). This may be a coincidence or it may suggest that cooperative breeding evolved very early in the history of this major passerine radiation. It also indicates that cooperative breeding not only has a strong historical component, but that it also may be rather resistant to change. The fact that, in some cases, most species of a genus are cooperative, with only one or few departing from that social pattern (e.g. New World jays, e.g. Genus *Aphelocoma*, *Campylorhychus* wrens, *Pomatostomus* babblers), while in other cases, most species are typical 'pair-only,' with only one cooperatively breeding species (the red-cockaded woodpecker within *Picoides*), demonstrates that, among living species, cooperative breeding may be either primitive or derived, depending on the specific case. As an example, most members of the wren Genus *Campylorhynchus* are cooperative breeders. In fact, only one of 11 species, the cactus wren, is known not to breed cooperatively. Factors associated with the kind of habitat occupied apparently have favoured loss of cooperative breeding in this one species (Farley 1995).

Table 14.1 Some genera in which cooperative breeding may be the primitive state

Genus (Family)	No. of species	No. of cooperative breeders	References
Merops (Meropidae)	21	6	Edwards and Naeem 1993
Phoeniculus (Phoeniculidae)	5	4	J. D. Ligon unpubl. ms.
Climacteris (Climacteridae)	6	4	Edwards and Naeem 1993
Malurus (Maluridae)	14	14	Edwards and Naeem 1993
Pomatostomus (Pomatostomidae)	5	5	Edwards and Naeem 1993
Lanius (Laniidae)[1]	26	2	Zack 1995
Aphelocoma (Corvidae)	5[2]	3	Peterson and Burt 1992
Prionops (Prionopidae)	8	8	Brown 1987; Zimmerman *et al.* 1996
Spreo (Sturnidae)	6	4	Edwards and Naeem 1993
Campylorhynchus (Troglodytidae)	11	8–10	Edwards and Naeem 1993; Farley 1995
Turdoides (Sylviidae)	26	25	Gaston 1977

[1]See Figure 14.3.
[2]Two scrub-jay populations recently elevated to species rank do not breed cooperatively, while the Florida scrub-jay does so.

Laniid shrikes are especially interesting in this regard in that while co-operative breeding seems to be the ancestral or primitive state, a large majority of contemporary species are not cooperative breeders (see Section 14.5.1.2).

14.5.1   Some examples of the role of the evolutionary history of cooperative breeding

14.5.1.1   New World jays

The New World jays are a distinctive sub-group of the traditional Family Corvidae. Most species are tropical or sub-tropical in distribution and exhibit cooperative breeding. One species of the Genus *Aphelocoma*, the Mexican jay, is a cooperative breeder throughout its geographically extensive, but eco-logically restricted, range. Another, the unicoloured jay, is likewise thought to be a cooperative breeder throughout its disjunct and comparatively small range. The third lineage, the 'scrub-jay' group, contains both cooperative breeders (Florida scrub-jay, plus a semi-cooperative population in southern Mexico) and non-cooperative breeders (scrub-jays of the western United States and northern Mexico).

The Genus *Aphelocoma* has been studied by Peterson and Burt (1992) with regard to the relationship between phylogenetic history (based on allozyme frequencies) and cooperative breeding. Their analyses produced some interest-ing conclusions, which are consistent with those of Edwards and Naeem (1993). First, they concluded that cooperative breeding predated the separation of *Aphelocoma* from the rest of the New World jays. Second, this finding led them to suggest that, for this group, the more useful question is, 'why has cooperative breeding been lost in scrub-jays of western North America?', rather than, 'Why are members of *Aphelocoma*, plus most other New World jays, cooperative breeders?' As Peterson and Burt point out, the evolutionary and ecological forces responsible for the transition from cooperative to non-cooperative breeding in the scrub-jay lineage may still be identifiable. The answer to the second question, in contrast, probably cannot be obtained by study of the ecology of living species of *Aphelocoma*.

With regard to the direction (polarity) of evolutionary change in helping behaviour in the Genus *Aphelocoma*, Brown and Li (1995) have recently questioned the conclusions of Peterson and Burt (1992). Based on combined allozyme and behavioural data, Brown and Li (1995) report that evolutionary change is greater for scrub-jays than for Mexican jays. They suggest that, unless evolutionary rates are considerably faster in Mexican jays, differentia-tion must have started earlier in the scrub-jays than in the other species of *Aphelocoma*. These authors then conclude that, 'If true, the Mexican Jay would have to be derived from the Scrub Jay and not vice versa' (Brown and Li 1995, p. 468).

Thus, we have two opposing views of the phylogenetic histories and evolutionary progression of helping behaviour within the Genus *Aphelocoma*. Peterson and Burt (1992) suggest that scrub-jays are descended from Mexican jays; Brown and Li (1995), on the other hand, argue just the opposite. The direction of change over evolutionary time is often not easy to ascertain. However, if we are clear on our starting point, and if we recognize that transitions from one direction to the other may occur repeatedly within a phylogenetic lineage (e.g. Szekely and Reynolds 1995; see Chapter 9), some progress can be made. First, the DNA–hybridization studies of Sibley and Ahlquist (1990) indicate that the Family Corvidae, including the New World jays, are derived from the great Australian radiation, certain lineages of which are virtually characterized by the prevalence of cooperatively breeding species (Russell 1989; Edwards and Naeem 1993). Thus, the ancestor of the contemporary New World jays may have brought cooperative breeding with them to the Western Hemisphere. This is consistent with the widespread occurrence of cooperative breeding (in one form or another) in this group.

Another hint at the evolutionary history of helping in scrub-jays is provided by the observation that young scrub-jays from the south-western United States sometimes attempt to feed begging juveniles of their own species or other jay species (Ligon 1985). This may suggest that the propensity to behave as a helper exists in western scrub-jays, for which helpers have never been recorded. If this speculative interpretation is correct, it suggests that these scrub-jays are descended from ancestors that exhibited helping behaviour. It also suggests that under appropriate environmental conditions helping behaviour could (re-)appear in western scrub-jays. The peculiar behaviour of scrub-jays in Oaxaca, Mexico, where non-breeders attempt to feed nestlings, but are prevented from doing so (Burt and Peterson 1993), also supports the suggestion that the direction of change in scrub-jays has been from cooperative to non-cooperative breeding.

With the differentiation of the genera and species that collectively make up the New World jay group, helping behaviour could have been lost and subsequently regained, or not regained, within each lineage recognized as a genus (e.g. *Aphelocoma*). For example, the Florida scrub-jay is one of the best-known examples of a cooperatively breeding species. However, about half of the reproductive units are composed of simple pairs (Woolfenden and Fitzpatrick 1984), which are reproductively effective, although somewhat less so, on average, than larger groups. Thus, the future evolutionary trajectory of cooperative breeding in this species cannot be confidently predicted. A change in climatic conditions leading to decreased environmental stability—greater aridity, for example, that made it impossible for territories to support a group year-round—could favour pairs-only units, with the eventual loss of cooperative breeding.

To conclude, an assessment of New World jays suggests that cooperative breeding may be primitive for the group as a whole. This is probably also true for the Genus *Aphelocoma* (Peterson and Burt 1992).

14.5.1.2   African fiscal shrikes

Another example of the importance of phylogenetic history in cooperative breeding systems comes from studies of fiscal shrikes in Kenya. A recent reassessment by Steve Zack (1995) of earlier studies on cooperative breeding in two species of *Lanius* fiscals illustrates the changes in 'conventional wisdom' that have occurred as a result of the inclusion of phylogenetic history into thinking on cooperative breeding. Briefly, Zack and Ligon (1985*a,b*) investigated the ecology and demography of two sympatric species of *Lanius* in an effort to identify the ecological factors favouring cooperative breeding in one of them, the grey-backed fiscal (Fig. 14.3), but not the other, the common fiscal. The common fiscal is a 'pairs-only' species, as are most of the other species in this large and widespread genus. Zack and Ligon (1985*a,b*) found that, in their study area, grey-backs occupied only acacia woodland. Within this habitat, higher perennial shrub cover was associated with more insect prey, which supported the idea that within the grey-back's habitat differences existed in quality of territories. In agreement with this, groups occupying territories containing more shrub cover were consistently larger and more productive than other groups. Moreover, in disagreement with the constraints models, low-cover areas were periodically occupied and vacated by pairs and small groups throughout the study. Thus, it appeared that extended natal philopatry

Fig. 14.3   Three cooperatively breeding grey-backed fiscals displaying cooperatively, or 'rallying', during a territorial dispute. Group defence of the territory is a conspicuous aspect of group-living in many cooperatively breeding species.

in this species resulted 'more from tactical restraint ... than ... constraints on dispersal ...' (Zack 1995, p. 35).

Zack and Ligon (1985*a,b*) assumed that cooperative breeding in the grey-backs was a derived trait for two reasons: (1) only two of the 26 members of the genus *Lanius* are cooperative breeders; and (2) these authors assumed that cooperative breeding in this species reflects a complex social system that must have evolved from a simpler, non-cooperative ancestral condition. Zack and Ligon erred on two counts. First, they did not clearly distinguish between origin and maintenance of cooperative breeding. Second, in their comparison of grey-backs and common fiscals, these authors failed to consider another species of *Lanius*, the long-tailed fiscal, which is closely related to grey-backs. The long-tailed fiscal (1) is a cooperative breeder and (2) occupies open country that is conspicuously different from the kind of habitat occupied by grey-backs. Long-tailed fiscals and common fiscals are often sympatric, and, as Zack (1995) points out, had the original comparative study included common fiscals and long-tailed fiscals, rather than grey-backed fiscals, the apparent relationship between habitat and cooperative breeding that Zack and Ligon described would not have existed.

Zack (1995) has now improved upon the original interpretation. He points out that grey-backed fiscals and long-tailed fiscals are sister taxa that are morphologically intermediate between the rest of the *Lanius* species (which do not breed cooperatively) and the shrike genera *Eurocephalus* and *Corvinella* (consisting of four species, all of which are cooperative breeders). The latter two

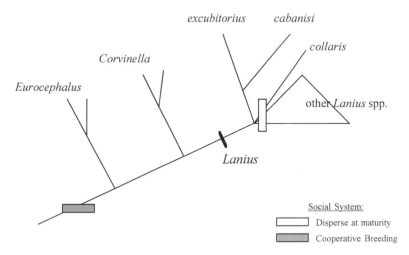

Fig. 14.4 A phylogeny of the Laniidae which suggests that cooperative breeding is the primitive state in this group, and that dispersal at maturity with pairs-only breeding is the derived condition in most members of the Genus *Lanius*. Reprinted with permission from Zack, S. (1995). Cooperative breeding in *Lanius* shrikes III. *Proceedings Western Foundation Vertebrate Zoology*, **6**, 34–8.

genera are considered by Sibley and Ahlquist (1990) to be ancestral to *Lanius*. Based on data from Sibley and Ahlquist (1990), Zack provides a phylogeny (Fig. 14.4) which suggests that cooperative breeding in this group preceded the evolution of non-cooperative breeding in all of the remaining extant species of *Lanius*.

In summary, at present there appears to be no good reason to assume that the habitats currently occupied by the two species of *Eurocephalus*, the two species of *Corvinella*, and the grey-backed and long-tailed fiscals (all endemic to sub-Saharan Africa) have any causal connection with the evolutionary origin of their cooperative breeding social systems.

14.6   Comparisons of the history of cooperative breeding in two woodpeckers

In this section I illustrate the importance of the historical perspective in attempts to unravel the evolutionary history of cooperative breeding of two species of woodpeckers. In one of these, the origin and maintenance of the social system may be accounted for by considering current adaptive significance, whereas in the other, origin and current adaptive significance (factors maintaining cooperative breeding) may not be closely related. A comparison of the acorn and red-cockaded woodpeckers (discussed above) illustrates how different systems labelled as cooperative breeding can be, even among related species. While the red-cockaded woodpecker is socially and genetically monogamous, as are some other cooperative breeders (see Table 12.1), the acorn woodpecker is polygynandrous (see Table 13.1).

In addition to the differences in their mating systems, comparison of each of these species with its congeners provides some insights into the possible role of phylogenetic history in the occurrence of cooperative breeding in these birds. (1) The genus *Melanerpes*, to which the acorn woodpecker belongs, is a large and widespread neotropical group of about 22 species. Many of these are cooperative breeders; thus, it is probable that the common ancestor of today's *Melanerpes* species was also a cooperative breeder. (2) The acorn woodpecker occurs from Colombia northward to the western and south-western United States and probably has expanded its range northward out of the neotropics comparatively recently. (3) This woodpecker is apparently a cooperative breeder throughout its extensive range. (4) Although in the United States the conspicuous behaviour of gathering and caching acorns in especially constructed 'granary' trees is closely associated with social living and cooperative breeding (Stacey and Koenig 1984; Koenig and Mumme 1987; Koenig and Stacey 1990), this is not the case in Colombia, where acorn woodpeckers occur and breed cooperatively in areas lacking oaks (Kattan 1988). Thus, because cooperative breeding is widespread in melanerpine woodpeckers, most of which are tropical in distribution, and because acorn woodpeckers breed cooperatively in South America in areas without oaks, it would probably be erroneous to attempt to account for the *origin* of the cooperative social system of North American populations of this species solely on the basis of the bird's behaviour and ecology in this region.

In contrast, a comparison between the red-cockaded woodpecker and its congeners leads to a very different conclusion concerning the evolutionary history of cooperative breeding in this species. First, among members of the large genus *Picoides*, cooperative breeding appears to be unique to this bird; i.e. it is a derived condition within the genus. Walters' (1990) reference to the red-cockaded woodpecker as a 'primitive' cooperative breeder reflects the relative recency of the evolution of cooperative breeding in this species. Second, as described previously, the ecology of red-cockaded woodpeckers is highly distinctive as compared to its congeners, particularly with regard to their dependence on living pines as cavity sites (Ligon 1970, Walters 1990). As discussed earlier, this dependence on a valuable and limited resource is consistent with ecologically-based explanations for the origin of cooperative breeding in this species (Walters *et al.* 1992*b*).

In short, although current adaptive explanations may account adequately for both the origin and maintenance of cooperative breeding in red-cockaded woodpeckers, historical factors are likely to be of major importance in any attempt to explain what appears superficially to be a similar social system in the acorn woodpecker. Current ecological factors may help to account for the maintenance of cooperative breeding in different parts of the acorn woodpecker's range, but for this species they are unlikely to explain fully the evolutionary origin of this form of sociality.

14.7   Communal breeding

Although the terms 'cooperative breeding' and 'communal breeding' have often been used interchangeably, it is worthwhile to distinguish between them. As emphasized at the beginning of this chapter, cooperative breeding is characterized by the presence and contributions of individuals that are not breeders: the 'helpers'. In contrast, communal breeding is identified by the presence of two or more mated pairs of birds, the females of which lay their eggs in a single nest. Mature, non-breeding nest helpers may be present or absent.

Communal breeding is characteristic of one subfamily of the Family Cuculidae, the neotropical Crotophaginae, which consists of the three species of anis and the guira cuckoo. Groove-billed anis live in social units composed of up to several mature pairs, which '... form conspicuous monogamous pair bonds within their groups' (Vehrencamp *et al.* 1986). Intragroup competition in the groove-billed ani takes the form of egg-tossing and egg-burying (Vehrencamp 1977). This is also true of the guira cuckoo, which, in addition, may practice infanticide (Macedo 1992). Quinn *et al.* (1994) employed

DNA fingerprinting to ascertain parentage of broods of young guira cuckoos. Although samples of relevant birds were incomplete, their data provided no evidence that genetic monogamy existed in the groups they studied. In some cases, nests contained half-siblings, indicating that polygamous (either polygynous or polyandrous) matings were of regular occurrence. Overall, it appears likely that the predominant mating system of the guira cuckoo may be polygynandry.

In view of the fact that in communal groups of groove-billed anis the dominant male conducts all nocturnal incubation, thereby doubling its chances of breeding-season mortality (Vehrencamp 1978), it will be of interest to learn the genetic parentage of broods of young anis.

14.8   Conclusions and summary

The phenomenon of cooperative breeding involves study of three major issues: nest helpers, delayed dispersal, and phylogenetic history. The significance of non-breeding helpers provided the initial impetus for study of cooperative breeding, largely because they appeared to pose a challenge to Darwin's theory of evolution. The effects of helpers on reproduction vary from species to species. In some cases, helpers increase the production of young birds, while in others no correlation has been detected between the presence or absence of helpers, or the number of helpers, and number of young birds produced.

Delayed dispersal is the behavioural factor that creates the non-breeding helpers, most of which are young birds that remain for an indefinite period in their natal territory. For many years it was thought that delayed dispersal was imposed on younger individuals by some sort of ecological constraint. The most frequently suggested constraint was 'habitat saturation', the scarcity or absence of unoccupied space suitable for establishing a territory and breeding. However, all avian species must contend with environmental constraints, yet few respond by remaining in their natal territory. An alternative suggestion is that under certain, uncommon conditions, there are critical resources that a young bird can obtain best or only in its natal territory, and that this, rather than a constraint on dispersal, such as an absence of occupiable space, has promoted natal philopatry.

It is becoming increasingly apparent that cooperative breeding provides a good example of the role of phylogenetic history in the complex social behaviour of contemporary species. For most of the modern history of the study of cooperative breeding, little or no thought was given to the historical aspects of the phenomenon. For more than twenty-five years students of cooperative breeding in birds have sought to identify its common adaptive theme, but with limited success. This quest has been hampered by the lack of recognition, until recently, of phylogenetic factors as important components of cooperative breeding. These components occur at two different levels. First, the behaviour that identifies cooperative breeding—the feeding of chicks by non-breeding helpers—is based on a general trait characteristic of altricial (and some precocial) bird species; it is not a special adaptation evolved anew in every cooperative breeder. Second, the distribution of cooperative breeding across a diverse array of species cannot be accounted for by broad ecological similarities among cooperative breeders.

A new insight is that cooperative breeding and its taxonomic, and even geographic, distribution appears to have been strongly influenced by phylogenetic history (Russell 1989; Edwards and Naeem 1993). The puzzle concerning the unusually high frequency of cooperative breeding in Australia largely disappears when historical factors are taken into account. It now appears that cooperative breeding was associated early on with the major passerine radiation on that continent (the Parvorder Corvida of Sibley and Ahlquist 1985). This may account, to a large extent, for the frequency of cooperative breeding in Australia and possibly elsewhere (e.g. the New World jays). In short, one reason that cooperative breeding may be so relatively common in Australia is simply because the common ancestors of many closely related Australian species may have practised it (e.g. fairy-wrens), with the trait being transmitted over evolutionary time to the species living today.

This suggestion does not deny the importance of current ecological conditions in Australia and elsewhere for maintaining cooperative breeding in the living descendants of early cooperative breeders. In some cases, the origin, and, perhaps in all cases, the maintenance of cooperative breeding is related to particular environmental circumstances. Clearly, a consideration of both current ecological and social factors, as well as phylogenetic history, is required to produce a comprehensive understanding of this social-breeding system.

15 Lek mating systems

15.1 Introduction

Lekking behaviour is one of the most interesting issues in sociobiology. Lek mating systems are characterized by clusters of males, often at a 'traditional' site, which females visit for mating. In contrast to the territorial behaviour of most male birds during the breeding season, males of typical lekking species come together and conduct their courtship activities in close proximity. Thus, the central question posed by lek mating systems is, why do males congregate at a single site? Lekking behaviour is also interesting because it is thought that 'The potential for sexual selection reaches full expression in species with lek mating systems' (Trail 1990, p. 1837). The clustering of males, which leads to high variance in male reproductive success and thus to intense sexual selection, is the reason that lek systems have been so attractive to students of sexual selection.

Recently, the books by Johnsgard (1994) and Höglund and Alatalo (1995) review and analyse these kinds of mating systems. Höglund and Alatalo recognize leks by the following limited criteria: 'Aggregated male display that females attend primarily for the purpose of fertilization'. Following Bradbury (1981), they discuss some additional aspects of leks. (1) Males of lekking species provide no parental investment in offspring beyond the sperm they contribute; i.e. there is no paternal care of young of any sort. (2) Males congregate at a specific site that is not used for other purposes, such as foraging or roosting. (3) Similarly, there are no resources for females at the display site (e.g. females do not nest at the sites). Thus, females come to the lek site to obtain only one 'resource', namely the sperm of one of more males. (4) Finally, females can choose among the males to select a preferred sperm donor. In a similar vein, Oring (1982) recommends that the term 'lek' be reserved for situations in which a group of males defends small, closely positioned courts used only for mating (i.e. the courts contain few, if any, resources other than the males themselves), females visit the arena only for purposes of choosing a mate and mating, and females subsequently nest elsewhere, where they provide all parental care.

Because males provide only sperm to inseminate females, it is thought that lek systems may be especially important to students of sexual selection for assessing the significance of female mate choice. Separate concerns, which can complicate interpretations of female choice, such as male parental care, matings by females with multiple males, quality of the male's territory, and overt male–male competition, are often absent in lekking systems (e.g. see Bradbury and Gibson 1983; Höglund and Alatalo 1995).

With regard to mating system terminology, how are lek systems to be classified? This is not an easy question, in part because, among the different lekking species, there is a great deal of variation in the details of their overall reproductive strategies. Höglund and Alatalo (1995) note that the terms 'lek-promiscuity', 'lek-polygyny', and 'lek mating systems' have been applied by different writers.

I have chosen to use lek-promiscuity to refer to the social mating systems of lekking species for the following reasons. First, it distinguishes the mating system seen at leks from other, less extreme polygynous mating systems. Second, the word 'polygyny' does not seem to be appropriate for lekking systems in which females briefly visit males solely for copulation. 'Polygyny' generally refers to an extended social relationship between a male and two or more females, which nest in the territory of their mate, with the male often providing some paternal care. Third, if the word 'promiscuity' indicates both multiple mates and an absence of mate discrimination by at least one sex, then males of at least some lekkers are clearly promiscuous. For example, in an especially well-studied lekking species, the black grouse, males readily mate with dummy females: 'males mount the dummies for up to 30 minutes and copulated repeatedly ...' (Höglund *et al.* 1995). In contrast, females of the black grouse and other lekking species (e.g. white-bearded and golden-headed manakins, Lill 1974, 1976) appear to be both highly discriminating and genetically monogamous. Characteristically, females of these species mate only one time or with only one male before deserting the lek and laying a clutch of eggs. In the ruff, however, females may mate with two or more males (see Oring 1982). Thus, female grouse and at least some manakins are genetically monogamous, while females of the ruff (reeves) may be genetically polyandrous.

Subsequent to the review of mating systems by Emlen and Oring (1977), avian lek systems have received much in-depth attention (e.g. Bradbury 1981; Wittenberger 1981; Oring 1982; Bradbury and Gibson 1983; Foster 1983; Beehler and Foster 1988; Gibson *et al.* 1991; Wiley 1991; Höglund and

Alatalo 1995), and the origins and adaptive significance of leks are far better understood than was the case just a few years ago.

15.2   The derivation of leks from other mating systems

Leks can be viewed as representing the endpoint on a continuum of mating–parenting dispersal strategies (Oring 1982), and it has been suggested that this endpoint is probably evolutionarily irreversible (van Rhijn 1990). If monogamy with biparental care was the evolutionary predecessor of most or all of the non-monogamous mating/parenting systems found in contemporary neognathous birds (see Section 10.6), lek-promiscuity can be viewed as the extreme outcome of evolutionary trajectories that first favoured the development of polygyny, along with the reduction and eventual loss of paternal care. That is, decreased necessity of paternal care, along with the evolution of male strategies designed to increase the number of mates, presumably led to polygyny. Polygyny, in turn, in association with little or no paternal care, may have been the intermediate condition in the evolutionary transformation from monogamy/biparental care to exploded and classical leks (see below and Thery 1992).

Polygynous mating systems, while rare as compared to monogamous ones, are much more common than lek mating systems, thus raising the question: if lekking is derived from polygyny, why is it so much rarer than polygyny? There may be several partial answers to this question. In the first place, since more than 90 per cent of all avian species are socially monogamous, this assures that shifts from monogamy to polygyny must occur more often than shifts from polygyny to lekking. Second, as a broad generalization, polygyny with a reduction in male parental care seems to favour the reproductive interests of breeding males over those of females. For example, in the highly polygynous yellow-rumped cacique, females lay two eggs and apparently must work hard to rear only one young. Presumably, if paternal care were available, the female's reproductive success could be doubled. Successful male caciques, on the other hand, may sire many young per year (Robinson 1986).

In contrast to polygyny, lek systems may reflect female control of mating strategies. When females obtain no parental aid from males, they are free to pursue exclusively their own reproductive interests (Gowaty 1996a,b). This may have 'forced' males into participating in lek mating systems. Because of the extremely high variance in mating success that may occur on leks, for the majority of males, lek-promiscuity will lead to less reproductive success than either monogamy or polygyny. It appears that, in the evolutionary history of classical leks, the tables were turned on males and the reproductive interests of females gained precedence over those of males. If these suggestions are true—that the average male benefits less from lek-promiscuity than from a polygynous system, and that lek systems evolved from polygyny—the obvious question is, how did this come about? Did males of lekking species lose control of the mating game?

One possible answer is as follows. Once care of offspring is provided solely by the female parent, and once females no longer obtain resources on a territory defended by males, they obviously do not have to accommodate the reproductive interests of males. When males become unimportant to females in terms of survival and the rearing of offspring, there is no motivation for females to mate with a male for reasons related to resources (including paternal care), as is seen in females of monogamous species, which typically 'allow' the social mate some or all paternity of the brood (see Section 12.4.1). In short, once paternal care becomes a resource unavailable and non-essential to females, then females cannot benefit by attempting to obtain it. On the other hand, both females and males still require copulations. (The fact that females of some lekking species (e.g. grouse) copulate only once per clutch may indicate that their mate choice decisions are unconstrained; see Section 12.4.1) A single male can potentially inseminate many females; in contrast, it is impossible for a female to produce young for many males. This may mean that, with regard to mating strategies, females are in the 'driver's seat'. One rare outcome of this situation is the evolution of leks.

15.3   The frequency and taxonomic distribution of lekking species

Because the overwhelming majority of bird species are socially monogamous with shared parental care, it follows that lek systems must be rare, and, indeed, this is the case. Of the approximately 9700 species of birds, about 100 (0.01 per cent) are known to lek (Höglund and Alatalo 1995, Table 2.2), and, of these, lekking is concentrated within a few families and genera. The taxonomic distribution of lekking indicates that it has evolved far fewer times than the number of species that mate on leks.

Höglund and Alatalo (1995, Table 2.2) list 15 families and sub-families of birds in which lekking has been recorded (see also Oring 1982, Table II for a more detailed listing of lekking species). For those taxonomic groups containing a number of species, the proportion of lekkers ranges from one per cent or less (Phasianinae, Psittacidae, Tyrannidae, Pycnonotidae) to about 50 per cent (Tetraoninae, Paradisaeidae). This compilation indicates that while in most cases the taxonomic distribution of lekking has a significant phylogenetic component, lekking can also appear in a single species in large, basically non-lekking groups. For example, there are more than 350 species of parrots (including cockatoos and lories), nearly all of which are thought to be monogamous with biparental care. Thus, possibly the most unlikely case of lekking is that of the kakapo of New Zealand, a flightless, terrestrial, nocturnal parrot (Merton *et al.* 1984). Even without the lekking behaviour, this is an unlikely combination of characteristics!

15.4 Types of leks

Leks occur in a variety of forms. Over the past two decades it has become clear that a single explanation for this phenomenon is probably as unlikely to be sufficient as a single explanation for other mating systems, such as polygyny, has proved to be. On the other hand, as with other mating systems, seeking generalizations to the extent that they provide insights into underlying causal factors is important. Two frequently used categories are *classical* and *exploded* leks.

15.4.1 Classical leks

Characteristics of classical leks were listed in the Introduction. The term 'classical leks' refers to sites where males are conspicuously clumped. Classical leks occur in a variety of avian groups, including several grouse (e.g. sage grouse, greater and lesser prairie chickens, sharp-tailed grouse, black grouse, capercaille), scolopacid shorebirds (ruff, great snipe, buff-breasted sandpiper, in part), birds of paradise (e.g. Raggiana bird of paradise), and cotingids (e.g. cock-of-the-rock, capuchinbird) and manakins. In most of these groups, the birds return to the same sites year after year, and, at least among some lekking grouse of North America, fidelity to particular lek locations can be extreme (Oring 1982).

In the grouse and cock-of-the-rock, males hold small territories on the ground, and inter-male aggression is a standard and conspicuous part of the activities that take place within the boundaries of the lek. Considerable aggression is also characteristic of the capuchinbird, an arboreal lekking cotengid (Trail 1990). In contrast, overt inter-male aggression at the lek is notably uncommon in the Raggiana and lesser birds of paradise (see Section 15.10).

15.4.2 Exploded leks

The term 'exploded lek' (Gilliard 1963; Foster 1983) refers to assemblages of displaying males that are separated by considerable distances. This spatial patterning of males is often detectable only as the result of careful mapping over a large area (Bradbury 1981). Typically, the individual males making up an exploded lek cannot see each other and they interact primarily by vocalizations. Although use of the males' territories by females is not thought to be a critical aspect of exploded-lek systems, because of the large distance between males, females can potentially forage and even nest within a male's territory (unlike the case for classical leks). Importantly, because of the mobility afforded by flight, female birds can visit several males within a brief period of time, thus paralleling a major advantage to females of classical leks.

Bradbury (1981, Fig. 9.5) and Foster (1983) view exploded leks as a compromise between the females' interest in having males close together (possible reasons for this are discussed below) and the disruptive effects of the close proximity of competitive males. Perhaps particularly for highly mobile, arboreal species, dispersed groups of displaying males allow females to compare a number of males before mating, while, at the same time, they also minimize male–male conflict and thus the potentially or actively disruptive effects of such interactions. However, Foster's (1983, Fig. 2) evolutionary scenario in which classical leks devolve into exploded leks in response to the disruptive behaviour of closely-spaced males seems unlikely, if it is true that classical leks evolved in response to female preferences for choosing a mate from within a group of males in close proximity. Although exploded leks may reflect a compromise between the interests of the female for male clustering and the interests of the male to avoid the extreme clumping seen in classical lekking systems, this pathway to exploded leks may be less likely than an evolutionary sequence in which exploded leks may be precursors of classical leks (Thery 1992).

15.5   Morphological and behavioural traits associated with lekking species

Among lekking species of birds are some of the most striking examples in existence of male morphological and behavioural traits apparently designed to capture the attention of females and to induce them to copulate. This, of course, is consistent with the assumption that, among birds, sexual selection is at its strongest in species where the only reproductive interaction between the sexes is copulation. In such species, this is the only avenue open to males to promote their fitness. Males control no resources of value to females or their offspring and they provide no parental care. Accordingly, males must go all out, so to speak, to attract females for mating.

The morphology, vocalizations, and behavioural movements and postures associated with courtship of several of the grouse, birds of paradise, and the ruff, along with some of the South American cotingids and manakins, are truly bizarre. It is easy to believe that in such species sexual selection has pushed development of male ornamental traits to their realizable limits.

15.5.1   Body size

In certain types of lekking species, sexual dimorphism in body size is typically greater than in their non-lekking relatives (Payne 1984). Presumably, this is due to the greater intensity of competition among males of such species. Höglund and Alatalo (1995) review the evidence concerning the relationship between lekking and male body size. In brief, they conclude that lekking is by no means always related to an increased difference in size of males and females (see also Höglund 1989; Oakes 1992; Höglund and Sillen-Tullberg

1994; and Section 7.2.5). In part, these differing results are due to the methods of analysis and, in part, they are due to the fact that there is a large phylogenetic component to this trait. In some lineages, sexual dimorphism of lekkers is indeed greater than in their non-lekking relatives, while in other lekking species, this is not the case. This makes sense: we would expect different male traits to be selected in different systems. For example, in some lekking species, males engage in aerial displays, and in these cases, males tend to be smaller than females.

In their survey across all lekking species, both vertebrate and invertebrate, Höglund and Alatalo (1995, Table 3.1) reported that male size and mating success were associated in only three species of vertebrates (an amphibian and two mammals). This is an interesting conclusion because Wiley (1974) reported that, in grouse, lekking correlates with large body size, strong sexual size dimorphism, and delayed maturation of males. This might suggest a relationship between male size and dominance, and between dominance and mating success. Wiley proposed that males of terrestrial grouse species delay breeding longer than females because they have a greater probability of subsequent survival due to their larger body size. Larger size could reduce the vulnerability of males both to predation and to low winter temperatures and snow cover, compared to females. Wiley (1974) suggested that these factors led to selection favouring sexual differences in the age of first reproduction, which, in turn, led to unbalanced sex ratios, with more ready-to-breed females than males, and that this led to polygyny (lek-promiscuity). Based on a consideration of degree of sexual size dimorphism and adult sex ratio in a number of species of grouse, Wittenberger (1978) disputed Wiley's scenario, suggesting instead that both the sexual size dimorphism and delayed breeding by males are consequences of intense male–male competition for mates. Wittenberger's comparative analysis implicated sexual selection (both male–male competition and female choice) as the causal factor in the sexual size dimorphism of polygynous grouse.

15.5.2   Bright, sexually dimorphic plumage colour

In many lekking forest species, such as birds of paradise, manakins, and cotingas, males are brightly coloured, while females are not. As with size dimorphism, it is not clear that sexual dimorphism in plumage is relatively more common in lekking species across-the-board than in their non-lekking relatives (Höglund and Alatalo 1995). Again, phylogenetic history needs to be taken into account. Males of several species of hummingbirds form leks; sexual dimorphism in plumage colour is minor or non-existent in these species, with males being duller than males of the average, non-lekking hummingbird.

In addition to their striking plumage and lekking behaviour, the reproductive biology of manakins is unusually interesting due to the complex 'team' displays of males of some species (e.g. McDonald and Potts 1994). With regard

to the various components of the overall display package, Prum's (1990) phylogenetic analyses indicate that in some cases the display behaviours of male
manakins evolved before the evolutionary development of the associated
ornamental plumage: '... derived male plumage traits in manakins have evolved
subsequent to the behavioural novelties in which they are prominently
featured' (Prum 1990, p. 225; see also Section 7.2.1).

If this is correct, there are very interesting implications. For some species,
including the long-tailed manakin, it appears that at least some plumage ornaments are of little current importance to females. Specifically, in this species,
tail length is not associated with female preference (McDonald 1989a). The
findings of McDonald (1989a) and Prum (1990), taken together, suggest that
the long tail of this manakin evolved to enhance the male's display performance, but that females do not gauge males by the lengths of these feathers (see
also Section 4.3.2).

Among lekking species, plumage traits (brightness, tail ornaments) usually
appear not to affect mate choice of females. In fact, according to the survey of
lekking species by Höglund and Alatalo (1995, Table 3.1), no relationship
between any aspect of male plumage colour and mating success has been
shown, with one exception. The exception cited by Höglund and Alatalo is the
great snipe, in which tail whiteness is related to male attractiveness (Höglund
et al. 1990; see also Section 6.3). However, feather traits of males of two additional lekking species are also important to females, namely the number of
ocelli in the peacock's train and the length of the tail of males of the Jackson's
widowbird (see Section 6.3).

15.5.3   Plumage monomorphism

It is not surprising that males of most lekking species are more brightly
coloured than females. What may be surprising, however, is that in a number
of species (18, approximately 20 per cent of the known lek-breeding bird
species), sexual dichromatism is lacking (Trail 1990, Table 1). Again, there is a
phylogenetic component to this pattern, in that seven of the 18 species are
hummingbirds, and four others are cotingas.

Based on his studies of lekking capuchinbirds, a species of cotinga in which
both sexes exhibit drab brown and black plumage colours, but with some morphological traits thought to be associated with sexual and/or social selection
(unfeathered, bare crown, elongated undertail coverts, and a cowl of upstand-
ing feathers at the back of the crown), Trail (1990) poses the question, why are
males and females of so many lekking birds alike in appearance, given the
strong sexual selection thought to be characteristic of such species?

The more general issue of plumage monomorphism, especially those cases
where both sexes are brightly coloured, is discussed in Section 4.7. Here I
summarize Trail's argument specifically for monomorphic lekking species.
Trail's thesis, in brief, is that monomorphism is the result of strong intrasexual

selection on both males and females. Trail found that, in capuchinbird leks, male–male competition was intense and that a single male obtained all of the matings. Female capuchinbirds were also intrasexually aggressive. Usually this took the form of threat displays and supplantings. These observations led Trail to suggest that the similar morphologies of the two sexes reflect similar social adaptations, and that maintenance of the monomorphism may be aided by a genetic correlation between the sexes.

Female capuchinbirds may benefit by intrasexual aggression because it stimulates male–male aggression, which, in turn, provides a mechanism by which females can evaluate males. In addition, in the social setting of the lek, females appear to benefit sometimes as a result of their male-like appearance and even male-like behaviour. According to Trail, these traits seemed to promote the females' access to the mating perch of the dominant male, as well as promoting dominance over other females.

15.5.4   Behavioural displays

In addition to specialized plumage ornaments, males of a majority of lekking species produce some of the most unusual behavioural displays known (e.g. the cooperative 'team' displays of long-tailed manakins, McDonald and Potts 1994, and the synchronous immobile display postures of the Raggiana bird of paradise, discussed below). Johnsgard (1994) provides an interesting, informative, and well-illustrated review of the display postures and movements of a wide variety of lekking species.

Another form of cooperation among males occurs in the ruff. Males are of two morphological and behavioural types. Independent or resident males compete for territorial space and, correspondingly, are large and aggressive. They vary in colour of the ruff feathers of the neck, but generally possess dark plumage. 'Satellite' males, in contrast, do not compete for territorial space, their ruffs are whitish in colour, and they are on average smaller in body size than the independent males (Fig. 15.1). Mating success depends first on drawing females to the lek, and the presence of satellite males is important in this regard. Females prefer larger leks (Lank and Smith 1992) and that satellite males are present; i.e. satellite males, together with the variously coloured independent males, apparently increase the attractiveness of leks to females flying over them (W. L. Hill 1991; Hugie and Lank 1997), as suggested by Stoner (1940) almost sixty years ago. Both the coloration and the distinctive behaviours of satellite males are genetically controlled. The presence of both independent and satellite males reflect a genetic polymorphism that is maintained by the tendency of females to mate regularly with each morphotype (Lank *et al.* 1995).

The displays of males of many lek species are noisy, as well as visually conspicuous. In most environments, sounds travel considerably farther than visual signals, and many of the sounds made by lekking males probably serve as

Fig. 15.1   Male ruffs at lek. The presence of both dark and light morphs enhances the attractiveness of a group of males to overflying females.

long-distance advertisement of their locations. Auditory signals of males may also be used by females at close range to assess males. The role of sound signals in mate choice has been studied in detail in only one lekking species. During courtship display, male sage grouse produce a double-popping sound. The time between the first and second pops, the 'inter-pop interval', correlates significantly with mate choice decisions by females (Gibson *et al.* 1991).

15.6   *Proposed benefits and costs of classical leks*

For the relatively few species of birds that exhibit lek mating systems, we assume that the net benefits of lekking must outweigh the net costs. Probably in no other type of mating system have the interests of the two sexes diverged more dramatically. Therefore, to understand leks it is necessary to consider lekking from the perspective of each sex. As suggested earlier, for the 'average' male, a lek mating system is a disaster: most males of classical lekking species probably die without leaving a single offspring. For females, too, lek systems may provide some problems.

In this section, some of the proposed benefits and costs directly related to classical lekking systems, in particular, are reviewed briefly. Some of the following points are relevant to the evolutionary origins of leks, while others are best considered as secondary benefits or costs once lekking has evolved. The issue of the evolutionary origins of leks is taken up in the next section.

15.6.1   Predator avoidance

One popular idea is that leks have evolved in response to predation pressures. That is, leks make the process of courtship and mating safer for all individuals of both sexes as they engage in mating activities while surrounded by many sharp-eyed conspecifics (see Höglund and Alatalo 1995, p. 152 for review). The safety-in-numbers phenomenon (Hamilton 1971) could explain, at least in part, why females prefer larger leks over smaller ones (Höglund and Alatalo 1995): a larger lek may simply be a safer place in which to assess males and to copulate.

In particular, the strong correlation between living in open country and lekking found in North American grouse has encouraged the idea that predation may be an important factor in the evolution of this kind of social organization. Although congregating in dense aggregations (e.g. sage grouse) to thwart predators is probably not the root cause for the evolution of lekking for any species, open-country grouse or others, it may well be the case that both courting males and their female observers are safer in concentrated groups. Oring (1982, p. 54) concludes '… that lek grouse definitely are less vulnerable than they would be if they displayed in solitary fashion in an open field' (see also Section 15.7.4).

Just how important the selective pressure of predation at leks has been over evolutionary time is difficult, if not impossible, to evaluate. Both Oring (1982) and Gibson and Bachman (1992) provide data suggesting that the species of North American open-country grouse they studied are rarely captured by avian predators at leks. The fact that the most likely predators rarely attack birds at leks, and even more rarely do so successfully, can be viewed as evidence for the effectiveness of leks in thwarting predators (see Oring 1982, pp. 53–4). However, even if this is so, it does not indicate that protection from predators is an evolved function of lekking behaviour.

Modern students of grouse leks have no way to assess the intensity of predation throughout the evolutionary history of most species, simply because where grouse occur—Eurasia and North America—many predators have been made rare or have been exterminated altogether by humans (see Oring 1982, p. 53). Golden eagles, for example, were probably important predators of the open-country grouse of North America, and in some areas this may still be the case (Hartzler 1974). For example, in Wyoming, an ornithology class watched a pair of golden eagles kill two adult male sage grouse (the 'master' cock and another strutting male) at their lek (Boyce 1990). Due to persecution, golden eagle numbers in the United States are much lower than when Europeans first began to occupy this area. (Although accurate censuses of birds during the last century are not available, data from egg collections indicate that in the western United States nesting densities of golden eagles were much greater even forty to fifty years ago than they are today; personal observation.) Thus, the fact that observations of predation attempts by eagles are currently infrequent in many areas (e.g. Gibson and Bachman 1992) provides no hint of the possible influence of eagles on the evolution and maintenance of sage grouse leks throughout the Pleistocene or their earlier evolutionary history. In fact, the spectacularly diverse mammalian and avian predator fauna of western North

America during the Pleistocene (e.g. Feduccia 1996), prior to the arrival of humans, indicates that open-country grouse were almost certainly subjected to strong predation pressures.

In brief, although grouse of both sexes may obtain increased safety from predators by gathering and displaying in groups, it is an open question whether this has been a causal factor in the evolution of lekking for any species. Predator pressure may be involved, however, in determining the specific location of leks in North American grassland grouse (Oring 1982). Lek sites either may move slowly (year by year) to safer spots or, alternatively, only those leks that are on safe sites persist for many years. For example, a large sage grouse lek in south-eastern Idaho was located on bare volcanic rock; i.e. there was little cover to obscure the birds' vision or to conceal the approach of terrestrial predators such as coyotes. At this site, one could find arrowheads of sizes used for birds, suggesting that the grouse had displayed here (and been hunted by humans) for many years (personal observation).

15.6.2   Stimulus pooling and increased per capita reproductive success on larger leks

In some cases, females are preferentially attracted to larger groups (Höglund and Alatalo 1995), and in two lekking birds, a positive relationship has been found between the number of males per lek (lek size) and per capita mating success of males. First, in black grouse, larger leks had more female visits, more copulations per male, and a higher probability that visiting females would mate; this last point indicates that females do indeed prefer to mate in larger leks (Alatalo *et al.* 1991; Höglund and Alatalo 1995). Similarly, Bradbury *et al.* (1989) found that, in sage grouse, female visiting rate per male increased up to a lek size of 25–50 males. Second, as described above, ruff leks contain two types of males, independent, territorial individuals and non-territorial, satellite males. Lank and Smith (1992) showed experimentally that females prefer larger leks, and Höglund *et al.* (1993) found that matings per territorial male increased with lek size. Likewise, there was a non-significant ($0.05 < P < 0.1$) tendency for satellite males to do better on larger leks. The combined signalling of males might stimulate females to choose at least some of them by lowering the threshold of the female's copulatory response.

A rare form of stimulus pooling occurs in several of the lekking manakins, where permanent 'teams' of two or more males display cooperatively. This cooperation is remarkable in view of the facts that (1) only one male of the team obtains copulations and (2) team members are not genetically related (McDonald and Potts 1994). The group displays of manakins provide some of the best evidence for the general proposition that multiple males may be more stimulating to a female than is a single male (Foster 1977, 1981).

Although stimulus pooling may provide a proximate explanation for the benefits to males occupying larger, rather than smaller, leks, it does not answer the ultimate question of *why* females tend to prefer males on leks in the first place. This central issue is addressed below.

15.6.3   Information sharing

De Vos (1979, in Oring 1982) showed that individual male black grouse at leks gain information about food location by following other males to food. This kind of benefit is probably not of general importance either to the evolution or maintenance of lek systems (see Höglund and Alatalo 1995, p. 156).

15.6.4   Mate copying

A potentially important benefit to females as they visit leks is that they may obtain information from other females concerning choice of mate; i.e. a female may copy the mate choice of another female (e.g. Gibson and Höglund 1992; Pruett-Jones 1992). Pruett-Jones (1992) lists a number of bird studies in which it has been suggested that females copy each other. Females might be expected to copy the mate choice decision of others for at least two reasons. (1) Costs of mate choice (time, energy) would be decreased by shortening the time required to make a decision, and (2) an inexperienced, young female might make a better mating decision by copying the choice of one or several more experienced individuals. The benefit of copying would be greater if experienced females choose first (Gibson and Höglund 1992) and, in a variety of species, older females do breed earlier than younger individuals. In addition to these benefits to females, mate copying has another effect: copying behaviour should increase the variance in mating success among males, thereby increasing the intensity of sexual selection (Pruett-Jones 1992).

Some of the best evidence for copying comes from two species of grouse. Gibson *et al.* (1991) found that some female sage grouse appeared to copy the mate choice of others. Within the same day, mate choice became more unanimous as the number of females mating at the same time increased. However, it is difficult to unequivocally demonstrate copying, because, following mating, individual male sage grouse increased their rate of display. This, in itself, could increase a male's attractiveness to females independent of copying *per se*. Gibson *et al.* (1991) recognize this and other alternative interpretations, and they point out that copying seems to occur not only during a single day, but also across days. Analysis of mating patterns over the course of several days led to the conclusion that the choices of some females were influenced by matings observed on a previous visit. In contrast, Spurrier *et al.* (1994) obtained no evidence that under aviary conditions female sage grouse copied the mate choice decisions of others.

Höglund *et al.* (1995; see also Höglund and Alatalo 1995) investigated the possibility of copying by female black grouse in two ways. First, they found that on five of 19 leks the number of sequential matings by a given male differed significantly from random mating by females. When the mating patterns on all 19 leks were combined, the total mating sequences also differed significantly from random matings. Second, these authors placed female dummies on leks, and males attempted prolonged copulations with the dummies. The sight of copulating males apparently increased the rates that females visited those males. However, six of the seven experimental males that mounted the dummies subsequently obtained no copulations from living females. This raises the possibility that females need to receive certain stimuli, lacking in the dummies, from mating females, as well as from the male, before copying another individual's mating decision.

Fiske *et al.* (1996) investigated the possibility that female great snipe copy each other's mate choice. These authors concluded that in this species copying is not a significant factor in mate selection by females.

15.6.5 Enhancement of mate choice via comparing males

One of the most frequent suggestions offered concerning benefits (and origins) of lekking is that leks offer females the opportunity to efficiently, quickly, and directly compare several or many males before making a critical mate choice decision (see below). For species in which females obtain no resources or other benefits from males, such as parental care, choice of mate may be particularly important, and in a number of well-studied species of classical lekkers, females do visit leks for several days before revealing a mating decision, moving from male to male, apparently making comparisons, before accepting copulation. The importance to a female of a considered choice of mate is probably further increased in those species in which a single copulation typically fertilizes the entire clutch of eggs (several, perhaps most, species of grouse; Höglund and Alatalo 1995), and in those species with a clutch size of one (most of the birds of paradise) and/or two (most manakins).

15.7 Costs of lekking

In this section costs of lekking for each sex are considered briefly. For reasons mentioned earlier, the view taken here is that classical lek systems are, in a sense, imposed on males. If we look at it from the point of view of the 'average' male, lekking, like a lottery, is unlikely to be rewarding. In discussing lek evolution, Bradbury (1981, p. 149) writes: '... clustering led to no net advantage for males in the absence of female preference and in many cases led to geometrical disadvantages... . Thus, for some of the males in all systems and all of the males in some systems, clustering is disadvantageous.'

This view leads to the idea that if leks do not promote the interests of most males, perhaps they promote the interests of females.

15.7.1   Intense competition for matings

For most males, the most conspicuous and critical cost of lek systems is that they will have little or no mating success. One of the key characteristics of leks, perhaps especially classical leks, is that a very small minority of males do very well indeed in terms of mating success, while the vast majority obtain zero matings, both per year and per lifetime. (This latter point might be especially true for species with high rates of annual mortality, such as grouse.) Such extreme skews in mating success are perhaps best known for the well-studied species of grouse (see Bradbury *et al.* 1985; Wiley 1991); however, they also occur in the leks of arboreal passerines. For example, in the lesser bird of paradise, one male in a lek of eight was overwhelmingly most successful in attracting and mating with females, obtaining 25 of 26 (96 per cent) copulations observed (Beehler 1983). Trail's (1990) study of lekking capuchinbirds yielded similar results: only one dominant individual of eight males making up a lek copulated with visiting females.

The absence of overt aggression at the lek among male lesser birds of paradise stands in striking contrast to the open conflict among males of the cock-of-the-rock. In this species, courtship disruption is a common cost of male aggregation. Disrupted females are likely to alter their mate choice decisions, and disrupting males receive a disproportionate share of the mating by those females (Trail 1985). Males that disrupted the same female several times apparently were '… able to focus the choice of disrupted females on themselves more effectively' (Trail 1985, p. 779). It appears that male disruptive behaviour is used by male cocks-of-the-rock to attract the attention of females and is used by females to assess males.

15.7.2   Extensive and sustained energetic output by males

Display performance can be a major cost for males, in terms both of time and energy. For some species, male success correlates with time spent on the lek (see Höglund and Alatalo 1995, Table 3.1). This is time that cannot be used for foraging, thus lek time has costs in terms of energy depletion. In addition, displays of males may be very costly. For the sage grouse, Vehrencamp *et al.* (1989) determined that the metabolic cost for vigorous displays was 14–17 times basal metabolic rate, an expense similar to that of flight. These authors also calculated that, on a 24-hour basis, actively displaying male sage grouse expend energy at the maximum rate typically sustainable by homeotherms. Thus, female grouse mate primarily with males capable of expending a lot of energy over extended periods of time.

15.7.3   Energetic costs to females of travel to leks

Concentrations of many males from over a large area make it necessary for females to travel long distances, and perhaps over unfamiliar ground, to mate. Using radio transmitters, Gibson and Bachman (1992) obtained direct measurements of the total distances that female sage grouse travelled per day during the period they were visiting leks. Gibson and Bachman concluded that the increased travel by these females amounted to only about one per cent of daily energy expenditure.

15.7.4   Increased risk of predation

It is possible that both travel to leks and attendance at leks increases the chances of an attack by a predator. Gibson and Bachmann (1992) investigated this potential risk to female sage grouse. Away from leks, they found no cost of travel, as measured by risk of predation. Potential predation costs were also found to be negligibly low at leks.

In the black grouse, predation rates at leks were also very low and no females were known to have been killed at a lek. Away from leks, however, the story was different. Twelve of 63 female black grouse were killed during the month of May, indicating severe predation during the lekking season (Höglund and Alatalo 1995, p. 109). One interpretation of these data is that they indirectly demonstrate just how effective a concentrated group—the lek—can be in promoting the safety of the individuals comprising it (Hamilton 1971).

Male lekkers typically focus their attention on female visitors to the lek. Thus, there is the possibility that males of some species suffer greater risk of predation while displaying on leks than when they are elsewhere. I am aware of no clear-cut demonstration that this is true, and the data for female black grouse suggest the opposite, i.e. that leks are relatively safe places to be (see also Oring 1982). Trail's (1987) study of the cock-of-the-rock indicates that males on leks do suffer low rates of predation; how this compares to predation away from the lek is not known.

15.8   Evolutionary origins of leks

The origins of avian lek systems pose an especially interesting problem, one that has generated many papers promoting one answer or another (e.g. Bradbury 1981; Oring 1982; Bradbury and Gibson 1983; Foster 1983; Beehler and Foster 1988; Höglund and Robertson 1990; Höglund and Alatalo 1995). None of the benefits to males mentioned in the previous section (e.g. reduced predation, stimulus pooling, etc.) can account for the origin of male clumping. When lek systems have developed, factors such as reduced predation, for example, may serve the interests of males. However, at most, these are secondary issues that can provide benefits to individual males after a classical lekking system has evolved.

Most recent studies of classical lek systems have indicated that, in one way or another, lek systems originate and are maintained in response to the reproductive strategies of females (see below). It appears that at leks a majority of males may be at a disadvantage with regard to their prospects for mating, partly because they are close to many other competitors for the females. If so, it could be argued that females have imposed their interests on males.

What factors brought male birds of certain species into tight aggregations that females visit for mating? Two alternative responses to the issue of selective pressures favouring lekking are: (1) leks originated as a result of special reproductive strategies of males (e.g. hotspot model; see below) or (2) the origins of leks were driven by the mating strategies of females (e.g. female preference model, hotshot model; see below). Before assessing these alternative scenarios, a point should be mentioned. It is unlikely that exactly the same explanation (i.e. same costs and benefits) will suffice for the diverse array of lekking species (see Oring 1982, Table II, and Höglund and Alatalo 1995, Table 2A); i.e. any attempt to generalize about the origin and maintenance of lekking behaviour will probably be no more than partially successful. All three of the factors emphasized in the models discussed below may have contributed in differing degrees to the evolution of lekking in one lineage or species or another. If this is correct, we would expect the explanations for the development of the lek systems of different species to vary.

The question I address here is seemingly a simple one: what factors have promoted the evolution of leks? Although some of the possible benefits and costs of lekking for each sex were mentioned above, this ultimate question has not been answered to the satisfaction of all students of the lek phenomenon. Here I review briefly three of the major theories proposed to explain how leks originate and the kinds of benefits obtained by each sex.

15.8.1   Female preference model

Emlen and Oring (1977) and Bradbury (1981) suggested that females can benefit from the clustering of males, because this makes it possible for the females to assess efficiently the relative quality of many males. Since females obtain only sperm from males, they are in a sense free to make 'demands' of males in exchange for paternity. Such a demand might be a simple one, e.g. that females will associate only with groups of males. If so, this factor alone could have initially promoted the evolution of leks. Bradbury's (1981) female preference model made several predictions about female movement patterns: (1) the diameters of the home ranges of most females should be less than inter-lek distances, and (2) most females should have breeding season home ranges that include only one lek.

The documented preference of females for larger leks is consistent with this scenario. In addition, several studies have demonstrated that within leks females do choose among the males (Höglund and Alatalo 1995). Thus, females make mating decisions at two levels; females prefer larger leks to smaller ones, and, within a lek, females prefer some males over others. However, it is not known, in an ultimate sense, why females prefer larger leks. Even so, for some species, the female preference hypothesis has been empirically supported, at least as an explanation for the maintenance of leks.

The female preference model also provides plausible answers to the question of why leks may have originated; namely, that under certain conditions females are able to 'force' males to cluster as a result of the female's preference for groups of males (i.e. lone males are unable to attract females). In short, the female preference model hypothesizes that leks originate and are maintained as a result of female mating strategies. Males must comply with the preferences of females for aggregations of males in order to have any hope of mating success. This model, however, does not consider factors affecting location of leks or their persistence over time.

In presenting their 'hotshot' hypothesis (see below), Bruce Beehler and Mercedes Foster (1988) provide a critique of the female preference model and make two points. (1) *Female preference.* Beehler and Foster state that male clustering because of female preference for groups has not received much empirical support. These authors contend that male clustering may impose a considerable cost on females, both in disruptions of matings and the effects of male dominance relationships on the female's freedom of mate choice. It should be recalled that there is now evidence, obtained since Beehler and Foster's proposal, that females of some lekking species do prefer larger groups (see Höglund and Alatalo 1995 for review), which indicates that leks do not impose a serious cost to the females. (2) *Freedom of female choice.* Beehler and Foster (1988) disagree with the critical assumption of the female preference model, that females have freedom of choice in lek situations. They point out that dominant males can inhibit copulations by subordinate males either by overt aggression or merely by their presence. The point that strong dominance relationships to the point of despotism can develop during the non-breeding season, and that the effects can be carried over into the breeding season, is an important one: rarity of overt aggression does not indicate the absence of a strong dominant–subordinate relationship among males within a lek. Because males may interact throughout the year, dominance relationships can continue from one breeding season to the next, even in the absence of active aggression at the lek site at the onset of each breeding period.

15.8.2   The hotspot model

Anticipating the hotspot model, Bradbury (1981, p. 149) suggests:

> … that males may be coaxed by females into aggregating at locations of maximal accessibility, e.g. at major traffic crossroads or minimally at traditional sites. The exact locations would be the result of the combined preferences and activity ranges of a number of females and hence might not be ideal for any one female'.

Subsequently, in contrast to female-driven explanations of leks, Bradbury and Gibson (1983) presented an explanation for the origin and spatial location of leks based, in part, on mating strategies of males. They suggested that individual males should gather at the places where, due to some aspect of the habitat, they will encounter the greatest number of females. These sites are 'hotspots'.

In the hotspot scenario, females have large home ranges with a high degree of overlap, which increases the likelihood that several or many females use the area within which the lek is sited. Two critical points are that foraging requirements are the basis for these large home ranges and that females are not constrained in their choice of mates. The hotspot model predicts that spacing between leks will be positively correlated with the size of female home ranges and that leks will be less than one female home-range diameter apart. Thus, females may visit several leks before mating without increasing their home ranges (Bradbury and Gibson 1983).

The hotspot model proposes an answer to the question of lek location, as well as to the question of which sex promotes male clumping in the first place. Males are attracted to specific sites where females will be most common. As more males are attracted to such sites, they are forced to either 'share' the site (i.e. to display in close proximity to other males) or settle singly at sites where females are less likely to be present.

The hotspot and female preference model differ in some important ways and are complementary in others. The female preference model is driven by the power of female choice: males are, in effect, forced to come together and line up for inspection. The hotspot model, on the other hand, is a male-initiated system. By using features of the habitat, males determine the location of highest density of females and display there. As the density of males drawn to such spots increases, each male eventually must accept the close proximity and the associated competition from other males, and a lek is born. Preferences by females for clustered males will strengthen the tendencies of males to aggregate, despite the disadvantages.

Together, these two models can provide explanations for leks that are adaptive for both sexes. The hotspot model argues that the placement of leks at specific locations is a male strategy, while the female preference model assumes that female choice promotes male clustering. In short, ecology (female spacing patterns based on resource distribution) can explain the initial aggregations of males at hotspots—leks—while female choice may be responsible for male–male interactions within the lek.

Beehler and Foster (1988) also question some of the assumptions of the hotspot model, as follows. (1) *Defining a home range.* The temporal stability of lek sites suggests that female homes ranges must also be stable over time. Beehler and Foster point out that female home ranges may vary seasonally and, more significantly, females may travel well out of their normal home ranges to visit a lek. Moreover, individual females may readily change their home ranges between successive nestings in a single year. For example, a female sharp-tailed grouse moved 20 km after losing her nest, and in greater prairie chickens, the new home ranges of females that have lost a nest overlap only five per cent, on average, with the original, pre-failure home range (Oring 1982). This behaviour indicates that attempts to elude nest predators are of major concern to female grouse, and that, as a result, home range site faithfulness can be low. (2) *Defining hotspots.* There may be a problem in defining hotspots in that aggregations of males may attract females, thereby increasing female traffic to the vicinity of the lek, rather than female traffic having initially attracted the males. In addition to the problem of which came first, hotspot or lek, Beehler and Foster suggest that the extreme stability of many leks over time (see below) contradicts the concept of female-defined hotspots. They argue that since habitat quality probably varies both in space and time, lekking groups of males would probably move more often than they do if males were cueing on centres of female activity. In addition, the fact that females of at least some species leave their normal home ranges to fly to distant leks refutes the generality of the hotspot concept.

15.8.3   The hotshot model

A third model to account for lek behaviour, the 'hotshot' model, was presented by Beehler and Foster (1988), who, as outlined above, question some of the basic assumptions of the female preference and hotspot models. Beehler and Foster suggest that the amazing skewedness in mating success of males of many lekking species (e.g. Beehler 1983) is more likely to be due to male–male social relationships, subtle though they might appear to be, than to the unanimity of choice among females visiting the lek (i.e. although the end result might be the same either way, the causal mechanisms are not). These points led Beehler and Foster (1988) to offer their hotshot model as an alternative explanation for lek mating systems.

The basic premise of the hotshot model is that in mating systems not influenced by resources, certain males are extremely successful at attracting mates as a result of 'conservative' female choice (Beehler and Foster 1988); such males are 'hotshots'. For lek development, the critical points are that other males are aware of the fact that certain males have a lot of drawing power, and that many males will occupy sites near an attractive male to take advantage of the fact that mate-seeking females will be drawn to the immediate vicinity of the hotshot.

Beehler and Foster (1988) believe that four factors can produce the spatial, demographic, and behavioural shifts necessary for the development of a lek system. These are: (1) male competition for dominance; (2) inequality of mating success among males; (3) simple and conservative female mating patterns, e.g. mate choice copying; and (4) secondary male reproductive strategies (e.g. intercepting females as they move to a hotshot male).

Evidence that Beehler and Foster provide for their hotshot model includes two points: (1) 'deterioration' of the lek following the experimental removal of dominant males (e.g. Robel and Ballard 1974), and (2) the fact that some secondary males obtain matings. This latter point suggested to Beehler and Foster that such males are attracted to the display sites of hotshot males in order to increase their own mating prospects.

Although it was not addressed specifically, the Beehler–Foster model also provides an answer to the question of why leks are where they are (founding hotshot males displayed there) and why they persist over time: once a site is established, young males, including hotshots-to-be, are attracted to the site because current hotshots are there and because females thus come there to mate. Beehler and Foster do not explicitly point out that all males start off as callow and probably unattractive youngsters. For their scheme to account for traditional sites that persist over many generations, young males that will eventually become hotshots must be attracted to sites originally developed by previous generations of hotshots. The costs and benefits to young, potential hotshots of joining an established lek versus displaying solitarily will probably vary from species to species. In some lekking species, individual males may remain apart from a lek (Beehler 1988), and lone males sometimes attract females and obtain matings (Kruijt *et al.* 1972). According to the perspective of Beehler and Foster, such loners would be predicted to be young hotshots-in-waiting. In contrast, lower 'quality' or less attractive males should always display at lekking areas.

15.9   An assessment of the models

All three models proposed to explain the origins of leks have received some empirical support. Both the female preference and hotshot models may be applicable for some species, but possibly not for others. For example, in some cases, female preference for clumped males is an important causal factor (e.g. black grouse, Höglund and Alatalo 1995), whereas in others, the attraction may be a particular 'hotshot' male. This appears to be the case, for example, for the capuchinbird (Trail 1990). Tests of specific models have produced negative results. A study of greater prairie chickens provided no support for the predictions of the female preference model that pertain to movement patterns of females (Schroeder 1991).

Species-to-species differences also make it difficult to generalize. For example, female ruffs visit a number of leks and may copulate at all of them. Thus, it is unlikely that the hotspot model can adequately explain this system. A recent consideration of the use of traditional lek sites by male ruffs concludes that a major benefit relates to the development of long-term (between years), as well as short-term, dominance relationships between males (Widemo 1997). Because females are likely to leave the lek in the face of high levels of male aggression, a stable hierarchy with little overt aggression is potentially beneficial to all males.

The hotspot model has received empirical support from a study by Thery (1992) of six species of manakins in French Guinea. As predicted by this model, in the five lek-breeding species, female home ranges were three to seven times larger than those of adult males. In contrast, adult males and females of the single non-lekking species exhibited similar home range sizes. Thery (1992) also found significant correlations between female home range size and male clustering, both in distances between neighbouring leks and in distances between neighbouring males. Larger female home ranges were associated with tighter clustering of males. Females did not expand their home ranges to visit leks. Instead, leks were located in areas of high female traffic, which was due to fruiting places and stream bathing sites. Although these findings are consistent with the hotspot model, the cause-and-effect issue may remain. That is, males may have aggregated in areas with a lot of female activity, or, alternatively, females may have increased their use of the area as a result of the presence of displaying males. For example, the statement '... leks were always situated near peaks of female feeding activity' (Thery 1992, p. 234) raises the question, are males aggregating where females are feeding or are females feeding near where males are aggregated?

The hotshot model, too, has received some empirical support. In some species, the alpha male is extremely despotic and the females appear to mate almost exclusively with that male (e.g. Beehler 1983; Trail 1990). However, I cannot agree with all aspects of Beehler and Foster's argument. For example, these authors believe that precise or careful discrimination between males by females is unlikely. They suggest, as an alternative, that the significant correlations between male mating success and regularity of lek attendance, display rate, and an acoustic component of display found in sage grouse (Gibson and Bradbury 1985) are behavioural traits that could correlate with male age, history of dominance on the lek, and/or past mating success, rather than being evidence of female discrimination. Although the first part of this statement may be true, this does not argue against careful discrimination by female sage grouse.

Subsequent to Beehler and Foster's paper, precise mate discrimination has been demonstrated in a number of species, both lekking and non-lekking (see Chapter 6). For the sage grouse, Vehrencamp *et al.* (1989; see also Gibson *et al.* 1991) determined that two of the traits mentioned by Beehler and Foster—display rate and regularity of lek attendance—are indicative of metabolic output, which is likely to be correlated with male condition.

To summarize, there is some empirical evidence to support each of these three models for lekking. It is probable that none of the models alone can provide a complete, general explanation for the origin and maintenance of lekking in any species, and certainly no single model is sufficient to account for all lek species. On the contrary, it appears that elements of at least two, and possibly all three, of the models are needed to provide a full explanation of the classical lekking phenomenon. Thery's (1992) comparative study of lekking in manakins, in which female movements were radio-tracked, provides some of the best evidence to date for the hotspot model. His study indicates that the other two models may also play roles in the development of lekking. The hotshot model may be best supported by those species in which male despotism seems to be extreme, e.g. lesser bird of paradise. The cock-of-the-rock provides support for the female preference model in that (1) mating skew is somewhat less extreme than in some apparently more despotic species, and (2) females use male aggression to make choice decisions.

The differing views of Bradbury and Gibson, on the one hand, and Foster and Beehler, on the other, may be due, at least in part, to differences in the biology of the lek species they studied. Bradbury and Gibson and their co-workers study sage grouse, while Foster and Beehler study manakins and birds of paradise. Annual mortality of adult male grouse is high, about 50 per cent, while annual mortality of manakins and birds of paradise probably is as low as 10 per cent per year (e.g. Lill 1976). This factor alone could create strong differences in the social relationships of males. We might expect, for example, that the high rate of turnover of male grouse would lead to more overt aggression on their leks than in species where the same males are present year after year, and where, moreover, the males interact socially throughout most of the year. This difference, based on male survivorship, could account for the emphasis that Beehler and Foster (1988) place on despotic males in their scenario of lek development.

15.10   A classically lekking bird of paradise

The lek systems of several species have been comprehensively investigated. Probably the best known are the black grouse, sage grouse, and ruff. Of the tropical lekkers, avian behavioural ecologists are probably most familiar with the cock-of-the-rock and several of the manakins (see Johnsgard 1994 and Höglund and Alatalo 1995 for reviews of these and other species). Data pertaining to female mate choice in several lekking species are described in Chapter 6. Because some of the birds of paradise are among the most extravagantly ornamented and behaviourally bizarre of all avian species, I have chosen to use one of them, the Raggiana bird of paradise (Genus *Paradisaea)*, to illustrate the interplay between lek behaviour and sexual selection (Beehler 1987*b*, 1988).

15.10.1   Male–male competition

LeCroy *et al.* (1980) and LeCroy (1981) suggested that lekking males of the Genus *Paradisaea* perform specific male–male dominance displays that are discrete from displays related to attraction and courtship of females, and that male dominance is the sole factor responsible for the distribution of matings among males. However, neither Frith (1981) nor Beehler (1988) observed any display interaction between males of the Raggiana bird of paradise that appeared to be related to the development of a dominance hierarchy. This is not to suggest, however, that males do not interact. There is some evidence that beneath the usually peaceful relationship among male lek members, a dominance hierarchy is present. Lek males may engage in 'scuffles', displacements, chases, and even major fights. In view of the documentation of male–male conflict, it seems contradictory that lek males rarely attempt to disrupt either pre-copulatory behaviour or copulations of rival males (e.g. Beehler 1988). Thus, in Raggiana, it appears that both a dominance hierarchy and a degree of mutual tolerance develops among regular members of the lek. This mutual tolerance contrasts strikingly with the aggressive behaviour and conflict of some other highly ornate tropical lekkers, such as the cock-of-the-rock. In support of the suggestion that male Raggiana exhibit tolerance, it also appears that rigid dominance hierarchies may not always be present. Two individually identifiable male Raggiana exhibited reversals of dominance; one male, 'A' displaced or chased 'B' four times, while the reverse was noted twice.

This apparently egalitarian approach by males to mating opportunities is not practised by all of Raggiana's close relatives. In the lesser bird of paradise (also a member of the Genus *Paradisaea*), one male in a lek of eight was overwhelmingly most successful in attracting and mating with females, obtaining 96 per cent (25 of 26) of the copulations observed (Beehler 1983). On two occasions this male was absent from his perch, which was not occupied by another male. In another instance, a male was solicited by a female but failed to respond; this suggests that he may have been inhibited by the presence of one or more higher ranking males. These observations of lek attendance and relative mating success suggest that despotism by a dominant male lesser bird of paradise is far more pronounced than is seen in Raggiana, and that it can be maintained without overt aggression.

In considering male–male interactions, some additional points need to be mentioned. First, for at least some species, several years are required before males attain adult plumage. Male Raggiana remain in female-like plumage for up to six years and may not acquire full adult male plumage for several years after that (Beehler 1988). Thus, acceptance and incorporation of a male into a lek may be a long-term process that takes place over a period of many years. Second, in this species, younger, female-plumaged males are usually driven from the lek by adult-plumaged males. In contrast, adult male lek members are mutually tolerant. The female-plumaged males maintain social contact with the lek males by counter-calling with them. This behaviour by young males may set the stage for their eventual acceptance into the lek (Beehler 1988). Third, adult male Raggiana display daily on their leks for eight months of the year, even though females nest over a period of only about three months. This extended occupancy of the lek site is probably related to the importance of male hierarchies within the lek (Le Croy 1981; Beehler 1988). Defence of status apparently forces competing males to be present at the lek while others are there. Thus, it is likely that each male must defend its position from other males throughout most or all of the year in order to maintain its rank.

Another factor that may make male–male competition at leks inconspicuous is the behaviour of females, which appears to dampen potential male aggression at the lek, as follows. Females come to leks and depart without mating far more often than they visit and mate (35 matings in 305 visits, Beehler 1988). Males that approach females aggressively, or even displaying males that approach females, cause them to leave, and active male–male aggression almost surely would do likewise. Therefore, it is probably to each male's benefit to be 'well-behaved' (Foster 1983); i.e. to behave in a manner that does not tend to drive females away from the lek (see also Widemo 1997). Aversion of females to male–male aggression at the lek may have promoted the development of rigid dominance relationships, etc., during the nine months of the year when females are not in breeding condition.

In short, although male–male conflict, both subtle and overt, occurs in classically lekking birds of paradise, there appears to be no hard evidence that special displays are associated solely with male aggression.

15.10.2   Female choice

Beehler (1988) found clear evidence of female choice in Raggiana's bird of paradise. Females usually visit the lek in groups of two to six. Typically, when one or more females enter the lek tree, the males begin to jump about their perches, giving the display call, and initiating an 'excitement display' (see LeCroy 1981, Beehler 1988 and Johnsgard 1994 for photographs of displays). All lek males within hearing distance, whether or not they were originally in the lek tree, immediately converge on the lek and join in this display. This 'convergence display' assembles all regular lek males and marks their initial interaction with the female. This is followed by the 'static display', in which the males are quiet and mostly immobile. At the end of this sequence, during which the wings are spread and the orange pectoral plumes are held over the back, the female begins closely examining males. If a male breaks his pose and moves towards the female, she retreats and may leave the lek entirely. Such behaviour presumably selects for all male lek members to remain at their display stations and to hold their poses.

When the female decides on a particular male, she often mounts him briefly: this is the male that subsequently mounts and copulates with her. Once the female has indicated her choice, only rarely do the other males interfere. At a lek in which four males were individually identifiable, all four received both female solicitations and matings. Thus, although skew in mating success does occur among males jointly occupying a lek, it appears that, in this species, most or all regular members of the lek will receive some matings (Beehler 1988). That is, unlike the situation in some other lekking species (e.g. lesser bird of paradise), all male lek participants are apparently likely to receive a genetic payoff during the course of a single breeding season. The fact that all lek males may benefit directly from visitations by females (i.e. all may receive copulations) may account, in part, for the lack of aggression by males.

In Raggiana, the behaviour of both males and females at the lek supports the view that female choice is being exercised. First, females move uninterruptedly among the males, observing them, and eventually solicit copulation from one of them. Second, the fact that all males in the lek copulate when solicited is important evidence that subtle forms of male dominance behaviour do not override (at least completely) the females' freedom of choice; i.e. unlike the leks of the lesser bird of paradise, described above, it appears that no male refrains from mating in deference to a nearby dominant individual.

Two other features of the biology of the typical birds of paradise may also have contributed to intensifying sexual selection, which presumably has promoted the extreme development of the plumage and display characters that have made these birds famous. First, in most of the strongly dimorphic, polygynous species, the clutch is composed of only a single egg. Small clutch sizes are common in tropical species; however, a one-egg clutch is obviously as far as selection can drive adaptive clutch size reduction. A clutch of one has several implications for both male–male competition and female mate choice. For one thing, it means that multiple paternity of broods is an impossibility, and thus that a single copulation will certainly monopolize all of a female's reproductive effort for that clutch; i.e. since there is no opportunity whatsoever for multiple paternity within a female's brood, all of a female's reproductive effort (nest-building, incubation, feeding of the nestling, etc.) will always benefit only a single male. This may, in a sense, force males to put an enormous amount of their time and effort into attempting to entice females to mate with them, and, as discussed above, this is a conspicuous aspect of the reproductive pattern of adult male birds of paradise. This point is reminiscent of the paternity pattern seen in the lekking grouse. Female grouse mate only once per clutch, thus, like the birds of paradise that lay only a single egg, each copulation is extremely valuable. The pattern of one sire per brood should increase the overall variance in male reproductive success beyond what would be the case if broods were typically sired by more than one male.

Second, male birds of paradise live for many years in immature plumage before beginning to attain the conspicuous ornamentation of adults. This implies (1) that the younger males are competitively inferior to older males and that they probably would be unable to take advantage (obtain matings) of showy adult plumage even if they possessed it, (2) that cryptic plumage is safer than bright plumage, and (3) that annual mortality rates are probably low (as is also implied by the one-egg clutches). These factors suggest that, for males, there may be no particular hurry to attain full breeding status. Moreover, as a result of (presumed) low mortality, it may require years for a young male to work its way into membership of a lek (Beehler 1988).

The delay in sexual maturation of male birds of paradise may be due ultimately to the benign environment occupied by most members of this group. That is, if birds of paradise have a dependable and high-quality food supply, if predation is low, if males are not required to participate in parental care activities, and if young males cannot compete successfully for lek membership, then selection may have favoured retardation of sexual maturity relative to females. In addition, over the annual cycle, these same benign environmental factors permit adult males to remain on or near their leks throughout the day for at least three-quarters of the year. This last point clearly suggests that male–male competition, either at the lek or for admission to the lek, or both, is an important component of sexual selection in these lekking birds of paradise.

To summarize, on the basis of the few well-studied cases of male and female sexual behaviour in lekking birds of paradise (Beehler 1988; the Pruett-Jones' (1990) study of sexual selection in the quasi-lekking Lawes' parotia is described in Section 6.3), it appears that both male–male competition and female choice are important, as Darwin intuited long ago. Male social relationships may be worked out to a large extent during the non-breeding season and over long periods of time, perhaps measured in years. As a result, overt aggression between males on leks, which apparently repels females, is rare, at least in the few species studied thus far.

15.11   Hidden leks in monogamous species?

Lek-like mating strategies occur in socially monogamous species with biparental care. Wagner (1997) has demonstrated that lek-like behaviours and mating patterns (i.e. strong skew in matings) occur in some colonial species that exhibit both social monogamy and biparental care (Fig. 15.2), and he argues that this may also be the case in many typical monogamous species that exhibit standard territorial behaviour (Fig. 15.3). Wagner's thesis, in brief, is that extra-pair copulations in socially monogamous species set the stage for two different reproductive strategies by both males and females, and that it is therefore important to distinguish between the genetic and the social parenting systems (see also Gowaty 1985, 1996a). Females of such species must have a social mate to provide care for their offspring; in addition, they will often attempt to obtain matings from males other than their social mates. From the males' perspective, obtaining both a social mate and EPCs is also important, although some males in the population will usually be unsuccessful in

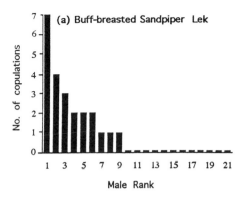
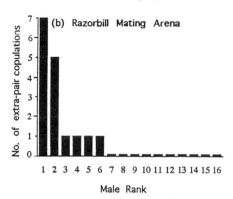

Fig. 15.2   The similarity between the distributions of (a) male copulation success in a lek of promiscuous buff-breasted sandpipers and (b) male extra-pair copulation success in a mating arena of monogamous razorbills. Reprinted with permission from Wagner, R. (1997). Hidden leks: sexual selection and the clumping of avian territories. In *Extra-pair mating tactics in birds*, Ornithological Monographs, No. 49 (ed. P. Parker and N. Burley), pp. 123–145. American Ornithologists' Union, Washington, DC.

achieving either, in part simply because overall sex ratios are typically male-biased (see Section 11.4.4).

The male-biased sex ratio has an important role in this scenario, because it provides leverage for females to obtain EPCs. For example, a young or otherwise competitively inferior male may have only two options—either to accept a mate on the female's terms, including being cuckolded, or to not form a pair bond at all. The facts that EPCs are so common in some socially monogamous species, and that it has been convincingly demonstrated that females are discriminating in mate choice (with preferred males also being high-quality males) together suggest that Wagner's 'hidden lek' hypothesis holds a great deal of promise for increasing our understanding of avian reproductive strategies.

Wagner (1997) predicts that when the genetic system is viewed separately from the social mating system, we will often find a 'lek' hidden among the territories of monogamous birds. Specifically, for territorial species, females will form a pair bond with a male that is positioned near an especially attractive, typically already-mated male. This is similar to the hotshot model, except that Wagner assumes that females actively discriminate among males; alternatively, the clumping of territories may be due to the preference of females to mate in an area where many males (potential EPC partners) are located, which parallels the female preference model for lekking.

The hidden lek hypothesis suggests that the relationship between high breeding density, as seen in colonial species, and EPC may reflect a strategy by females to obtain access to a number of males. This resembles the female

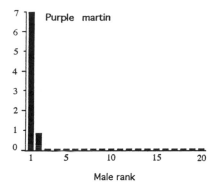

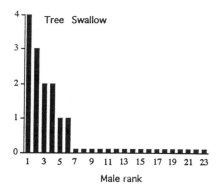

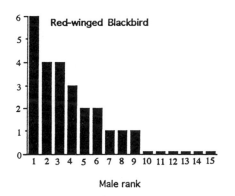

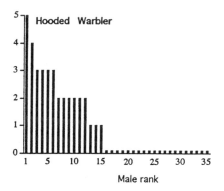

Fig. 15.3   Lek-like skews of extra-pair fertilization success of four species of North American passerine birds. Reprinted with permission from Wagner, R. (1997). Hidden leks: sexual selection and the clumping of avian territories. In *Extra-pair mating tactics in birds*, Ornithological Monographs, No. 49 (ed. P. Parker and N. Burley), pp. 123–145. American Ornithologists' Union, Washington, DC.

preference model of lek formation, rather than the more usual interpretation that the proximity imposed by coloniality increases opportunities for EPCs. In addition, the puzzling clumped distributions of territories within homogeneous habitat can potentially be explained by this hypothesis. Determining genetic parentage of offspring, together with careful behavioural studies, will be required to test the ability of the hidden lek hypothesis to explain breeding distributions in socially monogamous species (Wagner 1997).

One such test is a recent study by Hoi and Hoi-Leitner (1997), who found that extra-pair copulations and fertilizations were common in bearded tits that nested colonially. In contrast, extra-pair fertilizations were not detected in broods of pairs that nested solitarily. Females that nested colonially exhibited superior condition to solitary females, and their behaviour suggested that they competed actively for extra-pair copulations. Finally, in this species, it is the females that choose the nest site (i.e. choose to nest colonially or solitarily). All of these points accord nicely with Wagner's (1993) hidden lek hypothesis of colony formation.

In short, Wagner's hidden lek hypothesis indicates that, with regard to female choice of genetic mates, the reproductive strategies of lekking and non-lekking species may be less different from other mating systems than has usually been appreciated. It also re-emphasizes the importance of recognizing the distinctions between the social parenting systems and the genetic mating strategies followed by members of each sex.

15.12   Conclusions and summary

Although it is a rare mating system, lekking has evolved independently in a variety of avian species. Two categories of leks are commonly recognized, classical leks and exploded leks. In the former, males are tightly aggregated, sometimes with conspicuous male–male conflict, while in the latter, males are widely separated and therefore conflict is infrequent. Because of the often large number of individual males concentrated at a discrete site, classical leks are both more striking and more difficult to explain than exploded leks, which are similar to simple individual males territories with specific display sites. For these reasons, this chapter has focused primarily on classical leks.

With regard to their evolutionary derivation, it appears likely that classical leks represent an end-point on a continuum of mating systems. Strict monogamy (both social and genetic) and biparental care represent the other end of the spectrum. A necessary condition for the evolution of leks is the ability of females to rear offspring without male assistance. Thus, the intermediate steps from monogamy to lekking probably involved increasing polygyny, along with loss of male parental care. Once females obtained neither paternal care nor defended resources, as in male-held territories, they became 'emancipated' from having to compromise their reproductive interests with those of any particular male. That is, because individual males had nothing (neither parental care nor resources) to offer females, they lost their leverage with females. Under such conditions, females became free to follow strategies that maximized their own reproductive interests; i.e. a female had nothing to gain from any degree of fidelity to a particular male, with one exception, namely sperm from preferred individuals. The fact that females of some lekking species nevertheless show a high level of mating fidelity to one male strongly suggests that careful mate choice is important, and the most likely basis for this importance is differential genetic quality among males.

Three major hypotheses have been developed to account for classical leks: the female preference model, the hotspot model, and the hotshot model. The main premise of the first is that females prefer groups of males to solitary individuals because groups offer the opportunity for females to make comparisons among males. In the hotspot model, males are attracted to sites of maximal female traffic. The female preference and hotspot models can then act in a mutually reinforcing way, in that aggregations of males permit fuller female choice than would isolated males. The hotshot model suggests that females are attracted to certain males for conservative reasons, such as copying, and that other males are also attracted as a result of the presence of females. In this scheme, male despotism, rather than female choice *per se*, is the basis for the extreme variance in reproductive success. To reiterate, the female preference model is based on female choice, whereas male–male competition is the critical factor in the hotshot model. In contrast, in the hotspot model, males are concerned with the spatial distribution of resources required by females.

Empirical studies suggest that, across a variety of lekking species, each of these models may explain partially the adaptive significance, and possibly the evolutionary origins, of leks. An interspecific study of manakins by Thery (1992) provides good empirical evidence for the hotspot model, and it further suggests that all three models can contribute to explanations of lekking in this group. The extreme variance in male reproductive success characteristic of many lek species suggests either that relatively few males are strongly preferred by females or that such males are able to control the behaviour of other members of the lek. In some systems, the former interpretation appears to be correct (e.g. sage grouse, Gibson *et al.* 1991), while in others, the evidence suggests that despotic behaviour by a single male may be most important in determining which male mates with visiting females (e.g. Beehler 1983). Clearly, in many species, both male–male conflict and female preferences play important roles in the mate selection process (e.g. Trail 1985).

Because leks are a manifestation of extremely strong sexual selection, it comes as no surprise to learn that many birds of paradise, possibly the most spectacular and bizarre birds on Earth, exhibit lekking behaviour. (Johnsgard 1994 provides a review of the displays of several lekking species.) In the Raggiana bird of paradise, lekking behaviour is especially highly developed in that males appear to display in a collective or cooperative manner to enhance the group's overall attractiveness to females. Moreover, when a female signals a mating preference for copulation, the non-selected males typically do not interfere. This absence of overt competition by males is all the more striking in view of the fact that a single sperm (and thus a single copulation) will fertilize an entire 'clutch' (one egg) and thus monopolize the female's reproductive effort for most or all of the annual breeding season. The aversion of females to aggressive behaviour at the lek has apparently favoured the development of competitive restraint and mutual tolerance among males of this and some other lekking species.

From the perspective of sexual conflict, if leks pose major reproductive costs to the majority of males that obtain no matings, then it follows that leks primarily reflect reproductive strategies of females. This suggestion may be strengthened by the 'hidden lek' hypothesis of Richard Wagner. Wagner argues that lek-like mating systems may occur even in socially monogamous species with biparental care. In such species, females have a social mate with whom they share parental care. In addition, they often copulate with another male. In short, females share parental duties, as well as some parentage, with one male, while also obtaining sperm from another, possibly genetically superior, individual. Thus, the genetic mating system of some monogamous species may resemble that of lekking species.

16 Classical polyandry: the most puzzling of avian mating systems

16.1 Introduction

The term polyandry refers to a mating bond between a female and two or more males. Classical polyandry (Oring 1986) is the term used for social–mating systems where (1) a single female mates either sequentially or simultaneously with two or more males during a single breeding season, and (2) males typically form bonds with no more than one female per breeding cycle or per season. A key characteristic of classical polyandry is that each male has its own nest, where it incubates the eggs and later provides care for its (usually) precocial young, most typically with little or no help from the female. In classical polyandry, males not only provide virtually all parental care, females also compete aggressively for mating opportunities. Female–female competition is probably causally related to the reversed sexual dimorphism in body size and often colour seen in such species (e.g. Jehl and Murray 1986). Parental care exclusively by the male is a requisite for classical polyandry; however, monogamous mating systems with male-only care also occur (e.g. emus, kiwis). Thus, although exclusive paternal (as opposed to biparental or maternal) care is a requisite for classical polyandry, it is not synonymous with polyandry. I might also mention that females of both socially monogamous and socially polygynous species are often genetically polyandrous; in these cases, however, the concept of EPC is employed (see Chapter 12).

In contrast to classical polyandry, cooperative polyandry (see Chapter 13) is recognized by the presence of one female with two or more male mates at a single nest, and where all such males may contribute gametes and later provide parental care to eggs and chicks of that nest. These two categories—classical and cooperative polyandry—are fundamentally different kinds of systems in nearly every way (Oring 1986). The one similarity is that a single female may have social–mating bonds with two or more males either at the same time or sequentially.

In classically polyandrous species, the reproductive role assumed by each sex contrasts strikingly with the typical avian pattern of biparental care, as well as with the uniparental care provided by females of many polygynous species. The most conspicuous aspect of classical polyandry is parental care solely by the male. As soon as the clutch is laid (by one to several females in paleognathous species and by one female in polyandrous neognathous species), males assume all incubation duties. What combination of factors led males of such species down the evolutionary pathway to sole care of offspring, while at the same time producing females that seek to lay clutches for additional males? The benefits to females are readily envisioned; by laying a clutch for only one additional male, they can double their overall reproductive output. For males, on the other hand, the necessity of incubating the eggs means not only that mate guarding as the female produces a subsequent clutch may not be feasible, it also means that the male has severely reduced its prospects of copulating with additional females in the neighbourhood.

Thus, classical polyandry is probably the most interesting, and certainly is the least well understood, of the recognized avian mating systems. For example, Clutton-Brock (1991: p. 259) writes: 'No single explanation of avian polyandry is satisfactory, and its evolution continues to puzzle behavioral ecologists.' Basically, the problem has been based on the question: what are the selective forces that led males, with their great inherent potential for fertilizing females' gametes, to relinquish that potential benefit of maleness and to confine or restrict their reproductive efforts to one nest and its contents? Although paternal care is common in birds and is, in fact, almost the rule, parental care solely by the male definitely is not.

As discussed in detail in Chapter 10 and briefly below, sole paternal care may have been the first form of parenting to evolve in the Class Aves. In fact, it could have evolved before birds were birds; i.e. prior to the evolution of feathers. Incorporating a historical perspective into the issue of avian mating systems is essential, I believe, if we are ever going to understand classical polyandry.

16.2   Taxonomic distribution of exclusive paternal care and of classical polyandry

Exclusive male parental care occurs in most species of paleognathous birds, often in association with sequential classical polyandry (Table 16.1). As discussed in Chapter 10 and below, paternal care and classical polyandry in this group appear to have been strongly influenced by their phylogenetic history.

Table 16.1   Families of birds with parental care exclusively by males and in which classical polyandry is known or thought to occur

	Common name	Family	Order	Comments
Paleognathae				
	Tinamous	Tinamidae	Tinamiformes	Females of some species may lay eggs in the nests of several males.
	Rheas	Rheidae	Rheiformes	Males accept the eggs of several females into their nests; females may lay eggs in the nests of more than one male.
	Emu	Dromiceidae	Casuariiformes	Emus and cassowaries are thought to be primarily monogamous, with male care of eggs and chicks.
	Cassowaries	Casuariidae	Casuariiformes	
Neognathae				
	Buttonquail	Turnicidae	[Gruiformes][1]	Classical polyandry prevalent
	Jacanas	Jacanidae	Charadriiformes	Classical polyandry is thought to occur in 7 of the 8 living species.
	Painted-snipes	Rostratulidae	Charadriiformes	Classical polyandry is thought to occur in 1 of the 2 living species.
	Plains-wanderer	Pedionomidae	Charadriiformes	Classical polyandry is thought to occur in the sole living species.
	Sandpipers, phalaropes, etc.	Scolopacidae	Charadriiformes	Classical polyandry occurs commonly in the spotted sandpiper and less commonly in phalaropes.
	Plovers, lapwings, etc.	Charadriidae	Charadriiformes	Classical polyandry occurs in the Eurasian dotterel.
	Coucals	Cuculidae	Cuculiformes	Parental care conducted primarily by male in most species. Classical polyandry with exclusive parental care occurs in at least one species, the black coucal.

[1]The taxonomic position of the buttonquail is unknown (see text).

Classical polyandry is also of regular occurrence in one or more species of seven families of neognathous birds (Erckmann 1983; Oring 1986; Ligon 1993). Among neognathous birds, the single most striking fact about classical polyandry is its taxonomic or phylogenetic distribution: it is restricted almost exclusively to five families in one order of birds, the Charadriiformes (Table 16.1). Apparently, the only other examples of classical polyandry among neognathous species are the precocial buttonquail, Family Turnicidae, a group of uncertain taxonomic affinities (Sibley and Ahlquist 1990), and at least one altricial species of cuckoo, the black coucal (see below).

In addition, based on observations by Rand (1936), classical polyandry has been suggested for the subdesert mesite, a member of a primitive and taxonomically puzzling family of birds restricted to Madagascar. However, females of this species have subsequently been recorded incubating, and this is also true of at least one of the other two species (Rand 1951). Thus, whether classical polyandry actually occurs at all in mesites remains to be determined. In any case, the strict role reversal seen in other classically polyandrous species does not appear to be present in this group (see also Campbell and Lack 1985).

In some passerine species, some females apparently desert their mates and recently fledged offspring to form a mateship with a new male. This female strategy has been reported for several species of North American cardueline finches and has sometimes been labelled 'polyandry' (Middleton 1988; Seutin *et al.* 1991). Distinguishing between serial monogamy and sequential polyandry in cases such as these becomes a fine point. However, since females incubate the eggs and provide care for chicks at both nests, serial monogamy may be the more appropriate interpretation.

16.2.1   Classical polyandry in ratites and tinamous

As discussed in Chapter 10, exclusive paternal care is the rule in ratites and tinamous. Paternal care in these birds may be derived directly from the original form of parenting in the Class Aves. From paternal care, all major mating systems, including classical polyandry, can evolve in a comparatively parsimonious manner (van Rhijn 1984, 1990; see Fig. 10.2). In the greater rhea, ostrich, and at least three genera of tinamous, a group of females may lay eggs in the nest of a single male that presumably fertilized those eggs. After producing a full clutch, which the male then incubates alone, the females may then move to another male and repeat the process. Oring (1986) classified such systems as female defence polygyny (males), plus sequential polyandry (females).

Male mating systems
Males of the 10 living species of ratites are either monogamous or polygynous. Male kiwis are apparently monogamous, as are male emus (Coddington and Cockburn 1995) and cassowaries (Crome 1976). Male ostriches and rheas are simultaneously polygynous. Similarly, male tinamous may be either monogamous or polygynous.

Female mating systems
The mating systems of female ratites are either monogamous (e.g. kiwis) or sequentially polyandrous, or both may occur. The last is typical for the ostrich, the only ratite species in which females provide parental care (see also Section

10.5.1). In the ostrich, one of a group of females, the dominant 'major' hen, remains with the male and nest and shares incubation duties and the care of the young with the male; this female appears to be monogamous. The other females may move to a second male and lay eggs in his nest, thereby exhibiting sequential classical polyandry. Aside from ostriches, female ratites and tinamous apparently provide little or no parental care.

Several female rheas lay eggs in the nest of a male and, if environmental conditions are favourable, they may then leave the first male and produce eggs for the nest of a second. Thus, female rheas are facultatively sequentially polyandrous (Handford and Mares 1985). Sequential polyandry by females is also a common reproductive strategy in tinamous (see also Ridley 1978).

In emus, a single female forms a pair bond with a single male; thus, the female is initially socially monogamous (Davies 1976). After its mate begins incubation, the female may seek a second mate. If successful, female emus thus become sequentially polyandrous (Coddington and Cockburn 1995). Female emus compete vigorously for second mates, and the benefits to be obtained by successful competition are straightforward: mating with one additional male—sequential polyandry—potentially doubles annual reproductive output.

Like emus, cassowaries may also be monogamous at the beginning of the season, with the female sometimes obtaining a second mate after the first is occupied with parental duties (Crome 1976).

To conclude, in ratites and tinamous, parental care solely by the male, a trait that is probably primitive in this group (see Chapter 10), permits, and, in fact, has frequently encouraged, the development of sequential classical polyandry by females and single nest polygyny in males, a mating–parenting combination unknown in other birds.

16.2.2   Classical polyandry in shorebirds

Many authors have observed that the diversity of mating systems among the shorebirds exceeds that of any other comparable group (e.g. Pitelka *et al.* 1974; Oring 1982, 1986; Erckmann 1983; Jehl and Murray 1986; Szekely and Reynolds 1995; see also Lenington 1984; Walters 1984), and, in the minds of most avian behavioural ecologists, this is the group most closely identified with classical polyandry. Erckmann (1983) states that not only has classical polyandry been discovered in more species of shorebirds than in all other birds combined, but it has evolved in this group in more families and more species than has polygyny. Clearly, the diversity of mating systems and especially the question of why classical polyandry occurs in five distinctive shorebird families, while being almost non-existent in all other neognathous birds, has attracted a great deal of attention.

Polyandry may have come about in extant shorebirds in two different ways. First, as discussed in Chapter 10, in certain cases, exclusive paternal care might possibly have been the retained ancestral condition. Second, in other cases, the flexibility of mating systems in the shorebirds, as indicated by the frequency of transitions from one system to another (Szekely and Reynolds 1995), suggests that particular characteristics of this group may have promoted or permitted the evolution of classical polyandry. Thus, two approaches to the issue of polyandry in contemporary shorebirds can be taken. One can seek ecological explanations for the evolution and maintenance of polyandry or one can view the evolution of polyandry as one result of the phylogenetic history of this group. However, rather than viewing phylogenetic and ecological approaches as either–or alternatives, it is more productive (and also more complicated) to consider them as complementary, with each contributing to the development of the unusual diversity of mating systems, including polyandry, seen in shorebirds. That is, although the basic, conservative reproductive traits of the group have contributed to the evolutionary development of classical polyandry, it appears that in some cases ecological factors have also played major roles. If these suggestions prove to be correct, it would be futile to search for a single explanation to account for classical polyandry in shorebirds.

16.3   General traits of shorebirds that may promote polyandry

In this section I discuss the proposition that phylogenetic history is one critical aspect of the puzzle of mating system diversity seen in shorebirds. By 'phylogenetic history' I mean traits shared by shorebirds as a result of their common evolutionary history that may have contributed to the development and/or maintenance of classical polyandry. Several aspects of the biology of this group need to be considered in this context. These include characteristics of the eggs, clutch size, and patterns of parental care (who provides care, the ability of one parent to hatch and rear young). In addition, other traits, such as reversed size dimorphism and the ability of females to lay successive clutches, have probably played a role in the evolution and/or maintenance of polyandry in this group.

16.3.1   Reversed sexual dimorphism

Reversed sexual size dimorphism is the rule in scolopacids, irrespective of mating system. Jehl and Murray (1986, Appendix II) list weights by sex for 95 species; of these, the male-to-female size ratio was greater than one in only 16 species, while in 79 species, females were the larger sex. Jehl and Murray (1986) suggest that smaller size of males is related to aerial territorial and courtship displays. Large female body size relative to males, whatever its original adaptive significance, can be viewed as a factor that might help to promote the evolution and maintenance of polyandry.

16.3.2   Characteristics of the egg

The eggs of shorebirds are large relative to those of most other birds and thus are potentially costly to produce (Rahn *et al.* 1975). They also exhibit an unusual shape, with a great decrease in circumference from the large to the small end. Andersson (1978) analysed shorebirds eggs and showed that, as a result of their shape, the volume of the egg could be increased by 8 per cent over a spherical egg of the same diameter without increasing the egg area covered by the incubating bird. Andersson suggests that, in these birds, egg shape has resulted from selective pressures to maximize egg volume without increasing the area covered by the incubating bird, and that strong selection for maximal egg volume is probably why clutch size is limited to three or four. This scenario suggests that there has been strong selection on female shorebirds to produce clutches of the largest eggs that can be produced and that can be covered by the incubating bird. The fit of the bird to its clutch of eggs may be more precisely developed in shorebirds than in any other avian group.

16.3.3   Clutch size

Clutch size in shorebirds, from pantropical jacanas to the numerous arcticbreeding sandpipers to the dry country plains-wanderer of the Australian outback to the familiar killdeer of the United States, is strikingly invariable, usually numbering four eggs. Maclean (1972) first suggested that clutch size is a phylogenetically conservative trait in this group, pointing out that no species normally lays a clutch of more than four eggs, regardless of body size, zoogeographical distribution, type of habitat occupied, or ecological specialization. By use of extensive comparative data, Maclean (1972) also makes a case for the proposition that four eggs is the ancestral clutch size in this group, and that clutches of fewer than four eggs are adaptive deviations derived from the ancestral condition. Walters (1984) also provides arguments which support this perspective.

An obligate four-egg (or fewer) clutch places constraints on the reproductive potential of individuals in this group (Ligon 1993). This is not meant to suggest that the clutch size of four is completely immune to selection, as in several cases it has been reduced. Determinate clutch sizes do, however, suggest a strong developmental resistance to selection, based on ecological or demographic factors, for a clutch size of more than four. The fact that, among shorebirds, all of the few deviations from a four-egg clutch are reductions in number also supports the view that strong constraints serve to prevent an increase in clutch size. We cannot know why clutch size became so inflexible early in the history of this group, but the fact that this is the case suggests that knowledge of the ecology of contemporary species is insufficient to account for the remarkable consistency of clutch size across the diverse array of species and families that make up the shorebirds.

Earlier, I suggested that obligate four-egg clutches might be causally related to the diversity of mating systems, including polyandry, found in the shorebirds (Ligon 1993). In brief, the only available means of side-stepping (evolutionarily) the four egg/clutch limitation is by production of additional clutches. This is why the constraint of a small, fixed clutch size may be critical to the evolution of the great diversity of mating systems, including polyandry, found within this group. This alone, however, cannot account for the development of monogamy in species A, polygyny in species B, and polyandry in species C.

With regard specifically to classical polyandry, the key point is that, under certain conditions, one sex (here it is females) can benefit greatly by desertion, via increased reproduction. Such desertion, plus paternal care and the ability of the male to rear the brood unaided (see below), produces a sequentially polyandrous system. This scenario, however, leaves the issue of why males accept sole responsibility for the clutch incompletely resolved. A consideration of phylogenetic history is necessary to account for exclusive paternal care (see Section 10.5.2 and below).

Finally, fixed, restricted clutch sizes also occur in some other major groups that do not exhibit polyandry, such as the tube-nosed swimmers, loons, pigeons, doves, and hummingbirds, among others. However, in these, other kinds of constraints on clutch size (e.g. the feeding of young with crop milk in all pigeons and doves) appear to have limited, rather than promoted, diversity of mating systems and parental care. Clearly, factors in addition to a small, fixed clutch size must be present to account for the evolution of a diversity of mating systems within a group of closely related species.

16.3.4 Male parental care

Erckmann (1983) assumes that a mating system characterized by monogamy and biparental care was the ancestral condition from which all polygamous mating systems of shorebirds evolved (see also Szekely and Reynolds 1995; cf. van Rhijn 1984, 1990, and Chapter 10). Following this scenario, the involvement of the male in incubation and care of chicks can be viewed as an essential prerequisite for the evolution of sole male parental care, and thus also for classical polyandry. However, although male parental care is virtually the rule in birds, polyandry is rare and restricted to a few groups.

16.3.5 The need for only one parent to tend young

Under certain sets of conditions, one parent can rear as many young as two; in such species, selection to maintain biparental care might be expected to be weak. This is an important pre-adaptation for any sort of polygamy with uni-parental care, but, by itself, it does not help us to understand why the female parent remains with the clutch in some species, while in others, it is the male that does so.

16.3.6 Double-clutching

Some monogamous shorebird species exhibit a reproductive strategy that is often referred to as double-clutching. In this reproductive system, the female of a monogamous pair lays two clutches, the first being incubated by the male and the second by the female. Double-clutching is primarily a means employed by a mated pair to double their reproductive output within a single breeding season. (In addition to its occurrence in shorebirds, double-clutching has been recorded in a few galliforms; see Oring 1982, Table IV).

Most reviews of shorebird mating systems have suggested that double-clutching was the evolutionary precursor of classical polyandry in this group (e.g. Jenni 1974; Pitelka *et al.* 1974; Ridley 1978; Pienkowski and Greenwood 1979; Faaborg and Patterson 1981; Oring 1982). Erckmann (1983) refers to this evolutionary scenario, from typical single-nest monogamy to mono-gamous double-clutching to classical polyandry, as the stepping-stone model. The stepping-stone model has been appealing since it offers an explanation for how males come to provide sole parental care for broods and how females might gain the opportunity to mate with more than one male. Erckmann (1983) and Oring (1986) argue, however, that double-clutching and classical polyandry are unrelated phenomena and that each arose independently from single-nest monogamy with shared parental care.

Erckmann (1983) points out that double-clutching species differ from class-ically polyandrous species in several important ways. (1) Sexual dimorphism in double-clutchers is similar to that in related monogamous species, and the average annual gametic contributions per breeding individual is about the same for males and females. (2) Reversal of sex role has never been observed in any double-clutcher. (3) The contributions to parental care by the two sexes are nearly the same, as each bird incubates alone at its nest. In contrast, in classi-cally polyandrous shorebirds, females do not normally incubate the eggs. (4) Mate switching between clutches (sequential polygamy) is not a regular feature among double-clutching species. (5) There is a lack of a taxonomic relationship between double-clutching and polyandrous species. If double-clutching led, in an evolutionary sense, to true polyandry, some taxonomic groups would be expected to contain both mating systems. However, all double-clutching scolopacids are in the Subfamily Calidridinae, whereas all polyandrous scolopacids are in, or are closely related to, the Tringinae. In addition, except for the Charadriiformes, no order of birds includes both double-clutching and polyandrous species.

For the reasons listed above, both double-clutching and classical polyandry in some shorebird species (e.g. the spotted sandpiper; see below) probably evolved independently from monogamy with shared parental care (Erckmann 1983; Oring 1986). Double-clutching systems may have evolved by either of two different routes: through (a) double-brooding in monogamous species or (b) mate desertion and attempted polygamy by both sexes. Erckmann (1983) suggests that double-clutching in the mountain plover arose by the first route, while in the sandpipers it evolved via the second. Erckmann's suggestion that double-clutching arose in two very different ways is similar to the one offered here, namely that exclusive paternal care and classical polyandry in living shorebirds may also have arisen in two ways, via either ancestral exclusive paternal care or the derivation of exclusive paternal care from biparental care (see below).

Recognition that double-clutching and classical polyandry evolved independently and that species placed within each of these categories may have followed very different evolutionary pathways, together with the fact that these rare and peculiar systems occur almost exclusively within the shorebirds, lends additional weight to the suggestion that traits shared by shorebirds may be responsible for the diversity of mating systems seen within this group. As argued below, they also imply that current ecological factors cannot fully account for this diversity.

16.4   Ecological hypotheses for the evolution of classical polyandry in shorebirds

Nearly all previous discussions of the factors promoting classical polyandry in shorebirds have favoured ecological explanations to the virtual exclusion of phylogenetic history. Erckmann (1983) provides a comprehensive review of the evidence and logic pertaining to four ecological hypotheses for the evolution of polyandry in shorebirds; his conclusions are briefly summarized here. Erckmann's paper should be consulted for a more detailed assessment of these ideas.

16.4.1   The stressed female hypothesis for the evolution of sex-role reversal

Graul *et al.* (1977) suggest that food scarcity is the critical environmental variable favouring the initial evolution of sex-role reversal, with or without polyandry. The idea is that reduced energy reserves preclude female incubation following the energy-depleting task of producing eggs. Erckmann (1983) rejects this hypothesis because (1) there is no evidence that food scarcity generally prevents females from laying successive clutches, and (2) available evidence suggests that food scarcity selects for biparental incubation rather than exclusive male parental care.

16.4.2   The differential parental capacity hypothesis

According to this model, laying may deplete the energy reserves of females to such an extent that, although they can participate in incubation, they may be less able than males to increase investment in parental care if deserted (Graul 1974; Maynard-Smith 1977). By removing one member of 24 pairs of mono-gamous, single-nesting western sandpipers, Erckmann (1983) showed that males were indeed more capable than females of unassisted incubation, although all widowed sandpipers of both sexes abandoned their nests before any eggs hatched. The fact that the laying of a clutch does not produce large reductions in energy reserves of other arctic shorebirds also argues against this hypothesis (Erckmann 1983).

Erckmann (1983) suggests that although the difference in male and female capacity to be a single parent is probably small, the superior ability of males to incubate alone could have contributed to selection for female desertion and sex-role reversal in arctic shorebirds as a result of the necessity of a high level of nest attentiveness and an erratic food supply. However, since females incubate unassisted in nine of 24 arctic sandpipers, while males do so in only three (all double-clutchers; Pitelka *et al.* 1974), it appears doubtful that differential capacity of the sexes for single parenting was particularly important in the evolution of reproductive strategies, even in the arctic shorebirds.

16.4.3   The replacement clutch hypothesis

Frequent nest failure might select for emancipation of females so long as for-aging conditions were adequate for uniparental incubation by males (Jenni 1974; Emlen and Oring 1977). This hypothesis assumes that males should benefit by female emancipation from incubation duties, unless the female deserts them for a new mate.

Erckmann (1983) tested this idea by comparing mean nesting success and mean rate of replaceable nest loss in polyandrous species versus other shore-birds and found no significant differences. Moreover, among polyandrous species, there is a latitudinal trend in frequency of renesting that parallels the trend in monogamous species and the general trend in the frequency of nesting failures. Thus, there is no indication that recent production of a clutch sub-stantially impairs a female shorebird's ability to lay again immediately under normal conditions, or that emancipation from incubation allows females to replace clutches more rapidly than they would do otherwise.

16.4.4   The fluctuating food hypothesis

This model (Graul 1974; Parmalee and Payne 1973) suggests that double-clutch mating systems and classical polyandry may have evolved to facilitate facultative responses to annual variations in breeding conditions, specifically where food availability is highly variable within and between seasons as a con-

sequence of weather fluctuations. The basic argument is that females can capit-
alize on particularly good conditions by laying two clutches and that males
also benefit by fertilizing more than one clutch. This idea applies to the evolution of polyandry as well as double-clutching in that it assumes that periods of
food scarcity favoured assumption of all parental care by males and thus
established initial conditions conducive to the evolution of multiple-clutch
strategies (Graul *et al.* 1977; Erckmann 1983).

Erckmann's review indicates that this is not a generally applicable model
for the evolution of either double-clutching or polyandry. First, there is
insufficient annual variation in the productivity of females to indicate an
important influence for variation in food availability. Second, especially abundant food is not required for females to lay successive clutches. And third,
there is no evidence that production of multiple clutches generally influences
mortality of females. In short, according to Erckmann (1983, p. 160), none of
these ecological hypotheses has '... proved to be sufficient as a general model
for polyandry in shorebirds, but several are plausible for some species.'

16.5   Paternal care and classical polyandry as ancestral traits

For the past twenty-five to thirty years, avian mating systems have been con-
ceptualized, classified, and explained almost exclusively by use of a comparative ecological framework (e.g. Crook 1962, 1965; Lack 1968; Orians 1969;
Emlen and Oring 1977; Oring 1982, 1986; Murray 1984). Although this
approach has been productive for study of polygyny, it has not produced a satisfactory general model for classical polyandry (Erckmann 1983; Oring 1986;
Clutton-Brock 1991). This failure could be attributed either to inadequate ecological analyses (something critical may have been overlooked.) or to the possibility that ecological factors alone may be insufficient to account for the
appearance of polyandry. Although ecological–behavioural approaches are
essential to a full understanding of mating system diversity in the shorebirds,
this alone will probably not provide a satisfactory explanation for the evolution
of polyandry. In Chapter 10, I made a case for the proposition that exclusive
paternal care is the ancestral parenting system in birds. Here I suggest that this
may explain the origins of polyandry in some—but not all—of the shorebirds
that exhibit this mating system.

In any attempt to account for classical polyandry, the most critical issue is
the origin of exclusive paternal care (see Chapter 10). Once exclusive paternal
care has evolved, it is easy to see that females might frequently have more than
one mate, either simultaneously or sequentially. There are two ways by which
sole paternal care could have arisen in modern species of shorebirds, either
(1) paternal care was the first form of parenting in the 'transitional shorebirds,' suggested by Feduccia (1996) to represent the basal stock of all neognathous birds, or (2) paternal care arose from biparental care very early in the

history of shorebirds. The phylogenetic analysis of parenting by Szekely and Reynolds (1995) indicates that biparental care is ancestral in the Order Charadriiformes as a whole, but their results also indicate that paternal care is the ancestral condition in one of the two major subgroups, the Scolopacida, which includes almost all of the polyandrous species of shorebirds (see Section 10.5.2).

However, as discussed in Chapter 10, there may be reasons to question the conclusion that biparental care is ancestral in the group as a whole. Whether paternal care is primitive in all shorebirds (Charadriida plus Scolopacida) or, alternatively, is primitive only in the Scolopacida, the general point for purposes of this discussion is much the same, namely that exclusive paternal care (and classical polyandry) in extant species may have developed from two very different sources—(1) it may be a legacy from their ancestors or (2) it may have evolved from biparental care (see Fig. 10.3). Possible support for this latter suggestion comes from the fact that transitions from one type of parenting to another have apparently occurred frequently (see Section 10.5.2 and Szekely and Reynolds 1995).

16.6   Secondarily derived exclusive paternal care and classical polyandry

As mentioned above, polyandry in certain shorebirds is probably derived secondarily from biparental care. This suggestion is consistent with the analyses of Szekely and Reynolds (1995), which indicate that transitions from one form of parenting to another have occurred commonly in this group. For example, the sex-role reversal and polyandry of species such as the extensively studied spotted sandpiper are probably derived from biparental monogamy (Oring 1986). For this species, the most obvious and compelling evidence that male-only care was preceded by biparental care is the fact that female spotted sandpipers possess the ability and, under certain circumstances, the inclination to both incubate eggs and care for chicks. The challenge is to identify the set of factors that has promoted the development of paternal care and polyandry from biparental care and monogamy in species like the spotted sandpiper. Indirectly, this should also help us to understand why, among neognathous species, classical polyandry is so rare outside the Order Charadriiformes. The traits of shorebirds listed above—reversed sexual dimorphism in body size, small, fixed clutch size, ability of females to lay successive clutches, extensive paternal care, and the need for only one parent to tend young—may all contribute to the evolution of paternal care and polyandry in certain shorebirds, such as the spotted sandpiper.

Erckmann (1983) views these characteristics of shorebird biology as factors that may lead to a situation in which the potential for polygamy is similar in both sexes, e.g. the capacity in some species for one individual of either sex to incubate unassisted and the capacity for females to lay replacement clutches. Erckmann (1983) concludes that females need not benefit by male emancipation for polygyny to evolve, nor must males benefit by female emancipation for polyandry to evolve.

Erckmann also argues that, under such circumstances, polyandry may be as likely to evolve as polygyny from a monogamous system with shared parental care: females may benefit by desertion as much as males, and males may be as likely as females to accept all parental care. Are exclusive paternal care and polyandry actually as likely to evolve from biparental care as sole maternal care and polygyny, as Erckmann (1983) suggests? A consideration of the transitions from one type of parenting to another (Szekely and Reynolds 1995) suggests that an answer to this question depends on the starting point. According to Szekely and Reynolds, biparental care was the ancestral condition in shorebirds as a whole and remained so in the plover lineage, the Charadriida, and that, in this group, exclusive male care evolved from biparental care 1–5 times; in contrast, exclusive female care has not evolved. In the Scolopacida, male care was the ancestral condition; female care evolved 6–8 times from male care, and 0–3 times from biparental care. Thus, Szekely and Reynolds' analysis of the Charadriida does not support the proposition that polyandry and polygyny are equally likely to evolve from biparental care; it also suggests that without a consideration of phylogenetic history and of primitive and derived characteristics, it is not possible to evaluate Erckmann's suggestion.

Moreover, even in the Scolopacida, polyandry is considerably less common than typical monogamy and polygyny (except in the jacanas, painted-snipe, plains-wanderer, and phalaropes, where maternal care is usually absent and almost every species thus has potential for polyandry). This also raises some questions about Erckmann's suggestion that polyandry may be as likely to evolve as polygyny. Ignore, for the moment, the issue of phylogenetic history. If males are, for whatever reason, somewhat less 'willing' than females to accept sole parental care, an additional benefit or compensation of some sort might be required. What kinds of compensations might 'predispose' male shorebirds to accept sole parental care? Three issues, not discussed in most recent reviews of the evolution of polyandry, have been suggested (Ligon 1993). First, other things being equal (e.g. all clutches are the same size), the first (earliest) brood of the season is likely to be more valuable than later clutches, as measured by the probability of producing young that survive to maturity and attain breeding status. Second, certainty of paternity of its own clutch may be considerably greater for the first male mate of the season than for subsequent males. Third, the first male may sire more than four offspring by allowing its mate to promptly produce a second clutch that is cared for by a new male; i.e. second male mates sometimes incubate eggs fertilized by the female's previous mate. The latter two points have been documented by paternity analyses for males of the sequentially polyandrous spotted sandpiper (Oring *et al.* 1992).

In addition to the ecological and phylogenetic issues discussed above, there is another factor to consider in any attempt to understand the evolution of classical polyandry, which, it should be re-emphasized, is rare even among the shorebirds. This is the element of chance (Ligon 1993). Invoking chance as a factor in the evolution of polyandry in shorebirds may strike some as far-fetched; however, this suggestion has precedents. First, Erckmann (1983) implies as much when he argues that polygyny and polyandry may be equally likely to evolve from biparental care. (Even if they are not equally likely to evolve (see above), it appears that both polyandry and polygyny have developed from biparental care in some sandpipers.) Second, in discussing the evolutionary development of polyandry from biparental care, Selander (1972, p. 209) points out that which sex assumes the parental duties is, in part, due to chance, stating: 'Lewontin (1969) emphasizes that the determination of which alternative stable state a given dynamic system evolves to is a function of variation and the starting point.' Third, the 'desertion game' hypothesis (Maynard-Smith 1977; see also Ridley 1978) also considers which parent should cease to care for offspring, or desert, and leave this task for the other parent. As an example of the desertion game, in the snail kite, either the male or the female parent may desert, depending, in large part, on the opportunities available for remating (Beissinger 1986, 1987a,b).

The extraordinarily well-studied spotted sandpiper provides a possible example of the role of chance in the development of polyandry. Among extant species, spotted sandpipers are most closely related to the common sandpiper of Eurasia. According to Oring (1986), the common sandpiper, along with most of its closest relatives, exhibits monogamy with shared parental care. This probably represents the form of parenting from which the sex-role reversed polyandry of the spotted sandpiper evolved. Female common sandpipers are 9–15 per cent larger than males, while in spotted sandpipers, this difference is about 20 per cent (Jehl and Murray 1986). In summarizing his consideration of the factors promoting polyandry in the spotted sandpiper, Oring (1986, pp. 328–9) writes:

> All of the preadaptations for the evolution of male parental care and polyandry were present in ancestral *Actitis* populations. The fact that *A. hypoleucos* [the common sandpiper] is not polyandrous indicates that not all preadaptations have been operable across its Old World range, or that comparable selection pressures have been different than in the New World.

Although it is possible that ecological factors peculiar to North America promoted the evolution of sequential polyandry in the spotted sandpiper, it is also possible that chance played a role in the development of classical polyandry in this species, but not the common sandpiper; i.e. it is reasonable to assume that chance factors could have tipped the balance towards female desertion. In any case, once females became the 'deserting' sex, and once males became an extremely valuable resource to them, selection would promote traits associated with female–female competition, such as relatively larger body size and a high level of aggressiveness, along with an increase in clutch production. Until the specific ecological factors that initiated and established polyandry in the spotted sandpiper are identified, a role for chance in its development remains a viable possibility.

16.7   Proximate mechanisms of sex-role reversal: hormones

In considering sex-role reversed shorebirds, a question arises concerning the proximate mechanisms that may be responsible for 'female-like' behaviour of males and 'male-like' behaviour of females. Modifications of the sex hormones, or of responses to them, are the most obvious candidates. The hormonal control of sex-specific reproductive behaviour in typical birds is pretty well understood. Are alterations of hormones responsible for the role reversal seen in shorebirds? Early studies on sex-role reversed phalaropes did, in fact, report that females had higher levels of testosterone than males (e.g. Hohn 1970). However, more recent studies of the endocrine control of reproductive behaviour in the spotted sandpiper (Rissman and Wingfield 1984; Fivizzani and Oring 1986) and Wilson's phalarope (Fivizzani *et al.* 1986; Oring *et al.* 1988) failed to confirm the suggestion that females of either of these sex-role reversed species have unusually high levels of testosterone. In summarizing the data obtained for Wilson's phalaropes, Fivizzani *et al.* (1986:142) write:

> ... in spite of extreme sex-role reversal, there is no reversal of the typical avian pattern of androgen predominance in male plasma and estrogen predominance in the female. These results, plus those from spotted sandpipers (Rissman and Wingfield 1984) indicate that intense female aggression associated with mate acquisition in these role reversed species must be based upon mechanisms other than high female androgens and high male estradiol.

Fivizzani and Oring (1986) suggest that it is neural receptivity to sex-typical hormone profiles that have been altered in the evolution of the sex-role reversed breeding systems of these shorebirds.

In contrast to the sex-typical pattern of relative androgen levels, levels of the hormone prolactin, a principal regulator of the onset and maintenance of incubation, are consistently lower in prelaying and laying females of both spotted sandpipers and Wilson's phalaropes than in males at the same stage of the nesting cycle (Oring *et al.* 1986, 1988). In brief, the steroid hormones— testosterone and oestradiol—follow the typical 'ancestral' condition in these sex-role reversed species, whereas prolactin departs from the usual avian pattern in that it is higher in males than in females (Fivizzani *et al.* 1986).

16.8   The rarity of polyandry in phalaropes

Female phalaropes are larger and more brightly plumaged than males and maternal care is completely lacking; i.e. males provide all parental care. A female phalarope pairs with a male, promptly lays a clutch of four eggs, and then, as the male begins incubation, leaves and may attempt to court and mate with a second male. Despite this pronounced sex-role reversal and the absence of maternal care, polyandry is distinctly uncommon in phalaropes, with no more than 5–10 per cent of females acquiring a second mate (see Whitfield 1990 for references). In examining this issue, Whitfield (1990) considered a number of possible factors that might constrain polyandry in the red-necked phalarope.

At a behavioural level, polyandry is rare partly because males prefer not to mate with females that have already laid a clutch. Production of a first clutch does not appear to affect a female's ability to rapidly initiate a second clutch, nor are second-clutch eggs smaller than those of a female's first clutch, thus these factors cannot account for a male's reluctance to pair with a previously-mated female. Instead, it appears that the usual major factor leading males to reject a previously-mated female is the issue of genetic paternity; i.e. sperm placed in the female's reproductive tract by her first mate might fertilize some fraction of the clutch to which the second male would be committed. When male phalaropes did pair with previously-mated females, they copulated and attempted to copulate more often than first-mated males (suggesting that sperm competition might be great and, concomitantly, that certainty of paternity was lower), and males behaved more aggressively towards such females when the latter attempted to court them.

The infrequency of polyandry in phalaropes also appears to be affected by the 'operational sex ratio' (Emlen and Oring 1977). Most of the time during the breeding season, the sex ratio is biased towards females. This is the case for a variety of reasons, including (1) the fact that males disappeared to incubate as soon as they received a clutch of eggs, (2) asynchronous spring arrival schedules of the sexes, and (3) the effect of clutch failure on renesting opportunities (Colwell and Oring 1988). The upshot is that, at any point in time, there were likely to be more available females than available males and that, as a result, there was little opportunity for sequential polyandry. The relative shortage of males often led to intense female–female competition for a mate; Colwell and Oring recorded as many as 5–10 females pursuing a prospective mate.

Thus, although male phalaropes provide all of the parental care, leaving females free to obtain a second mate, females are rarely successful in doing so, for two very different reasons. First, males are often reluctant to mate with a female that has already produced a clutch of eggs for a male (Whitfield 1990). Second, even if males did not discriminate against such females, there are typically more unmated females than available males at any point during the short breeding season.

16.9   Paternal care and classical polyandry in buttonquail

Paternal care and speculation concerning possible phylogenetic relationships of buttonquail (Fig. 16.1) were discussed in Section 10.5.3. The 13 species

Fig. 16.1   In buttonquail, including the Madagascar buttonquail, the female exhibits the brighter plumage. In these birds, parental care is conducted by the male, sex-role reversal in body size, coloration, and aggressiveness is the rule, and classical polyandry is thought to be the predominant mating system.

(Johnsgard 1991) of buttonquail are apparently so far removed phylogenetically from all other living birds (Sibley and Ahlquist 1990) that they must have diverged very early in the history of neognathous birds. Because exclusive paternal care apparently occurs in all species of buttonquail, it is probably a primitive trait within the group. As discussed in Chapter 9, it even appears possible that the exclusive paternal care seen in buttonquail may have evolved directly from a condition of no parental care.

Because of the ubiquity of paternal care, and because so little is known of the behaviour of these birds in the wild, not much can be said about the possible role of ecological factors in either the evolution or maintenance of the mating systems of buttonquail. Virtually all that is known about their parental behaviour and mating strategies is based on studies of captive individuals (described and discussed by Johnsgard 1991). Females are larger, more vocal, and often more brightly plumaged than males, and they can be extremely aggressive. Correlated with this sex-role reversal, males of all species typically conduct almost all incubation and care of young. Some species may be monogamous, but most are probably polyandrous, either regularly or facultatively. In captivity, and presumably also in the wild, females may take a role in nest preparation, and they sometimes may contribute a bit of time to early incubation. However, after a clutch of eggs is laid, the major goal of the female is to attract a second male.

Aside from exclusive paternal care and polyandry, the most unusual aspect of the reproductive biology of these peculiar birds is their fecundity. Using conservative premises, Johnsgard (1991) calculated that a single female buttonquail might produce as many as 12 clutches per year. Assuming a 75 per cent rate of egg and chick mortality, the resulting female–young ratio at the end of the breeding season would be about 1:10. In addition, buttonquail reach sexual maturity at a tender age and have the ability, at least in captivity, to begin to lay eggs when they are about four months old. Thus, despite production of small clutches, female buttonquail have amazing reproductive potential, perhaps the greatest of any type of bird. Of course, this potential can only be reached if environmental conditions are ideal, and especially if sufficient males are present to accept responsibility for the female's clutches. Thus, males are a critical limiting resource for females. This last point indicates that sexual selection must be intense in female buttonquail, which probably accounts for their extreme aggressiveness (Johnsgard 1991).

16.10   Exclusive paternal care and classical polyandry in a cuckoo

Previous sections have considered paternal care and classical polyandry in ratites, tinamous, shorebirds, and buttonquail (see above and Chapter 10). All of these groups are characterized by precocial chicks. With regard to the development and maintenance of classical polyandry, this has been thought to be a noteworthy point, in that precocial chicks may require only a single parent, while altricial ones are more likely to require two. Classical polyandry, however, also apparently occurs in at least one altricial species, the black coucal (Vernon 1971; Irwin 1988; reviewed by Ligon 1993 and Andersson 1995).

Coucals (Fig. 16.2) are a distinctive group of cuckoos that are widely distributed in the Old World tropics. The approximately 30 species are especially interesting in that, with regard to their reproductive biology, sex-role reversal occurs, to varying degrees, in most or all species. Although the biology of most species is unstudied, Andersson (1995) provides a good review of what is known about the breeding biology of this unusual group. Because there are a large number of coucal species, which vary in the degree of reversed size dimorphism (Andersson 1995, Table 1) and probably sex-role reversal, this may potentially be the single best group of birds for study of the selective factors favouring the evolution of sex-role reversal.

Classical polyandry in the black coucal was discovered in southern Africa and described by C. J. Vernon (1971; also see Irwin 1988). Here I consider the evolution of polyandry in black coucals by examining two general categories of factors: (1) evolutionary history, as indicated by traits shared with other coucals, and (2) ecological aspects of the environment occupied by black coucals. Female black coucals are about 50 per cent larger than males (the most extreme case of reversed size dimorphism in the group), and it appears that only males engage in nest construction, incubation, and the feeding of nestlings and fledglings. Thus, except for copulation and the eggs donated by females, male black coucals are thought to conduct all reproductive activities alone. One social unit studied by Vernon (1971) consisted of three males that

Fig. 16.2 Senegal coucal. Sex-role reversal apparently occurs in all coucals; however, the black coucal is the only species of altricial bird for which classical polyandry has been documented. Reprinted with permission from Dover Publications, Inc.

remained spatially segregated and a single female, whose movements encompassed the ranges of all three males. On one occasion, the female was observed to move freely from one male home range to another in turn, calling each male that she visited to copulate with her, and possibly laying an egg in the nest of one of the males (Vernon 1971). Similar but less complete observations of these four birds were also made at other times. Thus, the evidence for classic polyandry in black coucals, although limited, is convincing.

16.10.1   Evolved characteristics of coucals that may promote polyandry

Like shorebirds, cuckoos as a whole exhibit a wide range of breeding systems. Many species are socially monogamous with biparental care, others exhibit communal breeding, still others are social parasites, and classical polyandry occurs in at least one species, the black coucal. (Polyandry has also has been recorded in the dwarf cuckoo of South America; (Ralph 1975). Although it is probably atypical in this species, this case illustrates the unusual flexibility in parental care systems of cuckoos.) The one major mating system apparently not found in this group is polygyny. Insofar as is known, in all non-parasitic species, the male incubates at night. This apparently obligate nocturnal incubation by males thus precludes the possibility of typical multi-nest polygyny (see Section 1.5.4).

Sex-role reversal is characteristic of coucals, and most species are thought to be monogamous with biparental care (Andersson 1995). Female coucals are larger than males, and males construct the nest, incubate, and feed the chicks with variable assistance from the female (Irwin 1988; Andersson 1995). This sex-role reversal appears to have set the stage for the development of polyandry in the black coucal; i.e. polyandry appears to be a derived condition. As Andersson (1995, pp. 179) points out: 'Regardless of why greater male parental care occurs in some monogamous birds, it should make polyandry more likely to evolve than in species with greater female parental care.'

In addition to traits associated with sex-role reversal, female coucals lay relatively large eggs. There is evidence that females of several of the monogamous species have difficulty in producing a clutch (Andersson 1995). For example, laying intervals of several days (up to nine) have been recorded. Unlike the shorebirds, coucals have variable clutch sizes, which are often small (e.g. two eggs). Like some shorebirds, such as the spotted sandpiper, they also apparently have the capacity, under favourable environmental conditions, to produce several clutches per season. This last point is significant in that it provides a clear adaptive basis for polyandry for female black coucals; i.e. females can benefit greatly by obtaining more than one mate.

Thus, it appears that some general traits of coucals, several of which are also shared with the shorebirds, have set the stage for the evolution of polyandry in the black coucal. These include: (1) sex-reversed size dimorphism, (2) male parental care, (3) the need for only one parent to tend the young (see Section 16.10.2, below), and (4) the ability of females to lay successive clutches. Black coucals exhibit all of these traits, which, together with certain, specific ecological conditions, have apparently promoted, or at least permitted, the evolutionary development of polyandry in this species (Ligon 1993).

Previously, I suggested that another trait might also have played an important role in the evolution of polyandry in coucals (Ligon 1993). That trait is nocturnal incubation by males, which is apparently characteristic of all species of non-parasitic cuckoos. My point was as follows: keeping in mind that classical polyandry could have evolved in black coucals only when all of the several factors listed above are present, nocturnal incubation by male cuckoos may reflect a tendency for females to contribute less parental care than males, as suggested by a lesser commitment to incubation duties, and, simultaneously, a greater physiological commitment of males to do so (Vehrencamp 1982). In other words, based on their phylogenetic history as cuckoos, male coucals have a long history of providing incubation and the physiological factors promoting incubation may be stronger in males than in females. Thus, the general sex-role pattern of incubation by non-parasitic cuckoos can be viewed as a pre-adaptation that, together with specific eco-logical and reproductive factors that permit uni-parental care, has favoured 'desertion' by female, rather than male, black coucals.

I speculated that the combination of male-biased care of young, plus poss-ibly chance (also discussed briefly for shorebirds), had tipped the balance from primarily male care to total male care in this one species of coucal. For example, theoretical treatments of the 'desertion game' (Maynard- Smith 1977; Ridley 1978) consider which parent should cease to care for offspring, or desert and leave this task to the other parent. (In snail kites, either the male or the female parent may desert; Beissinger 1986, 1987a,b.) In all coucals, it is females that have reduced parental care, which tips the balance for desertion towards females. This is seen in its most extreme form in black coucals, where, following egg laying, females apparently provide no care for their offspring.

16.10.2   Ecological factors influencing the evolution of polyandry in the black coucal

Thanks in part to Andersson's (1995) paper, I now think that I emphasized insufficiently the breeding environment in my earlier scenario for the evolution of polyandry in black coucals (Ligon 1993). Although coucals and possibly other non-parasitic tropical cuckoos possess a number of characteristics asso-ciated with classical polyandry, this mating system has not yet been detected in any other species of coucal (This group, however, is about as unstudied as it is interesting.); i.e. the conditions discussed above are necessary, but perhaps are not sufficient, to promote the evolutionary transition from monogamy–biparental care to polyandry–paternal care.

Unlike most other coucals, this species exhibits seasonal movements, migrating to breeding areas that may be unusually rich in food (Irwin 1988). During the breeding season, black coucals occupy moist grassland, especially areas that are seasonally inundated, and where insects, particularly grasshop-pers, are periodically abundant. Thus, the pre-breeding movement to such breeding sites may account for the ability of a single parent to rear a brood of young. This could have been the factor that tipped the balance from primarily paternal care, as appears to occur in other coucals, which are thought to be sedentary, to exclusive paternal care.

This suggestion specifically incorporates ecological factors—the abundance and perhaps kinds of food—plus behaviour perhaps novel among coucals, namely migration, into an explanation of polyandry in the black coucal (Andersson 1995). In addition, as in many other cuckoo species, the hatching of young is staggered, thus reducing the nestlings' peak food demands and thereby making it more likely that a single parent is able to provision the brood. Studies of additional species of coucals are needed to determine the relative contributions of phylogenetically-based traits and ecological circumstances in the evolution and maintenance of sex-role reversal in this group.

16.10.3   A unique proximate mechanism for sex-role reversal in coucals

As mentioned above, sex-role reversal apparently occurs to a greater or lesser extent in all coucals, with the male assuming a role generally considered to be more typical of female birds; i.e. in addition to being smaller, males construct the nest and carry out most or all incubation and provisioning duties. In some shorebirds, such sex-role reversal and polyandry is apparently not related to differential androgen production (see Section 16.7, above). Endocrine correlates of sex-role reversal in coucals are entirely unstudied.

Male coucals possess a unique gonadal condition that may proximately promote 'feminization' via a reduction by one-half of the primary source of testosterone (Ligon 1997). A morphological trait possibly related to sex-role reversal in this group is asymmetry in size and development of the testes. In at least some species, the left testis is 'atrophied', 'rudimentary', or even absent, while the right is of normal size (Rand 1933, 1936). Based on a large sample of 43 male specimens of the Madagascar coucal (considered to be a member of the same superspecies as the black coucal; Irwin 1988), Rand (1933, p. 219) reported that '... the right testis was always larger than the left, which was atrophied, never being firm and oval, and never showing any enlargement in the breeding season, even when the right was at its maximum size.' Rand (1933) and Chapin (1939) also reported the absence of the left testis in three other African species of coucals, including the black coucal; thus, this may be a feature typical of this group.

It appears reasonable to assume that loss of one of the two functional testes may have the effect of reducing the circulating testosterone level of male coucals, relative to the hormones important in nest-building, incubation, etc. One might speculate that atrophy of the left testis is the proximate means by which adaptive sex-role reversal has been effected in these birds. (Ligon 1997).

16.11   Conclusions and summary

The issue of classical polyandry leads to two questions: First, why has it evolved at all, and second, since it has evolved, why is it restricted to a few species in a few major groups: the ratites and tinamous, some shorebirds, the buttonquail, and at least one species of cuckoo? In this chapter I have suggested that this form of polyandry in living species may be based on two very different sorts of evolutionary histories. It should be explicitly recognized that the existence of parental care solely by the male is the key precursor of classical polyandry and, therefore, a consideration of the evolutionary origins of paternal care is a necessary first step in any attempt to understand classical polyandry.

In Chapter 10 I discussed the possibility that, in birds, exclusive paternal care is the 'ancestral' parenting system, and that it has been preserved in some lineages. Classical polyandry can be readily developed from a state of exclusive, or nearly exclusive, paternal care. A female simply leaves a clutch of eggs with a male and lays a second clutch for another male. The ratites and tinamous, buttonquail, and possibly some of the most distinctive shorebird families (jacanas, painted-snipe, plains-wanderer) may exhibit ancestral paternal care; i.e. paternal care derived directly from no parental care, rather than from biparental care. Overall, the case for paternal care being an ancestral, rather than a derived, trait is strongest for the ratites and tinamous, the most primitive of living birds, and for the buttonquail, all of which practise this form of parenting.

In addition, Feduccia's (1995) suggestion that the basal stock of all living neognathous birds are 'transitional shorebirds' is consistent with the idea that the paternal care–polyandry of some of the most distinctive shorebirds may reflect the original form of avian parenting. Szekely and Reynolds (1995) suggest that paternal care was primitive in one of the two major charadriiform lineages, the Scolopacida, which includes the jacanas, painted-snipe, plains-wanderer, phalaropes, and the polyandrous tringine sandpipers, although their reconstruction indicates that biparental care was the original form of parenting in the order Charadriiformes. How their choice of outgroups might affect this conclusion was discussed in Section 10.5.2.

Most readers probably have had the impression that avian polyandry is most closely associated with the shorebirds. This chapter is meant to demonstrate that although several shorebird lineages do exhibit exclusive paternal care and polyandry, this kind of mating system actually occurs more frequently in the ratites, tinamous, and buttonquail. The focus on shorebirds has been due, in large part, to the fact that they have received a great deal of study relative to the other groups, most of which are tropical in distribution.

Classical polyandry may have appeared in shorebirds for two basically different reasons. (1) In some species, sole paternal care with the potential for polyandry may be a primitive trait that reflects the original form of parenting in neognathous birds (possibly jacanas, painted-snipe, plains-wanderer). (2) Many features of shorebirds (frequency and extent of male parental care, reversed sexual dimorphism, small, fixed clutch size, etc.) seem to be necessary, but not in themselves sufficient, for the evolutionary development of sole paternal care-polyandry. (3) The phylogenetic analysis of parental care by Szekely and Reynolds (1995) indicates that in the largest extant group of shorebirds, the Scolopacida, sole paternal care was ancestral; i.e. that it evolved once in the common ancestor of jacanas, painted-snipe, the plains-wanderer, phalaropes, and sandpipers. Subsequent transitions from male care to female care or biparental care can account for the diversity of mating systems seen within this group. One can speculate that as semi-precocial and altricial young evolved in the descendants of the 'transitional shorebirds' (i.e. all other neognathous species), biparental care evolved concomitantly and remains the prevalent form of parenting in nearly every order of birds.

In other shorebirds, it appears that paternal care and classical polyandry evolved from a state of biparental care. Probable examples include the spotted sandpiper (Oring 1986) and the Eurasian dotterel (Nethersole-Thompson 1973), a species in which polyandry occasionally occurs. Depending on particular circumstances, females of the spotted sandpiper, the best-studied sex-role reversed shorebird, are hormonally and behaviourally equipped to competently perform all parental duties, which suggests a history of biparental care. Sex-role reversed female phalaropes, too, possess an endocrine system largely typical of 'normal' biparental birds. To summarize, polyandry in extant shorebirds may have come about in two fundamentally different ways—by retention of the ancestral condition in certain of the most distinctive groups (e.g. jacanas, plains-wanderer) and by evolution from a condition of biparental care in others.

A member of a little-known group of cuckoos, the black coucal, provides the only documented example of classical polyandry in an altricial species. Sex-role reversal with monogamy is thought to be the rule in coucals, although the degree of role reversal varies from species to species. Female coucals are larger than males and males conduct most of the parental duties, including nest construction, incubation, and care of the young. Thus, many of the prerequisites for polyandry are present throughout the group. Nevertheless, on the basis of current knowledge, it appears that only one of the approximately 30 species of coucals exhibits classical polyandry.

The annual cycle and breeding ecology of the black coucal is unusual, compared to other coucals, in that this species apparently migrates to especially productive sites (in terms of food availability) to breed. Breeding in an area where food may be especially abundant has two important effects that might promote the transition from monogamy to uniparental care and polyandry. First, abundant food makes it easier for females to lay many eggs (it appears that timely egg production sometimes may be difficult for coucals). Second, abundant food may make it possible for a single parent—the male—to rear a brood unassisted. Uniparental care is further aided by the staggered hatching pattern of chicks, which decreases the brood's demands at any point in time during the nesting cycle. Thus, because most species of coucals are thought to be monogamous with sex-role reversal, this may be the best group for comparative study of interactions of phylogenetic and ecological factors that predispose the development of classical polyandry and which, in at least one case, have provided the 'final push' to that rare mating system.

17 Conclusions: new research initiatives

17.1 Introduction

In this book I have attempted to review and synthesize the current literature dealing with two major aspects of the reproductive biology of birds, sexual selection and mating systems. In carrying out this project, I have gained a great deal of factual knowledge. I have also developed some impressions concerning the status of our understanding of these subjects, and where the field might go from here. These are discussed below.

One of the most satisfying aspects of this endeavour was seeing clear evidence that considerable scientific progress has been made in the two major areas covered by this book. Over the past three decades (about the length of my post-graduate student career), advances in our understanding of most of the topics treated here are truly impressive. It is difficult to imagine that they will be as great during the next thirty years.

Some of these advances can be illustrated by considering two statements made 30 years ago by the great British ornithologist David Lack in his book *Ecological adaptations for breeding in birds*, published in 1968. In this book, Lack briefly treated two topics that have been of great interest in contemporary avian evolutionary biology, sexual selection and the social/mating system labelled cooperative breeding. (1) On the subject of sexual selection theory, Lack (1968, p. 159) states: 'Clearly it is nowadays more meaningful to speak of the functions of secondary sexual characters of birds than of the theory of sexual selection.' To the contrary, most current students of sexual selection would probably argue that it is difficult to overstate the importance of theory in this area. (2) Concerning the phenomenon of cooperative breeding and nest helpers, Lack (1968, p. 72) writes: '… such behavior is uncommon, so is evidently selected against, and the extra helpers probably have no ecological significance.' The many studies of cooperative breeding conducted over the past twenty or more years clearly indicate that this conclusion was erroneous.

To be fair to Lack, I should emphasize that his views on both topics were actually more accurate than these two quotes, taken out of context, might suggest. The quotes are provided simply to show how risky it can be to predict future interpretations of topics that fall under the heading of evolutionary biology. Why did Lack appear to miss the mark? Clearly, in the 1960s his most critical handicap was the scarcity of well-developed ideas or theories. Empiricists need theory to suggest avenues for investigation, and, at the time Lack wrote the statements cited above, development of modern evolutionary theory for social behaviour was just getting underway. Key theoretical concepts especially relevant to cooperative breeding either were in their infancy (e.g. kin selection: Hamilton 1963, 1964; Maynard-Smith 1964; ecological constraints: Selander 1964) or had not appeared at all. Similarly, except for Darwin's original insights and Fisher's theory of runaway selection, both of which were long-neglected, most of the insightful ideas related to sexual selection had not yet appeared or were just beginning to appear at the time Lack wrote the comments quoted above. Clearly, those of us who study birds and what they do owe much to the theorists who have provided direction for empirical studies.

In short, thanks to the appearance of theories dealing with mating systems and sexual selection, and to the many outstanding empirical studies designed to test aspects of these theories, progress over the past twenty or so years has been impressively rapid. In the remainder of this short chapter, I offer some suggestions regarding the kinds of studies that may contribute to continuation of this progress.

17.2   Sexual selection: what do we know, and where do we go?

For most students of sexual selection, the most interesting ultimate issue, as well as the most controversial, has been related to the evolutionary significance of mate choice by female birds. Because most birds are socially monogamous and exhibit biparental care, direct selection is undoubtedly the single-most widespread factor influencing mate choice in this group. For those species, however, in which males provide neither resources nor parental care, other explanations have been sought.

Of the theories that attempt to explain mate choice in such systems, only good genes selection has received strong support based on empirical studies. For a number of well-studied species, the evidence indicates that females make mate choice decisions based on traits capable of providing information about 'condition', and thus about the relative genetic quality of males. If this is correct, then the key ultimate question pertaining to female mate choice has been resolved. It appears to me that the evidence supporting a good genes interpretation of mate choice in species without paternal contributions is sufficiently strong that we can consider that a 'verdict' has been reached in the debate over good genes versus alternatives such as Fisherian runaway (e.g. see Bradbury and Andersson 1987) or aesthetic mate choice. Even if other phenomena (Fisherian runaway; aesthetic mate choice, including sensory bias) also initially play a role, it appears that ultimately they would converge on male quality (good genes).

What remains to be tackled are related proximate and historical (phylogenetic) issues. Several topics addressed in this book require more study.

17.2.1   Phylogenetic studies of male ornaments and other display traits

In thinking about the 'evolution of avian breeding systems', it has become clear that we cannot safely use the biology of present-day species to infer the evolutionary origins of traits of interest. If we want to understand the origins of display traits—morphological ornaments, bizarre display behaviours, etc.— then phylogenetic analyses are required. Prum's (1990, 1994) studies of the display traits and mating systems of manakins suggest that within this group stereotyped displays evolved prior to at least some of the ornamental plumage (also see Kusmierski *et al.* 1997); i.e. plumage traits may have evolved to enhance the attractiveness of complex behavioural displays. They also indicate that lekking behaviour may have evolved several times in the manakins. Both of these conclusions are somewhat counter-intuitive and probably would never have been predicted or even suspected in the absence of phylogenetic studies.

Other phylogenetic analyses have also suggested unexpected evolutionary pathways. For example, strong sexual dichromatism in a number of species of New World blackbirds and orioles may have evolved from ancestors in which both sexes were similarly brightly coloured (Irwin 1994). With regard to parenting systems, Hughes' (1996) study of the evolutionary history of social parasitism in cuckoos suggests that the parasitic habit has been largely lost in the yellow-billed and black-billed cuckoos of North America.

These and other phylogenetic studies clearly indicate that many longstanding assumptions about the direction of evolutionary change may be in jeopardy. They also suggest that we have a lot to learn about the evolutionary origins of sexually-selected traits, as well as other traits associated with reproduction.

17.2.2   The significance of multiple ornaments

The evolution of diverse ornaments within closely related groups of species is an important issue. It has been suggested that ornamental traits should evolve more rapidly than non-ornamental traits, and that ornaments should be under less stringent genetic control (e.g. Møller 1993c). Differences in the underlying genetics of ornamental and non-ornamental traits are interesting issues, and ones that can help us to understand the role of sexual selection in the speciation process (see below).

Although most, if not all, birds possess more than a single ornament (here 'ornament' is broadly defined to include behaviours such as song and stereo-typed postures, as well as morphological characters), the significance of multiple ornaments is not well understood. In some species only one of a number of ornamental morphological traits appears to be important in female mate choice (e.g. Møller and Pomiankowski 1993a, Ligon et al. 1998). In some other systems, however, several ornaments may be used by both females and males (e.g. Mateos and Carranza 1995, 1997). It has been suggested that different ornaments may signal different messages about the male, or that multiple ornaments may provide additive information (e.g. Zuk et al. 1992). To date, however, there is little clear evidence that this is the case (see Møller and Pomiankowski 1993a and Ligon et al. 1998), although both ideas seem reasonable. Alternatively, it appears that females may shift their attention from one ornament to another over evolutionary time in an effort to utilize the trait that is currently most informative about male condition (Hill 1994b).

17.2.3   The multiple roles of testosterone in sexual selection

The point that testosterone is important in sexual selection might appear to be so obvious as to be uninteresting. That, however, would be a fallacious conclusion to draw. In addition to its effects on male condition and thus its importance in intra-sexual competition among males (influencing aggression, muscular development, etc.; e.g. Ligon et al. 1990), testosterone is also extremely important in inter-sexual interactions. To date, the multiple roles of testosterone in this context are incompletely understood. Testosterone is involved at least three distinct kinds of traits associated with male display and female mate choice.

First, the inverse relationship described between testosterone production and the effectiveness of the immune system (see Folstad and Karter 1992; Zuk et al. 1995) appears to provide a beautiful example of the relationship proposed by theory between the information value of a display influenced by testosterone and its cost. Clearly, we need to know more about the role of the testosterone–immunocompetence connection as it pertains to sexual selection.

Second, much male display behaviour is mediated by testosterone, and variation in such behaviour may provide a good indication of relative testosterone levels among males. Testosterone also influences the development of certain male ornaments. For example, in the red junglefowl and the domestic fowl (the junglefowl's domestic derivative), comb size precisely reflects testosterone level. In addition to the red junglefowl's comb, 'soft parts' of other species, including the wattles of pheasants and ptarmigan, the snood of turkeys, and bill colour in mallards are influenced by testosterone, to mention only a few. It is likely that all of these traits accurately signal aspects of male condition by revealing individual differences in levels of circulating testosterone. By identifying testosterone-dependent and testosterone-independent ornaments, we will gain a better understanding of the significance of multiple ornaments. As discussed in Chapter 4, a number of male morphological ornaments are apparently not used by females in assessing male quality. Is there a common proximate control of most characters used by females; e.g. do females usually rely on traits affected by testosterone?

Third, in contrast to its importance in the development of certain morphological characters, testosterone apparently does not influence the development of ornamental male plumage in most birds. For some strongly sexually-selected species in which the plumage of males is ornamented (e.g. mallard, turkey, red junglefowl), the evidence gathered to date indicates that male plumage is not important to females. Instead, when making choice decisions, females of these three species rely primarily on traits affected by testosterone (bill colour, snood length, and comb size, respectively). With regard to the issue of multiple ornaments, this suggests that females of such species may have used feather ornaments in mate choice decisions at an earlier time in their evolutionary history, but that they were superseded by other types of ornaments (Hill 1994*b*; Ligon and Zwartjes 1995a)—possibly testosterone-dependent ones—which may provide more precise and up-to-date information about male condition.

17.2.4   The relationship between carotenoid plumage pigments and male quality

The bright colours of the feathers of many kinds of birds are determined by carotenoid pigments. Because birds cannot synthesize carotenoids, this mode of feather coloration relies on the proper pigments being obtained in the diet. A relationship between diet and feather colour or brightness provides a means by which feather colour can signal the 'quality' of one male relative to others of its species. Interspecific comparisons indicate a relationship between the degree of sexual dichromatism and the use of plumage coloured by carotenoids, which suggests that carotenoid-based plumage colours may serve in sexual selection by providing an honest indicator of the phenotypic quality of males (Gray 1996).

This relationship between male condition and carotenoid-based feather colour has been comprehensively studied in the house finch. In this species, brighter males are more reproductively successful, are better parents, and survive better than males with less carotenoid pigmentation in their feathers (e.g. Hill 1991, Hill *et al.* 1994*a*). Although Gray's (1996) survey suggests that carotenoid pigments are important in sexual selection in a broad array of passerine species, single-species studies similar to those done by Hill on house finches will determine just how widespread the carotenoid–fitness relationship is.

17.2.5 The role of sensory bias in guiding the responses of females to new male ornaments

For obvious reasons, one of the least-studied aspects of the mate choice phenomenon, as currently understood, is how or why a new male colour or structural ornament gains the attention of females and attracts them to males bearing the new trait. That is, how and why does a newly mutated ornamental trait come to be important to females, and how does the trait and the preference for it become incorporated into the species as whole; i.e. how does a new mutation become fixed and how quickly might this occur? These issues are related both to perceptive abilities and ecology, as well as to the phylogenetic history of the species in question.

17.2.6 The role of sexual selection in speciation

One of the most exciting revelations produced by current studies of sexual selection pertains to its role in the process of speciation. Largely because of its historical nature, interpretation of this relationship currently must be somewhat speculative; however, it appears that sexual selection is causally related to the divergence of populations. This seems especially apparent in groups in which (1) sexual selection is strong, (2) males exhibit extreme ornamentation, and (3) the species making up the group are very closely related. The birds of paradise probably provide the most famous example of this interaction. Divergence of populations may be due, in part, simply to the fact that sexual selection accelerates differentiation of traits important in mate choice, and any factor that increases such divergence will play an important role in the speciation process.

17.3   Studies of mating systems

Partly because of their diversity, mating systems of birds have been of unusual interest to avian behavioural ecologists. In addition, the prevalence of monogamy—the mating system that characterizes the societies to which most human beings belong—provides a certain empathy between ourselves and the 'typical' bird. These factors, plus the fact that most aspects of the reproductive behaviour of birds are relatively easy to observe and quantify, can largely explain the great interest over the years in avian mating systems.

The subject of mating systems has long intrigued avian behavioural ecologists, and the discovery of a dichotomy between social and genetic mating systems provided a tremendous new impetus to mating system research. With the advent of molecular techniques, such as DNA fingerprinting, it became clear that the overall reproductive strategies of both male and female birds were much more complex, and therefore more interesting, than anyone had previously imagined. Hardly any researcher in this field would deny that the revelation of regular extra-pair copulations in a wide variety of birds was one of the most important advances ever made in the study of avian mating systems. Adding to the excitement of this discovery is the fact that this is one case where empirical facts led the development of theory, rather than vice versa. As discussed in Chapter 12, theory to account for extra-pair copulations that lead to fertilization is currently being actively developed.

Although a great deal is now known about the factors that contribute to the overall reproductive strategies of male and female birds, several areas of mating system research should yield significant additional advances in our thinking about this subject. Some of these are briefly considered below.

17.3.1   Evolutionary history of mating systems

We cannot know with certainty the kinds of mating systems exhibited by long-extinct species; however, this issue may be indirectly addressed by use of phylogenetic studies that deal both with mating systems and patterns of parental care. The form of parental care (i.e. biparental, exclusive paternal, exclusive maternal) is usually related to mating systems, and one of the most interesting questions about parental care pertains to its evolutionary origins.

Several authors have suggested that in birds parental care may originally have been conducted solely by males. This idea is less counter-intuitive than it initially seems, if we remember that paternal care is far better developed than maternal care in most other vertebrate groups, namely a variety of fishes and anuran amphibians; i.e. among vertebrates, exclusive maternal care is the rule (with some exceptions) only in mammals. Paternal care, either solely by the male or with the female, continues to be of critical importance in most contemporary avian species, including all of the most 'primitive' living species, the paleognathous birds, and in a strong majority of non-passerine neognathous species as well. Clearly, additional phylogenetic studies of the evolutionary sources of parental care of the sort conducted by Szekely and Reynolds (1995), will add greatly to our understanding of the origins of parental care patterns in various avian lineages.

Prum's (1990, 1994) studies of morphological and behavioural aspects of male displays in the manakins (Family Pipridae) and of the evolution of co-operative 'team' displays and lekking behaviour illustrate the kinds of approaches that should become more common in the future.

17.3.2   Reproductive strategies of female birds

Another area that is currently receiving a lot of attention pertains specifically to the reproductive strategies of females, e.g. the 'constrained female' hypothesis (see Gowaty 1996*b*) and related ideas. The potential benefits of extra-pair copulations to males have long been recognized. However, until recently, it has been less widely appreciated that in 'monogamous' birds, extra-pair copulations that lead to fertilization are also a widely-used reproductive strategy of females. As indicated in Chapter 12, ecological factors, together with the demands and effects of parental care (whether two parents produce more young than one), appear to be related to the frequency, and even the occurrence, of extra-pair copulations. Moreover, extra-pair copulations may even be responsible for the distinctive mating systems and strategies that characterize various species. For example, female grouse of the grasslands of North America receive no assistance from males in the rearing of young, either directly or indirectly. In this case, female choice of mating partner is essentially unconstrained. What do such 'unconstrained' females do? Unlike their male counterparts, they do not mate 'promiscuously.' Female grouse typically copulate only one time, and this single copulation fertilizes the entire clutch of eggs. Because females do not require male assistance in brood-rearing, they are presumably free to mate solely with the single male that they perceive to be of 'highest quality.'

Interestingly, in species in which the rearing of young birds absolutely requires the efforts of two parents, genetic monogamy also appears to be the rule, but for a different reason. In these cases, the female's reproductive success depends on the active contributions of her mate, and it appears that the male 'demands' a very high level of paternal certainty in exchange for his parenting efforts. In between these two extremes are those species in which females can rear some young, with or without her social mate's contributions. Here females often follow a strategy of mixed paternity—within a brood, some chicks are sired by the social mate, while others are sired by another, presumably 'superior', male or males.

Working out the main sub-themes of these kinds of female reproductive strategies is one of the most interesting challenges awaiting both empirically-oriented and theoretically-oriented students of monogamous and polygynous mating systems.

17.3.3   Classical polyandry

Another major theme requiring much additional work is classical polyandry. To date, no single explanation for the evolution of this mating system has proved to be adequate (Clutton-Brock 1991), and I have somewhat reluctantly become convinced that additional empirical studies of the ecology of polyandrous species will not really change this situation. Phylogenetic approaches, such as that of Szekely and Reynolds (1994), are essential for future progress, and will do far more to resolve the puzzle of the origins of classical polyandry than any number of field studies of polyandrous species.

Although they will probably not tell us why classical polyandry occurs almost exclusively in the Order Charadriiformes, comparative field studies may be very useful in certain specific cases. For example, classical polyandry is known to occur in one species of altricial bird, the black coucal. Additional field studies of this species, plus several other coucals, all of which exhibit some degree of sex-role reversal, should lead to a better understanding of the role of ecological factors in promoting the development of classical polyandry in this group.

17.4 Conclusions

One of the goals of scientific inquiry is to produce general explanations for the phenomenon being investigated. However, for some issues generality is often difficult to attain. Thus, it is satisfying to conclude that the 'why' of female mate choice is now pretty well understood. A major question posed by sexual selection—whether females obtain indirect benefits by choosing among males?—has been answered in the affirmative. Some of the specific mechanisms proposed for good genes selection have been supported (e.g. testosterone-dependent ornamentation, carotenoid pigments), while others, such as the proposed relationship between good genes and fluctuating asymmetry, are still in the testing stage. However, some other specific sorts of good genes scenarios, which have received a lot of attention, remain problematic. This appears to be true for the interspecific relationship between parasites and bright plumage proposed by Hamilton and Zuk (1982).

In contrast, generality about some of the other issues considered in this book remains more elusive. In part, this is because subjects or phenomena placed together under a single label are often actually rather different. This appears to be true for 'cooperative breeding', which actually encompasses a diversity of mating systems, helping patterns, effects of helpers, etc. (Edwards and Naeem 1993; McLennan and Brooks 1993). Viewed in this way, it is easy to see why there appears to be no single explanation for either the origin of cooperative breeding or the significance of helpers, and that a single all-inclusive explanation should not be expected.

The same can be said of each of the recognized mating systems (see Johnson and Burley 1997). Several factors promote monogamy and, depending on the species, members of each sex may mate monogamously either because this is the most productive breeding strategy for each sex (obligate monogamy; Murray 1984) or because one or more features of the environment may constrain an individual male, for example, to this mating system. A similar point can be made for polyandry. In the first place, avian 'polyandry' is composed of very different phenomena (cooperative and classical polyandry). Even within the category of classical polyandry the historical and selective backgrounds that have promoted polyandry probably vary from case to case. Determining the mix of factors that have contributed to the existence of classical polyandry in those few species exhibiting it provides an exciting challenge for future workers.

Appendix The common names and scientific classification of the species and groups mentioned in the text

Common Name	Scientific name	Family	Order
Accentor, alpine	*Prunella collaris*	Prunellidae	Passeriformes
Ani, groove-billed	*Crotophaga sulcirostris*	Cuculidae	Cuculiformes
Antbird, dusky	*Cercomacra tyrannina*	Thamnophilidae	Passeriformes
Auklet, crested	*Aethia cristatella*	Alcidae	Charadriiformes
Auks		Alcidae	Charadriiformes
Babblers, Australian	*Pomatostomus*	Pomatostomidae	Passeriformes
Barbets		Capitonidae	Piciformes
Bee-eater, European	*Merops apiaster*	Meropidae	Coraciiformes
Bee-eater, white-fronted	*Merops bullockoides*	Meropidae	Coraciiformes
Bellbird, bare-throated	*Procnias nudicollis*	Cotingidae	Passeriformes
Bird of paradise, blue	*Paradisaea rudolphi*	Paradisaeidae	Passeriformes
Bird of paradise, king	*Cicinnurus regius*	Paradisaeidae	Passeriformes
Bird of paradise, lesser	*Paradisaea minor*	Paradisaeidae	Passeriformes
Bird of paradise, Raggiana	*Paradisaea raggiana*	Paradisaeidae	Passeriformes
Bird of paradise, sicklebills	*Epimachus* spp.	Paradisaeidae	Passeriformes
Bird of paradise, twelve-wired	*Seleucidis melanoleuca*	Paradisaeidae	Passeriformes
Bishop, red	*Euplectes orix*	Ploceidae	Passeriformes
Blackbird, red-winged	*Agelaius ploeniceus*	Icteridae	Passeriformes
Blackbird, yellow-headed	*Xanthocephalus xanthocephalus*	Icteridae	Passeriformes
Bluebird, eastern	*Sialia sialis*	Turdidae	Passeriformes
Bobolink	*Dolichonyx oryzivorus*	Icteridae	Passeriformes
Booby, blue-footed	*Sula nebouxii*	Sulidae	Pelecaniformes
Bowerbird, Archbold's	*Archboldia papuensis*	Ptilonorhynchidae	Passeriformes
Bowerbird, fawn-breasted	*Chlamydera cerviniventris*	Ptilonorhynchidae	Passeriformes
Bowerbird, Macgregor's	*Amblyornis macgregoriae*	Ptilonorhynchidae	Passeriformes
Bowerbird, satin	*Ptilonorhynchus violaceus*	Ptilonorhynchidae	Passeriformes
Bowerbird, streaked	*Amblyornis subalaris*	Ptilonorhynchidae	Passeriformes
Bowerbird, spotted	*Chlamydera maculata*	Ptilonorhynchidae	Passeriformes
Bowerbird, Vogelkop	*Amblyornis inornatus*	Ptilonorhynchidae	Passeriformes
Brush-turkey, Australian	*Alectura lathami*	Megapodiidae	Galliformes
Budgerigar	*Melopsittacus undulatus*	Psittacidae	Psittaciformes
Bunting, corn	*Miliaria calandra*	Emberizidae	Passeriformes
Bunting, indigo	*Passerina cyanea*	Cardinalidae	Passeriformes
Bunting, lazuli	*Passerina amoena*	Cardinalidae	Passeriformes
Bunting, reed	*Emberiza schoeniclus*	Emberizidae	Passeriformes
Bunting, snow	*Plectrophenax nivalis*	Emberizidae	Passeriformes
Bushtit, common	*Psaltiparus minimus*	Aegithalidae	Passeriformes
Buttonquail, Madagascar	*Turnix nigricollis*	Turnicidae	Gruiformes

Appendix 1. Continued

Common Name	Scientific name	Family	Order
Cacique, yellow-rumped	*Cacicus cela*	Icteridae	Passeriformes
Capuchinbird	*Perissocephalus tricolor*	Cotingidae	Passeriformes
Canary, common (= European serin)	*Serinus serinus*	Fringillidae	Passeriformes
Capercaillie, (western)	*Tetrao urogallus*	Phasianidae	Galliformes
Cardinal, northern	*Cardinalis cardinalis*	Cardinalidae	Passeriformes
Cassowary, northern	*Casuarius unappendiculatus*	Casuariidae	Struthioniformes
Catbird (bowerbird), toothbilled	*Ailuroedus dentirostris*	Ptilonorhynchidae	Passeriformes
Cockatoo, pink	*Cacatua leadbetteri*	Psittacidae	Psittaciformes
Cock-of-the-rock (Guianan)	*Rupicola rupicola*	Cotingidae	Passeriformes
Colies (= mousebirds)		Coliidae	Coliiformes
Condor, Andean	*Vultur gryphus*	Cathartidae	Ciconiiformes
Condor, California	*Gymnogyps californianus*	Cathartidae	Ciconiiformes
Cormorants		Phalacrocoracidae	Pelecaniformes
Corvids		Corvidae	Passeriformes
Coucal, black	*Centropus grillii*	Cuculidae	Cuculiformes
Coucal, Madagascar	*Centropus toulou*	Cuculidae	Cuculiformes
Coucal, Senegal	*Centropus senegalensis*	Cuculidae	Cuculiformes
Coursers		Glareolidae	Charadriiformes
Cowbird, brown-headed	*Molothrus ater*	Iceridae	Passeriformes
Crane, whooping	*Grus americana*	Gruidae	Gruiformes
Cuckoo, black-billed	*Coccyzus erythrophthalmus*	Cuculidae	Cuculiformes
Cuckoo, common	*Cuculus canorus*	Cuculidae	Cuculiformes
Cuckoo, dwarf	*Coccyzus pumilus*	Cuculidae	Cuculiformes
Cuckoo, Great Spotted	*Clamator glandarius*	Cuculidae	Cuculiformes
Cuckoo, guira	*Guira guira*	Cuculidae	Cuculiformes
Cuckoo, yellow-billed	*Coccyzus americanus*	Cuculidae	Cuculiformes
Curassow, helmeted	*Pauxi pauxi*	Cracidae	Galliformes
Curassow, yellow-knobbed	*Crax daubentoni*	Cracidae	Galliformes
Divers (= loons)		Gaviidae	Gaviiformes
Dotterel, Eurasian	*Eudromias morinellus*	Charadriidae	Charadriiformes
Dove, rock	*Columba livia*	Columbidae	Columbiformes
Duck, black	*Anas rubripes*	Anatidae	Anseriformes
Duck, black-headed	*Heteronetta atricapilla*	Anatidae	Anseriformes
Duck, blue	*Hymenolaimus malacorhynchus*	Anatidae	Anseriformes
Duck, mallard	*Anas platyrhynchos*	Anatidae	Anseriformes
Duck, Mexican	*Anas diazi*	Anatidae	Anseriformes

Common name	Scientific name	Family	Order
Duck, mottled	*Anas fulvigula*	Anatidae	Anseriformes
Duck, northern pintail	*Anas acuta*	Anatidae	Anseriformes
Duck, wood	*Aix sponsa*	Anatidae	Anseriformes
Dunnock (= hedge accentor)	*Prunella modularis*	Prunellidae	Passeriformes
Eagle, golden	*Aquila chrysaetos*	Accipitridae	Falconiformes
Emu	*Dromaius novaehollandiae*	Dromiceidae	Casuariiformes
Fairy-wren, splendid	*Malurus splendens*	Maluridae	Passeriformes
Fairy-wren, superb	*Malurus cyaneus*	Maluridae	Passeriformes
Falcons		Falconidae	Falconiformes
Finch, double-barred	*Taeniopygia bichenovii*	Estrildidae	Passeriformes
Finch, house	*Carpodacus mexicanus*	Fringillidae	Passeriformes
Finch, large ground	*Geospiza magnirostris*	Emberizidae	Passeriformes
Finch, medium ground	*Geospiza fortis*	Emberizidae	Passeriformes
Finch, large cactus ground	*Geospiza conirostris*	Emberizidae	Passeriformes
Finch, zebra	*Taeniopygia guttata*	Estrildidae	Passeriformes
Finches, Darwin's		Emberizidae	Passeriformes
Fiscal, common	*Lanius collaris*	Laniidae	Passeriformes
Fiscal, grey-backed	*Lanius excubitoroides*	Laniidae	Passeriformes
Fiscal, long-tailed	*Lanius cabanisi*	Laniidae	Passeriformes
Flicker, northern	*Colaptes auratus*	Picidae	Piciformes
Flamingos		Phoenicopteridae	Phoenicopteriformes
Flycatcher, collared	*Ficedula albicollis*	Muscicapidae	Passeriformes
Flycatcher, fork-tailed	*Tyrannus savana*	Tyrannidae	Passeriformes
Flycatcher, pied	*Ficedula hypoleuca*	Muscicapidae	Passeriformes
Flycatcher, scissor-tailed	*Tyrannus forficatus*	Tyrannidae	Passeriformes
Fowl, domestic (chicken)	*Gallus gallus*	Phasianidae	Galliformes
Francolin, Erckel's	*Francolinus erckelii*	Phasianidae	Galliformes
Frigatebird, great	*Fregata minor*	Fregatidae	Pelecaniformes
Fulmar, northern	*Fulmarus glacialis*	Procellariidae	Procellariiformes
Geese		Anatidae	Anseriformes
Goldfinch, American	*Carduelis tristis*	Fringillidae	Passeriformes
Grackle, common	*Quiscalus quiscula*	Icteridae	Passeriformes
Grackle, great-tailed	*Quiscalus mexicanus*	Icteridae	Passeriformes
Grebes		Podicipedidae	Podicipediformes
Grosbeak, black-headed	*Pheucticus melanocephalus*	Emberizidae	Passeriformes
Grosbeak, rose-breasted	*Pheucticus ludovicianus*	Emberizidae	Passeriforme

Appendix 1. Continued

Common Name	Scientific name	Family	Order
Grouse, black	*Tetrao tetrix*	Phasianidae	Galliformes
Grouse, sage	*Centrocercus urophasianus*	Phasianidae	Galliformes
Grouse, sharp-tailed	*Tympanchus phasianellus*	Phasianidae	Galliformes
Guineafowl, helmeted	*Numida meleagris*	Numididae	Galliformes
Gull, black-headed	*Larus ridibundus*	Laridae	Charadriiformes
Gull, herring	*Larus argentatus*	Laridae	Charadriiformes
Hawk, Galapagos	*Buteo galapogensis*	Accipitridae	Falconiformes
Hawk, Harris'	*Parabuteo unicinctus*	Accipitridae	Falconiformes
Herons		Ardeidae	Ciconiiformes
Honeyguide, orange (= yellow)-rumped	*Indicator xanthonotus*	Indicatoridae	Piciformes
Hornbill, southern ground	*Bucorvus cafer*	Bucerotidae	Coraciiformes
Hummingbird, booted racket-tail	*Ocreatus underwoodii*	Trochiidae	Apodiformes
Hummingbird, purple-throated carib	*Eulampis jugularis*	Trochiidae	Trochiliformes
Hummingbird, streamertail (red-billed)	*Trochilus polytmus*	Troachilidae	Apodiformes
Ibis, scarlet	*Eudocimus ruber*	Threskiornithidae	Ciconiiformes
Ibis, white	*Eudocimus albus*	Threskiornithidae	Ciconiiformes
Indigobirds	*Vidua* spp.	Estrildidae	Passeriformes
Jacamars		Galbulidae	Piciformes
Jacanas		Jacanidae	Charadriiformes
Jay, green	*Cyanocorax yncas*	Corvidae	Passeriformes
Jay, Mexican	*Aphelocoma ultramarina*	Corvidae	Passeriformes
Jay, pinyon	*Gymnorhinus cyanocephalus*	Corvidae	Passeriformes
Jay, scrub (incl. Florida scrub-jay)	*Aphelocoma coerulescens*	Corvidae	Passeriformes
Jay, unicolored	*Apelocoma unicolor*	Cordivae	Passeriformes
Junco, dark-eyed	*Junco hyemalis*	Fringillidae	Passeriformes
Junglefowl, Ceylon	*Gallus lafayetti*	Phasianidae	Galliformes
Junglefowl, green	*Gallus varius*	Phasianidae	Galliformes
Junglefowl, grey	*Gallus sonneratii*	Phasianidae	Galliformes
Junglefowl, red	*Gallus gallus*	Phasianidae	Galliformes
Kakapo	*Strigops hapbroptilus*	Psittacidae	Psittaciformes
Kestrel, American	*Falco sparverius*	Falconidae	Falconiformes

Kestrel, common	*Falco tinnunculus*	Falconidae	Falconiformes
Killdeer	*Charadrius vociferus*	Charadriidae	Charadriiformes
Kingfisher (= common or European)	*Alcedo atthis*	Alcedinidae	Coraciiformes
Kingfisher, pied	*Ceryle rudis*	Alcedinidae	Coraciiformes
Kite, snail	*Rostrhamus sociabilis*	Accipitridae	Falconiformes
Kiwis	*Apteryx* spp.	Apterygidae	Dinornithiformes
Longspur, Smith's	*Calcarius pictus*	Emberizidae	Passeriformes
Loon, common	*Gavia immer*	Gaviidae	Gaviiformes
Lyrebird, superb	*Menura novaehollandiae*	Menuridae	Passeriformes
Magpie, black-billed	*Pica pica*	Corvidae	Passeriformes
Magpie, yellow-billed	*Pica nuttalli*	Corvidae	Passeriformes
Malleefowl	*Leipoa ocellata*	Megapodiidae	Galliformes
Manakin, golden-collared	*Manacus vitellinus*	Pipridae	Passeriformes
Manakin, golden-headed	*Pipra erythrocephala*	Pipridae	Passeriformes
Manakin, helmeted	*Antilophia galeata*	Pipridae	Passeriformes
Manakin, long-tailed	*Chiroxiphia linearis*	Pipridae	Passeriformes
Manakin, white-bearded	*Manacus nanacus*	Pipridae	Passeriformes
Manakin, white-collared	*Manacus candei*	Pipridae	Passeriformes
Manucodes	*Manucodie* spp.	Paradisaeidae	Passeriformes
Martin, purple	*Progne subis*	Hirundinidae	Passeriformes
Megapode (= scrubfowl), Melanesian	*Megapodius eremita*	Megapodiidae	Galliformes
Mesite, subdesert (monia)	*Monias benschi*	Mesoenatidae	Gruiformes
Mimids		Mimidae	Passeriformes
Mockingbird, northern	*Mimus polyglottos*	Mimidae	Passeriformes
Moorhen, dusky	*Gallinula tenebrosa*	Rallidae	Gruiformes
Moorhen (common)	*Gallinula chloropus*	Rallidae	Gruiformes
Motmots		Momotidae	Coraciiformes
Muscovy (duck)	*Cairina moschata*	Anatidae	Anseriformes
Myna, common	*Acridotheres tristis*	Mimidae	Passeriformes
Native hen, Tasmanian	*Gallinula mortierii*	Rallidae	Gruiformes
Nightjar, pennant-winged	*Macrodipteryx vexillarius*	Caprimulgidae	Caprimulgiformes
Nightjar, standard-winged	*Macrodipteryx longipennis*	Caprimulgidae	Caprimulgiformes

Appendix 1. Continued

Common Name	Scientific name	Family	Order
Oriole, Baltimore	*Icterus galbula*	Icteridae	Passeriformes
Oropendolas	*Psarocolius* sp. and *Gymnostinops* sp.	Iceridae	Passeriformes
Ostrich	*Struthio camelus*	Struthionidae	Struthioniformes
Owl, eastern screech	*Otus asio*	Strigidae	Strigiformes
Owl, Tengmalm's	*Aegolius funereus*	Strigidae	Strigiformes
Owls		Strigidae and Tytonidae	Strigiformes
Oxpecker, yellow-billed	*Buphagus africanus*	Sturnidae	Passeriformes
Oxpecker, red-billed	*Buphagus erythrorhynchus*	Sturnidae	Passeriformes
Oystercatcher, European (= Eurasian)	*Haematopus ostralegus*	Haematopodidae	Charadriiformes
Painted-snipe	*Rostratula* sp.	Rostratulidae	Charadriiformes
Palila	*Loxioides bailleui*	Drepanididae	Passeriformes
Paradise-flycatcher, African	*Terpsiphone viridis*	Monarchidae	Passeriformes
Paradise-flycatcher, Asian	*Terpsiphone paradisi*	Monarchidae	Passeriformes
Parotia, Lawes'	*Parotia lawesii*	Paradisaeidae	Passeriformes
Parrot, eclectus	*Eclectus roratus*	Psittacidae	Psittaciformes
Partridge, grey	*Perdix perdix*	Phasianidae	Galliformes
Peacock-pheasant, grey	*Polyplectron bicalcaratum*	Phasianidae	Galliformes
Peafowl, Indian (= common or blue)	*Pavo cristatus*	Phasianidae	Galliformes
Peafowl, Congo	*Afropavo congensis*	Phasianidae	Galliformes
Penguin, emperor	*Aptenodytes forsteri*	Spheniscidae	Sphenisciformes
Penguin, rockhopper	*Eudyptes chrysocome*	Spheniscidae	Sphenisciformes
Phalarope, red-necked (= northern)	*Phalaropus lobatus*	Scolopacidae	Charadriiformes
Phalarope, Wilson's	*Steganopus tricolor*	Scolopacidae	Charadriiformes
Pheasant, Bulwer's (wattled)	*Lophura bulweri*	Phasianidae	Galliformes
Pheasant, golden	*Chrysolophus pictus*	Phasianidae	Galliformes
Pheasant, great argus	*Argusianus argus*	Phasianidae	Galliformes
Pheasant, Kalij	*Lophura leucomelanos*	Phasianidae	Galliformes
Pheasant, Lady Amherst's	*Chrysolophus amherstiae*	Phasianidae	Galliformes
Pheasant, Reeve's	*Syrmaticus reevesii*	Phasianidae	Galliformes
Pheasant, ring-necked	*Phasianus colchicus*	Phasianidae	Galliformes
Pigeons (and doves)		Columbidae	Columbiformes
Pigeon, feral (rock dove)	*Columba livia*	Columbidae	Columbiformes
Pigeon, Victoria crowned	*Goura victoria*	Columbidae	Columbiformes
Pipit, meadow	*Anthus pratensis*	Motacillidae	Passeriformes
Plains-wanderer	*Pedionomus torquatus*	Pedionomidae	Charadriiformes
Plover, mountain	*Charadrius montanus*	Charadriidae	Charadriiformes
Poor-will	*Phalaenoptilus nuttallii*	Caprimulgidae	Caprimulgiformes
Prairie chicken, greater	*Tympanuchus cupido*	Phasianidae	Galliformes
Prairie chicken, lesser	*Tympanuchus pallidicinctus*	Phasianidae	Galliformes

Common name	Scientific name	Family	Order
Pratincoles		Glareolidae	Charadriiformes
Procellariiforms			Procellariiformes
Ptarmigan, willow	*Lagopus lagopus*	Phasianidae	Galliformes
Puffbirds		Bucconidae	Piciformes
Puffin, tufted	*Fratercula cirrhata*	Alcidae	Charadriiformes
Pukeko (= purple swamphen)	*Porphyrio porphyrio*	Rallidae	Gruiformes
Quail, coternix	*Coternix* spp.	Phasianidae	Galliformes
Quail, bobwhite	*Colinus virginianus*	Odontophoridae	Galliformes
Quail, Gambel's	*Callipepla gambelii*	Odontophoridae	Galliformes
Quail plover (= lark buttonquail)	*Ortyxelos meiffrenii*	Turnicidae	[Gruiformes]
Quetzal, resplendent	*Pharomachrus mocinno*	Trogonidae	Trogoniformes
Razorbill	*Alca torda*	Alcidae	Charadriiformes
Redpoll, greater	*Carduelis flammea*	Fringillidae	Passeriformes
Reed-warbler, greater	*Acrocephalus arundinaceus*	Sylviidae	Passeriformes
Reed-warbler, Henderson	*Acrocephalus vaughni*	Sylviidae	Passeriformes
Redstart, American	*Setophaga ruticilla*	Parulidae	Passeriformes
Rhea, greater (= common)	*Rhea americana*	Rheidae	Rheiformes
Rifleman	*Acanthisitta chloris*	Acanthisittidae	Passeriformes
Roadrunner, greater	*Geococcyx californianus*	Cuculidae	Cuculiformes
Robin, American	*Turdus migratorius*	Turdidae	Passeriformes
Robin, European	*Erithacus rubecula*	Turdidae	Passeriformes
Rook	*Corvus frugilegus*	Corvidae	Passeriformes
Rosellas	*Platycercus* spp.	Psittacidae	Psittaciformes
Ruff	*Philomachus pugnax*	Scolopacidae	Charadriiformes
Sandgrouse		Pteroclidae	Charadriiformes
Sandpiper, buff-breasted	*Tryngites subruficollis*	Scolopacidae	Charadriiformes
Sandpiper, common	*Tringa hypoleucos*	Scolopacidae	Charadriiformes
Sandpiper, purple	*Calidris maritima*	Scolopacidae	Charadriiformes
Sandpiper, spotted	*Tringa macularia*	Scolopacidae	Charadriiformes
Sandpiper, western	*Calidris mauri*	Scolopacidae	Charadriiformes
Sapsucker, Williamson's	*Sphyrapicus thyroideus*	Picidae	Piciformes
Sapsucker, yellow-bellied	*Sphyrapicus varius*	Picidae	Piciformes
Scrub-birds	*Atrichornis* spp.	Atrichornithidae	Passeriformes
Seedsnipe		Thinocoridae	Charadriiformes
Shrike, bull-headed	*Lanius bucephalus*	Laniidae	Passeriformes
Shrikes (see also see fiscals)		Laniidae	Passeriformes
Silver-eye	*Zosterops citrinellus*	Zosteropidae	Passeriformes
Snipe, common	*Gallinago gallinago*	Scolopacidae	Charadriiformes

Appendix 1. Continued

Common Name	Scientific name	Family	Order
Snipe, great	*Gallinago media*	Scolopacidae	Charadriiformes
Solitaire, Townsend's	*Myadestes townsendi*	Turdidae	Passeriformes
Sparrow, house	*Passer domesticus*	Passeridae	Passeriformes
Sparrow, savannah	*Passerculus sandwichensis*	Emberizidae	Passeriformes
Sparrow, seaside	*Ammodramus maritimus*	Emberizidae	Passeriformes
Sparrow, song	*Melospiza melodia*	Emberizidae	Passeriformes
Sparrow, swamp	*Melospiza georgiana*	Emberizidae	Passeriformes
Sparrow, white-crowned	*Zonotrichia leucophrys*	Emberizidae	Passeriformes
Sparrow, white-throated	*Zonotrichia albicollis*	Emberizidae	Passeriformes
Sparrowhawk, Eurasian	*Accipiter nisus*	Arcipitridae	Falconiformes
Starling, European	*Sturnus vulgaris*	Sturnidae	Passeriformes
Stork, saddle-billed		Ciconiidae	Ciconiiformes
Sunbird, scarlet-tufted malachite (red-tufted)	*Ephippiorhynchus senegalensis*	Nectarinidae	Passeriformes
Swallow, barn	*Nectarinia johnstoni*	Hirundinidae	Passeriformes
Swallow, tree	*Hirundo rustica*	Hirundinidae	Passeriformes
Swifts	*Tachycineta bicolor*	Apodidae	Apodiformes
Sylviids		Sylviidae	Passeriformes
tanagers		Thraupidae	Passeriformes
Tern, common	*Sterna hirundo*	Laridae	Charadriiformes
Tinamous		Tinamidae	Tinamiformes
Tinamou, elegant crested	*Eudromia elegans*	Tinamidae	Tinamiformes
Tit, bearded (parrotbill)	*Panurus biarmicus*	Sylviidae	Passeriformes
Tit, blue	*Parus caeruleus*	Paridae	Passeriformes
Tit, great	*Parus major*	Paridae	Passeriformes
Tit, penduline (Eurasian)	*Remis pendulinus*	Remizidae	Passeriformes
Tit, willow	*Parus montanus*	Paridae	Passeriformes
Toucan, keel-billed	*Ramphastos sulfuratus*	Rhamphastidae	Piciformes
Touracos		Musophagidae	Musophagiformes
Tragopan, Temminck's	*Tragopan temminckii*	Phasianidae	Galliformes
Treeswifts	*Hemiprocne sp.*	Hemiprocnidae	Apodiformes
Trogons		Trogonidae	Trogoniformes
Trumpeter, white-winged (pale-winged)	*Psophia leucoptera*	Psophiidae	Gruiformes
Tube-nosed swimmers			Procellariiformes
Turkey (= common or wild)	*Meleagris gallopavo*	Phasianidae	Galliformes

Common name	Scientific name	Family	Order
Turkey, ocellated	*Agriocharis ocellata*	Phasianidae	Galliformes
Tyrant-manakin, dwarf	*Tyranneutes stolzmanni*	Pipridae	Passeriformes
Vireo, red-eyed	*Vireo olivaceus*	Vireonidae	Passeriformes
Vulture, black	*Coragyps atratus*	Cathartidae	Ciconiiformes
Vulture, greater yellow-headed	*Cathartes melanbrotus*	Cathartidae	Ciconiiformes
Vulture, king	*Sarcoramphus papa*	Cathartidae	Ciconiiformes
Vulture, lesser yellow-headed	*Cathartes burrovianus*	Cathartidae	Ciconiiformes
Vulture, turkey	*Cathartes aura*	Cathartidae	Ciconiiformes
Warbler, sedge	*Acrocephalus schoenobaenus*	Sylviidae	Passeriformes
Warbler, Seychelles	*Acrocephalus (Bebrornis) sechellensis*	Sylviidae	Passeriformes
Warbler, willow	*Phylloscopus trochilus*	Sylviidae	Passeriformes
Warbler, wood	*Phylloscopus sibilatrix*	Sylviidae	Passeriformes
Warbler, yellow	*Dendroica petechia*	Parulidae	Passeriformes
Waxwing, cedar	*Bombycilla cedrorum*	Bombycillidae	Passeriformes
Weaver, parasitic	*Anomalospiza imberbis*	Ploceidae	Passeriformes
Wheatear, black	*Oenanthe leucura*	Turdicae	Passeriformes
Whydah, northern paradise	*Vidua orientalis*	Estrildidae	Passeriformes
Whydah, eastern paradise	*Vidua paradisaea*	Estrildidae	Passeriformes
Whydah, pin-tailed	*Vidua macroura*	Estrildidae	Passeriformes
Whydah, shaft-tailed (= queen)	*Vidua regia*	Estrildidae	Passeriformes
Widowbird, Jackson's	*Euplectes jacksoni*	Ploceidae	Passeriformes
Widowbird, long-tailed	*Euplectes progne*	Ploceidae	Passeriformes
Woodcock, American	*Scolopax minor*	Scolopacidae	Charadriiformes
Woodhoopoe, green	*Phoeniculus purpureus*	Phoeniculidae	Upupiformes
Woodpecker, acorn	*Melanerpes formicivorus*	Picidae	Piciformes
Woodpecker, downy	*Picoides pubescens*	Picidae	Piciformes
Woodpecker, hairy	*Picoides villosus*	Picidae	Piciformes
Woodpecker, red-cockaded	*Picoides borealis*	Picidae	Piciformes
Wood-warblers		Parulidae	Passeriformes
Wren (winter)	*Troglodytes troglodytes*	Troglodytidae	Passeriformes
Wren, bicolored	*Campylorhynchus griseus*	Troglodytidae	Passeriformes
Wren, Carolina	*Thryothorus ludovicianus*	Troglodytidae	Passeriformes
Wren, house	*Troglodytes aedon*	Troglodytidae	Passeriformes
Wren, marsh	*Cistothorus palustris*	Trogoldytidae	Passeriformes
Wren, stripe-backed	*Campylorhynchus nuchalis*	Troglodytidae	Passeriformes
Yellowhammer	*Emberiza citrinella*	Emberizidae	Passeriformes

Generic and specific names follow Sibley and Monroe (1990); families and orders follow the more traditional classifications of Howard and Moore (1991) and Gill (1995).

References

Alatalo, R. V., Carlson, A., Lundberg, A., and Ulfstrand, S. (1981). The conflict between male polygamy and female monogamy: the case of the pied flycatcher *Ficedula hypoleuca. American Naturalist*, **117**, 738–52.

Alatalo, R. V., Lundberg, A., and Ståhlbrandt, K. (1982). Why do pied flycatcher females mate with already-mated males? *Animal Behaviour*, **30**, 585–93.

Alatalo, R. V., Lundberg, A., and Glynn, C. (1986). Female pied flycatchers choose territory quality and not male characteristics. *Nature, London*, **323**, 152–3.

Alatalo, R. V., Göttlander, K., and Lundberg, A. (1987). Extra-pair copulations and mate guarding in the polyterritorial pied flycatcher, *Ficedula hypoleuca. Behaviour*, **101**, 139–55.

Alatalo, R. V., Höglund, J., and Lundberg, A. (1988). Patterns of variation in tail ornament size in birds. *Biological Journal of the Linnean Society*, **34**, 363–74.

Alatalo, R. V., Glynn, C., and Lundberg, A. (1990a). Singing rate and female attraction in the pied flycatcher: an experiment. *Animal Behaviour*, **39**, 601–3.

Alatalo, R. V., Eriksson, D., Gustafsson, L., and Lundberg, A. (1990b). Hybridization between pied and collared flycatchers—sexual selection and speciation theory. *Journal of Evolutionary Biology*, **3**, 375–89.

Alatalo, R. V., Höglund, J., and Lundberg, A. (1991). Lekking in black grouse—a test of male viability. *Nature, London*, **352**, 155–6.

Alatalo, R. V., Höglund, J., Lundberg, A., Rintamäki, P. T., and Silverin, B. (1996). Testosterone and male mating success on black grouse leks. *Proceedings of the Royal Society of London, B*, **263**, 1697–1702.

Alcock, J. (1975). *Animal behavior: an evolutionary approach*. Sinauer Associates, Sunderland, Massachusetts.

Alexander, R. D. (1974). The evolution of social behavior. *Annual Review of Ecology and Systematics*, **5**, 324–83.

Allee, W. C., Collias, N. E., and Lutherman, C. Z. (1939). Modifications of the social order in flocks of hens by the injection of testosterone propionate. *Physiological Zoology*, **12**, 412–39.

Altenburg, W., Daan, S., Starkenburg, J., and Zijlstra, M. (1982). Polygamy in the marsh harrier, *Circus aeruginosus*: individual variation in hunting performance and number of mates. *Behaviour*, **79**, 272–312.

Altmann, S. A., Wagner, S. S., and Lenington, S. (1977). Two models for the evolution of polygyny. *Behavioral Ecology and Sociobiology*, **2**, 397–410.

Alvarez, H. (1975). The social system of the green jay in Columbia. *Living Bird*, **14**, 5–44.

American Ornithologists' Union (1995). Fortieth supplement to the American Ornithologists' Union Checklist of North American birds. *Auk*, **112**, 819–30.

Andersson, M. (1978). Optimal egg shape in waders. *Ornis Fennica*, **55**, 105–9.

Andersson, M. (1982*a*). Sexual selection, natural selection and quality advertisement. *Biological Journal of the Linnean Society*, **17**, 375–93.

Andersson, M. (1982*b*). Female choice selects for extreme tail length in a widowbird. *Nature, London*, **299**, 818–20.

Andersson, M. (1994). *Sexual selection*. Princeton University Press.

Andersson, M. (1995). Evolution of reversed sex roles, sexual size dimorphism, and mating systems in coucals (Centropodidae, Aves). *Biological Journal of the Linnean Society*, **54**, 173–81.

Andersson, S. (1989). Sexual selection and cues for female choice in leks of Jackson's widowbird *Euplectes jacksoni*. *Behavioral Ecology and Sociobiology*, **25**, 403–10.

Andersson, S. (1991). Bowers on the savanna: display courts and mate choice in a lekking widowbird. *Behavioral Ecology*, **2**, 210–18.

Andersson, S. (1992). Female preference for long tails in lekking Jackson's widowbirds: experimental evidence. *Animal Behaviour*, **43**, 379–88.

Andersson, S. and Andersson, M. (1994). Tail ornamentation, size dimorphism and wing length in the genus *Euplectes* (Ploceinae). *Auk*, **111**, 80–6.

Arcese, P. (1989). Intrasexual competition and the mating system in primarily monogamous birds: the case of the song sparrow. *Animal Behaviour*, **38**, 96–111.

Armstrong, E. A. (1965). *Bird display and behaviour: an introduction to the study of bird psychology*, Revised edn. Dover Publications, Inc., New York.

Arnold, S. J. (1983). Sexual selection: the interface of theory and empiricism. In *Mate choice* (ed. P. Bateson), pp. 67–107. Cambridge University Press.

Arnold, S. J. (1994). Is there a unifying concept of sexual selection that applies to both plants and animals? *American Naturalist*, **144**, (**supplement**) S1–S12.

Baker, M. C. (1983). The behavioral response of female Nuttall's white-crowned sparrows to song of the natal dialect and alien dialect. *Behavioral Ecology and Sociobiology*, **12**, 309–15.

Baker, M. C. (1988). Sexual selection and size of repertoire in songbirds. In *Acta XIX Congressus Internationalis Ornithologica* (ed. H. Ouellet), pp. 1358–65. University of Ottawa Press.

Baker, M. C. and Baker, A. E. M. (1988). Vocal and visual stimuli enabling copulation behavior in female buntings. *Behavioral Ecology and Sociobiology*, **23**, 105–8.

Baker, M. C., Spitler-Nabors, K. J., and Bradley, D. C. (1981). Early experience determines song dialect responsiveness of female sparrows. *Science*, **214**, 819–20.

Baker, M. C., Spitler-Nabors, K. J., and Bradley, D. C. (1982). The response of female mountain white-crowned sparrows to songs from their native dialect and an alien dialect. *Behavioral Ecology and Sociobiology*, **10**, 175–9.

Baker, M. C., Bjerke, T. K., Lampe, H. U., and Espmark, Y. O. (1986). Sexual response of female great tits to variation in size of males' song repertoires. *American Naturalist*, **128**, 491–8.

Baker, M. C., Bjerke, T. K., Lampke, H. U., and Espmark, Y. O. (1987*a*). Sexual response of female yellowhammers to differences in regional song dialects and repertoire sizes. *Animal Behaviour*, **35**, 395–401.

Baker, M. C., Spitler-Nabors, K. J., Thompson, A. D., Jr., and Cunningham, M. A. (1987*b*). Reproductive behaviour of female white-crowned sparrows: effect of dialects and synthetic hybrid songs. *Animal Behaviour*, **35**, 1766–74.

Balaban, E. 1997. Changes in multiple brain regions underlie species differences in a complex congenital behavior. *Proceedings of the National Academy of Science, USA*, **94**, 2001–6.

Balda, R. R. and Bateman, G. C. (1972). The breeding biology of the pinyon jay. *Living Bird*, **11**, 5–42.

Balmford, A., Jones, I. L., and Thomas, A. L. R. (1993). On avian asymmetry: evidence of natural selection for symmetrical tails and wings in birds. *Proceedings of the Royal Society of London B*, **252**, 245–51.

Baptista, L. F. and Morton, M. L. (1982). Song dialects and mate selection in montane white-crowned sparrows. *Auk*, **99**, 537–47.

Baptista, L. F. and Morton, M. L. (1988). Song learning in montane white-crowned sparrows: from whom and when. *Animal Behaviour*, **36**, 1753–64.

Barnard, P. (1989). Territoriality and the determinants of male mating success in the southern African whydahs (Vidua). *Ostrich*, **60**, 103–17.

Barnard, P. (1990). Male tail length, sexual display intensity and female sexual response in a parasitic African finch. *Animal Behaviour*, **39**, 652–6.

Barnard, P. (1991). Ornament and body size variation and their measurement in natural populations. *Biological Journal of the Linnean Society*, **42**, 379–88.

Barraclough, T. G., Harvey, P. H., and Nee, S. (1995). Sexual selection and taxonomic diversity in passerine birds. *Proceedings of the Royal Society of London B*, **259**, 211–15.

Bart, J. and Tornes, A. (1989). Importance of monogamous male birds in determining reproductive success: evidence for house wrens and a review of male-removal experiments. *Behavioral Ecology and Sociobiology*, **24**, 109–16.

Basolo, A. L. (1990). Female preferences predates the evolution of the sword in swordtail fish. *Science*, **250**, 808–10.

Basolo, A. L. (1995*a*). A further examination of a pre-existing bias favouring a sword in the genus *Xiphophorus*. *Animal Behaviour*, **50**, 365–75.

Basolo, A. L. (1995*b*). Phylogenetic evidence for the role of a pre-existing bias in sexual selection. *Proceedings of the Royal Society of London B*, **259**, 307–11.

Bateman, A. J. (1948). Intra-sexual selection in *Drosophila*. *Heredity*, **2**, 349–68.

Bateson, P. (1983). Optimal outbreeding. In *Mate choice*. (ed. P. Bateson), pp. 257–79. Cambridge University Press, Cambridge.

Baylis, J. R. (1982). Avian vocal mimicry: its function and evolution. In *Acoustic communication in birds*, Vol. 2 (ed. D. E. Kroodsma and E. H. Miller), pp. 51–83. Academic Press, New York.

Beani, L. and Dessi-Fulgheri, F. (1995). Mate choice in the grey partridge, *Perdix perdix*: role of physical and behavioural male traits. *Animal Behaviour*, **49**, 347–56.

Bednarz, J. C. (1987). Pair and group reproductive success, polyandry, and cooperative breeding in Harris' hawks. *Auk*, **104**, 393–404.

Bednarz, J. C. (1988). Cooperative hunting in Harris' hawks (*Parabuteo unicintus*). *Science*, **239**, 1525–7.

Beebe, W. (1990). *A monograph of the pheasants*, Vol. 1–4. Dover Publications, Inc., New York.

Beehler, B. M. (1983). Lek behavior of the lesser bird of paradise. *Auk*, **100**, 992–5.

Beehler, B. M. (1987b). Birds of paradise and mating system theory—predictions and observations. *Emu*, **87**, 78–89.

Beehler, B. M. (1987a). Ecology and behavior of the buff-tailed sicklebill. *Auk*, **104**, 48–55.

Beehler, B. M. (1988). Lek behavior of the Raggiana bird of paradise. *National Geographic Research*, **4**, 343–58.

Beehler, B. M. (1989). The birds of paradise. *Scientific American*, **261**, 117–23.

Beehler, B. M. and Foster, M. S. (1988). Hotshots, hotspots, and female preference in the organization of lek mating systems. *American Naturalist*, **131**, 203–19.

Beehler, B. and Pruett-Jones, S. G. (1983). Display, dispersion and diet of birds of paradise: a comparison of nine species. *Behavioral Ecology and Sociobiology*, **13**, 229–38.

Beehler, B. M., Pratt, T. K., and Zimmerman, D. A. (1986). *Birds of New Guinea*. Princeton University Press.

Beissinger, S. R. (1986). Demography, environmental uncertainty, and the evolution of mate desertion in the snail kite. *Ecology*, **68**, 1445–59.

Beissinger, S. R. (1987a). Anisogamy overcome: female strategies in snail kites. *American Naturalist*, **129**, 486–500.

Beissinger, S. R. (1987b). Mate desertion and reproductive effort in the snail kite. *Animal Behaviour*, **35**, 1504–19.

Beissinger, S. R. (1990). Experimental brood manipulations and the monoparental threshold in snail kites. *American Naturalist*, **136**, 20–38.

Beissinger, S. R. and Snyder, N. F. R. (1987). Mate desertion in the snail kite. *Animal Behaviour*, **35**, 477–87.

Beletsky, L. D., Gori, D. F., Freeman, S., and Wingfield, J. C. (1995). Testosterone and polygyny in birds. *Current Ornithology*, **12**, 1–41.

Belthoff, J. R., Dufty, A. M. J., and Gauthreaux, S. A. J. (1994). Plumage variation, plasma steroids and social dominance in male house finches. *Condor*, **96**, 614–25.

Bennett, A. T. D., Cuthill, I. C., and Norris, K. J. (1994). Sexual selection and the mismeasure of color. *American Naturalist*, **144**, 848–60.

Bennett, A. T. D., Cuthill, I. C., Partridge, J. C., and Maler, E. J. (1996). Ultraviolet vision and mate choice in zebra-finches. *Nature, London*, 380, 433–5.

Bensch, S. and Hasselquist, D. (1992). Evidence for active female choice in a polygynous warbler. *Animal Behaviour*, **44**, 301–12.

Bertram, B. C. R. (1979). Ostriches recognize their own eggs and discard others. *Nature, London*, **279**, 233–4.

Bertram, B. C. R. (1992). *The ostrich communal nesting system*. Princeton University Press.

Birkhead, T. R. and Goodburn, S. F. (1989). Magpie. In *Lifetime reproduction in birds* (ed. I. Newton), pp. 173–82. Academic Press, London.

Birkhead, T. R. and Møller, A. P. (1992). *Sperm competition in birds: evolutionary causes and consequences*. Academic Press, London.

Birkhead, T. R. and Møller, A. P. (1996). Monogamy and sperm competition in birds. In *Partnerships in birds, the study of monogamy* (ed. J. M. Black), pp. 323–43. Oxford University Press.

Birkhead, T. R., Burke, T., Zann, R. A., Hunter, F. M., and Krupa, A. P. (1990). Extra-pair paternity and intraspecific brood parasitism in wild zebra finches, *Taeniopygia guttata*, revealed by DNA fingerprinting. *Behavioral Ecology and Sociobiology*, **27**, 315–24.

Björklund, M. (1990). A phylogenetic interpretation of sexual dimorphism in body size and ornament in relation to mating systems in birds. *Journal of Evolutionary Biology*, **3**, 171–83.

Björklund, M. (1991). Evolution, phylogeny, sexual dimorphism and mating system in the grackles (*Quiscalus* spp.: Icterinae). *Evolution*, **45**, 608–21.

Blackburn, D. G. and Evans, H. E. (1986). Why are there no viviparous birds? *American Naturalist*, **128**, 165–90.

Blum, M. S. and Blum, N. A. (ed.) (1979). *Sexual selection and reproductive competition in insects*. Academic Press, New York.

Boag, P. T. and Grant, P. R. (1981). Intense natural selection in a population of Darwin's finches (Geospizinae) in the Galápagos. *Science*, **214**, 82–5.

Bock, W. J. (1979). A synthetic explanation of macroevolutionary change—a reductionistic approach. *Bulletin of the Carnegie Museum of Natural History*, **13**, 20–69.

Borgia, G. (1985*a*). Bower quality, number of decorations, and mating success of male satin bowerbirds (*Ptilonorhynchus violaceus*): an experimental analysis. *Animal Behaviour*, **33**, 266–71.

Borgia, G. (1985*b*). Bower destruction and sexual competition in the satin bowerbird (*Ptilonorhynchus violaceus*). *Behavioral Ecology and Sociobiology*, **18**, 91–100.

Borgia, G. (1986*a*). Sexual selection in bowerbirds. *Scientific American*, **254**, 92–100.

Borgia, G. (1986*b*). Satin bowerbird parasites: a test of the bright male hypothesis. *Behavioral Ecology and Sociobiology*, **19**, 355–8.

Borgia, G. (1987). A critical review of sexual selection models. In *Sexual selection: testing the alternatives* (ed. J. W. Bradbury and M. B. Andersson), pp. 55–66. John Wiley and Sons, Chichester, New York.

Borgia, G. (1995*a*). Complex male display and female choice in the spotted bowerbird: specialized functions for different bower decorations. *Animal Behaviour*, **49**, 1291–301.

Borgia, G. (1995*b*). Why do bowerbirds build bowers? *American Scientist*, **83**, 542–47.

Borgia, G. (1995*c*). Threat reduction as a cause of differences in bower structure, bower decoration and male display in two closely related bowerbirds, *Chlamydera nuchalis* and *C. maculata*. *Emu*, **95**, 1–12.

Borgia, G. and Collis, K. (1989). Female choice for parasite-free male satin bowerbirds and the evolution of bright male plumage. *Behavioral Ecology and Sociobiology*, **25**, 445–54.

Borgia, G. and Collis, K. (1990). Parasites and bright male plumage in the satin bowerbird (*Ptilonorhynchus violaceus*). *American Zoologist*, **30**, 279–85.

Borgia, G. and Gore, M. A. (1986). Feather stealing in the satin bowerbird (*Ptilonorhynchus violaceus*): male competition and the quality of display. *Animal Behaviour*, **34**, 727–38.

Borgia, G., Pruett-Jones, S. G., and Pruett-Jones, M. A. (1985). The evolution of bower-building and the assessment of male quality. *Zeitschrift für Tierpsychologie*, **67**, 225–36.

Boss, W. R. (1943). Hormonal determination of adult characters and sex behavior in herring gulls (*Larus argentatus*). *Journal of Experimental Zoology*, **94**, 181–209.

Bossema, I. and Kruijt, J. P. (1982). Male activity and female mate acceptance in the mallard (*Anas platyrhynchos*). *Behaviour*, **79**, 313–24.

Bowman, R. and Bird, D. M. (1987). Behavioral strategies of American kestrels during mate replacement. *Behavioral Ecology and Sociobiology*, **20**, 129–35.

Boyce, M. S. (1990). The red queen visits sage grouse leks. *American Zoologist*, **30**, 263–70.

Bradbury, J. W. (1981). The evolution of leks. In *Natural selection and social behavior: research and new theory* (ed. R. D. Alexander and D. W. Tinkle), pp. 138–69. Chiron, New York.

Bradbury, J. W. and Andersson, M. B. (ed.) (1987). *Sexual selection: testing the alternatives*. Wiley, Chichester.

Bradbury, J. W. and Gibson, R. M. (1983). Leks and mate choice. In *Mate choice* (ed. P. Bateson), pp. 109–38. Cambridge University Press.

Bradbury, J. W., Vehrencamp, S. L., and Gibson, R. M. (1985). Leks and the unanimity of female choice. In *Evolution: essays in honour of John Maynard Smith* (ed. P. F. Greenwood, P. H. Harvey, and M. Slatkin), pp. 301–14. Cambridge University Press.

Bradbury, J. W., Gibson, R. M., and Vehrencamp, S. L. (1989). Dispersion of displaying male sage grouse. I. Patterns of temporal variation. *Behavioral Ecology and Sociobiology*, **24**, 1–14.

Bray, O. E., Kennelly, J. J., and Guarino, J. L. (1975). Fertility of eggs produced on territories of vasectomized red-wing blackbirds. *Wilson Bulletin*, **87**, 187–95.

Breitwisch, R. (1988). Sex differences in defense of eggs and nestlings by northern mockingbirds (*Mimus polyglottos*). *Animal Behaviour*, **36**, 62–72.

Breitwisch, R. (1989). Mortality patterns, sex ratios, and parental investment in monogamous birds. In *Current ornithology*, Vol. 6 (ed. D. M. Power), pp. 1–49. Plenum Press, New York.

Breitwisch, R., and Whitesides, G. H. (1987). Directionality of singing and non-singing behaviour of mated and unmated northern mockingbirds, *Mimus polyglottos*. *Animal Behaviour*, **35**, 331–39.

Briskie, J. V. (1992). Copulation patterns and sperm competition in the polygynandrous Smith's longspur. *Auk*, **109**, 563–75.

Briskie, J. V. (1993). Anatomical adaptations to sperm competition in Smith's longspurs and other polygynandrous passerines. *Auk*, **110**, 875–88.

Briskie, J. V. (1996). Lack of sperm storage by female migrants and the significance of copulations en route. *Condor*, **98**, 417.

Brodkorb, P. (1960). How many species of birds have existed? *Bulletin of the Florida State Museum*, **4**, 349–71.

Brodsky, L. M. (1988). Ornament size influences mating success in male rock ptarmigan. *Animal Behaviour*, **36**, 662–7.

Brodsky, L. M. and Weatherhead, P. J. (1984). Behavioral and ecological factors contributing to American black duck–mallard hybridization. *Journal of Wildlife Management*, **48**, 846–52.

Brodsky, L. M., Ankney, C. D., and Dennis, D. G. (1988). The influence of male dominance on social interactions in black ducks and mallards. *Animal Behaviour*, **36**, 1371–8.

Brooke, M. D. and Hartley, I. (1995). Nesting Henderson reed-warblers (*Acrocephalus vaughani tati*) studied by DNA fingerprinting: unrelated coalitions in a stable habitat. *Auk*, **112**, 77–86.

Brooker, M. G., Rowley, I., Adams, M., and Bauerstock, P. R. (1990). Promiscuity: an inbreeding avoidance mechanism in a socially monogamous species? *Behavioral Ecology and Sociobiology*, **26**, 191–9.

Brown, J. L. (1974). Alternate routes to sociality in jays with a theory for the evolution of altruism and communal breeding. *American Zoologist*, **14**, 63–80.

Brown, J. L. (1975). *The evolution of behavior*. W. W. Norton, New York.

Brown, J. L. (1978). Avian communal breeding systems. *Annual Review of Ecology and Systematics*, **9**, 123–55.

Brown, J. L. (1987). *Helping and communal breeding in birds: ecology and evolution.* Princeton University Press.

Brown, J. L., Brown, E. R., and Dow, D. D. (1982). Helpers: effects of experimental removal on reproductive success. *Science*, **215**, 421–2.

Brown, J. L. and Eklund, A. (1994). Kin recognition and the major histocompatability complex: an integrative review. *American Naturalist*, **143**, 435–61.

Brown, J. L. and Li, S.-H. (1995). Phylogeny of social behavior in *Aphelocoma* jays: a role for hybridization? *Auk*, **112**, 464–72.

Brown, J. L., Brown, E. R., and Dow, D. D. (1982). Helpers: effects of experimental removal on reproductive success. *Science*, **215**, 421–2.

Bruce, J. P., Quinn, J. S., Sloane, S. A., and White, B. N. (1996). DNA fingerprinting reveals monogamy in the bushtit, a cooperative breeding species. *Auk*, **113**, 511–16.

Bruning, D. (1974). Social structure and reproductive behavior in the greater rhea. *Living Bird*, **13**, 251–94.

Bruning, D. (1985). Buttonquail. In *A dictionary of birds* (ed. B. Campbell and E. Lack), pp. 76–7. Buteo Books, Vermillion, SD.

Brush, A. H. (1990). Metabolism of carotenoid pigments in birds. *FASEB Journal*, **4**, 2969–77.

Brush, A. H. and Power, D. M. (1976). House finch pigmentation: carotenoid metabolism and the effect of diet. *Auk*, **93**, 725–39.

Buchholz, R. (1991). Older males have bigger knobs: correlates of ornamentation in two species of curassow. *Auk*, **108**, 153–60.

Buchholz, R. (1995). Female choice, parasite load and male ornamentation in wild turkeys. *Animal Behaviour*, **50**, 929–43.

Buchholz, R. (1996). Thermoregulatory role of the unfeathered head and neck in male wild Turkeys. *Auk*, **113**, 310–8.

Buchholz, R. (1997). Male dominance and variation in fleshy head ornamentation in wild turkeys. *Journal of Avian Biology*, **28**, 223–30.

Bunzel, M. and Drüke, J. (1989). Kingfisher. In *Lifetime reproduction in birds* (ed. I. Newton), pp. 107–16. Academic Press, London.

Burke, T., Davies, N. B., Bruford, M. W., and Hatchwell, B. J. (1989). Parental care and mating behaviour of polyandrous dunnocks *Prunella modularis* related to paternity by DNA fingerprinting. *Nature, London*, **338**, 249–51.

Burley, N. (1981). The evolution of sexual indistinguishability. In *Natural selection and social behavior: recent research and new theory*. (ed. R. D. Alexander and D. W. Tinkle), pp 121–137. Chiron Press, New York.

Burley, N. (1985). The organization of behavior and the evolution of sexually selected traits. In *Avian monogamy*, Ornithological Monographs, No. 37 (ed. P. A. Gowaty and D. Mock), pp. 22–44. American Ornithologists' Union, Washington.

Burley, N. (1986*a*). Sexual selection for aesthetic traits in species with biparental care. *American Naturalist*, **127**, 415–45.

Burley, N. (1986*b*). Comparison of the band-colour preferences of two species of estrildid finches. *Animal Behaviour*, **34**, 1732–41.

Burley, N. (1988). The differential-allocation hypothesis: an experimental test. *American Naturalist*, **132**, 611–28.

Burley, N. (1991). Mate choice by multiple criteria in a monogamous species. *American Naturalist*, **117**, 515–28.

Burley, N. and Coopersmith, C. B. (1987). Bill color preferences of zebra finches. *Ethology*, **76**, 133–51.

Burley, N. T. and Price, D. K. (1991). Extrapair copulation and attractiveness in zebra finches. *Proceedings of the International Ornithological Congress*, **20**, 1367–72.

Burley, N., Prantzberg, G., and Radman, P. (1982). Influence of colour banding on the conspecific preferences of zebra finches. *Animal Behaviour*, **30**, 444–5.

Burley, N., Tidemann, S. C., and Halupka, K. (1991). Bill colour and parasite levels of zebra finches. In *Bird–parasite interactions: ecology, evolution, and behaviour* (ed. J. E. Loye and M. Zuk), pp. 359–76. Oxford University Press.

Burt, D. B. and Peterson, A. T. (1993). Biology of cooperative-breeding scrub jays (*Aphelocoma coerulescens*) of Oaxaca, Mexico. *Auk*, **110**, 207–14.

Cade, T. J. and Maclean, G. L. (1967). Transport of water by adult sandgrouse to their young. *Condor*, **69**, 323–43.

Campbell, B. (ed.) (1972). *Sexual selection and the descent of man, 1871–1971*. Aldine, Chicago, Illinois.

Campbell, B. and Lack, E. (1985). *A dictionary of birds*. Buteo Books, Vermillion, South Dakota.

Carey, M. and Nolan, V., Jr. (1979). Population dynamics of indigo buntings and the evolution of polygyny. *Evolution*, **33**, 1180–92.

Carlsson, B. G., Hornfeldt, B., and Lofgren, O. (1987). Bigyny in Tengmalm's owl *Aegolius funereus*: effect of mating strategy on breeding success. *Ornis Scandinavica*, **18**, 237–43.

Catchpole, C. K. (1980). Sexual selection and the evolution of complex songs among European warblers of the genus *Acrocephalus*. *Behaviour*, **74**, 146–66.

Catchpole, C. K. and Slater, P. J. B. (1995). *Bird song*. Cambridge University Press.

Catchpole, C. K., Dittami, J., and Leisler, B. (1984). Differential responses to male song repertoires in female songbirds implanted with oestradiol. *Nature, London*, **312**, 563–4.

Cezilly, F. and Nager, R. G. (1995). Comparative evidence for a positive association between divorce and extra-pair paternity in birds. *Proceedings of the Royal Society of London B*, **262**, 7–12.

Chapin, J. P. (1939). The birds of the Belgian Congo. II. *Bulletin of the American Museum of Natural History*, **75**, 1–632.

Chappell, M. A., Zuk, M., Kwan, T., and Johnsen, T. S. (1995). Energy cost of an avian vocal display: crowing in red junglefowl. *Animal Behaviour*, **49**, 254–6.

Charlesworth, B. (1987). The heritability of fitness. In *Sexual selection: testing the alternatives* (ed. J. W. Bradbury and M. B. Andersson), pp. 21–40. Wiley, Chichester.

Chiappe, L. M. (1995). The first 85 million years of avian evolution. *Nature*, **378**, 349–55.

Choudhury, S. (1995). Divorce in birds: a review of the hypotheses. *Animal Behaviour*, **50**, 413–29.

Clark, G. A., Jr. (1964). Life histories and the evolution of megapodes. *Living Bird*, **3**, 149–67.

Clayton, D. H. (1990). Mate choice in experimentally parasitized rock doves: lousy males lose. *American Zoologist*, **30**, 251–62.

Clutton-Brock, T. H. (ed.) (1988). *Reproductive success*. University of Chicago Press, Chicago, Illinois.

Clutton-Brock, T. H. (1991). *The evolution of parental care*. Princeton University Press, Princeton.

Clutton-Brock, T. H. and Godfray, C. (1991). Parental investment. In *Behavioural ecology: an evolutionary approach* (ed. J. R. Krebs and N. B. Davies), pp. 234–62. Blackwell Scientific Publications, Oxford.

Clutton-Brock, T. H. and Harvey, P. H. (1979). Comparison and adaptation. *Proceedings of the Royal Society of London B*, **205**, 547–65.

Coates, B. J. (1990). *The birds of Papua New Guinea. Vol. II, Passerines*. Dove Publications, Alderley, Queensland.

Coddington, C. L. and Cockburn, A. (1995). The mating system of free-living emus. *Australian Journal of Zoology*, **43**, 365–72.

Collias, N. E. (1943). Statistical analysis of factors which make for success in initial encounters between hens. *American Naturalist*, **77**, 519–38.

Collias, N. E. and Collias, E. C. (1985). Social behavior of unconfined red jungle fowl. *Zoonooz*, **58**, 4–11.

Collias, N. E., Collias, E. C., Hunsaker, D., and Minning, L. (1966). Locality fixation, mobility and social organization within an unconfined population of red jungle fowl. *Animal Behaviour*, **14**, 550–9.

Collins, S. A. and ten Cate, C. (1996). Does beak colour affect female preference in zebra finches? *Animal Behaviour*, **52**, 105–12.

Collis, K. and Borgia, G. (1992). Age-related effects of testosterone, plumage, and experience on aggression and social dominance in juvenile male satin bowerbirds (*Ptilonorhynchus violaceus*). *Auk*, **109**, 422–34.

Colwell, M. A. and Oring, L. W. (1988). Sex ratios and intrasexual competition for mates in a sex-role reversed shorebird, Wilson's phalarope (*Phalaropus tricolor*). *Behavioral Ecology and Sociobiology*, **22**, 165–73.

Cooper, A. and Penny, D. (1997). Mass survival of birds across the Cretaceous–Tertiary boundary: molecular evidence. *Science*, **275**, 1109–13.

Cooper, W. T. and Forshaw, J. M. (1977). *The birds of paradise and bowerbirds*. Collins, Sydney.

Cox, F. E. G. (1989). Parasites and sexual selection. *Nature, London*, **341**, 289.

Coyne, J. A. (1992). Genetics and speciation. *Nature, London*, **355**, 511–5.

Cracraft, J. (1974). Phylogeny and evolution of the ratite birds. *Ibis*, **116**, 494–521.

Cracraft, J. (1986). Origin and evolution of continental biotas: speciation and historical congruence within the Australian avifauna. *Evolution*, **40**, 977–96.

Craig, J. L. and Jamieson, I. G. (1990). Pukeko: different approaches and some different answers. In *Cooperative breeding in birds: long-term studies of ecology and behavior* (ed. P. B. Stacey and W. D. Koenig), pp. 385–412. Cambridge University Press.

Crockett, A. B. (1975). Ecology and behavior of the Williamson sapsucker in Colorado. Ph.D. dissertation, University of Colorado.

Crockett, A. B. and Hansley, P. L. (1977). Coition, nesting, and postfledging behavior of Williamson's Sapsucker in Colorado. *Living Bird*, **16**, 7–20.

Crome, F. H. J. (1976). Some observations on the biology of the Cassowary in northern Queensland. *Emu*, **76**, 8–14.

Cronin, E. W. J. and Sherman, P. W. (1976). A resource-based mating system: the orange-rumped honeyguide. *Living Bird*, **15**, 5–32.

Cronin, H. (1991). *The ant and the peacock*. Cambridge University Press.

Crook, J. H. (1962). The adaptive significance of pair formation types in weaverbirds. *Symposia, Zoological Society of London*, **8**, 57–70.

Crook, J. H. (1965). The adaptive significance of avian social organization. *Symposia, Zoological Society of London*, **14**, 181–218.

Crowe, T. M., Keith, S., and Brown, L. H. (1986). Order Galliformes. In *The birds of Africa*, Vol. 2 (ed. E. K. Urban, C. H. Fry, and S. Keith), pp. 1–75. Academic Press, London.

Cuthill, I. C. and MacDonald, W. A. (1990). Experimental manipulation of the dawn and dusk chorus in the blackbird *Turdus merula*. *Behavioral Ecology and Sociobiology*, **26**, 209–16.

Dahlgren, J. (1990). Females choose vigilant males: an experiment with the monogamous grey partridge *Perdix perdix*. *Animal Behaviour*, **39**, 646–51.

Dale, S. and Slagsvold, T. (1990). Random settlement of female pied flycatchers, *Ficedula hypoleuca*: significance of male territory size. *Animal Behaviour*, **39**, 231–43.

Dale, S., Rinden, H., and Slagsvold, T. (1992). Competition for a male restricts mate search of female pied flycatchers. *Behavioral Ecology and Sociobiology*, **30**, 165–76.

Darwin, C. (1859). *On the origin of species. Facsimile of the first edition. 1964.* Harvard University Press, Cambridge, Massachusetts.

Darwin, C. (1871). The descent of man and selection in relation to sex. Murray, London.

Davidar, P. and Morton, E. S. (1993). Living with parasites: prevalence of a blood parasite and its effect on survivorship in the purple martin. *Auk*, **110**, 109–16.

Davies, N. B. (1989). Sexual conflict and the polygyny threshold. *Animal Behaviour*, **38**, 226–34.

Davies, N. B. (1990). Dunnocks: cooperation and conflict among males and females in a variable mating system. In *Cooperative breeding in birds: long-term studies of ecology and behavior* (ed. P. B. Stacey and W. D. Koenig), pp. 455–85. Cambridge University Press.

Davies, N. B. (1991). Mating systems. In *Behavioural ecology: an evolutionary approach* (ed. J. R. Krebs and N. B. Davies), pp. 263–94. Blackwell Science Publications, Oxford.

Davies, N. B. (1992). *Dunnock behaviour and social evolution*. Oxford University Press.

Davies, N. B., Hartley, I. R., Hatchwell, B. J., Desrochers, A., Skeer, J., and Nebel, D. (1995). The polygynandrous mating system of the alpine accentor, *Prunella collaris*. I. Ecological causes and reproductive conflicts. *Animal Behaviour*, **49**, 769–88.

Davies, S. J. J. F. (1976). The natural history of the emu in comparison with that of other ratites. In *Proceedings of the XVI International Ornithological Congress, Canberra, Australia, 1974* (ed. H. J. Frith and J. H. Calaby), pp. 109–20. Australia Academy of Science.

Davison, G. W. H. (1985). Avian spurs. *Journal of Zoology*, **206**, 353–66.

Decker, M. D., Parker, P. G., Minchella, D. J., and Rabenold, K. N. (1993). Monogamy in black vultures: genetic evidence from DNA fingerprinting. *Behavioral Ecology*, **4**, 29–35.

de Kiriline, L. (1954). The voluble singer of the tree-tops. *Audubon Magazine*, **56**, 109–11.

Delacour, J. (1977). *The pheasants of the world*, 2nd edn. Saiga Ltd., Surrey, England.

DeLay, L. S., Faaborg, J., Naranjo, J., Paz, S. M., de Vries, T., and Parker, P. G. (1996). Paternal care in the cooperatively polyandrous Galapagos hawk. *Condor*, **98**, 300–11.

Delehanty, D. J., Fleischer, R. C., Colwell, M. A., and Oring, L. W. (1998). Sex-role reversal and the absence of extra-pair fertilization in Wilson's phalaropes. *Animal Behaviour*, **55**, 995–1002.

Dewsbury, D. A. (1978). *Comparative animal behavior*. McGraw-Hill, New York.

Diamond, J. M. (1972). *Avifauna of the eastern highlands of New Guinea*. Nuttall Ornithology Club, Cambridge, Massachusetts.

Diamond, J. M. (1982). Evolution of bowerbirds' bower: animal origins of the aesthetic sense. *Nature, London*, **297**, 99–102.

Diamond, J. M. (1986*a*). Animal art: variation in bower decorating style among male bowerbirds *Amblyornis inornatus*. *Proceedings of the National Academy of Sciences, U.S.A.*, **88**, 3042–6.

Diamond, J. M. (1986*b*). Biology of birds of paradise and bowerbirds. *Annual Review of Ecology and Systematics*, **17**, 17–37.

Diamond, J. M. (1987). Bower building and decoration by the bowerbird *Amblyornis inornatus*. *Ethology*, **74**, 177–204.

Diamond, J. M. (1988). Experimental study of bower decoration by the bowerbird *Amblyornis inornantus*, using colored poker chips. *American Naturalist*, **131**, 631–53.

Dilger, W. C. (1962). The behavior of lovebirds. *Scientific American*, **206**, 88–98.

Dixon, A., Ross, D., O'Malley, S. L. C., and Burke, T. (1994). Paternal investment inversely related to degree of extra-pair paternity in the reed bunting. *Nature, London*, **371**, 698–700.

Domm, L. V. (1939). Modifications in sex and secondary sexual characters. In *Sex and internal secretions: a survey of recent research* (ed. E. Allen), pp. 227–327. Williams and Wilkins, Baltimore, Maryland.

Dow, D. D. (1980). Communally breeding Australian birds with an analysis of distributional and environmental factors. *Emu*, **80**, 121–40.

Downhower, J. F. (1976). Darwin's finches and the evolution of sexual dimorphism in body size. *Nature, London*, **263**, 558–63.

Dufty, A. M. J. (1986), Singing and the establishment and maintenance of dominance hierarchies in captive brown-headed cowbirds. *Behavioral Ecology and Sociobiology*, **19**, 49–55.

Dunn, P. O. and Hannon, S. J. (1989). Evidence for obligate male parental care in black-billed magpies. *Auk*, **106**, 635–44.

Dunn, P. O., Cockburn, A., and Mulder, R. A. (1995). Fairy-wren helpers often care for young to which they are unrelated. *Proceedings of the Royal Society of London B*, **259**, 339–43.

du Plessis, M. A. (1989). Behavioural ecology of the redbilled woodhoopoe *Phoeniculus purpureus* in South Africa. Ph.D. dissertation., University of Cape Town.

du Plessis, M. A. (1991). The role of helpers in feeding chicks in cooperatively breeding green (red-billed) woodhoopoes. *Behavioral Ecology and Sociobiology*, **28**, 291–5.

du Plessis, M. A. and Williams, J. B. (1994), Communal cavity roosting in green woodhoopoes: consequences for energy expenditure and the seasonal pattern of mortality. *Auk*, **111**, 292–9.

Eason, P. K. and Sherman, P. T. (1995). Dominance status, mating strategies and copulation success in cooperatively polyandrous white-winged trumpeters, *Psophia leucopter* (Aves: Psophiidae). *Animal Behaviour*, **49**, 725–36.

Eberhardt, L. S. (1994). Oxygen consumption during singing by male Carolina Wrens. *Auk*, **111**, 124–30.

Eberhardt, L. S. (1996). Energy expenditure during singing: a reply to Gaunt *et al.* *Auk*, **113**, 721–3.

Edwards, S. V. and Naeem, S. (1993). The phylogenetic component of cooperative breeding in perching birds. *American Naturalist*, **141**, 754–89.

Elbin, S. B., Crowe, T. M., and Graves, H. B. (1986). Reproductive behavior of the helmeted guinea fowl (*Numida meleagris*): mating system and parental care. *Applied Animal Behaviour Science*, **16**, 179–97.

Emlen, S. T. (1978). The evolution of cooperative breeding in birds. In *Behavioural ecology: an evolutionary approach* (ed. J. R. Krebs and N. B. Davies), pp. 245–81. Blackwell Scientific Publications, Oxford.

Emlen, S. T. (1982a). The evolution of helping. I. An ecological constraints model. *American Naturalist*, **119**, 29–39.

Emlen, S. T. (1982b). The evolution of helping. II. The role of behavioral conflict. *American Naturalist*, **119**, 40–53.

Emlen, S. T. (1990). The white-fronted bee-eater: helping in a colonially nesting species. In *Cooperative breeding in birds: long-term studies of ecology and behavior* (ed. P. B. Stacey and W. D. Koenig), pp. 489–526. Cambridge University Press.

Emlen, S. T. (1991). Evolution of cooperative breeding in birds and mammals. In *Behavioural ecology: an evolutionary approach* (ed. J. R. Krebs and N. B. Davies), pp. 301–37. Blackwell Scientific Publications, Oxford.

Emlen, S. T. (1994). Benefits, constraints and the evolution of the family. *Trends in Ecology and Evolution*, **9**, 282–5.

Emlen, S. T. and Oring, L. W. (1977). Ecology, sexual selection, and the evolution of mating systems. *Science*, **187**, 215–23.

Emlen, S. T., Demong, N. J., and Demong, D. J. (1989). Experimental induction of infanticide in female wattled jacanas. *Auk*, **106**, 1–7.

Emlen, S. T., Reeve, H. K., Sherman, P. W., Wrege, P. H., Ratnieks, F. L. W., and Shellman-Reeve, J. (1991). Adaptive versus nonadaptive explanations of behavior: the case of alloparental helping. *American Naturalist*, **138**, 259–70.

Enquist, M. and Arak, A. (1994). Symmetry, beauty and evolution. *Nature, London*, **372**, 169–72.

Ens, B. J., Safriel, U. N., and Harris, M. P. (1993). Divorce in the long-lived and monogamous oystercatcher, *Haematopus ostralegus*: incompatability or choosing the better option? *Animal Behaviour*, **45**, 1199–217.

Ens, B. J., Choudhury, S., and Black, J. M. (1996). Mate fidelity and divorce in monogamous birds. In *Partnerships in birds: the study of monogamy* (ed. J. M. Black), pp. 344–95. Oxford University Press.

Erckmann, W. J. (1983). The evolution of polyandry in shorebirds: an evaluation of hypotheses. In *Social behavior of female vertebrates* (ed. S. K. Wasser), pp. 113–68. Academic Press, New York.

Eriksson, D. and Wallin, L. (1986). Male bird song attracts females—a field experiment. *Behavioral Ecology and Sociobiology*, **19**, 297–9.

Evans, M. R. (1991). The size of adornments of male scarlet-tufted malachite sunbirds varies with environmental conditions, as predicted by handicap theory. *Animal Behaviour*, **42**, 797–805.

Evans, M. R. (1993). Fluctuating asymmetry and long tails: the mechanical effects of asymmetry may act to enforce honest advertisement. *Proceedings of the Royal Society of London B*, **253**, 205–9.

Evans, M. R. (1997). Nest building signals male condition rather than age in wrens. *Animal Behaviour*, **53**, 749–55.

Evans, M. R. and Hatchwell, B. J. (1993). New slants on ornament asymmetry. *Proceedings of the Royal Society of London B*, **251**, 171–7.

Evans, M. R., Martins, T. L. F., and Haley, M. (1994). The asymmetrical cost of tail elongation in red-billed streamertails. *Proceedings of the Royal Society of London B*, **256**, 97–103.

Ezlanowski, A. (1985). The evolution of parental care in birds with reference to fossil embryos. In *Acta XVIII Congressus Internationalis Ornithologici* (ed. V. D. Ilyichev and V. M. Gavriolov), pp. 178–83. Nauka, Moscow.

Faaborg, J. and Bednarz, J. C. (1990). Galápagos and Harris' hawks: divergent causes of sociality in two raptors. In *Cooperative breeding in birds: long-term studies of ecology and behavior* (ed. P. B. Stacey and W. D. Koenig), pp. 359–83. Cambridge University Press.

Faaborg, J. and Patterson, C. B. (1981). The characteristics and occurrence of cooperative polyandry. *Ibis*, **123**, 477–84.

Faaborg, J., Parker, P. G., DeLay, L., Vries, T. de, Bednarz, J. C., Paz, S. M., *et al.* (1995). Confirmation of cooperative polyandry in the Galapagos hawk (*Buteo galapagoensis*). *Behavioral Ecology and Sociobiology*, **36**, 83–90.

Farley, G. H. (1995). Thermal, social and distributional consequences of nighttime cavity roosting in *Campylorhynchus* wrens. Ph.D. dissertation, The University of New Mexico, Albuquerque.

Feduccia, A. (1995). Explosive evolution in tertiary birds and mammals. *Science*, **267**, 637–8.

Feduccia, A. (1996). *The origin and evolution of birds*. Yale University Press, New Haven, Connecticut.

Fernández, G. J., and Reboreda, J. C. (1998). Effects of clutch size and timing of breeding on reproductive success of greater rheas. *Auk*, **115**, 340–8.

Fisher, R. A. (1915). The evolution of sexual preference. *Eugenics Review*, **7**, 184–92.

Fisher, R. A. (1930). *The genetical theory of natural selection*. Clarendon Press, Oxford.

Fisher, R. A. (1958). *The genetical theory of natural selection*, 2nd edn. Dover Publications, Inc., New York.

Fiske, P., Kalas, J. A., and Saether, S. A. (1996). Do female great snipe copy each other's mate choice? *Animal Behaviour*, **51**, 1355–62.

Fitzpatrick, J. W. and Woolfenden, G. E. (1988). Components of lifetime reproductive success in the Florida scrub jay. In *Reproductive success*. (ed. Clutton-Brock, T. H.), pp. 305–20. University of Chicago Press, Chicago.

Fivizzani, A. J. and Oring, L. W. (1986). Plasma steroid hormones in relation to behavioral sex role reversal in the spotted sandpiper, *Actitis macularia*. *Biology of Reproduction*, **35**, 1195–201.

Fivizzani, A. J., Colwell, M. A., and Oring, L. W. (1986). Plasma steroid hormone levels in free-living Wilson's phalaropes, *Phalaropus tricolor*. *General and Comparative Endocrinology*, **62**, 137–44.

Fleischer, R. C., Tarr, C. L., and Pratt, T. K. (1994). Genetic structure and mating system in the palila, an endangered Hawaiian honeycreeper, as assessed by DNA fingerprinting. *Molecular Ecology*, **3**, 383–92.

Fleischer, R. C., Tarr, C. L., Morton, E. S., Sangmeister, A., and Derrickson, K. C. (1997). Mating system of the dusky antbird, a tropical passerine, as assessed by DNA fingerprinting. *Condor*, **99**, 512–14.

Fleishman, L. J. (1992). The influence of the sensory system and the environment on motion patterns in the visual displays of anoline lizards and other vertebrates. *American Naturalist*, **139 (supplement)**, S36–61.

Folstad, I. and Karter, A. J. (1992). Parasites, bright males, and the immunocompetence handicap. *American Naturalist*, **139**, 603–22.

Ford, H. A. (1989). *Ecology of birds—an Australian perspective*. Surrey Beatty, Chipping Norton, NSW.

Ford, H. A., Bell, H., Nias, R., and Noske, R. (1988). The relationships between ecology and the incidence of cooperative breeding in Australian birds. *Behavioral Ecology and Sociobiology*, **22**, 239–49.

Ford, N. L. (1983). Variation in mate fidelity in monogamous birds. In *Current ornithology*, Vol. 1 (ed. R. F. Johnston), pp. 329–56. Plenum, New York.

Foster, M. S. (1977). Odd couples in manakins: a study of social organization and cooperative breeding in *Chiroxiphia linearis*. *American Naturalist*, **111**, 845–53.

Foster, M. S. (1981). Cooperative behavior and social organization of the swallow-tailed manakin (*Chiroxiphia caudata*). *Behavioral Ecology and Sociobiology*, **9**, 167–77.

Foster, M. S. (1983). Disruption, dispersion, and dominance in lek-breeding birds. *American Naturalist*, **122**, 53–71.

Fox, D. L. (1976). Animal biochromes and structural colors. University of California Press, Berkeley.

Frederick, P. C. (1987*a*). Extrapair copulations in the mating system of white ibis (*Eudocimus albus*). *Behaviour*, **100**, 170–201.

Frederick, P. (1987*b*). Responses of male white ibises to their mate's extra-pair copulations. *Behavioral Ecology and Sociobiology*, **21**, 223–8.

Freed, L. A. (1987). The long-term pair bond of tropical house wrens: advantage or constraint? *American Naturalist*, **130**, 505–25.

Frith, C. B. (1981). Displays of Count Raggi's bird-of-paradise *Paradisaea raggiana* and congeneric species. *Emu*, **81**, 193–201.

Frith, C. B. and Frith, D. W. (1993). Courtship display of the tooth-billed bowerbird *Scenopoeetes dentirostris* and its behavioural and systematic significance. *Emu*, **93**, 129–36.

Fry, C. H. (1969). Structural and functional adaptation to display in the standard-winged nightjar *Macrodipteryx longipennis*. *Journal of Zoology*, **157**, 19–24.

Fry, C. H. and Harwin, R. M. (1988). Caprimulgidae, nightjars. In *Birds of Africa*, Vol. 3 (ed. C. H. Fry, S. Keith, and E. K. Urban), pp. 155–97. Academic Press, London.

Fry, C. H., Fry, K., and Harris, A. (1992). *Kingfishers, bee-eaters and rollers*. Princeton University Press.

Fumihito, A., Miyake, T., Sumi, S.-I., Takada, M., Ohno, S., and Kondo, N. (1994). One subspecies of the red junglefowl (*Gallus gallus gallus*) suffices as the matriarchic ancestor of all domestic breeds. *Proceedings of the National Academy of Sciences*, **91**, 12505–09.

Fusani, L., Beani, L., Lupo, C., and Dessi-Fulgheri, F. (1997). Sexually selected vigilance behaviour of the grey partridge is affected by plasma androgen levels. *Animal Behaviour*, **54**, 1013–18.

Futuyma, D. J. (1986). *Evolutionary biology*, 2nd edn. Sinauer Associates, Sunderland, Massachusetts.

Futuyma, D. J. (1989). In *Speciation and its consequences* (ed. D. Otte and J. Endler), pp. 557–8. Sinauer Press, Sunderland, Massachusetts.

Galeotti, P., Saino, N., Sacchi, R., and Møller, A. P. (1997). Song correlates with social context, testosterone and body condition in male barn swallows. *Animal Behaviour*, **53**, 687–700.

Garnett, S. T. (1980). The social organization of the dusky moorhen, *Gallinula tenebrosa* Gould (Aves: Rallidae). *Australian Wildlife Research*, **7**, 103–12.

Garson, P. T., Pleszczynska, W. K., and Holm, C. H. (1981). The 'polygyny threshold' model: a reassessment. *Canadian Journal of Zoology*, **59**, 902–10.

Gaston, A. J. (1977). Social behaviour within groups of jungle babblers (*Turdoides striatus*). *Animal Behaviour*, **25**, 828–48.

Gaunt, A. S., Bucher, T. L., Gaunt, S. L. L., and Baptista, L. F. (1996). Is singing costly? *Auk*, **113**, 718–21.

Gavin, T. A. and Bollinger, E. R. (1985). Multiple paternity in a territorial passerine: the bobolink. *Auk*, **102**, 550–5.

Gayou, D. C. (1986). The social system of the Texas green jay. *Auk*, **103**, 540–7.

Gehlbach, F. R. (1994). The eastern screech owl: life history, ecology, and behavior in the suburbs and countryside. Texas A & M University Press.

George, T. L. (1987). Behavior of territorial male and female Townsend's solitaires (*Myadestes townsendi*) in winter. *American Midland Naturalist*, **118**, 121–7.

Gibbs, H. L., Weatherhead, P. J., Boag, P. T., White, B. N., Tabak, L. M., and Hoysak, D. J. (1990). Realized reproductive success of polygynous red-winged blackbirds revealed by DNA markers. *Science*, **250**, 1394–7.

Gibbs, H. L., Goldizen, A. W., Bullough, C., and Goldizen, A. R. (1994). Parentage analysis of multimale social groups of Tasmanian native hens (*Tribonyx mortierii*): genetic evidence for monogamy and polyandry. *Behavioral Ecology and Sociobiology*, **35**, 363–71.

Gibson, R. M. (1990). Relationships between blood parasites, mating success and phenotypic cues in male sage grouse (*Centrocercus urophasianus*). *American Zoologist*, **30**, 271–8.

Gibson, R. M. and Bachman, G. C. (1992). The costs of female choice in a lekking bird. *Behavioral Ecology*, **2**, 300–9.

Gibson, R. M. and Bradbury, J. W. (1985). Sexual selection in lekking sage grouse: phenotypic correlates of male mating success. *Behavioral Ecology and Sociobiology*, **18**, 117–23.

Gibson, R. M. and Höglund, J. (1992). Copying and sexual selection. *Trends in Ecology and Evolution*, **7**, 229–32.

Gibson, R. M., Bradbury, J. W., and Vehrencamp, S. (1991). Mate choice in lekking sage grouse revisited: the roles of vocal display, female site fidelity, and copying. *Behavioral Ecology*, **2**, 165–80.

Gill, F. B. (1995). *Ornithology*, 2nd edn. W. H. Freeman and Company, New York.

Gilliard, E. T. (1963). The evolution of bowerbirds. *Scientific American*, **209**, 38–46.

Gilliard, E. T. (1969). *Birds of paradise and bower birds*. Natural History Press, Garden City, New York.

Gittleman, J. L. (1981). The phylogeny of parental care in fishes. *Animal Behaviour*, **29**, 936–41.

Gladstone, D. E. (1979). Promiscuity in monogamous colonial birds. *American Naturalist*, **114**, 545–57.

Goldizen, A. W., Goldizen, A. R., Putland, D. A., Lambert, D. M., Millar, C. D., and Buchan, J. C. (1998). "Wife-sharing" in the Tasmanian native hen (*Gallinula mortierii*): is it caused by a male-biased sex ratio? *Auk*, **115**, 528–32.

Göransson, G., von Schantz, T., Fröberg, I., Helgée, A., and Wittzell, H. (1990). Male characteristics, viability and harem size in the pheasant, *Phasianus colchicus*. *Animal Behaviour*, **40**, 89–104.

Gould, S. J. and Lewontin, R. (1979). The spandrels of San Marco and the Panglossian paradigm: a critique of the adaptationist programme. *Proceedings of the Royal Society of London B*, **205**, 581–98.

Gould, S. J. and Vrba, E. S. (1982). Exaptation—a missing term in the science of form. *Paleobiology*, **8**, 4–15.

Gowaty, P. A. (1981). An extension of the Orians–Verner–Willson model to account for mating systems besides polygyny. *American Naturalist*, **118**, 851–9.

Gowaty, P. A. (1985). Multiple parentage and apparent monogamy in birds. In *Avian monogamy*, Ornithological Monographs, No. 37 (ed. P. A. Gowaty and D. W. Mock), pp. 11–21. The American Ornithologists' Union, Washington, DC.

Gowaty, P. A. (1996a). Battles of the sexes and origins of monogamy. In *Partnership in birds: the study of monogamy* (ed. J. M. Black), pp. 21–52. Oxford University Press.

Gowaty, P. A. (1996*b*). Field studies of parental care in birds. *Advances in the Study of Behavior*, **25**, 477–531.

Gowaty, P. A. and Mock, D. (ed.) (1985). *Avian monogamy*, Ornithological Monographs, No. 37. American Ornithologists Union, Washington, DC.

Grafen, A. (1990). Biological signals as handicaps. *Journal of Theoretical Biology*, **144**, 517–46.

Grahn, M. and von Schantz, T. (1994). Fashion and age in pheasants: age differences in mate choice. *Proceedings of the Royal Society of London B*, **255**, 237–41.

Grant, B. R. (1984). The significance of song variation in a population of Darwin's finches. *Behaviour*, **89**, 90–116.

Grant, P. R. and Grant, B. R. (1992). Hybridization of bird species. *Science*, **256**, 193–7.

Graul, W. D. (1974). Adaptive aspects of the mountain plover social system. *Living Bird*, **12**, 69–74.

Graul, W. D., Derrickson, S. R., and Mock, D. W. (1977). The evolution of avian polyandry. *American Naturalist*, **111**, 812–16.

Graves, J., Whiten, A., and Henzi, P. (1984). Why does the herring gull lay three eggs? *Animal Behaviour*, **32**, 798–805.

Gray, D. A. (1996). Carotenoids and sexual dichromatism in North American passerine birds. *American Naturalist*, **148**, 453–80.

Gray, D. A. and Hagelin, J. C. (1996). Song repertoires and sensory exploitation: reconsidering the case of the common grackle. *Animal Behaviour*, **52**, 795–800.

Greenlaw, J. S. and Post, W. (1985). Evolution of monogamy in seaside sparrows, *Ammodramus maritimus*: tests of hypotheses. *Animal Behaviour*, **33**, 373–83.

Gross, M. R. and Sargent, R. C. (1985). The evolution of male and female parental care in fishes. *American Zoologist*, **25**, 807–22.

Grossman, C. J. (1985). Interactions between the gonadal steroids and the immune system. *Science*, **227**, 257–61.

Gyllensten, U. B., Jakobsson, S., and Temrin, H. (1990). No evidence for illegitimate young in monogamous and polygynous warblers. *Nature, London*, **343**, 168–70.

Haig, S. M., Walters, J. R., and Plissner, J. H. (1994). Genetic evidence for monogamy in the cooperatively breeding red-cockaded woodpecker. *Behavioral Ecology and Sociobiology*, **34**, 295–303.

Halliday, T. R. (1983). The study of mate choice. In *Mate choice* (ed. P. Bateson), pp. 3–32. Cambridge University Press.

Hamilton, W. D. (1963). The evolution of altruistic behaviour. *American Naturalist*, **97**, 354–6.

Hamilton, W. D. (1964). The evolution of social behavior. *Journal of Theoretical Biology*, **7**, 1–52.

Hamilton, W. D. (1971). Geometry for the selfish herd. *Journal of Theoretical Biology*, **31**, 295–311.

Hamilton, W. D. and Zuk, M. (1982). Heritable true fitness and bright birds: a role for parasites. *Science*, **218**, 384–7.

Hamilton, W. D. and Zuk, M. (1989). Reply to Cox: parasites and sexual selection. *Nature, London*, **341**, 289–90.

Hamilton, W. J., III and Orians, G. H. (1965). Evolution of brood parasitism in altricial birds. *Condor*, **67**, 361–82.

Handford, P. and Mares, M. A. (1985). The mating system of ratites and tinamous: an evolutionary perspective. *Biological Journal of the Linnean Society*, **25**, 77–104.

Hansen, A. J. and Rowher, S. (1986). Coverable badges and resource defense in birds. *Animal Behaviour*, **34**, 69–76.

Hardesty, M. (1931). The structural basis for the response of the comb of the brown leghorn fowl to the sex hormones. *American Journal of Anatomy*, **47**, 277–323.

Harding, C. F. (1983). Hormonal influences on avian aggressive behavior. In *Hormones and aggressive behavior* (ed. B. B. Savre), pp. 435–67. Plenum Press, New York.

Hartley, I. R. and Davies, N. B. (1994). Limits to cooperative polyandry in birds. *Proceedings of the Royal Society of London B*, **257**, 67–73.

Hartley, I. R. and Shepherd, M. A. (1995). A random female settlement model can explain polygyny in the corn bunting. *Animal Behaviour*, **49**, 1111–8.

Hartley, I. R., Davies, N. B., Hatchwell, B. J., Desrochers, A., Nebel, D., and Burke, T. (1995). The polygynandrous mating system of the alpine accentor, *Prunella collaris*. II. Multiple paternity and parental effort. *Animal Behaviour*, **49**, 789–803.

Hartmann, W. (1985). The effect of selection and genetic factors on resistance to disease in fowl: a review. *World Poultry Science Journal*, **41**, 20–35.

Hartzler, J. E. (1972). An analysis of sage grouse lek behavior. Ph.D. dissertation, University of Montana.

Hartzler, J. E. (1974). Predation and the daily timing of sage grouse leks. *Auk*, **91**, 532–6.

Hartzler, J. E. and Jenni, D. A. (1988). Mate choice by female sage grouse. In *Adaptive strategies and population ecologies of northern grouse*, Vol. 1 (ed. A. T. Bergerud and M. W. Gratson), pp. 240–69. University of Minnesota Press.

Harvey, P. H. and Bradbury, J. W. (1991). Sexual selection. In *Behavioral ecology: an evolutionary approach* (ed. J. R. Krebs and N. B. Davies), pp. 203–33. Blackwell Scientific Publications, Oxford.

Harvey, P. H. and Pagel, M. D. (1991). *The comparative method in evolutionary biology*. Oxford University Press.

Harvey, P. H., Greenwood, P. J., and Campbell, B. (1984). Timing of laying by the pied flycatcher in relation to age of male and female parent. *Bird Study*, **31**, 57–60.

Hasselquist, D., Bensch, S., and von Schantz, T. (1995). Low frequency of extra-pair paternity in the polygynous great reed warbler, *Acrocephalus arundinaceus*. *Behavioral Ecology*, **6**, 27–38.

Hasselquist, D., Bensch, S., and von Schantz, T. (1996). Correlation between male song repertoire, extra-pair paternity and offspring survival in the great reed warbler. *Nature, London*, **381**, 229–32.

Haydock, J., Parker, P. G., and Rabenold, K. N. (1996). Extra-pair paternity uncommon in the cooperatively breeding bicolored wren. *Behavioral Ecology and Sociobiology*, **38**, 1–16.

Hedges, S. B., Parker, P. H., Sibley, C. G., and Kumar, S. (1996). Continental break-up and the ordinal diversification of birds and mammals. *Nature, London*, **381**, 226–9.

Heg, D., Ens, B. J., Burke, T., Jenkins, L., and Kruijt, J. P. (1993). Why does the typically monogamous oystercatcher (*Haematopus ostralegus*) engage in extra-pair copulations? *Behaviour*, **126**, 247–89.

Hensley, M. M. and Cope, J. B. (1951). Further data on removal and repopulation of the breeding birds in a spruce-fir forest community. *Auk*, **68**, 483–93.

Hill, G. E. (1986). Severe aggression between female black-headed grosbeaks. *Wilson Bulletin*, **98**, 486–8.

Hill, G. E. (1990). Female house finches prefer colourful males: sexual selection for a condition dependent trait. *Animal Behaviour*, **40**, 563–72.

Hill, G. E. (1991). Plumage coloration is a sexually selected indicator of male quality. *Nature, London*, **350**, 337–9.

Hill, G. E. (1992). The proximate basis of variation in carotenoid pigmentation in male house finches. *Auk*, **109**, 1–12.

Hill, G. E. (1993a). Male mate choice and the evolution of female plumage coloration in the house finch. *Evolution*, **47**, 1515–25.

Hill, G. E. (1993b). Geographic variation in the carotenoid plumage pigmentation of male house finches (*Carpodacus mexicanus*). *Biological Journal of the Linnean Society*, **49**, 63–86.

Hill, G. E. (1994a). Geographic variation in male ornamentation and female mate preference in the house finch: a comparative test of models of sexual selection. *Behavioral Ecology*, **5**, 64–73.

Hill, G. E. (1994b). Trait elaboration via adaptive mate choice: sexual conflict in the evolution of signals of male quality. *Ethology Ecology and Evolution*, **6**, 351–70.

Hill, G. E. (1995a). Evolutionary inference from patterns of female preference and male display. *Behavioral Ecology*, **6**, 350–1.

Hill, G. E. (1995b). Ornamental traits as indicators of environmental health: condition-dependent display traits hold promise as potent biomonitors. *BioScience*, **45**, 25–31.

Hill, G. E. (1995c). Review of A. P. Møller's *Sexual selection and the barn swallow*. *Quarterly Review of Biology*, **70**, 347–8.

Hill, G. E. and Montgomerie, R. (1994). Plumage colour signals nutritional condition in the house finch. *Proceedings of the Royal Society of London B*, **258**, 47–52.

Hill, G. E., Montgomerie, R., Inouye, C. Y., and Dales, J. (1994a). Influence of dietary carotenoids on plasma and plumage colour in the house finch: intra- and intersexual variation. *Functional Ecology*, **8**, 343–50.

Hill, G. E., Montgomerie, R., Roeder, C., and Boag, P. (1994b). Sexual selection and cuckoldry in a monogamous songbird: implications for sexual selection theory. *Behavioral Ecology and Sociobiology*, **35**, 193–9.

Hill, W. L. (1991). Correlates of male mating success in the ruff *Philomachus pugnax*, a lekking shorebird. *Behavioral Ecology and Sociobiology*, **29**, 367–72.

Hillgarth, N. (1990a). Parasites and female choice in the ring-necked pheasant. *American Zoologist*, **30**, 227–33.

Hillgarth, N. (1990b). Pheasant spurs out of fashion. *Nature, London*, **345**, 119–20.

Hinde, R. A. (1970). *Animal behaviour: a synthesis of ethology and comparative psychology*. McGraw-Hill, New York.

Höglund, J. (1989). Size and plumage dimorphism in lek-breeding birds: a comparative analysis. *American Naturalist*, **134**, 72–87.

Höglund, J. and Alatalo, R. V. (1995). *Leks*. Princeton University Press.

Höglund, J. and Lundberg, A. (1987). Sexual selection in a monomorphic lek-breeding bird: correlates of male mating success in the great snipe *Gallinago media*. *Behavioral Ecology and Sociobiology*, **21**, 211–16.

Höglund, J. and Robertson, J. G. M. (1990). Female preferences, male decision rules and the evolution of leks in the great snipe *Gallinago media*. *Animal Behaviour*, **40**, 15–22.

Höglund, J. and Sillen-Tullberg, B. (1994). Does lekking promote the evolution of male-biased size dimorphism in birds? On the use of comparative approaches. *American Naturalist*, **144**, 881–9.

Höglund, J., Eriksson, M., and Lindell, L. E. (1990). Females of the lek-breeding great snipe, *Gallinago media*, prefer males with white tails. *Animal Behaviour*, **40**, 23–32.

Höglund, J., Montgomerie, R., and Widemo, F. (1993). Costs and consequences of variation in the size of ruff leks. *Behavioral Ecology and Sociobiology*, **32**, 31–9.

Höglund, J., Alatalo, R. V., Gibson, R. M., and Lundberg, A. (1995). Mate-choice copying in black grouse. *Animal Behaviour*, **49**, 1627–33.

Hohn, E. O. (1970). Gonadal hormone concentrations in northern phalaropes in relation to nuptial plumage. *Canadian Journal of Zoology*, **48**, 400–1.

Hoi, H. and Hoi-Leitner, M. (1997). An alternate route to coloniality in the bearded tit: females pursue extra-pair fertilizations. *Behavioral Ecology*, **8**, 113–9.

Holmberg, K., Edsman, L., and Klint, T. (1989). Female mate preferences and male attributes in mallard ducks *Anas platyrhynchos*. *Animal Behaviour*, **38**, 1–7.

Horn, A. G., Leonard, M. L., and Weary, D. M. (1995). Oxygen consumption during crowing by roosters: talk is cheap. *Animal Behaviour*, **50**, 1171–5.

Hötker, H. (1989). Meadowpipit. In *Lifetime reproduction in birds* (ed. I. Newton), pp. 119–33. Academic Press, London.

Houde, P. (1988). *Paleognathous birds from the early Tertiary of the northern hemisphere*. Nuttall Ornithological Club, Cambridge, Massachusetts.

Howard, R. D. (1974). The influence of sexual selection and interspecific competition on mockingbird song (*Mimus polyglottos*). *Evolution*, **28**, 428–38.

Howard, R. and Moore, A. (1991). *A complete checklist of the birds of the world*. Academic Press, San Diego.

Hoysack, D. J. and Ankney, C. D. (1996). Correlates of behavioural dominance in mallards and American black ducks. *Animal Behaviour*, **51**, 409–19.

Hughes, J. M. (1996). Phylogenetic analysis of the Cuculidae (Aves: Cuculiformes) using behavioral and ecological characters. *Auk*, **113**, 10–22.

Hugie, D. M. and Lank, D. B. (1997). The resident's dilemma: a female choice model for the evolution of alternative mating strategies in lekking male ruffs (*Philomachus pugnax*). *Behavioral Ecology*, **8**, 218–25.

Hunter, F. M., Burke, T., and Watts, S. E. (1992). Frequent copulation as a method of paternity assurance in the northern fulmar. *Animal Behaviour*, **44**, 149–56.

Hutt, F. B. (1949). *Genetics of the fowl*. McGraw-Hill, New York.

Huxley, J. S. (1938). Darwin's theory of sexual selection and the data subsumed by it, in the light of recent research. *American Naturalist*, **72**, 416–33.

Irwin, M. P. S. (1988). Cuculiformes. In *Birds of Africa*, Vol. 3 (ed. C. H. Fry, S. Keith, and E. K. Urban), pp. 58–104. Academic Press, London.

Irwin, R. E. (1990). Directional sexual selection cannot explain variation in song repertoire size in the New World Blackbirds (Icterinae). *Ethology*, **85**, 212–24.

Irwin, R. (1994). The evolution of plumage dichromatism in the new world blackbirds: social selection on female brightness? *American Naturalist*, **144**, 890–907.

Jacob, F. (1977). Evolution and tinkering. *Science*, **196**, 1161–6.

Jamieson, I. G. (1989*a*). Levels of analysis or analyses at the same level. *Animal Behaviour*, **37**, 696–7.

Jamieson, I. G. (1989*b*). Behavioral heterochrony and the evolution of birds' helping at the nest: an unselected consequence of communal breeding. *American Naturalist*, **133**, 394–406.

Jamieson, I. G. and Craig, J. L. (1987). Critique of helping behavior in birds: a departure from functional explanations. In *Perspectives in ethology*, Vol. 7 (ed. P. Bateson and P. Klopfer), pp. 79–98. Plenum Press, New York.

Jamieson, I. G., Quinn, J. S., Rose, P. A., and White, B. N. (1994). Shared paternity among non-relatives is a result of an egalitarian mating system in a communally breeding bird, the pukeko. *Proceedings of the Royal Society of London B*, **257**, 271–7.

Järvi, T., Radesäter, T., and Jakobsson, S. (1980). The song of the willow warbler *Phylloscopus trochilus* with special reference to singing behavior in agonistic situations. *Ornis Scandinavica*, **11**, 236–42.

Jehl, J. R., Jr. and Murray, B. G., Jr. (1986). The evolution of normal and reverse sexual size dimorphism in shorebirds and other birds. In *Current ornithology*, Vol. 3 (ed. R. F. Johnston), pp. 1–86. Plenum Press, New York.

Jenni, D. A. (1974). Evolution of polyandry in birds. *American Zoologist*, **14**, 129–44.

Jenni, D. A. and Betts, B. J. (1978). Sex differences in nest construction, incubation, and parental behaviour in the polyandrous American Jacana (*Jacana spinosa*). *Animal Behaviour*, **26**, 207–18.

Johns, J. E. (1964). Testosterone-induced nuptial feathers in phalaropes. *Condor*, **66**, 449–55.

Johnsgard, P. A. (1960). A quantitative study of sexual behavior of mallards and black ducks. *Wilson Bulletin*, **72**, 135–55.

Johnsgard, P. A. (1986). *The pheasants of the world*. Oxford University Press.

Johnsgard, P. A. (1991). *Bustards, hemipodes, and sandgrouse: birds of dry places*. Oxford University Press.

Johnsgard, P. A. (1994). *Arena birds: sexual selection and behavior*. Smithsonian Institution Press, Washington, DC.

Johnson, K. (1988*a*). Sexual selection in piñon jays I: female choice and male–male competition. *Animal Behaviour*, **36**, 1038–47.

Johnson, K. (1988*b*). Sexual selection in piñon jays II: male choice and female–female competition. *Animal Behaviour*, **36**, 1048–53.

Johnson, K. and Burley, N. T. (1997). Mating tactics and mating systems of birds. In *Extra-pair mating tactics in birds*, Ornithological Monographs, No. 49 (ed. P. Parker and N. Burley), pp. 21–60. American Ornithologists' Union, Washington, DC.

Johnson, K. and Marzluff, J. M. (1990). Some problems and approaches in avian mate choice. *Auk*, **107**, 296–304.

Johnson, K., Dalton, R., and Burley, N. (1993*a*). Preferences of female American goldfinches (*Carduelis tristis*) for natural and artificial male traits. *Behavioral Ecology*, **4**, 138–43.

Johnson, K., Ligon, J. D., Thornhill, R., and Zuk, M. (1993*b*). The direction of mothers' and daughters' preference and the heritability of male ornaments in red junglefowl (*Gallus gallus*). *Behavioral Ecology*, **4**, 254–59.

Johnson, L. L. and Boyce, M. S. (1991). Female choice of males with low parasite loads in sage grouse. In *Bird–parasite interactions: ecology, evolution, and behaviour* (ed. J. E. Loye and M. Zuk), pp. 377–88. Oxford University Press.

Johnson, L. S. and Kermott, L. H. (1993). Why is reduced male parental assistance detrimental to the reproductive success of secondary female house wrens. *Animal Behaviour*, **46**, 1111–20.

Johnson, L. S., Kermott, L. H., and Lein, M. R. (1993). The cost of polygyny in the house wren *Troglodytes aedon*. *Journal of Animal Ecology*, **62**, 669–82.

Johnson, S. G. (1991). Effects of predation, parasites, and phylogeny on the evolution of bright coloration in North American male passerines. *Evolutionary Ecology*, **5**, 52–62.

Johnstone, R. A. (1994). Female preference for symmetrical males as a by-product of selection for mate recognition. *Nature, London*, **372**, 172–5.

Jones, C. S., Lessells, C. M., and Krebs, J. R. (1991). Helpers-at-the-nest in European bee-eaters (*Merops apiaster*): a genetic analysis. In *DNA fingerprinting approaches and applications* (ed. T. Burke, G. Dolf, A. J. Jeffreys, and R. Wolff), pp. 169–92. Birkhauser, Basel, Switzerland.

Jones, D. N., Dekker, R. W. R. J., and Roselaar, C. S. (1995). *The megapodes*. Oxford University Press.

Jones, I. L. and Hunter, F. M. (1993). Mutual sexual selection in a monogamous seabird. *Nature, London*, **362**, 238–9.

Jones, I. L. and Montgomerie, R. (1992). Least auklet ornaments: do they function as quality indicators? *Behavioral Ecology and Sociobiology*, **30**, 43–52.

Joste, N., Ligon, J. D., and Stacey, P. B. (1985). Shared paternity in the acorn woodpecker (*Melanerpes formicivorus*). *Behavioral ecology and sociobiology*, **17**, 39–41.

Kattan, G. (1988). Food habits and social organization of acorn woodpeckers in Colombia. *Condor*, **90**, 100–6.

Keast, A. (1958). Infraspecific variation in the Australian finches. *Emu*, **58**, 219–46.

Keck, W. N. (1934). The control of the secondary sex characters in the English sparrow *Passer domesticus* (Linnaeus). *Journal of Experimental Zoology*, **67**, 315–41.

Kemp, A. (1995). *The hornbills*. Oxford University Press.

Kempenaers, B. (1995). Polygyny in the blue tit: intra- and inter-sexual conflicts. *Animal Behaviour*, **49**, 1047–64.

Kempenaers, B., Verheyen, G. R., Van den Broeck, M., Burke, T., Van Broeckhoven, C., and Dhondt, A. A. (1992). Extra-pair paternity results from female preference for high-quality males in the blue tit. *Nature, London*, **357**, 494–6.

Kennedy, M., Spencer, H. G., and Gray, R. D. (1996). Hop, step and gape: do social displays of the Pelecaniformes reflect phylogeny? *Animal Behaviour*, **51**, 273–91.

Ketterson, E. D. and Nolan, V. J. (1992). Hormones and life histories: an integrative approach. *American Naturalist*, **140** (**supplement**), S33–62.

Kimball, R. T. (1995). Sexual selection in house sparrows, *Passer domesticus*. Ph.D. dissertation, The University of New Mexico.

Kimball, R. T. (1996). Female choice for male morphological traits in house sparrows, *Passer domesticus*. *Ethology*, **102**, 639–48.

Kimball, R. T., Ligon, J. D., and Merola-Zwartjes, M. (1997). Fluctuating asymmetry in red junglefowl. *Journal of Evolutionary Biology*, **10**, 441–57.

King, A. P., West, M. J., and Eastzer, D. H. (1980). Song structure and song development as potential contributors to reproductive isolation in cowbirds (*Molothrus ater*). *Journal of Comparative Physiology and Psychology*, **94**, 1028–39.

Kirkpatrick, C. E., Robinson, S. K., and Kitron, U. D. (1991). Phenotypic correlates of blood parasitism in the common grackle. In *Bird–parasite interactions: ecology, evolution and behaviour* (ed. J. E. Loye and M. Zuk), pp. 344–58. Oxford University Press.

Kirkpatrick, M. (1982). Sexual selection and the evolution of mate choice. *Evolution*, **36**, 1–12.

Kirkpatrick, M. (1987). The evolutionary forces acting on female mating preferences in polygynous animals. In *Sexual selection: testing the alternatives* (ed. J. W. Bradbury and M. B. Andersson), pp. 67–82. Wiley, Chichester.

Kirkpatrick, M. and Rosenthal, G. G. (1994). Symmetry without fear. *Nature, London*, **372**, 134–5.

Kirkpatrick, M. and Ryan, M. J. (1991). The paradox of the lek and the evolution of mating preferences. *Nature, London*, **350**, 33–8.

Klint, T. (1980). Influence of male nuptial plumage on mate selection in the female mallard (*Anas platyrhynchos* L.). *Animal Behaviour*, **28**, 1230–8.

Kodric-Brown, A. and Brown, J. H. (1984). Truth in advertising: the kinds of traits favored in sexual selection. *American Naturalist*, **124**, 309–23.

Koenig, W. D. and Mumme, R. L. (1987). *Population ecology of the cooperatively breeding acorn woodpecker*. Princeton University Press.

Koenig, W. D. and Stacey, P. B. (1990). Acorn woodpeckers: group-living and food storage under contrasting ecological conditions. In *Cooperative breeding in birds: long-term studies of ecology and behavior* (ed. P. B. Stacey and W. D. Koenig), pp. 413–53. Cambridge University Press.

Koenig, W. D., Pitelka, F. A., Carmen, W. J., Mumme, R. L., and Stanback, M. T. (1992). The evolution of delayed dispersal in cooperative breeders. *The Quarterly Review of Biology*, **67**, 111–50.

Komdeur, J. (1992). Importance of habitat saturation and territory quality for evolution of cooperative breeding in the Seychelles warbler. *Nature, London*, **360**, 768.

Komdeur, J. (1994). Experimental evidence for helping and hindering by previous offspring in the cooperative breeding Seychelles warbler *Acrocephalus sechellensis*. *Behavioral Ecology and Sociobiology*, **34**, 175–86.

Komdeur, J. (1996). Influence of helping and breeding experience on reproductive performance in the Seychelles warbler: a translocation experiment. *Behavioral Ecology*, **7**, 326–33.

Komdeur, J., Huffstadt, A., Prast, W., Castle, G., Mileto, R., and Wattel, J. (1995). Transfer experiments of Seychelles warblers to new islands: changes in dispersal and helping behaviour. *Animal Behaviour*, **49**, 695–708.

Korpimaki, E. (1991). Poor reproductive success of polygynously mated female Tengmalm's owls: are better options available? *Animal Behaviour*, **41**, 37–48.

Krebs, J. R., Ashcroft, R., and Webber, M. I. (1978). Song repertoires and territory defense in the great tit. *Nature, London*, **271**, 539–42.

Krimbas, C. B. (1984). On adaptation, neo-Darwinian tautology, and population fitness. *Evolutionary Biology*, **17**, 1–57.

Kroodsma, D. E. (1976). Reproductive development in a female songbird: differential stimulation by quality of male song. *Science*, **192**, 574–5.

Kroodsma, D. E. (1977). Correlates of song organization among North American wrens. *American Naturalist*, **111**, 995–1008.

Kroodsma, D. E. (1979). Vocal dueling among male marsh wrens: evidence for ritualized expressions of dominance/subordinance. *Auk*, **96**, 506–15.

Kroodsma, D. E. and Byers, B. E. (1991). The function(s) of bird song. *American Zoologist*, **31**, 318–28.

Kroodsma, D. E., Bereson, R. C., Byers, B. E., and Minear, E. (1989). Use of song types by the chestnut-sided warbler: evidence for both intra- and inter-sexual functions. *Canadian Journal of Zoology*, **67**, 447–56.

Kruijt, J. P., de Vos, G. J., and Bossema, I. (1972). The arena system of Black Grouse. *Proceedings of the International Ornithological Congress*, **15**, 399–423.

Kusmierski, R., Borgia, G., Crozier, R. H., and Chan, B. H. Y. (1993). Molecular information on bowerbird phylogeny and the evolution of exaggerated male characteristics. *Journal of Evolutionary Biology*, **6**, 737–52.

Kusmierski, R., Borgia, G., Uy, A., and Crozier, R. H. (1997). Labile evolution of display traits in bowerbirds indicates reduced effects of phylogenetic constraints. *Proceedings of the Royal Society of London B*, **264**, 307–13.

Lack, D. (1940). Pair formation in birds. *Condor*, **42**, 269–86.

Lack, D. (1968). *Ecological adaptations for breeding in birds*. Chapman and Hall, London.

Lande, R. (1981). Models of speciation by sexual selection on polygenetic traits. *Proceedings of the National Academy of Sciences, U.S.A.*, **78**, 3721–5.

Lande, R. (1982). Rapid origin of sexual isolation and character divergence in a cline. *Evolution*, **36**, 213–23.

Lank, D. B. and Smith, C. M. (1992). Females prefer larger leks: field experiments with ruffs (*Philomachus pugnax*). *Behavioral Ecology and Sociobiology*, **30**, 323–9.

Lank, D. B., Smith, C. M., Hanotte, O., Burke, T., and Cooke, F. (1995). Genetic polymorphism for alternative mating behavior in lekking male ruff *Philomachus pugnax*. *Nature, London*, **378**, 59–62.

Lanyon, S. M. (1992). Interspecific brood parasitism in blackbirds (Icterinae): a phylogenetic perspective. *Science*, **255**, 77–9.

Lawless, S. G., Ritchison, G., Klatt, P. H., and Westneat, D. F. (1997). The mating strategies of eastern screech owls: a genetic analysis. *Condor*, **99**, 213–17.

Lawrence, L. de K. (1967). *A comparative life-history study of four species of woodpeckers*, Ornithological Monographs, No. 5. American Ornithologists Union, Washington, DC.

LeCroy, M. (1981). The genus Paradesia—display and evolution. *American Museum Novitiates*, **2714**, 1–52.

LeCroy, M., Kulupi A., and Peckover, W. S. (1980). Goldie's bird of paradise: display, natural history and traditional relationships of people to the birds. *Wilson Bulletin*, **92**, 289–301.

Lemon, R. E., Weary, D. M., and Norris, K. J. (1992). Male morphology and behavior correlate with reproductive success in the American Redstart (*Setophaga ruticilla*). *Behavioral Ecology and Sociobiology*, **29**, 399–403.

Lenington, S. (1984). The evolution of polyandry in shorebirds. In *Shorebirds: breeding behavior and populations* (ed. J. Burger and B. L. Olla), pp. 149–65. Plenum Press, New York.

Leonard, M. L. (1990). Polygyny in marsh wrens: asynchronous settlement as an alternative to the polygyny-threshold model. *American Naturalist*, **136**, 446–58.

Leonard, M. L. and Horn, A. G. (1995). Crowing in relation to status in roosters. *Animal Behaviour*, **49**, 1283–90.

Leonard, M. L., Horn, A. G., and Eden, S. F. (1989). Does juvenile helping enhance breeder reproductive success? A removal experiment on moorhens. *Behavioral Ecology and Sociobiology*, **25**, 357–61.

Leroi, A. M., Rose, M. R., and Lauder, G. V. (1994). What does the comparative method reveal about adaptation? *American Naturalist*, **143**, 381–402.

Lifjeld, J. T. and Slagsvold, T. (1991). Sexual conflict among polygynous pied flycatchers feeding young. *Behavioral Ecology*, **2**, 106–15.

Lifjeld, J. T., Slagsvold, T., and Lampe, H. M. (1991). Low frequency of extra-pair paternity in pied flycatchers revealed by DNA fingerprinting. *Behavioral Ecology and Sociobiology*, **29**, 94–101.

Lifjeld, J. T., Dunn, P. O., Robertson, R. J., and Boag, P. T. (1993). Extra-pair paternity in monogamous tree swallows. *Animal Behaviour*, **45**, 213–29.

Ligon, J. D. (1968*a*). Sexual differences in foraging behavior in two species of *Dendrocopos* woodpeckers. *Auk*, **85**, 203–15.

Ligon, J. D. (1968*b*). The biology of the elf owl, *Micrathene whitneyi*. University of Michigan Museum of Zoology, Miscellaneous Publication No. 136.

Ligon, J. D. (1970). Behavior and breeding biology of the red-cockaded woodpecker. *Auk*, **87**, 255–78.

Ligon, J. D. (1978). Reproductive interdependence of piñon jays and piñon pines. *Ecological Monographs*, **48**, 11–126.

Ligon, J. D. (1983). Cooperation and reciprocity in avian social systems. *American Naturalist*, **121**, 336–84.

Ligon, J. D. (1985). *Review* The Florida scrub jay: demography of a cooperative-breeding bird: Woolfenden, G. E., Fitzpatrick, J. W. *Science*, **227**, 1573–4.

Ligon, J. D. (1993). The role of phylogenetic history in the evolution of contemporary avian mating and parental care systems. In *Current ornithology*, Vol. 10 (ed. D. M. Power), pp. 1–46. Plenum Press, New York.

Ligon, J. D. (1997). A single functional testis as a unique proximate mechanism promoting sex-role reversal in coucals. *Auk*, **114**, 800–1.

Ligon, J. D. and Ligon, S. H. (1978). The communal social system of the green woodhoopoe in Kenya. *Living Bird*, **17**, 159–98.

Ligon, J. D. and Ligon, S. H. (1988). Territory quality: key determinant of fitness in the group living green woodhoopoe. In *The ecology of social behavior* (ed. C. N. Slobochikoff), pp. 229–53. Academic Press, San Diego, California.

Ligon, J. D. and Ligon, S. H. (1989). Green woodhoopoe. In *Lifetime reproduction in birds* (ed. I. Newton), pp. 219–32. Academic Press, London.

Ligon, J. D. and Ligon, S. H. (1990). Green woodhoopoes: life history traits and sociality. In *Cooperative breeding in birds: long-term studies of ecology and behavior* (ed. P. B. Stacey and W. D. Koenig), pp. 33–65. Cambridge University Press.

Ligon, J. D. and Stacey, P. B. (1989). On the significance of helping behavior in birds. *Auk*, **106**, 700–5.

Ligon, J. D. and Stacey, P. B. (1991). The origin and maintenance of helping behavior in birds. *American Naturalist*, **138**, 254–8.

Ligon, J. D. and White, J. L. (1974). Molt and its timing in the piñon jay, *Gymnorhinus cyanocephalus*. *Condor*, **76**, 274–87.

Ligon, J. D. and Zwartjes, P W (1995*a*). Ornate plumage of male red jungle fowl does not influence female mate choice. *Animal Behaviour*, **49**, 117–25.

Ligon, J. D. and Zwartjes, P. W. (1995*b*). Female red junglefowl choose to mate with multiple males. *Animal Behaviour*, **49**, 127–35.

Ligon, J. D., Carey, C., and Ligon, S. H. (1988). Cavity roosting, philopatry, and co-operative breeding in the green woodhoopoe may reflect a physiological trait response. *Auk*, **105**, 123–7.

Ligon, J. D., Thornhill, R., Zuk, M., and Johnson, K. (1990). Male–male competition and the role of testosterone in sexual selection in red jungle fowl. *Animal Behaviour*, **40**, 367–73.

Ligon, J. D., Ligon, S. H., and Ford, H. A. (1991). An experimental study of the bases of male philopatry in the cooperatively breeding superb fairy-wren *Malurus cyaneus*. *Ethology*, **87**, 134–48.

Ligon, J. D., Kimball, R. T., and Merola-Zwartjes, M. (1998). Mate choice by female red junglefowl: the issues of multiple ornaments and flucuating asymmetry. *Animal Behaviour*, **55**, 41–50.

Liker, A., and Székely, T. (1997). Aggression among female lapwings. *Animal Behaviour*, **54**, 797–802.

Lill, A. (1974). Sexual behavior of the lek forming white-bearded manakin. *Zeitschrift für Tierpsychologie*, **36**, 1–36.

Lill, A. (1976). Lek behavior in the golden-headed manakin, *Pipra erythrocephala*, in Trinidad (West Indies). *Zeitschrift für Tierpsychologie, Suppl.*, **18**, 1–83.

Lill, A. (1979). An assessment of male parental investment and pair bonding in the polygamous superb lyrebird. *Auk*, **96**, 489–98.

Lill, A. (1986). Time–energy budgets during reproduction and the evolution of single parenting in the superb lyrebird. *Australian Journal of Zoology*, **34**, 351–71.

Livezey, B. C. (1991). A phylogenetic analysis and classification of recent dabbling ducks (Tribe Anatini) based on comparative morphology. *Auk*, **108**, 471–508.

Loftredo, C. A. and Borgia, G. (1986*a*). Sexual selection, mating systems, and the evolution of avian acoustical displays. *American Naturalist*, **128**, 773–94.

Loffredo, C. A. and Borgia, G. (1986*b*). Male courtship vocalizationa as cues for mate choice in the satin bowerbird (*Ptilonorhynchus violaceus*). *Auk*, **103**, 189–95.

Lofts, B. and Murton, R. K. (1973). Reproduction in birds. In *Avian biology*, Vol. 3 (ed. D. S. Farner, J. R. King, and K. C. Parkes), pp. 1–107. Academic Press, New York.

Lorenz, K. Z. (1941). Vergleichende Bewegungsstudien an Anatinen. *Journal fur Ornithologie*, **89**, 194–294.

Loye, J. E. and Zuk, M. (ed.) (1991). *Bird–parasite interactions: ecology, evolution, and behaviour*. Oxford University Press.

Lozano, G. A. and Lemon, R. E. (1996). Male plumage, paternal care and reproductive success in yellow warblers, Dendroica petechia. *Animal Behaviour*, **51**, 225–72.

Lundberg, A. and Alatalo, R. V. (1992). *The pied flycatcher*. T. & A. D. Poyser, London.

Macedo, R. H. (1992). Reproductive patterns and social organization of the communal guira cuckoo (*Guira guira*) in central Brazil. *Auk*, **109**, 786–99.

Maclean, G. L. (1972). Clutch size and evolution in the Charadrii. *Auk*, **89**, 299–324.

Mader, W. J. (1975). Biology of the Harris' hawks in southern Arizona. *Living Bird*, **14**, 59–85.

Mader, W. J. (1979). Breeding behavior of a polyandrous trio of Harris' hawks in southern Arizona. *Auk*, **96**, 776–88.

Manning, A. (1979). *An introduction to animal behavior.* Addison-Wesley, Reading, Massachusetts.

Manning, J. T. (1989). Age-advertisement and the evolution of the peacock's train. *Journal of Evolutionary Biology*, **2**, 379–84.

Manning, J. T. and Hartley, M. A. (1991). Symmetry and ornamentation are correlated in the peacock's train. *Animal Behaviour*, **42**, 1020–1.

Marini, M. A. and Cavalcanti, R. B. (1992). Mating systems of the helmeted manakin (*Antilophia galeata*) in central Brazil. *Auk*, **109**, 911–3.

Marshall, A. J. (1954). *Bower-birds.* Clarendon, Oxford.

Marshall, A. J. (1961). Breeding seasons and migration. In *Biology and comparative physiology of birds*, Vol. II (ed. A. J. Marshall), pp. 307–39. Academic Press, New York.

Marshall, A. J. and Coombs, C. J. F. (1957). The interaction of environmental, internal and behavioural factors in the rook, *Corvus f. frugilegus* Linnaeus. *Proceedings of the Royal Society of London B*, **128**, 545–89.

Martin, S. G. (1974). Adaptations for polygynous breeding in the bobolink, *Dolichonyx oryzivorus*. *American Zoologist*, **14**, 109–19.

Marzluff, J. M. and Balda, R. P. (1992). *The pinyon jay.* T. & A. D. Poyser, London.

Mateos, C. and Carranza, J. (1995). Female choice for morphological features of male ring-necked pheasants. *Animal Behaviour*, **49**, 737–48.

Mateos, C. and Carranza, J. (1996). On the intersexual selection for spurs in the ring-necked pheasant. *Behavioral Ecology*, **7**, 362–9.

Mateos, C. and Carranza, J. (1997). Signals in intra-sexual competition between ring-necked pheasant males. *Animal Behaviour*, **53**, 471–85.

Mather, M. H. and Robertson, R. J. (1992). Honest advertisement in flight displays of bobolinks (*Dolichonyx oryzivorus*). *Auk*, **109**, 869–73.

Maynard-Smith, J. (1964). Group selection and kin selection. *Nature, London*, **201**, 1145–7.

Maynard-Smith, J. (1977). Parental investment: a prospective analysis. *Animal Behaviour*, **25**, 1–9.

Maynard-Smith, J. and Ridpath, M. G. (1972). Wife sharing in the tasmanian native hen, *Tribonyx mortierii*: a case of kin selection. *American Naturalist*, **106**, 447–52.

Mayr, E. (1963). *Animal species and evolution.* Harvard University Press, Cambridge, Massachusetts.

Mayr, E. (1970). *Populations, species, and evolution.* The Belknap Press, Cambridge, Massachusetts.

Mayr, E. (1983). How to carry out the adaptationist program? *American Naturalist*, **121**, 324–34.

McCleery, R. H. and Perrins, C. M. (1989). Great tit. In *Lifetime reproduction in birds* (ed. I. Newton), pp. 35–53. Academic Press, London.

McDonald, D. B., (1989a). Correlates of male making success in a lekking bird with male-male competition. *Animal Behaviour*, **37**, 1007–22.

McDonald, D. B. (1989*b*). Cooperation under sexual selection: age-graded changes in a lekking bird. *American Naturalist*, **134**, 709–30.

McDonald, D. B. and Potts, W. K. (1994). Cooperative display and relatedness among males in a lek-mating bird. *Science*, **266**, 1030–2.

McDonald, M. V. (1989). Function of song in Scott's seaside sparrow, *Ammodramus maritinus peninsulae*. *Animal Behaviour*, **38**, 468–85.

McGregor, P. K. and Krebs, J. R. (1982). Mating and song types in the great tit. *Nature, London*, **297**, 60–1.

McGregor, P. K. and Krebs, J. R. (1984). Song learning and deceptive mimicry. *Animal Behaviour*, **32**, 280–7.

McGregor, P. K. and Krebs, J. R. (1989). Song learning in adult great tits (*Parus major*): effects of neighbours. *Behaviour*, **108**, 139–59.

McGregor, P. K., Krebs, J. R., and Perrins, C. M. (1981). Song repertoires and lifetime reproductive success in the great tit (*Parus major*). *American Naturalist*, **118**, 149–59.

McKinney, F. (1985). Primary and secondary male reproductive strategies of dabbling ducks. In *Avian monogamy*, Ornithological Monographs, No. 37 (ed. P. A. Gowaty and D. Mock), pp. 68–82. American Ornithologists' Union, Washington, DC.

McKinney, F. (1986). Ecological factors influencing the social systems of migratory dabbling ducks. In *Ecological aspects of social evolution* (ed. D. I. Rubenstein and R. W. Wrangham), pp. 153–71. Princeton University Press.

McKinney, F., Cheng, K. M., and Bruggers, D. J. (1984). Sperm competition in apparently monogamous birds. In *Sperm competition and the evolution of animal mating systems* (ed. R. L. Smith), pp. 523–45. Academic Press, Orlando.

McKitrick, M. C. (1992). Phylogenetic analysis of avian parental care. *Auk*, **109**, 828–46.

McKitrick, M. C. (1993). Phylogenetic constraint in evolutionary history: has it any explanatory power? *Annual Review of Ecology and Systematics*, **24**, 307–30.

McKitrick, M. C. and Zink, R. M. (1988). Species concepts in ornithology. *Condor*, **90**, 1–14.

McLennan, D. A., and Brooks, D. R. (1993). The phylogenetic component of cooperative breeding in perching birds: a commentary. *American Naturalist*, **141**, 790–5.

McLennan, D. A., Brooks, D. R., and McPhail, J. D. (1988). The benefits of communication between comparative ethology and phylogenetic systematics: a case using gasterosteid fishes. *Canadian Journal of Zoology*, **66**, 2177–90.

McPherson, J. M. (1988). Preferences of cedar waxwings in the laboratory for fruit species, colour and size: a comparison with field observations. *Animal Behaviour*, **36**, 961–9.

Meek, S. B., Robertson, R. J., and Boag, P. T. (1994). Extrapair paternity and intraspecific brood parasitism in eastern bluebirds revealed by DNA fingerprinting. *Auk*, **111**, 739–44.

Merton, D. V., Morris, R. V., and Atkinson, I. A. E. (1984). Lek behaviour in a parrot: the kakapo, *Strigops habroptilus*, of New Zealand. *Ibis*, **126**, 277–83.

Meyer, A., Morrissey, J. M., and Schartl, M. (1994). Recurrent origin of a sexually selected trait in *Xiphophorus* fishes inferred from a molecular phylogeny. *Nature, London*, **368**, 539–41.

Middleton, A. L. A. (1988). Polyandry in the mating system of the American goldfinch, *Carduelis tristis. Canadian Journal of Zoology*, **66**, 296–9.

Miles, D. B. and Dunham, A. E. (1993). Historical perspectives in ecology and evolutionary biology: the use of phylogenetic comparative analyses. *Annual Review of Ecology and Systematics*, **24**, 587–619.

Millington, S. J. and Price, T. D. (1985). Song inheritance and mating patterns in Darwin's finches. *Auk*, **102**, 342–6.

Mitra, S., Landel, H., and Pruett-Jones, S. (1996). Species richness covaries with mating system in birds. *Auk*, **113**, 544–51.

Mock, D. W. (1985). An introduction to the neglected mating system. In *Avian monogamy*, Ornithological Monographs, No. 37 (ed. P. A. Gowaty and D. Mock), pp. 1–10. American Ornithologists' Union, Washington, DC.

Møller, A. P. (1986). Mating systems among European passerines: a review. *Ibis*, **128**, 234–50.

Møller, A. P. (1987*a*). Variation in badge size in male house sparrows *Passer domesticus*: evidence for status signalling. *Animal Behaviour*, **35**, 1637–44.

Møller, A. P. (1987*b*). Social control of deception among status signalling house sparrows *Passer domesticus. Behavioral Ecology and Sociobiology*, **20**, 307–11.

Møller, A. P. (1988). Badge size in the house sparrow *Passer domesticus*: effects of intra- and intersexual selection. *Sociobiology*, **22**, 273–8.

Møller, A. P. (1989). Natural and sexual selection on a plumage signal of status and on morphology in house sparrows, *Passer domesticus. Journal of Evolutionary Biology*, **2**, 125–40.

Møller, A. P. (1990*a*). Effects of a haematophagous mite on the barn swallow (*Hirundo rustica*): a test of the Hamilton and Zuk hypothesis. *Evolution*, **44**, 771–84.

Møller, A. P. (1990*b*). Male tail length and female mate choice in the monogamous swallow *Hirundo rustica. Animal Behaviour*, **39**, 458–65.

Møller, A. P. (1991). Sexual ornament size and the cost of fluctuating asymmetry. *Proceedings of the Royal Society of London B*, **243**, 59–62.

Møller, A. P. (1992*a*). Parasites differentially increase the degree of fluctuating asymmetry in secondary sexual characters. *Journal of Evolutionary Biology*, **5**, 691–9.

Møller, A. P. (1992*b*). Female swallow preference for symmetrical male sexual ornaments. *Nature, London*, **357**, 238–40.

Møller, A. P. (1992*c*). Patterns of fluctuating asymmetry in weapons: evidence for reliable signaling of quality in beetle horns and bird spurs. *Proceedings of the Royal Society of London B*, **248**, 199–206.

Møller, A. P. (1993*a*). Patterns of fluctuating asymmetry in sexual ornaments predict female choice. *Journal of Evolutionary Biology*, **6**, 481–91.

Møller, A. P. (1993*b*). Morphology and sexual selection in the barn swallow *Hirundo rustica* in Chernobyl, Ukraine. *Proceedings of the Royal Society of London B*, **252**, 51–7.

Møller, A. P. (1993*c*). Developmental stability, sexual selection and speciation. *Journal of Evolutionary Biology*, **6**, 493–509.

Møller, A. P. (1994). *Sexual selection and the barn swallow*. Oxford University Press, Oxford.

Møller, A. P. and Birkhead, T. R. (1993). Certainty of paternity covaries with paternal care in birds. *Behavioral Ecology and Sociobiology*, **33**, 261–8.

Møller, A. P. and Birkhead, T. R. (1994). The evolution of plumage brightness in birds is related to extrapair paternity. *Evolution*, **48**, 1089–1100.

Møller, A. P. and Höglund, J. (1991). Patterns of fluctuating asymmetry in avian feather ornaments: implications for models of sexual selection. *Proceedings of the Royal Society of London B*, **245**, 1–5.

Møller, A. P. and Pomiankowski, A. (1993*a*). Why have birds got multiple sexual ornaments? *Behavioral Ecology and Sociobiology*, **32**, 167–76.

Møller, A. P. and Pomiankowski, A. (1993*b*). Fluctuating asymmetry and sexual selection. *Genetica*, **89**, 267–79.

Morton, E. S., Forman, L., and Braun, M. (1990). Extra-pair fertilizations and the evolution of colonial breeding in purple martins. *Auk*, **107**, 275–83.

Mountjoy, D. J. and Lemon, R. J. (1996). Female choice for complex song in the European starling: a field experiment. *Behavioral Ecology and Sociology*, **38**, 65–71.

Mountjoy, D. J. and Robertson, R. J. (1988). Why are waxwings waxy? Delayed plumage maturation in the cedar waxwing. *Auk*, **105**, 61–9.

Mulder, R. A. and Cockburn, A. (1993). Sperm competition and the reproductive anatomy of male superb fairy-wrens. *Auk*, **110**, 588–93.

Mulder, R. A., Dunn, P. O., Cockburn, A., Lazenby-Cohen, K. A., and Howell, M. J. (1994). Helpers liberate female fairy-wrens from constraints on extra-pair mate choice. *Proceedings of the Royal Society of London B*, **255**, 223–9.

Muma, K. E. and Weatherhead, P. J. (1989). Male traits expressed in females: direct or indirect sexual selection? *Behavioral Ecology and Sociobiology*, **25**, 23–31.

Mumme, R. L. (1992*a*). Delayed dispersal and cooperative breeding in the Seychelles warbler. *Trends in Ecology and Evolution*, **7**, 330–1.

Mumme, R. L. (1992*b*). Do helpers increase reproductive success: an experimental analysis in the Florida scrub jay. *Behavioral Ecology and Sociobiology*, **31**, 319–28.

Mumme, R. L. and Koenig, W. D. (1983). Reproductive competition in the communal acorn woodpecker: sisters destroy each other's eggs. *Nature, London*, **306**, 583–4.

Mumme, R. L., Koenig, W. D., Zink, R. M., and Marten, J. A. (1985). Genetic variation and parentage in a California population of acorn woodpeckers. *Auk*, **102**, 305–12.

Mundinger, P. C. (1972). Annual testicular cycle and bill color change in the eastern American goldfinch. *Auk*, **89**, 403–19.

Murray, B. G., Jr. (1984). A demographic theory on the evolution of mating systems as exemplified by birds. In *Evolutionary biology*, Vol. 18 (ed. M. K. Hecht, B. Wallace, and G. T. Prance), pp. 71–140. Plenum Press, New York.

Nelson, D. A. (1984). Communication of intentions in agnostic contexts by the pigeon guillemot, *Cepphus columba*. *Behaviour*, **88**, 145–89.

Nethersole-Thompson, D. (1973). *The dotterel*. Collins, London.

Newton, I. (1986). *The sparrowhawk*. T & A. D. Poyser, Calaton.

Newton, I. (ed.) (1989). *Lifetime reproduction in birds*. Academic Press, London.

Nilsson, J. (1994). Energetic stress and the degree of fluctuating asymmetry: implications for a long-lasting, honest signal. *Evolutionary Ecology*, **8**, 248–55.

Nisbet, I. (1973). Courtship-feeding, egg-size and breeding success in common terns. *Nature, London*, **241**, 141–2.

Norell, M. A., Clark, J. M., Chiappe, L. M., and Dashzeveg, D. (1995). A nesting dinosaur. *Nature, London*, **378**, 774–6.

Norris, K. J. (1990a). Female choice and the evolution of the conspicuous plumage coloration of monogamous male great tits. *Behavioral Ecology and Sociobiology*, **26**, 129–38.

Norris, K. J. (1990b). Female choice and the quality of parental care in the great tit *Parus major*. *Behavioral Ecology and Sociobiology*, **27**, 275–81.

Norris, K. J. (1993). Heritable variation in a plumage indicator of viability in male great tits. *Nature, London*, **362**, 537–9.

Oakes, E. J. (1992). Lekking and the evolution of sexual dimorphism in birds: comparative approaches. *American Naturalist*, **140**, 665–84.

Oakes, E. J. and Barnard, P. (1994). Fluctuating asymmetry and mate choice in paradise whydahs, *Vidua paradisaea*: an experimental manipulation. *Animal Behaviour*, **48**, 937–43.

O'Donald, P. (1973). Models of sexual and natural selection in polygynous species. *Heredity*, **31**, 145–56.

O'Donald, P. (1983). Sexual selection by female choice. In *Mate choice* (ed. P. Bateson), pp. 53–66. Cambridge University Press.

Olson, S. L. and Steadman, D. W. (1981). The relationships of the Pedionomidae (Aves: Charadriiformes). *Smithsonian Contributions to Zoology*, **337**, 1–25.

Omland, K. E. (1996a). Female mallard mating preferences for multiple male ornaments: I. natural variation. *Behavioral Ecology and Sociobiology*, **39**, 353–60.

Omland, K. E. (1996b). Female mallard mating preferences for multiple male ornaments: II. experimental variation. *Behavioral Ecology and Sociobiology*, **39**, 361–6.

Oppenheim, R. W. (1972). Pre-hatching and hatching behavior in birds: a comparative study of altricial and precocial species. *Animal Behaviour*, **20**, 644–55.

Orell, M., Rytkönen, S., and Koivula, K. (1994). Causes of divorce in the monogamous willow tit, *Parus montanus*, and consequences for reproductive success. *Animal Behaviour*, **48**, 1143–54.

Orians, G. H. (1961). The ecology of blackbird social systems. *Ecological Monographs*, **31**, 285–312.

Orians, G. H. (1969). On the evolution of mating systems in birds and mammals. *American Naturalist*, **103**, 589–603.

Orians, G. H. (1980). *Some adaptations of marsh-nesting blackbirds*. Princeton University Press.

Oring, L. W. (1982). Avian mating systems. In *Avian biology*, Vol. 6 (ed. D. S. Farner, J. S. King, and K. C. Parkes), pp. 1–92. Academic Press, New York.

Oring, L. W. (1986). Avian polyandry. In *Current ornithology*, Vol. 3 (ed. R. J. Johnston), pp. 309–51. Plenum Press, New York.

Oring, L. W., Fivizzani, A. J., El Halawani, M. E., and Goldsmith, A. (1986). Seasonal changes in prolactin and leutenizing hormone in the polyandrous spotted sandpiper, *Actitis macularia*. *General and Comparative Endocrinology*, **62**, 394–403.

Oring, L. W., Fivizzani, A. J., Colwell, M. A., and El Halawani, M. E. (1988). Hormonal changes associated with natural and manipulated incubation in the sex-role reversed Wilson's phalarope. *General and Comparative Endocrinology*, **72**, 247–56.

Oring, L. W., Fleischer, R. C., Reed, J. M., and Marsden, K. E. (1992). Cuckoldry through stored sperm in the sequentially polyandrous spotted sandpiper. *Nature, London*, **359**, 631–3.

Oring, L. W., Reed, J. M., and Alberico, J. A. R. (1994). Mate acquisition tactics in polyandrous spotted sandpipers (*Actitis macularia*): the role of age and experience. *Behavioral Ecology*, **5**, 9–16.

Owens, I. P. F. and Short, R. V. (1995). Hormonal basis of sexual dimorphism in birds: implications for new theories of sexual selection. *Trends in Ecology and Evolution*, **10**, 44–7.

Owens, P. F., Dixon, A., Burke, T., and Thompson, Des. B. A. (1995). Strategic paternity assurance in the sex-role reversed Eurasian dotterel (*Charadrius morinellus*): behavioral and genetic evidence. *Behavioral Ecology*, **6**, 14–26.

Palmer, A. R. and Strobeck, C. (1986). Fluctuating asymmetry: measurement, analysis, patterns. *Annual Review of Ecology and Systematics*, **17**, 391–421.

Palokangas, P., Korpimäki, E., Hakkarainen, H., Huhta, E., Tolonen, P., and Alatalo, R. V. (1994). Female kestrels gain reproductive success by choosing brightly ornamented males. *Animal Behaviour*, **47**, 443–8.

Parmelee, D. F. and Payne, R. B. (1973). On multiple broods and the breeding strategy of arctic sandpipers. *Ibis*, **115**, 218–26.

Parsons, P. A. (1990). Fluctuating asymmetry: an epigenetic measure of stress. *Biological Review*, **65**, 131–45.

Parsons, T. J., Olson, S. L., and Braun, M. J. (1993). Unidirectional spread of secondary sexual plumage traits across an avian hybrid zone. *Science*, **260**, 1643–6.

Pärt, T. (1995). Does breeding experience explain increased reproductive success with age? An experiment. *Proceedings of the Royal Society of London, B*, **360**, 113–17.

Partridge, L. and Halliday, T. (1984). Mating patterns and mate choice. In *Behavioral ecology* (ed. J. R. Krebs and N. B. Davies), pp. 222–50. Sinauer Associates, Sunderland, Massachusetts.

Patterson, C. B., Erckmann, W. J., and Orians, G. H. (1980). An experimental study of parental investment and polygyny in male blackbirds. *American Naturalist*, **116**, 757–69.

Payne, R. B. (1973). Individual laying histories and the clutch size and number of eggs of parasitic cuckoos. *Condor*, **75**, 414–38.

Payne, R. B. (1979). Sexual selection and intersexual differences in variation of mating success. *American Naturalist*, **114**, 447–52.

Payne, R. B. (1981). Population structure and social behavior: models for testing the ecological significance of song dialects in birds. In *Natural selection and social behavior: research and new theory* (ed. R. D. Alexander and D. W. Tinkle), pp. 108–20. Chiron, New York.

Payne, R. B. (1982). Ecological consequences of song matching: breeding success and intraspecific song mimicry in indigo buntings. *Ecology*, **63**, 401–11.

Payne, R. B. (1983). Bird songs, sexual selection and female mating strategies. In *Social behavior of female vertebrates* (ed. S. K. Wasser), pp. 55–90. Academic Press, New York.

Payne, R. B. (1984). *Sexual selection, lek and arena behavior, and sexual size dimorphism in birds*, Ornithological Monographs, No. 33. American Ornithologists Union, Washington, DC.

Payne, R. B., Payne, L. L., and Doehlart, S. M. (1987). Song, mate choice and the question of kin recognition in a migratory songbird. *Animal Behaviour*, **35**, 35–47.

Peek, F. W. (1972). An experimental study of the territorial function of vocal and visual display in the male red-winged blackbird (*Agelaius phoeniceus*). *Animal Behaviour*, **20**, 112–18.

Persson, O. and Öhrström, P. (1989). A new avian mating system: ambisexual polygamy in the penduline tit *Remiz pendulinus*. *Ornis Scandinavica*, **20**, 105–11.

Peterson, A. T. and Burt, D. B. (1992). Phylogenetic history of social evolution and habitat use in the *Aphelocoma* jays. *Animal Behaviour*, **44**, 859–66.

Petrie, M. (1983). Female moorhens compete for small, fat males. *Science*, **220**, 413–15.

Petrie, M. (1992). Peacocks with low mating success are more likely to suffer predation. *Animal Behaviour*, **44**, 585–6.

Petrie, M. (1993). Do peacock's trains advertise age? *Journal of Evolutionary Biology*, **6**, 443–8.

Petrie, M. (1994). Improved growth and survival of offspring of peacocks with more elaborate trains. *Nature, London*, **371**, 598–9.

Petrie, M. and Halliday, T. (1994). Experimental and natural changes in the peacock's (*Pavo cristatus*) train can affect mating success. *Behavioral Ecology and Sociobiology*, **35**, 213–17.

Petrie, M. and Williams, A. (1993). Peahens lay more eggs for peacocks with larger trains. *Proceedings of the Royal Society of London B*, **251**, 127–31.

Petrie, M., Halliday, T., and Sanders, C. (1991). Peahens prefer peacocks with elaborate trains. *Animal Behaviour*, **41**, 323–31.

Petrinovich, L. and Baptista, L. F. (1984). Song dialects, mate selection, and breeding success in white-crowned sparrows. *Animal Behaviour*, **32**, 1078–88.

Pienkowski, M. W. and Greenwood, J. J. D. (1979). Why change mates? *Biological Journal of the Linnean Society*, **12**, 85–94.

Pierce, E. P. and Lifjeld, J. T. (1998). High paternity without paternity-assurance behavior in the purple sandpiper, a species with high paternal investment. *Auk*, **115**, 602–12.

Pierotti, R. and Annett, C. A. (1993). Hybridization and male parental investment in birds. *Condor*, **95**, 670–9.

Pinxten, R., Hanotte, O., Eens, M., Verheyen, R. F., Dhondt, A. A., and Burke, T. (1993). Extra-pair paternity and intraspecific brood parasitism in the European starling, *Sturnus vulgaris*: evidence from DNA fingerprinting. *Animal Behaviour*, **45**, 795–809.

Piper, W. D., Ever, D. C., Meyer, M. W., Tischler, K. B., Kaplan, J. D., and Fleischer, R. C. (1997). Genetic monogamy in the common loon (*Gavia immer*). *Behavioral Ecology and Sociobiology*, **41**, 25–31.

Pitelka, F. A., Holmes, R. T., and MacLean, S. F. J. (1974). Ecology and evolution of social organization in arctic sandpipers. *American Zoologist*, **14**, 184–204.

Pizzey, G. (1980). *A field guide to the birds of Australia*. Princeton University Press.

Pomiankowski, A. (1994). Swordplay and sensory bias. *Nature, London*, **368**, 494–5.

Pomiankowski, A. and Møller, A. P. (1995). A resolution of the lek paradox. *Proceedings of the Royal Society of London B*, **360**, 31–9.

Price, D. K. and Burley, N. T. (1994). Constraints on the evolution of attractive traits: selection in male and female zebra finches. *American Naturalist*, **144**, 908–34.

Pruett-Jones, S. G. (1992). Independent versus non-independent mate choice: do females copy each other? *American Naturalist*, **140**, 1000–9.

Pruett-Jones, S. G. and Lewis, M. J. (1990). Habitat limitation and sex ratio promote delayed dispersal in superb fairy-wrens. *Nature, London*, **348**, 541–2.

Pruett-Jones, M. A. and Pruett-Jones, S. G. (1982). Spacing and distribution of bowers in MacGregor's bowerbird (*Amblyornis macgregoriae*). *Behavioral Ecology and Sociobiology*, **11**, 25–32.

Pruett-Jones, M. A. and Pruett-Jones, S. G. (1983). The bowerbird's labor of love. *Natural History*, **9**, 49–55.

Pruett-Jones, S. G. and Pruett-Jones, M. A. (1986). Altitudinal distribution and seasonal activity patterns of birds of paradise. *National Geographic Research*, **2**, 87–105.

Pruett-Jones, S. G. and Pruett-Jones, M. A. (1988). The use of court objects by Lawes' parotia. *Condor*, **90**, 538–45.

Pruett-Jones, S. G. and Pruett-Jones, M. A. (1990). Sexual selection through female choice in Lawes' parotia, a lek-mating bird of paradise. *Evolution*, **44**, 486 501.

Pruett-Jones, S. G., Pruett-Jones, M. A., and Jones, H. I. (1990). Parasites and sexual selection in birds of paradise. *American Zoologist*, **30**, 287–98.

Pruett-Jones, S. G., Pruett-Jones, M. A., and Jones, H. I. (1991). Parasites and sexual selection in a New Guinea avifauna. In *Current Ornithology*, Vol. 8, pp. 213–45.

Prum, R. O. (1990). Phylogenetic analysis of the evolution of display behavior in the neotropical manakins (Aves: Pipridae). *Ethology*, **84**, 202–31.

Prum, R. O. (1994). Phylogenetic analysis of the evolution of alternative social behavior in the manakins (Aves: Pipridae). *Evolution*, **48**, 1657–75.

Quay, W. B. (1989). Insemination of Tennessee warblers during spring migration. *Condor*, **91**, 660–70.

Quinn, J. S., Macedo, R., and White, B. N. (1994). Genetic relatedness of communally breeding *Guira* cuckoos. *Animal Behaviour*, **47**, 515–29.

Rabenold, K. N. (1985). Cooperation in breeding by nonreproductive wrens: kinship, reciprocity, and demography. *Behavioral Ecology and Sociobiology*, **17**, 1–17.

Rabenold, P. P., Rabenold, K. N., Piper, W. H., Haydock, J., and Zack, S. W. (1990). Shared paternity revealed by genetic analysis in cooperatively breeding tropical wrens. *Nature, London*, **348**, 538–40.

Rahn, H., Paganelli, C. V., and Ar, A. (1975). Relation of avian egg weight to body weight. *Auk*, **92**, 750–65.

Ralph, C. P. (1975). Life style of *Coccyzus pumilus*, a tropical cuckoo. *Condor*, **77**, 60–72.

Rand, A. L. (1933). Testicular asymmetry in the Madagascar coucal. *Auk*, **50**, 219–20.

Rand, A. L. (1936). Distribution and habits of Madagascar birds. *Bulletin of the American Museum of Natural History*, **72**, 143–499.

Rand, A. L. (1951). The nests and eggs of *Mesoenas unicolor* of Madagascar. *Auk*, **68**, 23–6.

Read, A. F. (1987). Comparative evidence supports the Hamilton and Zuk hypothesis on parasites and sexual selection. *Nature, London*, **327**, 68–70.

Read, A. F. (1990). Parasites and the evolution of host sexual behaviour. In *Parasitism and host behaviour* (ed. C. J. Barnard and J. M. Behnke), pp. 117–57. Taylor and Francis, London.

Read, A. F. and Harvey, P. H. (1989). Reassessment of comparative evidence for Hamilton and Zuk theory on the evolution of secondary sexual characters. *Nature, London*, **339**, 618–20.

Read, A. F. and Weary, D. M. (1990). Sexual selection and the evolution of bird song: a test of the Hamilton–Zuk hypothesis. *Behavioral Ecology and Sociobiology*, **26**, 47–56.

Reeve, H. K. and Sherman, P. W. (1993). Adaptation and the goals of evolutionary research. *The Quarterly Review of Biology*, **68**, 1–32.

Regelmann, K. and Curio, E. (1986). Why do great tits (*Parus major*) males defend their brood more than females do? *Animal Behaviour*, **34**, 1206–14.

Reyer, H.-U. (1990). Pied kingfishers: ecological causes and reproductive consequences of cooperative breeding. In *Cooperative breeding in birds: long-term studies of ecology and behavior* (ed. P. B. Stacey and W. D. Koenig), pp. 527–57. Cambridge University Press.

Ridley, M. (1978). Paternal care. *Animal Behaviour*, **26**, 904–32.

Ridpath, M. G. (1972). The Tasmanian native hen, *Tribonyx mortierii*. *Commonwealth Scientific and Industrial Research Organization, Wildlife Research*, **17**, 53–118.

Rissman, E. F. and Wingfield, J. C. (1984). Hormonal correlates of polyandry in the spotted sandpiper, *Actitis macularia*. *General and Comparative Endocrinology*, **56**, 401–5.

Ritchison, G., Klatt, P. H., and Westneat, D. F. (1994). Mate guarding and extra-pair paternity in northern cardinals. *Condor*, **96**, 1055–63.

Robel, R. J. and Ballard, W. B. (1974). Lek social organization and reproductive success in the greater prairie chicken. *American Zoologist*, **14**, 121–8.

Robertson, B. C. and Kikkawa, J. (1994). How do they do it?—Monogamy in *Zosterops lateralis*. *Journal of Ornithology*, **135**, 459.

Robinson, S. K. (1986). The evolution of social behavior and mating systems in the blackbirds (Icterinae). In *Ecological aspects of social evolution: birds and mammals* (ed. D. I. Rubenstein and R. W. Wrangham), pp. 175–200. Princeton University Press.

Rowley, I. C. R. (1965). The life history of the superb blue wren, *Malurus cyanea*. *Emu*, **64**, 251–97.

Rowley, I. C. R. (1974). Co-operative breeding in Australian birds. *Proceedings of the International Ornithological Congress*, **16**, 657–66.

Rowley, I. C. R. (1981). The communal way of life in the splendid wren, *Malurus splendens*. *Zeitschrift für Tierpsychologie*, **55**, 228–67.

Rowley, I. C. R. (1983). Re-mating in birds. In *Mate choice* (ed. P. Bateson), pp. 331–60. Cambridge University Press.

Rowley, I. C. R. (1991). Petal-carrying by fairy-wrens of the genus *Malurus*. *Australian Bird Watcher*, **14**, 75–81.

Rowley, I. C. R. and Russell, E. (1990). Splendid fairy-wrens: demonstrating the importance of longevity. In *Cooperative breeding in birds: long-term studies of ecology and behavior* (ed. P. B. Stacey and W. D. Koenig), pp. 1–30. Cambridge University Press.

Rowley, I., Russell, E., and Brooker, M. (1986). Inbreeding: benefits may outweigh costs. *Animal Behaviour*, **34**, 939–41.

Rowley, I., Russell, E., Brown, R., and Brown, M. (1988). The ecology and breeding biology of the red-winged fairy-wren *Malurus elegans*. *Emu*, **88**, 161–76.

Russell, E. M. (1989). Co-operative breeding—a Gondwanan perspective. *Emu*, **89**, 61–2.

Ryan, M. J. and Keddy-Hector, A. (1992). Directional patterns of female mate choice and the role of sensory biases. *American Naturalist*, **139 (supplement)**, S4–S35.

Ryan, M. J. and Rand, A. S. (1993). Sexual selection and signal evolution: the ghost of biases past. *Philosophical Transactions of the Royal Society of London, Series B*, **340**, 187–95.

Ryan, M. J. and Rand, A. S. (1995). Female responses to ancestral advertisement calls in Tungara frogs. *Science*, **269**, 390–2.

Ryan, M. J., Fox, J. H., Wilczynski, W., and Rand, A. S. (1990). Sexual selection for sensory exploitation in the frog *Physalaemus pustulosus*. *Nature, London*, **343**, 66–7.

Sætre, G.-P., Dale, S., and Slagsvold, T. (1994). Female pied flycatchers prefer brightly coloured males. *Animal Behaviour*, **48**, 1407–16.

Saino, N. and Møller, A. P. (1994). Secondary sexual characters, parasites and testosterone in the barn swallow, *Hirundo rustica*. *Animal Behaviour*, **48**, 1325–33.

Saino, N. and Møller, A. P. (1995). Testosterone correlates of mate guarding, singing and aggressive behaviour in male barn swallows, *Hirundo rustica*. *Animal Behaviour*, **49**, 465–72.

Santee, W. R. and Bakken, G. S. (1987). Social displays in red-winged blackbirds (*Agelaius phoeniceus*): sensitivity to thermoregulatory costs. *Auk*, **104**, 413–20.

Savalli, V. M. (1989). Female choice. *Nature*, **339**, 432.

Schneider, J and Lamprecht, J. (1990). The importance of biparental care in a precocial, monogamous bird, the bar-headed goose (*Anser indicus*). *Behavioral Ecology and Sociobiology*, **27**, 415–19.

Schoech, S. J., Mumme, R. L., and Wingfield, J. C. (1996). Prolactin and helping behaviour in the cooperatively breeding Florida scrub-jay, *Aphelocoma coerulescens*. *Animal Behaviour*, **52**, 445–56.

Schroeder, M. A. (1991). Movement and lek visitation by female greater prairie-chickens in relation to predictions of Bradbury's female preference hypothesis of lek evolution. *Auk*, **108**, 896–903.

Searcy, W. A. (1979). Female choice of mates: a general model for birds and its application to red-winged blackbirds (*Agelaius phoeniceus*). *American Naturalist*, **114**, 77–100.

Searcy, W. A. (1984). Song repertoire size and female preferences in song sparrows. *Behavioral Ecology and Sociobiology*, **14**, 281–6.

Searcy, W. A. (1988). Dual intersexual and intrasexual functions of song in red-winged blackbirds. In *Acta XIX Congressus Internationalis Ornithologica* (ed. H. Ouellet), pp. 1373–81. University of Ottawa Press.

Searcy, W. A. (1992). Song repertoire and mate choice in birds. *American Zoologist*, **32**, 71–80.

Searcy, W. A. and Andersson, M. (1986). Sexual selection and the evolution of song. *Annual Review of Ecology and Systematics*, **17**, 507–33.

Searcy, W. A. and Marler, P. (1981). A test for responsiveness to song structure and programming in female sparrows. *Science*, **213**, 926–8.

Searcy, W. A. and Yasukawa, K. (1983). Sexual selection and red-winged blackbirds. *American Scientist*, **71**, 166–74.

Searcy, W. A. and Yasukawa, K. (1989). Alternative models of territorial polygyny in birds. *American Naturalist*, **134**, 323–43.

Searcy, W. A. and Yasukawa, K. (1995). *Polygyny and sexual selection in red-winged blackbirds*. Princeton University Press.

Searcy, W. A., Searcy, M. H., and Marler, P. (1982). The response of swamp sparrows to acoustically distinct song types. *Behaviour*, **80**, 70–83.

Selander, R. K. (1964). Speciation in wrens of the genus *Campylorhynchus*. *University of California Publications in Zoology*, **74**, 1–305.

Selander, R. K. (1965). On mating systems and sexual selection. *American Naturalist*, **99**, 129–41.

Selander, R. K. (1972). Sexual selection and dimorphism in birds. In *Sexual selection and the descent of man, 1871–1971* (ed. B. Campbell), pp. 180–230. Aldine, Chicago.

Seutin, G., Boag, P. T., White, B. N., and Ratcliffe, L. M. (1991). Sequential polyandry in the common redpoll (*Carduelis flammea*). *Auk*, **108**, 166–70.

Sharpe, R. S. and Johnsgard, P. A. (1966). Inheritance of behavioral characters in F2 mallard pintail (*Anas platyrhynchos* L. *Anas acuta* L.) hybrids. *Behaviour*, **27**, 259–72.

Sheldon, B. C. (1994). Male phenotype, fertility, and the pursuit of extra-pair copulations by female birds. *Proceedings of the Royal Society of London, B*, **257**, 25–30.

Sheldon, F. H. and Whittingham, L. A. (1997). *Phylogeny in studies of bird ecology, behavior, and morphology* (ed. D. Mindell), pp. 279–99. Academic Press, New York.

Sherley, G. (1989). Benefits of courtship feeding for rifleman (*Acanthisitta chloris*) parents. *Behaviour*, **109**, 303–18.

Sherman, P. T. (1995*a*). Breeding biology of white-winged trumpeters (*Psophia leucoptera*) in Peru. *Auk*, **112**, 285–95.

Sherman, P. T. (1995*b*). Social organization of cooperatively polyandrous white-winged trumpeters (*Psophia leucoptera*) in Peru. *Auk*, **112**, 296–309.

Sherman, P. W. (1988). The levels of analysis. *Animal Behaviour*, **36**, 616–19.

Sherman, P. W. (1989). The clitoris debate and the levels of analysis. *Animal Behaviour*, **37**, 697–8.

Sherman, P. W., Seeley, T. D., and Reeve, H. K. (1988). Parasites, pathogens, and polyandry in social Hymenoptera. *American Naturalist*, **131**, 602–10.

Shutler, D. and Weatherhead, P. J. (1990). Target of sexual selection: song and plumage of wood warblers. *Evolution*, **44**, 1967–77.

Sibley, C. G. (1957). The evolutionary and taxonomic significance of sexual dimorphism and hybridization in birds. *Condor*, **59**, 166–91.

Sibley, C. G. and Ahlquist, J. E. (1985). The phylogeny and classification of the Australo–Papuan passerine birds. *Emu*, **85**, 1–14.

Sibley, C. G. and Ahlquist, J. E. (1990). *Phylogeny and classification of birds: a study in molecular evolution*. Yale University Press, New Haven, Conn.

Sibley, C. G. and Monroe, B. L. J. (1990). *Distribution and taxonomy of birds of the world*. Yale University Press, New Haven, Conn.

Silver, R., Andrews, H., and Ball, G. F. (1985). Parental care in an ecological perspective: a quantitative analysis of avian subfamilies. *American Zoologist*, **25**, 823–40.

Simmons, R. E. (1988). Food and the deceptive acquisition of mates by polygynous male harriers. *Behavioral Ecology and Sociobiology*, **23**, 83–92.

Skutch, A. F. (1976). *Parent birds and their young.* University of Texas Press.

Slagsvold, T. (1986). Nest site settlement by the pied flycatcher: does the female choose her mate for the quality of his house or himself? *Ornis Scandinavica,* **17,** 210–20.

Slagsvold, T. and Lifjeld, J. T. (1994). Polygyny in birds: the role of competition between females for male parental care. *American Naturalist,* **143,** 59–94.

Smith, D. G. (1979). Male singing ability and territory integrity in red-winged blackbirds (*Agelaius phoeniceus*). *Behaviour,* **68,** 193–206.

Smith, H. G. and Montgomerie, R. (1991). Sexual selection and the tail ornaments of North American barn swallows. *Behavioral Ecology and Sociobiology,* **28,** 195–201.

Smith, H. G., Montgomerie, R., Pöldmaa, T., White, B. N., and Boag, P. T. (1991). DNA fingerprinting reveals relation between tail ornaments and cockoldry in barn swallows, *Hirundo rustica. Behavioral Ecology,* **2,** 90–8.

Smith, H. G., Sandell, M. I., and Bruun, M. (1995). Paternal care in the European starling, *Sturnus vulgaris:* incubation. *Animal Behaviour,* **50,** 323–31.

Smith, S. M. (1988). Extra-pair copulations in black-capped chickadees: the role of the female. *Behaviour,* **107,** 15–23.

Smith, W. D. (1977). *The behavior of communicating: an ethological approach.* Harvard University Press, Cambridge, Massachusetts.

Snow, D. W. (1961). The displays of the manakins *Pipra pipra* and *Tyranneutes virescens. Ibis,* **103a,** 110–13.

Soler, M., Soler, J. J., Møller, A. P., Moreno, J., and Linden, M. (1996). The functional significance of sexual display: stone carrying in the black wheatear. *Animal Behaviour,* **51,** 247–54.

Sonerud, G. A. (1992). Nest predation may make the 'deception hypothesis' unnecessary to explain polygyny in the Tengmalm's owl. *Animal Behaviour,* **43,** 871–4.

Spurrier, M. F., Boyce, M. S., and Manly, B. F. J. (1991). Effects of parasites on mate choice by captive sage grouse. In *Bird–parasite interactions: ecology, evolution, and behaviour* (ed. J. E. Loye and M. Zuk), pp. 389–98. Oxford University Press.

Spurrier, M. F., Boyce, M. S., and Manly, B. F. J. (1994). Lek behavior in captive sage grouse, *Centrocercus urophasianus. Animal Behaviour,* **47,** 303–10.

Stacey, P. B. (1979). Kinship, promiscuity, and communal breeding in the acorn woodpecker. *Behavioral Ecology and Sociobiology,* **6,** 53–66.

Stacey, P. B. (1982). Female promiscuity and male reproductive success in social birds and mammals. *American Naturalist,* **120,** 51–64.

Stacey, P. B. and Bock, C. E. (1978). Social plasticity in the acorn woodpecker. *Science,* **202,** 1298–300.

Stacey, P. B. and Koenig, W. D. (1984). Cooperative breeding in the acorn woodpecker. *Scientific American,* **251,** 114–21.

Stacey, P. B. and Koenig, W. D. (ed.) (1990). *Cooperative breeding in birds: long-term studies of ecology and behavior.* Cambridge University Press.

Stacey, P. B. and Ligon, J. D. (1987). Territory quality and dispersal options in the acorn woodpecker, and a challenge to the habitat-saturation model of cooperative breeding. *American Naturalist,* **130,** 654–76.

Stacey, P. B. and Ligon, J. D. (1991). The benefits-of-philopatry hypothesis for the evolution of cooperative breeding: variation in territory quality and group-size effects. *American Naturalist*, **137**, 831–46.

Staddon, J. E. R. (1975). A note on the evolutionary significance of 'supernormal stimuli'. *American Naturalist*, **109**, 541–5.

Stenmark, G., Slagsvold, T., and Lifjeld, J. T. (1988). Polygyny in the pied flycatcher, *Ficedula hypoleuca*: a test of the deception hypothesis. *Animal Behaviour*, **36**, 1646–57.

Stevens, L. (1991). *Genetics and evolution of the domestic fowl*. Cambridge University Press.

Stewart, R. E. and Aldrich, J. W. (1951). Removal and repopulation of breeding birds in a spruce-fir forest community. *Auk*, **68**, 471–82.

Stokes, A. W. (1971). Parental and courtship feeding in red junglefowl. *Auk*, **88**, 21–9.

Stokes, A. W. and Williams, H. W. (1971). Courtship feeding in gallinaceous birds. *Auk*, **88**, 543–59.

Stokkan, K. A. (1979*a*). Testosterone and day length-dependent development of comb size and breeding plumage of male willow ptarmigan (*Lagopus lagopus lagopus*). *Auk*, **96**, 106–15.

Stokkan, K. A. (1979*b*). The effect of permanent short days and castration on plumage and comb growth in male willow ptarmigan (*Lagopus lagopus*). *Auk*, **96**, 682–7.

Stoner, C. R. (1940). *Courtship and display among birds*. Country Life, London.

Strain, J. G. and Mumme, R. L. (1988). Effects of food supplementation, song playback, and temperature on vocal territorial behavior of Carolina wrens. *Auk*, **105**, 11–16.

Strauch, J. G., Jr. (1978). The phylogeny of the Charadriiformes (Aves): a new estimate using the method of character compatability analysis. *Transactions of the Zoological Society of London*, **34**, 269–345.

Stutchbury, B. J. (1991*a*). The adaptive significance of male subadult plumage in purple martins: plumage dyeing experiments. *Behavioral Ecology and Sociobiology*, **29**, 297–306.

Stutchbury, B. J. (1991*b*). Floater behaviour and territory acquisition in male purple martins. *Animal Behaviour*, **42**, 435–44.

Stutchbury, B. J. and Morton, E. S. (1995). The effect of breeding synchrony on extra-pair mating systems in songbirds. *Behaviour*, **132**, 675–90.

Sullivan, M. S. and Hillgarth, N. (1993). Mating system correlates of tarsal spurs in the Phasianidae. *Journal of Zoology*, **231**, 203–14.

Sundberg, J. and Larsson, C. (1994). Male coloration as an indicator of parental quality in the yellowhammer, *Emberiza citrinella*. *Animal Behaviour*, **48**, 885–92.

Swaddle, J. P. and Cuthill, I. C. (1994*a*). Female zebra finches prefer males with symmetric chest plumage. *Proceedings of the Royal Society of London B*, **258**, 267–71.

Swaddle, J. P. and Cuthill, I. C. (1994*b*). Preference for symmetric males by female zebra finches. *Nature, London*, **367**, 165–6.

Swaddle, J. P. and Witter, M. S. (1994). Food, feathers and fluctuating asymmetries. *Proceedings of the Royal Society of London B*, **255**, 147–52.

Szekely, T. and Reynolds, J. D. (1995). Evolutionary transitions in parental care in shorebirds. *Proceedings of the Royal Society of London B*, **262**, 57–64.

Taber, R. D. (1949). Observations on the breeding behavior of the ring-necked pheasant. *Condor*, **51**, 153–75.

Taylor, P. B. (1986). Turnicidae, button-quail. In *Birds of Africa*, Vol. 2 (ed. E. K. Urban, C. H. Fry, and S. Keith), pp. 76–83. Academic Press, London.

Temrin, H. and Tullberg, B. S. (1995). A phylogenetic analysis of the evolution of avian mating systems in relation to altricial and precocial young. *Behavioral Ecology*, **6**, 296–307.

Terborgh, J. W. (1971). Distribution of environmental gradients: theory and a preliminary interpretation of distributional patterns of the Cordillera Vilcabamba, Peru. *Ecology*, **53**, 23–40.

Terborgh, J. W. and Weske, J. S. (1975). The role of competition in the distribution of Andean birds. *Ecology*, **56**, 562–75.

Thapliyal, J. P. and Saxena, R. N. (1961). Plumage control in Indian weaver bird (*Ploceus philippinus*). *Naturwissenschaften*, **48**, 741–2.

Thery, M. (1992). The evolution of leks through female choice: differential clustering and space utilization in six sympatric manakins. *Behavioral Ecology and Sociobiology*, **30**, 227–37.

Thomas, A. L. R. and Balmford, A. (1995). How natural selection shapes birds' tails. *The American Naturalist*, **146**, 848–68.

Thornhill, R. (1988). The jungle fowl hen's cackle incites male competition. *Verhandlungen. der Deutschent. Zoologischen Gesellschaft*, **81**, 145–54.

Thornhill, R. (1990). The study of adaptation. In *Explanation and interpretation in the study of animal behavior*, Vol. 2 (ed. M. Bekoff and D. Jamieson). Westview Press, Boulder, Colorado.

Tinbergen, N. (1948). Social releasers and the experimental method required for their study. *Wilson Bulletin*, **60**, 6–51.

Tinbergen, N. (1951). *The study of instinct*. Claredon Press, Oxford.

Tinbergen, N. (1952). Derived activities: their causation, biological significance, origin and emancipation during evolution. *The Quarterly Review of Biology*, **27**, 1–32.

Tinbergen, N. (1953). *Social behaviour in animals*. Methuen, London.

Tinbergen, N. (1963). On aims and methods of ethology. *Zeitschrift für Tierpsychologie*, **20**, 410–33.

Tomback, D. F. and Baker, M. C. (1984). Assortative mating by white-crowned sparrows at song dialect boundaries. *Animal Behaviour*, **32**, 465–9.

Tordoff, H. B. (1967). An interesting case of mortality in the Gouldian finch. *Auk*, **84**, 604–5.

Trail, P. W. (1985). Courtship disruption modifies mate choice in a lek-breeding bird. *Science*, **227**, 778–80.

Trail, P. W. (1987). Predation and anti-predator behavior at Guianan cock-of-the-rock leks. *Auk*, **104**, 496–507.

Trail, P. W. (1990). Why should lek-breeders be monomorphic? *Evolution*, **44**, 1837–52.

Triggs, S., Williams, M., Marshall, S., and Chambers, G. (1991). Genetic relationships within a population of blue duck, *Hymenolaimus malacorhynchos*. *Wildfowl*, **42**, 87–93.

Trivers, R. L. (1971). The evolution of reciprocal altruism. *The Quarterly Review of Biology*, **46**, 35–57.

Trivers, R. L. (1972). Parental investment and sexual selection. In *Sexual selection and the descent of man, 1871–1971* (ed. B. Campbell), pp. 136–79. Aldine, Chicago.

Trivers, R. L. (1974). Parent–offspring conflict. *American Zoologist*, **14**, 249–64.

Turner, C. D. and Bagnara, J. T. (1971). *General endocrinology*. W. B. Saunders, Philadelphia, Penn.

van Oordt, G. J. and Junge, G. C. A. (1933). The influence of the testis hormone on the development of ambosexual characters in the blackheaded gull (*Larus ridibundus*). *Acta Brevia Neerlandica de Physiologia, Pharmacologia, Microbiologia*, **3**, 15–17.

van Oordt, G. J. and Junge, G. C. A. (1934). The relation between the gonads and the secondary sexual characters in the ruff (*Philomachus pugnax*). *Bulletin de la Societe de Biologie de Lettonie*, **4**, 141–6.

van Rhijn, J. G. (1984). Phylogenetical constraints in the evolution of parental care strategies of birds. *Netherlands Journal of Zoology*, **34**, 103–22.

van Rhijn, J. G. (1990). Unidirectionality in the phylogeny of social organization, with special reference to birds. *Behaviour*, **115**, 153–73.

van Rhijn, J. G. (1991). Mate guarding as a key factor in the evolution of parental care in birds. *Animal Behaviour*, **41**, 963–70.

Van Tyne, J. and Berger, A. J. (1959). *Fundamentals of ornithology*. Wiley, New York.

Van Valen, L. (1962). A study of fluctuating asymmetry. *Evolution*, **16**, 125–42.

Vehrencamp, S. L. (1977). Relative fecundity and parental effort in communally nesting anis, (*Crotophaga sulcirostris*). *Science*, **197**, 403–5.

Vehrencamp, S. L. (1978). The adaptive significance of communal nesting in groove-billed anis (*Crotophaga sulcirostris*). *Behavioral Ecology and Sociobiology*, **4**, 1–33.

Vehrencamp, S. L. (1982). Body temperatures of incubating versus nonincubating roadrunners. *Condor*, **84**, 203–7.

Vehrencamp, S. L. and Bradbury, J. W. (1984). Mating systems and ecology. In *Behavioural ecology: an evolutionary approach* (ed. J. R. Krebs and N. B. Davies), pp. 251–78. Blackwell Scientific Publications, Oxford.

Vehrencamp, S. L., Bowen, B. S., and Koford, R. R. (1986). Breeding roles and pairing patterns within communal groups of groove-billed anis. *Animal Behaviour*, **34**, 347–66.

Vehrencamp, S. L., Bradbury, J. W., and Gibson, R. M. (1989). The energetic cost of display in male sage grouse. *Animal Behaviour*, **38**, 885–96.

Veiga, J. P. (1990). Sexual conflict in the house sparrow: interference between poly-gynously mated females versus asymmetric male investment. *Behavioral Ecology and Sociobiology*, **27**, 345–50.

Veiga, J. P. (1992). Why are house sparrows predominantly monogamous? A test of hypotheses. *Animal Behaviour*, **43**, 361–70.

Vellenga, R. E. (1980). Moults of the satin bowerbird *Ptilonorhynchus violaceus*. *Emu*, **80**, 49–54.

Verbeek, N. A. M. (1973). The exploitation system of the yellow-billed magpie. *University of California Publications in Zoology*, **99**, 1–58.

Verner, J. (1964). Evolution of polygamy in the long-billed marsh wren. *Evolution*, **18**, 252–61.

Verner, J. (1965). Breeding biology of the long-billed marsh wren. *Condor*, **67**, 6–30.

Verner, J. and Engelsen, G. H. (1970). Territories, multiple nest building, and polygyny in the long-billed marsh wren. *Auk*, **87**, 557–67.

Verner, J. and Willson, M. F. (1966). The influence of habitats on mating systems of North American passerine birds. *Ecology*, **47**, 143–7.

Verner, J. and Willson, M. F. (1969). *Mating systems, sexual dimorphism, and the role of male North American passerine birds in the nesting cycle*, Ornithological Monographs, No. 9. American Ornithologists Union, Washington, DC.

Vernon, C. J. (1971). Notes on the biology of the black coucal. *Ostrich*, **42**, 242–58.

Vevers, H. G. (1962). The influence of the ovaries on secondary sexual characters. In *The Ovary*, Vol. 2 (ed. S. Zuckerman), pp. 263–89. Academic Press, New York.

Vleck, C. M., Mays, N. A., Danson, J. W., and Goldsmith, A. (1991). Hormonal correlates of parental and helping behavior in cooperatively breeding Harris' hawks (*Parabuteo unicinctus*). *Auk*, **108**, 638–48.

von Hartmann, L. (1969). Nest-site and evolution of polygamy in European passerine birds. *Ornis Fennica*, **46**, 1–12.

von Schantz, T., Göransson, G., Andersson, G., Froberg, I., Grahn, M., Helgee, A., *et al.* (1989). Female choice selects for a viability-based male trait in pheasants. *Nature, London*, **337**, 166–9.

von Schantz, T., Grahn, M., and Göransson, G. (1994). Intersexual selection and reproductive success in the pheasant *Phasianus colchicus*. *American Naturalist*, **144**, 510–27.

Wagner, R. H. (1991). The use of extrapair copulations for mate appraisal by razorbills, *Alca torda*. *Behavioral Ecology*, **2**, 198–203.

Wagner, R. H. (1992*a*). Confidence of paternity and parental effort in razorbills. *Auk*, **109**, 556–62.

Wagner, R. H. (1992*b*). The pursuit of extra-pair copulations by monogamous female razorbills: how do females benefit? *Behavioral Ecology and Sociobiology*, **29**, 455–64.

Wagner, R. H. (1992*c*). Extra-pair copulations in a lek: the secondary mating system of monogamous razorbills. *Behavioral Ecology and Sociobiology*, **31**, 63–71.

Wagner, R. H. (1993). The pursuit of extra-pair copulations by female birds: a new hypothesis of colony formation. *Journal of Theoretical Biology*, **163**, 333–46.

Wagner, R. H. (1997). Hidden leks: sexual selection and the clumping of avian territories. In *Extra-pair mating tactics in birds*, Ornithological Monographs, No. 49 (ed. P. Parker and N. Burley), pp. 123–45. American Ornithologists' Union, Washington, DC.

Wagner, R. H., Schug, M. D., and Morton, E. S. (1996*a*). Condition dependent control of paternity by female purple martins: implications for coloniality. *Behavioral Ecology and Sociobiology*, **38**, 379–89.

Wagner, R. H., Schug, M. D., and Morton, E. S. (1996*b*). Confidence of paternity, actual paternity and parental effort by purple martins. *Animal Behaviour*, **52**, 123–32.

Wallace, A. R. (1889). *Darwinism*. Macmillan, London.

Walters, J. R. (1984). The evolution of parental behavior and clutch size in shorebirds. In *Breeding behavior and populations* (ed. J. Burger and B. L. Olla), pp. 243–87. Plenum Publishing, New York.

Walters, J. R. (1990). Red-cockaded woodpeckers: a 'primitive' cooperative breeder. In *Cooperative breeding in birds: long-term studies of ecology and behavior* (ed. P. B. Stacey and W. D. Koenig), pp. 69–101. Cambridge University Press.

Walters, J. R., Copeyon, C. K., and Carter, J. H. (1992*a*). Test of the ecological basis of cooperative breeding in red-cockaded woodpeckers. *Auk*, **109**, 90–7.

Walters, J. R., Doerr, P. D., and Carter, J. H. (1992*b*). Delayed dispersal and reproduction as a life-history tactic in cooperative breeders: fitness calculations from red-cockaded woodpeckers. *American Naturalist*, **139**, 623–43.

Wanntorp, H.-E., Brooks, D. R., Nilsson, T., Nylin, S., Ronquist, F., Stearns, S. C., and Wedell, N. (1990). Phylogenetic approaches in ecology. *Oikos*, **57**, 119–32.

Watson, P. J. and Thornhill, R. (1994). Fluctuating asymmetry and sexual selection. *Trends in Ecology and Evolution*, **9**, 21–5.

Weatherhead, P. J. (1979). Ecological correlates of monogamy in tundra-breeding savannah sparrows. *Auk*, **96**, 391–401.

Weatherhead, P. J. (1990). Secondary sexual traits, parasites and polygyny in red-winged blackbirds, *Agelaius phoeniceus*. *Behavioral Ecology*, **1**, 125–30.

Weatherhead, P. J. and Boag, P. T. (1995). Pair and extra-pair mating success relative to male quality in red-winged blackbirds. *Behavioral Ecology and Sociobiology*, **37**, 81–91.

Weatherhead, P. J., Bennett, G. F., and Shutler, D. (1991). Sexual selection and parasites in wood-warblers. *Auk*, **108**, 147–52.

Webster, M. S. (1992). Sexual dimorphism, mating system and body size in New World blackbirds (Icterinae). *Evolution*, **46**, 1621–41.

Wedekind, C. and Folstad, I. (1994). Adaptive or non-adaptive immunosuppression by sex hormones? *American Naturalist*, **143**, 936–8.

Weidmann, U. (1990). Plumage quality and mate choice in mallards (*Anas platyrhynchos*). *Behaviour*, **115**, 127–41.

Wesolowski, T. (1994). On the origin of parental care and the early evolution of male and female parental roles in birds. *American Naturalist*, **143**, 39–58.

West, M. J., King, A. P., and Eastzer, D. H. (1981*a*). Validating the female bioassay of cowbird song: relating differences in song potency to mating success. *Animal Behaviour*, **29**, 490–501.

West, M. J., King, A. P., and Eastzer, D. H. (1981*b*). The cowbird: reflections on development from an unlikely source. *American Scientist*, **69**, 56–66.

West-Eberhard, M. J. (1975). The evolution of social behavior by kin selection. *The Quarterly Review of Biology*, **50**, 1–33.

West-Eberhard, M. J. (1983). Sexual selection, social competition and speciation. *The Quarterly Review of Biology*, **58**, 155–83.

West-Eberhard, M. J. (1984). Sexual selection, competitive communication and species-specific signals in insects. In *Insect communication* (ed. T. Lewis), pp. 283–324. Academic Press, London.

Westneat, D. F. (1987*a*). Extra-pair copulations in a predominantly monogamous bird: observations of behavior. *Animal Behaviour*, **35**, 865–76.

Westneat, D. F. (1987*b*). Extra-pair fertilizations in a predominantly monogamous bird: genetic evidence. *Animal Behaviour*, **35**, 877–86.

Westneat, D. F. (1988*a*). Male parental care and extrapair copulations in the indigo bunting. *Auk*, **105**, 149–60.

Westneat, D. (1988*b*). The relationships among polygyny, male parental care, and female breeding success in the indigo bunting. *Auk*, **105**, 372–4.

Westneat, D. F. (1990). Genetic parentage in the indigo bunting: a study using DNA fingerprinting. *Behavioral Ecology and Sociobiology*, **27**, 67–76.

Westneat, D. F. (1993). Polygyny and extrapair fertilizations in eastern red-winged blackbirds. *Behavioral Ecology*, **4**, 49–60.

Westneat, D. F. (1995). Paternity and paternal behaviour in the red-winged blackbird, *Agelaius phoeniceus*. *Animal Behaviour*, **49**, 21–35.

Westneat, D. F., Sherman, P. W., and Morton, M. L. (1990). The ecology and evolution of extra-pair copulations in birds. In *Current ornithology*, Vol. 7 (ed. D. M. Power), pp. 331–69. Plenum Press, New York.

Wetton, J. H. and Parkin, D. T. (1991). An association between fertility and cuckoldry in the house sparrow, *Passer domesticus*. *Proceedings of the Royal Society of London B*, **245**, 227–33.

Whitfield, D. P. (1990). Male choice and sperm competition as constraints on polyandry in the red-necked phalarope, *Phalaropus lobatus*. *Behavioral Ecology and Sociobiology*, **27**, 247–54.

Whittingham, L. A., Kirkconnell, A., and Ratcliffe, L. M. (1992). Differences in song and sexual dimorphism between Cuban and North American red-winged blackbirds (*Agelaius phoeniceus*). *Auk*, **109**, 928–33.

Whittingham, L. A., Dunn, P. O., and Robertson, R. J. (1993). Confidence of paternity and male parental care: an experimental study in tree swallows. *Animal Behaviour*, **46**, 139–47.

Widemo, F. (1997). The social implications of traditional use of lek sites in the ruff *Philomachus pugnax*. *Behavioral Ecology*, **8**, 211–17.

Wiley, R. H. (1973). Territoriality and non-random mating in sage grouse, *Centrocercus urophasianus*. *Animal Behaviour Monographs*, **6**, 85–169.

Wiley, R. H. (1974). Evolution of social organization and life-history patterns among grouse. *The Quarterly Review of Biology*, **49**, 201–27.

Wiley, R. H. (1976*a*). Affiliation between the sexes in common grackles I. Specificity and seasonal progression. *Zeitschrift für Tierpsychologie*, **40**, 59–79.

Wiley, R. H. (1976*b*). Affiliation between the sexes in common grackles II. Spatial and vocal coordination. *Zeitschrift für Tierpsychologie*, **40**, 244–64.

Wiley, R. H. (1991). Lekking in birds and mammals: behavioral and evolutionary issues. *Advances in the Study of Behavior*, **20**, 201–91.

Williams, D. M. (1983). Mate choice in the mallard. In *Mate choice* (ed. P. P. G. Bateson), pp. 297–309. Cambridge University Press.

Williams, G. C. (1966). *Adaptation and natural selection*. Princeton University Press.

Williams, G. C. (1975). *Sex and evolution*. Princeton University Press.

Williams, G. C. (1992). *Natural selection: domains, levels, and challenges*. Oxford University Press.

Williams, P. A., Lawton, M. F., and Lawton, R. O. (1994). Population growth, range expansion, and competition in the cooperatively breeding brown jay, *Cyanocorax morio*. *Animal Behaviour*, **48**, 309–22.

Wilson, E. O. (1971). *The insect societies*. Harvard University Press, Cambridge, Massachusetts.

Wilson, E. O. (1975). *Sociobiology*. Belknap/Harvard University Press, Cambridge, Massachusetts.

Wingfield, J. C. (1984). Androgens and mating systems: testosterone-induced polygyny in normally monogamous birds. *Auk*, **101**, 665–71.

Wingfield, J. C., Ball, G. F., Dufty, A. M. J., Hegner, R. E., and Ramenofsky, R. E. (1987). Testosterone and aggression in birds. *American Scientist*, **75**, 602–8.

Wingfield, J. C., Hegner, R. E., and Dufty, A. M. J. (1990). The 'challenge hypothesis': theoretical implications for patterns of testosterone secretion, mating systems, and breeding strategies. *American Naturalist*, **136**, 829–46.

Winkler, D. W. and Sheldon, F. H. (1993). Evolution of nest construction in swallows (Hirundinidae): a molecular phylogenetic perspective. *Proceedings of the National Academy of Sciences, USA*, **90**, 5705–7.

Winquist, T. and Lemon, R. E. (1994). Sexual selection and exaggerated male tail length in birds. *American Naturalist*, **143**, 95–116.

Witmer, M. C. (1996). Consequences of an alien shrub on the plumage coloration and ecology of cedar waxwings. *Auk*, **113**, 735–43.

Witschi, E. (1935). Seasonal sex characters in birds and their hormonal control. *Wilson Bulletin*, **47**, 177–88.

Witschi, E. (1961). Sex and secondary sexual characters. In *Biology and comparative physiology of birds*, Vol. 2 (ed. A. J. Marshall), pp. 115–68. Academic Press, London.

Wittenberger, J. F. (1976). The ecological factors selecting for polygyny in altricial birds. *American Naturalist*, **110**, 779–99.

Wittenberger, J. F. (1978). The evolution of mating systems in grouse. *Condor*, **80**, 126–37.

Wittenberger, J. F. (1979). The evolution of mating systems in birds and mammals. In *Handbooks of behavioral neurobiology: Social behavior and communication* (ed. P. Marler and J. Vandenbergh), pp. 197–232. Plenum, New York.

Wittenberger, J. F. (1980). Feeding of secondary nestlings by polygynous male bobolinks in Oregon. *Wilson Bulletin*, **92**, 330–40.

Wittenberger, J. F. (1981). *Animal social behavior*. Duxbury Press, Boston, Massachusetts.

Wittenberger, J. F. and Tilson, R. L. (1980). The evolution of monogamy: hypotheses and evidence. *Annual Review of Ecology and Systematics*, **11**, 197–232.

Witter, M. S. and Swaddle, J. P. (1994). Fluctuating asymmetries, competition and dominance. *Proceedings of the Royal Society of London B*, **256**, 299–303.

Wittzell, H. (1991). Directional selection on morphology in the pheasant, *Phasianus colchicus*. *Oikos*, **61**, 394–400.

Wolf, L. L. (1975). 'Prostitution' behavior in a tropical hummingbird. *Condor*, **77**, 140–4.

Wolf, L., Ketterson, E. D., and Nolan, V., Jr. (1988). Paternal influence on growth and survival of dark-eyed junco young: do parental males benefit? *Animal Behaviour*, **36**, 1601–18.

Wood-Gush, D. G. M. (1954). The courtship of the brown leghorn cock. *British Journal of Animal Behavior*, **2**, 95–102.

Woolfenden, G. E. and Fitzpatrick, J. W. (1984). *The Florida scrub jay: demography of a cooperative-breeding bird*. Princeton University Press.

Yamagishi, S., Nishiumi, I., and Shimoda, C. (1992). Extrapair fertilizations in monogamous bull-headed shrikes revealed by DNA fingerprinting. Auk, 109, 711–21.

Yasmin, S. and Yahya, H. S. A. (1996). Correlates of mating success in Indian peafowl. *Auk*, **113**, 490–2.

Yasukawa, K. (1981a). Song repertoires in the red-winged blackbird (*Agelaius phoeniceus*): a test of the Beau Geste hypothesis. *Animal Behaviour*, **25**, 475–8.

Yasukawa, K. (1981b). Male quality and female choice of mate in the red-winged blackbird (*Agelaius phoeniceus*). *Ecology*, **62**, 922–9.

Yezerinac, S. M. and Weatherhead, P. J. (1997). Extra-pair mating, male plumage coloration and sexual selection in yellow warblers (*Dendroica petechia*). *Proceeding of the Royal Society of London, B*, **264**, 527–32.

Young, J. R., Hupp, J. W., Bradbury, J. W., and Braun, C. E. (1994). Phenotypic divergence of secondary sexual traits among sage grouse, *Centrocercus urophasianus*, populations. *Animal Behaviour*, **47**, 1353–62.

Zack, S. (1990). Coupling delayed breeding with short-distance dispersal in cooperatively breeding birds. *Ethology*, **86**, 265–86.

Zack, S. (1995). Cooperative breeding in *Lanius* shrikes. III. a reply in hindsight to Zack and Ligon I, II (1985). *Proceedings Western Foundation Vertebrate Zoology*, **6**, 34–8.

Zack, S. and Ligon, J. D. (1985a). Cooperative breeding in *Lanius* shrikes. I. Habitat and demography of two sympatric species. *Auk*, **102**, 754–65.

Zack, S. and Ligon, J. D. (1985b). Cooperative breeding in *Lanius* shrikes. II. Maintenance of group-living in a nonsaturated habitat. *Auk*, **102**, 766–73.

Zack, S. and Rabenold, K. N. (1989). Assessment, age and proximity in dispersal contests among cooperative wrens: field experiments. *Animal Behaviour*, **38**, 235–47.

Zahavi, A. (1975). Mate selection—a selection for a handicap. *Journal of Theoretical Biology*, **53**, 205–14.

Zahavi, A. (1977). The cost of honesty (further remarks on the handicap principle). *Journal of Theoretical Biology*, **67**, 603–5.

Zahavi, A. (1991). On the definition of sexual selection, Fisher's model, and the evolution of waste and of signals in general. *Animal Behaviour*, **42**, 501–3.

Zimmerman, D. A., Turner, D. A., and Pearson, D. J. (1996). *Birds of Kenya and northern Tanzania*. Princeton University Press.

Zimmerman, J. L. (1966). Polygyny in the dickcissel. *Auk*, **83**, 534–46.

Zuk, M. (1991a). Sexual ornaments as animal signals. *Trends in Ecology and Evolution*, **6**, 228–31.

Zuk, M. (1991b). Parasites and bright birds: new data and a new prediction. In *Bird–parasite interactions: ecology, evolution, and behaviour* (ed. J. E. Loye and M. Zuk), pp. 317–27. Oxford University Press.

Zuk, M. (1992). The role of parasites in sexual selection: current evidence and future directions. *Advances in the Study of Behavior*, **21**, 39–68.

Zuk, M., Thornhill, R., Ligon, J. D., Johnson, K., Austad, S., Ligon, S. H., *et al.* (1990a). The role of male ornaments and courtship behavior in female mate choice of red jungle fowl. *American Naturalist*, **136**, 459–73.

Zuk, M., Thornhill, R., Ligon, J. D., and Johnson, K. (1990b). Parasites and mate choice in red jungle fowl. *American Zoologist*, **30**, 235–44.

Zuk, M., Johnson, K., Thornhill, R., and Ligon, J. D. (1990c). Mechanisms of female choice in red jungle fowl. *Evolution*, **44**, 477–85.

Zuk, M., Ligon, J. D., and Thornhill, R. (1992). Effects of experimental manipulation of male secondary sex characters on female mate preference in red jungle fowl. *Animal Behaviour*, **44**, 999–1006.

Zuk, M., Johnsen, T. S., and Maclarty, T. (1995). Endocrine-immune interactions, ornaments and mate choice in red jungle fowl. *Proceedings of the Royal Society of London B*, **260**, 205–10.

Index

Birds are indexed by their common names, in inverted form, e.g. blackbirds, red-winged. Scientific names and classifications are listed in the Appendix (page 435)

The alphabetical arrangement of the index is letter-by-letter.

DATE DUE